Medical Primatology

Medical Primatology

History, biological foundations and applications

Eman P. Fridman

formerly Primate Information Center
Institute of Experimental Pathology and Therapy
(Sukhumi Primate Center)
USSR Academy of Medical Sciences

Edited by

Ronald D. Nadler

Yerkes Regional Primate Research Center
Emory University, Atlanta

CRC Press
Taylor & Francis Group
Boca Raton London New York

CRC Press is an imprint of the
Taylor & Francis Group, an **informa** business
A TAYLOR & FRANCIS BOOK

First published 2002 by Taylor & Francis

Published 2019 by CRC Press
Taylor & Francis Group
6000 Broken Sound Parkway NW, Suite 300
Boca Raton, FL 33487-2742

© 2002 by Taylor & Francis Group, LLC
CRC Press is an imprint of Taylor & Francis Group, an Informa business

First issued in paperback 2019

No claim to original U.S. Government works

ISBN 13: 978-0-367-45501-9 (pbk)
ISBN 13: 978-0-415-27583-5 (hbk)

Visit the Taylor & Francis Web site at
http://www.taylorandfrancis.com

and the CRC Press Web site at
http://www.crcpress.com

Typeset in Garamond 3 by Graphicraft Limited, Hong Kong

British Library Cataloguing in Publication Data
A catalogue record for this book is available from the British Library

Library of Congress Cataloging in Publication Data
A catalogue record has been requested

To my wife Lina
Emman P. Fridman

Contents

PART III
Biological foundations and applications of medical primatology

Eman P. Fridman, D. Biol.

Acknowledgements

The author would like to thank Roman Selivanov and Evgeniy Morozov for translating the text from Russian, Nina and Michael Rabinovich for word processing the text, and Alexander Avgustinov for digital manipulation of the illustrations.

The images in this book have been reproduced with permission from the following sources:

Institute of Experimental Pathology and Therapy (Sukhumi Primate Center) (S. Nikitenko), former USSR

Institute of Medical Primatology of the Russian Academy of Medical Sciences (Sochi-Adler Primate Center), Russia

Cologne Zoo (E. Kullman, G. Nogge, U. Hick, R. Shlosser, W. Spiess), Germany

Duke University Primate Center, USA

Yerkes Regional Primate Research Center, USA

and other centers and journals.

Acknowledgements

Part I

History

Chapter 1

General comments on the history of primatology

The history of medical primatology is, naturally, related to the history of primatology proper. The history of primatology, moreover, is rather dramatic. The science of primatology to which humans belong (together with apes, monkeys, and prosimians), and the biological links between humans and other primates, have always been caught in the maelstrom of ideological controversy. Naturally enough, this controversy has had a negative impact on the evolution of the science of humans and their nearest relatives. The development of medical primatology has also been adversely affected by the difficulty of conducting scientific studies with these animals. Their habitats, as a rule, were beyond the scientist's reach and their adaptation to captivity was quite poor. As a result, primatology was actually established as a science only in the 1960s, considerably later than the other biological or zoological sciences (Schultz, 1966; Nesturch, 1968).

In my opinion, there is no need in this monograph on medical primatology to give thorough consideration to the history of primatology proper. This was done previously in the book, *Primates* (Fridman, 1979b), although that volume does not contain a great deal of detail. Nevertheless, a brief review of the development of primatology is in order.

A relatively large number of authors, usually prominent primatologists and other scientists, have written treatises on the history of this science, including T. Huxley (1863), R. M. and A. W. Yerkes (1929), W. McDermott (1938), H. Janson (1952), M. F. Nesturch (1960), A. Schultz (1966), R. and D. Morris (1966), V. Reynolds (1967), and W. Hill (1969). These scholars collected and reported a great number of valuable facts on the history of human knowledge of primates, from the time of Aristotle, Pliny, and Galen to Tyson, Buffon, and Darwin to modern times. It is part of their great service to science. Yet, their works present only facts; there is no theoretical analysis which traces the development of primatology, defines the periods of its history, or describes its ideological and conceptual basis. A. Schultz alone attempted to relate our knowledge of primates to the conception and affirmation of the theory of evolution (Schultz, 1966). More consideration is given nowadays to the theoretical aspects of the history of primatology (Spencer, 1995). The first symposium of more than 30 speakers on the history of primate research, *Ape, Man, Apeman: Changing Views Since 1600* (Corbey and Theuissen, eds, 1995), which took place in Leiden, the Netherlands, in the summer of 1993, was an extraordinary event. It was followed a year later by the *First Symposium on the History of Primatology* presented at the XVth Congress of the International Primatological Society which was held in Bali, Indonesia (Nadler and Dukelow, 1996).

Primatological studies employ historical surveys of individual countries (Nesturch, 1968; Fridman, 1967b; Lapin, 1994) and reviews of separate primate species or divisions of science (Hill, 1969; Simons and Covert, 1981; Coolidge, 1984; Schaller, 1988; Fedigan and Strum, 1999; Mitchell, 1999). In accordance with my purposes, I distinguish three periods in the history of primatology.

EARLY INFORMATION ON SIMIANS
(FROM ANCIENT TIMES TO THE 17TH CENTURY)

Our first acquaintance with the other primates dates from the Paleolithic period (Indonesia, South France). The earliest evidence in the history of our knowledge of primates is found in cultural artifacts depicting monkeys, made by early Asians and Africans, especially the Egyptians. The Egyptians regarded some monkeys (baboons) as consecrated to gods (Ra or Thoth, at different times) and used them for ritualistic purposes (Figure 1.1). The ancient world was well aware of the primates. It may be assumed that not only monkeys, but also apes were known in ancient Greece, to say nothing of other countries which were the natural habitats of primates.

Some evidence of primates may be found in the works of Aristotle, Pliny the Elder, and C. Galen (whose studies of primate anatomy were quite extensive, though often discrepant, a fact which was later pointed out by Vesalius). The Greek grammarian, Horapollo (4th century A.D.) remarkably once made a valuable observation on the periodic bleeding (menstruation) of female primates. Albert Magnus (13th century) considered simians 'the connecting link' between humans and the rest of the animal world and even attempted to classify them. Noteworthy also are A. Vesalius' (1543) works on monkey anatomy and an archaic compilation of facts on primates taken from

Figure 1.1 Sacred baboon (*Papio hamadryas*) between man and God in Ancient Egypt (Egyptian papyrus, about 1000 B.C.) (from Ploog, 1972).

Figure 1.2 'Devilish' reputation of monkeys and apes in the Middle Ages: Confrontation of people with devil and simian in Judeo-Christian tradition (Stuttgart's Psalter – after Morris and Morris, 1966).

other authors which was published by K. Gesner in 1551. That is essentially all we know about primates from the first period. It should be noted that in the 12th century and probably earlier, monkeys were called 'simia' in Latin, due to their likeness to humans (*Monkeys*, 1980, and references therein). H. Janson (1952) showed that this word was related to the Latin 'similitudo'. It is apparent that the word is also reminiscent of the English 'similarity', in association with the monkey, much the same as the word 'monkey' itself is related to the English 'man' (Yerkes and Yerkes, 1929).

At the end of the first period in the history of primatology, the science from today's perspective was extremely primitive. Being relatively well-known to, and even deified in, the ancient heathen (olympic) mythology, and possibly for that very reason, simians came to be regarded in Europe as 'satanic' creatures that were 'hostile' to humans. This conception developed after monotheism became predominant in the Dark Ages, the time of the Inquisition. Only when the Renaissance flowered and ideological pressure was somewhat reduced did the monkey begin to lose its 'devilish' reputation, but was seen nevertheless as a caricature of a human; a clown and a simpleton (Figure 1.2) (Morris 1966; Reynolds, 1967). We should agree with Yerkes and Yerkes (1929), therefore, who wrote that the Middle Ages not only failed to enrich the ancient's knowledge of primates, but in fact, caused some of what had been known to be lost. During those times, monkeys and apes were no longer differentiated, although a vague notion of the difference had existed in ancient times. Almost completely unknown (judging from the literature) were the platyrrhine ('broad-nosed') primates or, the New World monkeys. Knowledge of prosimians was also nonexistent. Descriptions of catarrhine monkeys and hominoids were confused and mingled with legends. We can describe the initial scientific data only in terms of anatomy.

The first period in the development of primatology, which ended at the beginning of the 17th century, was admittedly a period of the most primitive, rudimentary knowledge of primates. A similar stage in the history of general biology was essentially completed in the 15th century, and the 16th century saw scholastic works in the natural sciences on fish (G. Rondele, I. Salvini), birds (P. Belon), and insects (T. Mofette, U. Aldrovandi) which went down in the history of natural sciences.

EXPANDING THE KNOWLEDGE OF PRIMATES AND THE PROBLEMS OF PRIMATE CLASSIFICATION IN PRE-DARWINIAN TIMES (FROM THE 17TH CENTURY TO THE FIRST HALF OF THE 19TH CENTURY)

This was a time of rapid growth and systematization of data and of the subsequent formation, in the first half of the 19th century, of all the fundamental biological sciences. The science of zoology began in the 18th century. Anthropology was formed in the second half of the century and the first anthropological society was established in Paris by the end of this period (1859). At the same time, the concept of transformationism became more and more popular. The proposal by the classical or creationist biologist, G. Cuvier, that naturalists should 'name, describe, and classify' appeared too limited by the turn of the century. 18,000 to 20,000 animals had been studied by this time, but the task that Cuvier had set his contemporaries was apparently completed; biology was becoming a science which would discover the evolutionary laws of nature.

The state of affairs was different in primatology. The need to 'name, describe, and classify' persisted until as late as the first half of the 19th century. The science of our nearest relatives, however, was threatened by something far more significant than a lack of factual data. At the time of the French revolution, the biological affinity of humans and the apes was a rather perturbing subject. The idea of a common progenitor of all primates was openly discussed by the French Enlighteners and later by J. Lamarck. This could not fail to provoke controversy. As a result, the uniform order Primates, as defined by C. Linnaeus in 1758, was soon divided into two distinct orders: Bimana ('the two-handed', in which only humans were placed), and Quadrumana ('the four-handed', which included the apes, monkeys, and prosimians). This division persisted for over a hundred years. The term 'primates', as such, vanished from scientific papers. The order was reunified as one taxonomic group only in 1863 by T. Huxley.

In the 17th and 18th centuries a multitude of primate descriptions emerged, contributed by adventurers, seafarers, and naturalists. These descriptions were generally confused, not very reliable, and rather apocryphal. It was during this period, however, that some genuinely scientific papers were written as well. The period began with the very first scholarly publication on chimpanzee anatomy written in 1641 by Nicolas Tulpius, the prominent Dutch anatomist (Figure 1.3). Although the title included such words as 'Indian satyr', 'Homo silvestris', and 'Orang-Outang' (the last two terms both mean 'man of the woods'), the author went on to describe the pygmy chimpanzee or bonobo (Reynolds, 1967). Tulpius discovered certain examples of resemblance in the animal's body to that of the human and described them in his work. Thomas Willis, another famous European anatomist, found the similarity of human and non-human primate brains remarkable and described them in his paper of 1664. According to W. Bynum (1973), Willis was wary of theologians, and therefore explained the

Figure 1.3 Nicolas Tulp(ius). Detail from Rembrant's picture 'Professor Tulp (the anatomy lesson)', 1632. (See colour plate section)

incredible fact he had established by reference to the then fashionable principle of 'non-material complementation'. A study by one of the predecessors of modern systematics, J. Ray (1693), is also worth mentioning. Ray was the first to draw a taxonomic distinction between lower and higher simians (monkeys and apes).

A book on chimpanzee anatomy (1699) by the outstanding English anatomist, Edward Tyson (1650–1708) was a brilliant finale to the 17th century, establishing another landmark in the history of primatology (Figure 1.4). This classic work (many times reprinted; last edition released in 1969) qualified its author as the founder of primatology. In spite of the fact that its title mentions the 'orang-outang' and uses some currently unacceptable terminology which was common at the time, the book presents a detailed description of chimpanzee anatomy, drawing an explicit comparison between human and nonhuman primates, and pointing out many similarities and differences. Tyson considered his 'Pygmie' to be 'the missing link' in the chain connecting humans and monkeys (the word 'chain' was underlined by him) (Figure 1.5).

I have already mentioned the incredible contribution to primatology made by C. Linnaeus (1708–1778). It was he who joined humans, apes, monkeys, and half-monkeys in one taxonomic order and gave this order its present name, Primates (1758). It should be noted that it is the only mammalian order, which, despite violent disagreement, retained its order status, as defined by Linnaeus, together with its original name.

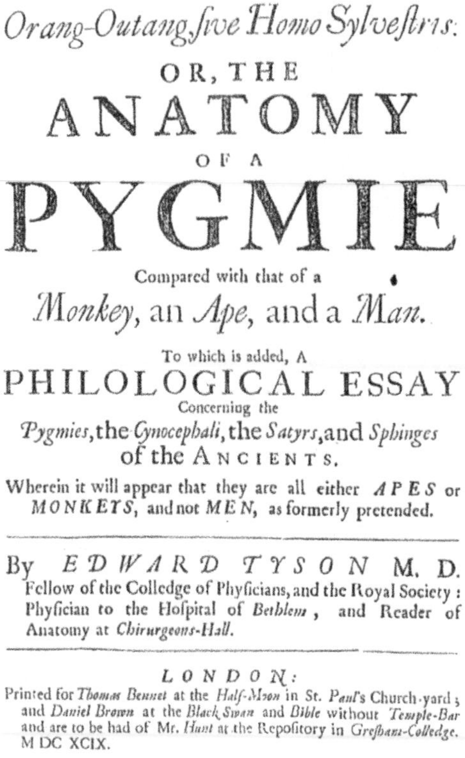

Orang-Outang,ſive Homo Sylveſtris:

OR, THE

ANATOMY

OF A

PYGMIE

Compared with that of a

Monkey, an *Ape,* and a *Man.*

To which is added, A

PHILOLOGICAL ESSAY

Concerning the

Pygmies, the *Cynocephali,* the *Satyrs,* and *Sphinges*
of the A N C I E N T S.

Wherein it will appear that they are all either *A P E S* or
M O N K E Y S, and not *M E N,* as formerly pretended.

By *E D W A R D   T Y S O N* M. D.

Fellow of the Colledge of Phyſicians, and the Royal Society :
Phyſician to the Hoſpital of *Bethlem,* and Reader of
Anatomy at *Chirurgeons-Hall.*

L O N D O N:

Printed for *Thomas Bennet* at the *Half-Moon* in St. *Paul's* Church-yard ;
and *Daniel Brown* at the *Black Swan* and *Bible* without *Temple-Bar*
and are to be had of Mr. *Hunt* at the Repoſitory in *Greſham-Colledge.*
M DC XCIX.

Figure 1.4 Title page of the book by E. Tyson, 1699.

Figure 1.5 E. Tyson's 'Pygmie' (from his book of 1699).

The anatomy of three contemporary species of apes had already been described by the end of the 18th century: chimpanzee (Tulpius, Tyson), orangutan (P. Camper, A. Fosmaer), and gibbon (J. van Iperen and F. Schouman). None of these species, however, was identified taxonomically and their nomenclature was confusing. The attempt by C. Hoppius, Linnaeus' student from St. Petersburg, to organize and systematize 'anthropomorpha' in his thesis of 1760 (Russian translation published in 1777) was not very successful. The task of identification and classification of the three species in question was undertaken in the first quarter of the 19th century. The gorilla, as we know, was discovered only in the 1840s (Savage and Wyman, 1847).

During the second period of our history, data were rapidly accumulating and numerous species of monkey and prosimian were being inventoried. Many scholars contributed to this endeavor. A significant contribution was made by the early evolutionists (transformationists) who were Darwin's predecessors. G. Buffon (1707–1788) wrote his famous *Histoire Naturelle* (1749–1788), which included a special article on simians (Buffon, 1766), and E. Geoffroy Saint-Hilaire (1772–1844) described 18 new primate genera and introduced the classic division of Anthropoidea into catarrhines and platyrrhines (Saint-Hilaire, 1812).

Thus, by the middle of the 19th century, the majority of primate species had been inventoried by scholars and preliminary classification had been given to all the major groups. Primatology of that time, however, could not yet be called a science. There existed no synthesis of data in primatology and it was long before the first papers on phylogenesis were written and full-scale comparisons made of monkeys and apes with humans, beyond the first few anatomical ones. The affinity of human and nonhuman primates, however, had been established; sometimes it was even exaggerated. Most remarkably, as mentioned above, the order Primates was split in two for no substantial reason: Bimana (humans) and Quadrumana (apes, monkeys, and prosimians). This distinction was said to account for the gap between humans and other primates, although, as is shown below, it was made on the basis of a quite far-fetched criterion.

THE REVOLUTION IN PRIMATOLOGY DUE TO EVOLUTIONARY THEORY; THE FORMATION AND DEVELOPMENT OF PRIMATOLOGY; THE BURGEONING OF MODERN PRIMATOLOGY (FROM THE SECOND HALF OF THE 19TH CENTURY TO THE PRESENT TIME)

The third period in the history of primatology encompasses modern times; from Charles Darwin and T. Huxley to the end of the 20th century. This is the time that the theory of evolution and, subsequently, the theory of the descent of man were established and substantiated. Triumphant expansion of evolutionary ideas began, despite the strong resistance of anti-Darwinists, and persisted until the end of the 19th century. There was also, however, a time of decline in acceptance of Darwin's theory, from the beginning of the 20th century to the 1930s. After that period, a new upsurge of evolutionism took place; neo-Darwinism was fortifying its position and outstanding discoveries of the 20th century were being made, such as deciphering the genetic code, defining DNA's role and structure, and obtaining new data on the genealogy of humans. All of these discoveries appreciably influenced the development of primatology, a science which is organically connected to Darwinism. Progress did not follow an upward sloping curve, however, once a sufficient amount of data was compiled in primate biology, but instead took a rather uncertain path and advanced at times with great difficulty. Only toward the end of the 1950s did a dynamic upswing begin which crowned the final stages of the formation of primatology as a *science*. Primatology was thriving, aided greatly by the above mentioned favorable conditions and an incredibly rapid and large-scale development of *medical primatology*, which began in the 1960s.

After Charles Darwin's (1809–1882) work of genius on the origin of species had been published in 1859, Thomas Huxley's (1825–1895) book on the status of humans in the animal kingdom (Huxley, 1863) was a logical sequel from the scientific point of view (but not in light of the world view at that time). The significance of this latter book in the history of primatology cannot be overestimated. Based on a great amount of factual data, the conclusion at which Huxley arrived was that humans and apes were of closer resemblance and kinship to each other than were apes and monkeys! As mentioned above, Huxley provided the scientific foundation for the single order Primates, introduced earlier by Linnaeus.

Restoration of the order Primates was convincingly confirmed on the basis of the natural system by Darwin in his book on the descent of man (Darwin, 1871). (It is noteworthy that Darwin criticized Huxley for placing humans in one taxonomic *family* with the apes, rather than positioning them together in one *sub-family*, in accordance with the scientific data). After that time, it was impossible to separate humans from the other primates of the system. This and other publications by Darwin contained a wealth of material on primates. Darwin provided scientific data to support anthropogenesis, introduced the phylogenetic synthesis of the order Primates, and demonstrated that the source of the affinity between humans and the apes lies in their common origin. He was the first to point out the special value of the simians as subjects in research and the similarities between human and simian pathology. It was the latter similarities which opened new vistas for modern medical primatology.

In the second half of the 19th century and the beginning of the 20th century the number of discoveries and descriptions of primates increased dramatically, especially with respect to the platyrrhine monkeys and prosimians. At the beginning of the century, a new mountain species of gorilla was discovered (*Gorilla gorilla beringei*) and then another species of chimpanzee (pygmy chimpanzee or bonobo). Confusion in the classification and in the nomenclature of the primates, however, persisted for some time, essentially to the present time.

The first nursery for great apes was built in 1906 by Madame Rosalia Abreu in Cuba. Although it was not designed for scientific purposes, it played a prominent role in the history of primatology. It was here that the first live chimpanzee was born in captivity. Madame Abreu also corresponded with such distinguished scientists as Mechnikov (1845–1916) and Robert M. Yerkes (1876–1956), who at that time were themselves entertaining the idea of establishing primate centers. Yerkes eventually obtained his first apes (among other primates) from Madame Abreu, with which he established the first American primate center in 1930 in Orange Park, Florida. The first primate center in the Soviet Union was built in 1927 in Sukhumi; it played an important role in the development of other such centers throughout the world. Beginning in the 1960s, seven national primate research centers were established in the United States and others were built in other countries.

At the beginning of the 20th century, medical primatology was conceived at the junction of the rapidly developing science of primatology and experimental medicine. In primatology proper, a rapid accumulation of knowledge occurred in the first half of the 20th century in the fields of primate anatomy and paleontology (R. Pocock, W. Gregory, W. Le Gros Clark, A. Schultz, W. Hill, and others), systematics (W. Gregory, H. Coolidge, G. Simpson, M. Nesturch, and others), primate behavior (N. N. Ladigina-Kohts, W. Köhler, R. M. Yerkes, I. Pavlov), and ecology (C. R. Carpenter, S. Zuckerman, H. W. Nissen, S. Washburn, and others). There was a gradual consolidation of various types of data on primates into a single science (R. M. Yerkes, T. Ruch, H. Hofer, A. Schultz, D. Stark, and others). The *Bibliographia Primatologica* (1941) was published, the result of many years of work by the outstanding primatologist, T. Ruch, a follower of Yerkes. In the Introduction, the famous physiologist, J. Fulton, recorded his astonishment at the small number of works which had been written on primate biology. In the entire period from ancient times to 1939, only 463 references were located. (By the beginning of the 1970s, this number was published in one year (Fridman, 1974).) By the end of the 1930s and beginning of 1940s, the contemporary term 'primatology' had been coined (E. Hooten, as quoted by J. Erwin (1983), G. Bonch-Osmolovsky (1940), and T. Ruch (1941)). It should be noted that 'anthropology', a cognate term, had been used more than 400 years earlier in 1516 (Blanckaert, 1989). Nevertheless, even in the 1940s, prominent scholars were still in doubt as to whether they should use the word 'primate' (Montagu, 1941).

By the end of the 1950s, due to the favorable conditions mentioned above, there was a new and powerful upsurge in the development of primatology which reached a pinnacle between the 1960s and 1990s. Topical periodicals and annual publications on primates appeared for the first time. The Japanese journal 'Primates' was established in 1959. The International Primatological Society was established in 1964 (Preuschoft, 1996). Biannual international congresses and annual national meetings and symposia were organized regularly. One indication of the rise in primatology is worth noting. Only two theses on nonhuman primates were defended in the field of anthropology in the United States during the period from 1929 to 1960, while from 1960 to 1971, a total of 161 academic degrees in primatology were awarded (Gilmore, 1981). Having traveled far on a long and winding road, primatology has become a *science*. The highly dynamic development of primatology, beginning during the 1960s, has no precedent in history. As I noted above, this development has been stimulated and supported considerably by primate research in the biomedical area.

*   *   *

Chapter 2

History of medical primatology

I use the term 'medical primatology' to denote the contemporary trend in the medical and biological sciences to conduct biomedical studies on nonhuman primates with the purpose of solving problems of human pathology and biology (*Comprehensive Medical Encyclopedia*, Lapin and Fridman, 1983). This trend incorporates comparative biology, anatomy, physiology, biochemistry, primate pathology, and some more specific data on the order Primates, primarily evolutionary, phylogenetic and taxonomic data. Medical primatology also deals with the traits of affinity between human physiological systems and organs and those of other primates, essentially in terms of their similarities and common functions. In contrast to all other experimental animals, such as rodents, cats, and dogs, primates represent the same biological order as the human species and are thus, the nearest relatives of humans. This relationship to humans requires that the medical primatologist possess not only the knowledge of his specific area of research, but also a combination of strictly medical skills for the maintenance and breeding of primates in captivity.

Usage of the approximately comparable term 'experimental primatology' may be considered correct in many cases (Fridman, 1967a; Firsov, 1982). Medical primatology has developed to the point where this trend, in many respects, is now concerned with matters which traditionally pertained to primatology proper, including anatomy, behaviour, ecology, and genetics, to name just a few. The combining of these and other data within the domain of medical primatology is inevitable and quite reasonable. On the whole, it facilitates the progress of research on both humans and their nearest relatives.

The literature on the history of medical primatology is rather sparse. There is, of course, the paper by the prominent British primatologist, S. Zuckerman (1963), in which he made the first attempt to collect all the medical research on primates from Galen to the 1960s. That paper cites valuable data, reinforced by documentary sources. It was a concise compilation, however, thereby omitting many examples of primate research as well as the historical problems which have confronted the field.

An interesting chapter in a collective volume was published quite recently on the history of primate use in medical research (Johnsen, 1995). Without differentiating between medical primatology with apes and monkeys from primatology proper, the author presents an historical account of primate studies, describes the first primatological centers of the world, provides an extensive picture of virological and other medical research on primates, and describes the American regional primate research centers and the rather complicated objectives of contemporary research in the United States. The United States, perhaps, serves as the best example of the successful use of primates in biomedical research in the second half of the 20th century. Johnsen, however, believes that experiments on primates will become outmoded in the future due to the development of biotechnology. I am compelled to disagree with this author. Should such a change occur (which is doubtful, if we refer to all types of experiments and primates), it will probably be a long time from now. Studies on the history of medical primatology performed by the author of the present book, which have been published since the middle of the 1960s, will be discussed further in due course.

A contribution of considerable significance to the history of medical primatology has been made by prominent primatologists, scholars, and scientists who included in their papers introductory historical reviews of primate research (Carmichael, 1969; Lapin, 1988). Also of value are the rather numerous descriptions of the world's primatological centers. These descriptions include the first primate center in Cuba (Yerkes, 1925), the Yale Laboratories of Primate Biology in the United States (subsequently renamed the Yerkes Regional Primate Research Center) (Yerkes, 1943; Nissen, 1944; Bourne, 1965, 1971; King and Yarbrough, 1994; Nadler, 1994a), the Sukhumi Center in the Soviet Union (Botchkarev, 1932; Lapin and Fridman, 1966, 1988), the Japan Primate Center (Simonds, 1962), the Caribbean Primate Center (Carpenter, 1940, 1972; Kessler, 1989), the descriptions of the United States and world primate research centers (Regional . . . , 1968; Kuhn, 1970; Goodwin, 1972; Vaitukaitis, 1994).

Of special value are historical reviews in separate divisions of science concerned with primate use, such as virology (Kalter, 1969; Heberling and Kalter, 1974), oncology (O'Connor, 1969), and physiology (Voronin and Firsov, 1967), and descriptions of certain discoveries in the field of medicine and biology, to which I refer in due course.

Quite indispensable are the historical archives obtained by the author from the Sukhumi Primate Center (Soviet Union), the Russian Academy of Sciences (St. Petersburg), the Russian Ministry of Health (Moscow), the Institute of Experimental Medicine (Russian Academy of Medical Sciences), the Old and Rare Books Departments of the Russian State Library and the Moscow State University (Moscow), Washington University (Seattle, Washington, USA), the Regional Primate Research Center at the University of Wisconsin (Madison, Wisconsin, USA), and the Robert M. Yerkes Foundation of the Yerkes Regional Primate Research Center at Emory University (Atlanta, Georgia, USA).

PERIODS IN THE HISTORY OF MEDICAL PRIMATOLOGY

I divide the history of medical primatology into two main periods:

(i) The period of spontaneous and empirical use of primates in experiments, the pre-history of medical primatology (from Hippocrates to the end of the 19th century).
(ii) The period of scientifically substantiated medical primatology (from the beginning of the 20th century to the present time).

The following four stages may be defined within the second, main period:

a. The birth of medical primatology (turn of the century).
b. The initiation of biomedical research based on large primate centers located in regions outside the natural habitats of monkeys and apes (the 1920s).

c.   The wide use of primates in medical and biological research (since the 1960s).

d.   Research on primates, imported or specially bred in research centers, in, as well as distant from, their natural habitats (since the 1970s).

My further account of medical primatology may deviate at times from this division of the science into specific periods in consideration of the ease of presentation.

THE BIRTH OF MODERN MEDICAL PRIMATOLOGY

Medical experiments on primates have been known since the age of ancient Egypt (2nd millennium B.C.). Egyptian priests (who were also doctors), as already mentioned, dissected baboon cadavers for ritualistic purposes. There is evidence in the literature that Hippocrates, 'the father of medicine', dissected monkeys in order to find out how gall is excreted. Alexandrian doctors (3rd century B.C.) also carried out anatomical studies of monkeys' inner organs, as did the aforementioned C. Galen. There is also evidence of one Rufus of Ethes, who studied the anatomy of simians even before Galen (see Zuckerman, 1963). Great doctors of the Orient, Avicenna, Al-Razi, and Juhanna-ibn-Messavai (9th and 10th centuries A.D.) also conducted research using monkeys to study 'miasms'. I have already discussed A. Vesalius. Evidently, experiments on primates, a rather advanced stage of the research enterprise, were done at those very locations to where the centers of world civilization subsequently moved.

Complete experiments on syphilis in primates are known to have been performed in the 19th century, e.g., by Dovass in 1845, Langleberta in 1864, Turrenne in 1874, and Clebs in 1878, followed by a whole group of syphilidologists, including E. Sperk, the famous researcher from St. Petersburg (1893–1898). The majority of these references were taken from reviews by E. Mechnikov and E. Roux (1903), and M. Chlenov (1902), but I believe the situation was much the same in other fields of medical research. Lebert (1874), a Breslav professor, studied tuberculosis using monkeys as early as 1869. W. Carter and then R. Koch made a study of relapsing fever in primates in India (1879) and L. Pasteur and his colleagues conducted rabies and cholera experiments on monkeys in the 1880s.

The majority of these early experiments on primates were conducted in a haphazard manner. The ape or monkey was used not as an experimental subject of special qualities, but rather, as an ordinary laboratory species such as mice, guinea pigs and other animals. At that time (and even considerably later) medicine was completely out of touch with the ideas of evolution. As a matter of fact, experimental medicine itself was only beginning to take shape; it was founded as late as the 1880s. No consideration was given to the biological foundations for experiments on primates. No one seemed to be concerned with that issue; even today it remains an unresolved problem in many respects. In any event, by that time it was imperative that experimenters had some elementary knowledge of primates. In fact, there was some information available on the taxonomy of the experimental subjects. Nevertheless, all the studies conducted during this period were characterized either by ignorance of the species or by omission of the relevant considerations, a fact first pointed out by E. Mechnikov, and then by Osman Hill. As a result of these combined 'negative' (as it seems today) conditions, the most significant characteristic of this period is clear: with little or no exception, there was no actual success in the experiments with primates. In particular, scientists failed to reproduce in primates syphilis, enteric fever, poliomyelitis, or any other infections, although such attempts were made. Models of these and other diseases, of course, were subsequently induced in monkeys or apes. Thus, it may be concluded that sporadic experimenting on primates in the 19th century indeed corresponds to a period of spontaneous and empirical research.

A separate comment must be made regarding the early *physiological* experiments on primates, which were first conducted in the 1870s and which produced some significant results. Before that time none of the prominent physiologists (C. Bell, C. Bernard, E. Gering, C. Ludwig, F. Majandi, J. P. Muller) had regarded the simian as a subject for experimentation. C. Bernard wrote in his *Introduction to the Study of Experimental Medicine* (1866) that a medical researcher should be interested primarily in dogs, cats, horses, rabbits, oxen, sheep, pigs, and poultry. Special attention was given to the frog, without which 'physiology would be impossible even now' (ibid, p. 150). It is apparent that there was no place for a monkey in the studies of this great physiologist (who, not surprisingly, had denounced Darwinism).

I have already called attention to this schism between physiology and Darwinism (Fridman and Khassabov, 1972). As early as 1873, however, the outstanding Russian physiologist, I. M. Sechenov, wrote that 'all physiological processes must be regarded in light of Darwin's great teachings'. He proposed a system of principles based on the evolution of neuropsychological activity (Karamyan, 1980). The problem, however, was not resolved before the 1930s, and even then the schism could not be entirely overcome. Was it not for this very reason that by the beginning of the 1960s, incredibly, we still were content to have experimental data obtained from cats, dogs, rodents, and other animals, despite the availability of primates, the nearest relatives of humans (Montagna, 1968)?

The first physiological studies of primates seem to have begun as a result of the discovery, by D. Fritch and E. Hitzig in 1870, of the reaction of the dog's cortex to stimulation by galvanic current. Hitzig himself published one of the first studies (1874) in primate physiology. According to E. Ewarts (1982), V. Betz, a Russian anatomist, discovered very large neurons in the motor cortex of humans and nonhuman primates. Nowadays these neurons are called 'Betz cells'. These are the cells that 'descend' from the brain and form direct links with spinal cord motor neurons, including those that control the muscle movements which effect human speech and precise control of the fingers. It is readily understood that no traditional laboratory animals would satisfy such an experiment; the magnitude of such direct links is clearly much greater in primates.

There were several other scientists who lived during this period and experimented on nonhuman primates (e.g., L. Luciani), but the most prominent of them all was David Ferrier (1843–1928), a professor of legal medicine and a doctor of the Royal College clinic in London. At the beginning of the 1870s he began experimenting on primates to study the rather pressing problem of the distribution of specific functions within the brain (Ferrier, 1874). His studies were combined in one monograph 'The Functions of the Brain', the last chapter of which is devoted to the similarity of human and nonhuman primate brains. Ferrier's works became a celebrated part of history. They have been discussed with great interest, especially those on stimulation and ablation of the auditory cortex (Glickstein, 1985; Heffner, 1987). In this field, Ferrier was indeed ahead of his time. I. Sechenov expressed special appreciation for one of Ferrier's discoveries in particular, namely, his finding that removing the brain loci, which causes paralysis of the skeletal muscles, has the weakest effect in dogs, a much greater effect in monkeys, and the greatest effect in humans. This finding was of great interest for experimental physiologists.

From the perspective of medical primatology, Ferrier's works are of great value for yet another reason. A thread of continuity may be traced from them to other pioneering studies with primates. Especially noteworthy are the studies of Horsley and Schaffer (1883), who investigated cortical functions, and those of Beevor and Horsley (1890), who followed the same course to become the first physiologists, it appears, to experiment with a great ape (the orangutan). These authors drew an important conclusion with regard to the difference between brain function in apes and monkeys. A connection may be traced from Ferrier's research to the first experiments on primates by the classical physiologist, C. Sherrington (1889). S. Brown of New York University, who worked in England with E. Schaffer, probably published the first experimental study on primates in America in 1888. Finally, Ferrier's influence extends to a multitude of experiments on primates performed between 1886 and 1906 by the prominent Russian physiologist and neurologist, V. M. Bechterew. Bechterew's research resulted in a fundamental seven-volume work *Foundations of Brain Functions Study* (Bechterew, 1906). The last two volumes are devoted to experiments on primates.

Even these important studies, however, failed to provide the impetus for using the nonhuman primate as a special experimental subject. Evolutionary concepts were alien to many scientists. The anthropogenic status of simians and the close phylogenetic position of them in relation to the higher primate – the human species – still had not attracted the attention of physiologists. Neither they nor their numerous anatomist predecessors recognized any scientific value in the discovery of brain similarities among close representatives of the order. During this period the monkey was 'just another' laboratory animal. Of course, it may be argued that sometimes a scientist could make a momentous discovery without either subscribing to the theory of evolution or having any theoretical knowledge of the experimental subject. This would be a good example, since there have been similar precedents in history. The subject of the present discussion, however, concerns experiments on simians, which are the closest relatives of humans. Can such a cardinal factor be neglected in the course of an experiment which has as its ultimate subject the human species itself?

All of the above factors affected experimental work directly. Let us demonstrate the connection by using an example from D. Ferrier's experiments. A few references in his publications and the legends to his figures establish that he experimented on macaques (and probably, baboons), although it is impossible to determine exactly which species of this large primate genus (about 20 species) he chose. In his extensive journal publications of 1875 concerned with the primate brain, macaques are mentioned once in the first series of experiments (Ferrier, 1875a). In the second series of 25 experiments, only experiment #5 has a reference to 'a macaque of large size' and experiment #16 includes 'a large monkey, a kind of baboon'. In the remaining 23 experiments on monkeys, not even the genus is indicated (Ferrier, 1875b). Of course, the scientist cannot be blamed for what he 'failed to accomplish'. Such were the times and the physiologist's knowledge of primate taxonomy. Nevertheless, taking into account the primate species, to say nothing of the genus, sometimes proved of great importance to the experimental results.

In order to form a genuinely scientific and theoretically substantiated justification for using primates in biomedical research, at least the following specific prerequisites were required (in addition to the more general factors necessary for the development of any science):

A corresponding level of development in medicine and biology. As a result of the research of F. Majandi, C. Bernard and other scientists (according to I. P. Pavlov) and subsequently,

L. Pasteur, R. Koch, and J. Lister, this correspondence was reached in the 1880s. The invention of vaccines and serum therapy techniques opened up unprecedented prospects in the struggle for improved public health. Hazardous diseases such as syphilis, typhoid fever, poliomyelitis, dysentery, pneumonia, measles, pox, and many others were placed on the agenda of experimental medicine. A number of 'new' illnesses were soon discovered whose nature had been theretofore unknown. All these diseases were impossible to induce in ordinary laboratory animals. The involvement of the nearest relatives of humans, i.e., monkeys and apes, was inevitable.

The second imperative prerequisite for the development of medical primatology was a *sufficient scientific knowledge of simians and other primates*, as well as a certain amount of comparative primatology, i.e., knowledge of the relationship between the anatomy and physiology of humans and those of other animals. As noted in Chapter 1, this knowledge, 'in preliminary approximation', was obtained in the second half of the 19th century. Obviously, most primate species had been identified and inventoried by that time and many of them had been described quite intelligibly, even though their classification remained rather confused. As far as papers and books in comparative biology are concerned, they began to appear with ever increasing frequency. Most important among these were the previously mentioned classic works by Darwin and Huxley, books by R. Owen, E. Haeckel, and K. Foght, and a number of comparative anatomical studies on separate organ systems in the fields of neurology (Vrolik, Graciole, Tidemann, Brocä) and embryology (Deniker, Zelenka), plus hematology studies at the beginning of the 20th century (Nuttal, Friedenthal, Grunbaum).

The third prerequisite for the development of medical primatology was the *adoption of Darwinian concepts in medical science*. The experimental research on primates demonstrated that Darwin's theory was gaining a foothold in medicine and biology. This is apparent if we examine the field of physiology, which, as noted above, did not accept the theory of evolution until the beginning of the 1930s. The reasons for this were the rejection of Darwinism in the 19th century (sometimes an ideological rejection) and the crisis of Darwinism in the first third of the 20th century (due to discoveries in the field of genetics and the onset of fundamentalism), which were overcome with the introduction of the synthetic theory of evolution or neo-Darwinism. As soon as the convergence of physiology and evolutionary theory became a reality, there was a conspicuous increase in physiological research on primates. In 1942, two prominent American physiologists, Theodore Ruch and John Fulton, published the results of many years of difficult and meticulous research with primates, ushering in the era of medical primatology. It is unlikely these authors were aware of the crucial issues in the history of medical primatology; they merely presented their data in a careful and thorough manner (Ruch and Fulton, 1942). A graph borrowed from their article gives a clear numerical outline of the history of primate physiology which is consistent with the discussion presented above (Figure 2.1). It is appropriate to add that the original photocopy of this graph was generously given by Prof. Ruch himself to the author of this book during a conference on primates held at the Sukhumi Primate Center in 1966.

Much earlier than it was accepted in physiology, Darwinism was accepted in microbiology and infectious disease pathology. In these areas the threat of dangerous diseases, implying interactions between macro- and micro-organisms and immunological interdependencies, called for a more decisive evolutionary approach. By the end of the 19th century, medical science required not only experiments on primates for its further growth, but also the theoretical foundation for such experiments, i.e., a combination of medicine

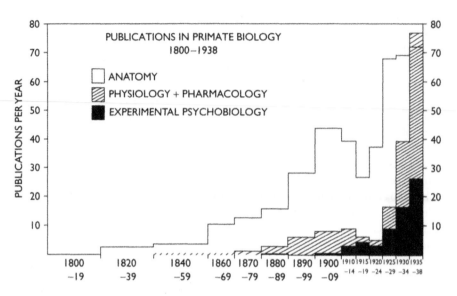

Figure 2.1 The growth of primate studies from 1800 to 1938 (from Ruch and Fulton 1942).

with Darwinism. R. Virchow, an opponent of Darwinism (as were most medical pundits of that time), lectured in London shortly before the end of the 19th century in commemoration of T. Huxley. While discussing the impact of his famous book (Huxley, 1863), Virchow declared that 'One may have any kind of opinion about the descent of the human species, but the conviction of a complete correlation between human and animal organizations has become ubiquitous. Therefore, all biological sciences, especially physiology and pathology, have been provided a great stimulus for adopting corresponding methods of study. In particular, everything that is based on experiment should be investigated on animals first . . . Suffice it to say, the boundary between man and animal in biological sciences is becoming less and less definite' (Virchow, 1899, p. 8). Even though he gave Huxley his due, Virchow did not call explicitly for a unification of medicine and Darwin's theory. This unification was brought about by the outstanding Russian scientist and researcher, Elie Mechnikov (1845–1916).

Unfortunately, the English-speaking reader is not sufficiently aware of this man's contributions to science – he became one of the first Nobel Prize winners (1908) – possibly because he published his main studies in popular books as well as professional ones, both in Russian and in French (after 1888 he lived in Paris and worked at the Pasteur Institute for the rest of his life). I quote mainly from the academic editions of Mechnikov's works, in Russian and less often, in French. Mechnikov was a well-rounded biologist and pathologist. Having also worked successfully in the fields of zoology, embryology, anthropology, gerontology, and microbiology, and being one of the founders of modern immunology and comparative pathology, he remained a talented Darwinist for his entire life. Mechnikov studied primates thoroughly and possessed extensive knowledge of them. This was reflected in his books, *Etudes on Human Nature* and *Etudes of Optimism* and a multitude of experimental studies. Mechnikov provided important data summaries on the relation of humans to other primates which were derived from research that was conducted after Darwin, Huxley, and Haeckel. His most remarkable contribution in terms of medical primatology, however, was the introduction of evolutionary concepts into medicine and his demonstration of the fruitfulness of this unification by reference to his own life's work.

A simple enumeration of the titles of some of Mechnikov's works gives some idea of his scientific aspirations: *Anthropology and Darwinism* (1875), *Darwinism and Medicine* (1910), and *World Outlook and Medicine* (1910). Mechnikov emphasized the significance of Darwinism for medicine and biology in Cambridge at a celebration in Darwin's honour (1909), in his speech at his own jubilee (1915), in annual reviews of infectious disease pathology from 1909–1912, and in his many other works. 'I believe', Mechnikov wrote, 'that for the success of the science of the descent of species and for the benefit of medicine, a unification of these two domains appears extremely important' (Mechnikov, 1943, p. 219). Such was the theoretical background Mechnikov possessed even before he initiated his research with simians.

It should be recognized that Mechnikov encountered aggressive opposition along the way. The well-known chronicler of Mechnikov's research, R. I. Belkin wrote, 'Mechnikov was the first scholar who was well-equipped to begin restructuring medicine on the basis of Darwin's theory and to bring about drastic changes in medical views . . . Mechnikov came up against the serried ranks of anti-Darwinists in medicine' (Belkin, 1958, p. 347). The concept of phagocytosis, for which Mechnikov was awarded the Nobel Prize, was the first great success of evolutionary theory. In order to support his concept of immunity, Mechnikov began his investigations in the field of comparative pathology, namely, studies of infections. As a result of this research, he arrived at the idea of directly modeling human diseases using nonhuman primates. Thereafter, he began using monkeys and apes for tackling more general problems in biology (gerontology) as well.

In May 1885, in connection with his concept of phagocytosis, Mechnikov attempted to conduct research with monkeys on the problems of relapsing fever, a 'genuine blood disease' which was widespread in Russia at that time. The attempt was not successful, however, because he could obtain no monkeys for the experiments. The next year he managed to obtain three rhesus macaques (*Macaca mulatta*) and three guenons (*Cercophithecus spp.*). He was able to induce relapsing fever in the monkeys, thereby creating one of the first models of this disease (Mechnikov, 1887). This and other studies on monkeys, carried out by Mechnikov in his laboratory in Odessa, became the starting point for many immunologists who subsequently investigated the reticuloendothelial system by means of splenectomy or blocking.

After Mechnikov moved to Paris (1888) on the invitation of L. Pasteur, his interest in primate experiments, especially on apes, grew greater still. Mechnikov was particularly interested in infectious diseases, which, according to his assumption, killed 'noble tissue cells' and caused premature wear (tightening) of vessel walls. Even the Pasteur Institute, however, had no financial means to procure the chimpanzees needed for the experiment (they cost from 1,000 to 2,000 francs each at that time). In 1903, Mechnikov was awarded the prize of the Madrid Medical Congress (5,000 francs) and decided to use it to buy the apes. Learning of this, Emil Roux, then head of the Pasteur Institute and a future Nobel Prize winner who had received the Ifla-Oziris prize of 100,000 francs, also donated that money for the experiments[1].

---

1 Informed of the selflessness of these scientists, the wealthy Ifla-Oziris bequeathed an enormous fortune to the Pasteur Institute.

Mechnikov consented to accept the donation on one condition: they would conduct the experiments together. As is known, the publications on syphilis were signed by both Mechnikov and Roux, whereas all of the summaries and reviews belong to Mechnikov alone.

Truly sensational studies on primates began in 1903. For over a hundred years, beginning with Hanter's research in 1788, scientists had attempted to reproduce a model of syphilis in various animals, even in monkeys, but all to no avail[2]. This was finally accomplished by Mechnikov and Roux (1903). Intensive studies of syphilis also began after this time in Germany, France, Russia, and Austro-Hungary. In 1905, Shaudin and Hoffman discovered *Spirocheta pallida*, the pathogen of syphilis (Shaudin requested Mechnikov to expedite the publications on modeling the disease, which facilitated recognition of the discovery of the pathogen). The first effective means of fighting syphilis, Salvarsan, and later, Neosalvarsan, were soon found (P. Ehrlich). I should note that even today there is no better model of syphilis than that in nonhuman primates (Musher *et al.*, 1976). Mechnikov's studies of syphilis continued for 4 years (1903–1906). More than 270 monkeys were used in the experiments, including 60 hominoids (Fridman, 1967a).

By that time, Mechnikov was already a true medical primatologist, an interpretation which is supported by his early studies. The first article on experimental syphilis states (as do his later works) that the study of infections to which other animals were not susceptible, 'quite naturally' should be done on primates, since the idea of 'species continuity' and the descent of humans from a nonhuman primate had become commonly accepted. Mechnikov alluded to Huxley and more recent investigations by Grunbaum and Nutall into the 'hemolytic, agglutinative and precipitable properties of serums', which revealed the same characteristics in humans as those found in apes (Mechnikov, 1959, p. 227). Special attention was called to the taxonomy of the primates he used. In a speech on May 25 1905 in the Society of Parisian Hospital Doctors, Mechnikov directly attributed the failure of past scientists to model syphilis to their dismissal of the relevant characteristics of primate systematics. Mechnikov himself discovered that there were different degrees of susceptibility to syphilis depending upon the species of primate which was chosen, from greatest in the apes to some totally immune monkeys, e.g., *Mandrillus* (ibid, p. 238). Knowledge of these species variations allowed researchers to apply the original technique of pathogen passage, with the purpose of weakening its virulence, from a susceptible species to an immune one and then in the opposite direction (for example, from chimpanzee to macaque and then back to chimpanzee). This technique was used in attempts to create vaccines, in utilizing less expensive monkeys as controls, in reducing the incubation period of infections, and in other cases as well.

Because of his views on premature aging, Mechnikov developed a special interest in intestinal infections. In 1908, many medical scientists were trying to determine the cause of 'baby cholera' (baby diarrhea), which was killing children and which was at the time defined as a non-infectious disease by leading specialists. Mechnikov modeled the disease in a chimpanzee and proved it to be an infection caused by *Proteus*. This study is important, apart from its historic value for medical practice. In the framework of this chapter, it shows the indispensability of primates for studying diseases of unknown etiology, but assuming an infectious nature (Mechnikov, 1955).

---

2 In the 1890s E. Sperk (St. Petersburg) was close to achieving a model of syphilis in baboons, but he died shortly thereafter and the research was not continued (Sperk, 1896).

Mechnikov achieved another scientific triumph in 1910 with his pupil and colleague, A. Besredka, by creating a model of typhoid/enteric fever. By that time more than 30 years had passed since the day Ebert discovered the pathogen of this disease. More than 20 vaccines had been developed against this infection and prophylactic serum had been prepared, but none of these could be tested because there was no experimental model for the tests. Attempts were made to reproduce this disease in various animals, but to no avail. On March 3 1910, Mechnikov presented to the Paris Academy of Medicine his research on modeling typhoid fever in the chimpanzee by adding the feces of an infected person to the animal's food. Even today there is no model of this dangerous disease except the hominoid simia (ape) model (Mechnikov and Bezredka, 1910). It is known that Mechnikov used chimpanzees to study cholera, obtained a model of diphtheria, and even attempted to induce experimental cancer in primates. Interestingly, Mechnikov's laboratory in Paris at that time (1903) was also the work place of N. Petrov. Petrov, the founder of modern Russian oncology, subsequently went on to work at the Sukhumi Primate Center, where he became the first investigator to induce tumors in simians (Petrov et al., 1951).

Mechnikov was also the first investigator to begin extensive gerontological studies with chimpanzees. It was in connection with this work that Mechnikov proposed the construction of primate nurseries for captive rearing of monkeys and apes, the future primate centers for biomedical research. Such nurseries, of course, subsequently played a decisive role in the history of medical primatology. Mechnikov made this suggestion on his 70th birthday at a celebration in Paris on May 16 1915 (Mechnikov, 1915), and then reiterated the idea in the preface to the 5th edition of his book, *Studies on Human Nature*. A similar suggestion with respect to primate nurseries was made in the context of comparative psychology by another strong proponent of primatology, R. M. Yerkes (1915, 1916).

Thus, it is clear that Elie Mechnikov made a fundamental contribution to the theoretical foundations of modern medical primatology. He demonstrated pragmatically the important role of theory in experimental science, particularly evolutionary ideas, by developing models of human diseases in primates which were impossible with other laboratory animals. In the course of these experiments, he formulated several proposals in the area of medical primatology which continue to be relevant even today. These include the proposal of using primates to clarify the etiology of diseases of an unknown, but presumably infectious character, the proposal that taxonomic identification of experimental monkeys and apes was crucial to the outcome of an experiment, and the proposal for the combined experimental use of various primate species of different biological affinity to humans. Finally, it is in Mechnikov's numerous articles and archives that we find the first proposal to construct a laboratory in which primates are bred and reared for the specific purpose of conducting biomedical research (Mechnikov, 1959, pp. 347–387).

I must also not fail to mention the fundamental impact that Elie Mechnikov's work produced on the further development of science, especially of medical primatology. In the final years of his life Mechnikov enjoyed incredible scientific recognition. The resolution by the London Royal Society to award Mechnikov the Albert Medal stated, 'Professor Mechnikov's discoveries, more than any other living man's work, ensure the control of contagious diseases and gradual health improvement for European nations . . .' (Amlinsky, 1964, p. 332). Influenced and assisted by Mechnikov, dozens of scientists in Russia and other countries began experimenting on primates to study relapsing fever, syphilis, and other diseases. Some scientists received primates from him. C. Landsteiner and E. Popper, who were in scientific correspondence with Mechnikov, not only modeled poliomyelitis in the macaque (1908), but determined the viral character of this neural infection one year later. The victory over this disease in the 1950s and 1960s became yet another triumph of medical primatology. (D. O. Johnsen [1995] is mistaken when he stated that the two authors mentioned above won the Nobel prize for the discovery of the polio virus. Although they may have deserved the award, they did not receive it. Landsteiner won the Nobel Prize for the discovery of human blood groups in 1930, after studying on a great number of nonhuman primates [Landsteiner and Miller, 1925], thus making it yet another success of medical primatology.) At Mechnikov's request, the Pasteur Institute sent chimpanzees to C. Nicolle in Tunis, who failed to obtain a model of typhus fever in monkeys, but was successful in inducing this human disease in the chimpanzees he received. Nicolle's studies of typhus fever were also crowned with the Nobel Prize in 1928.

Evaluating Mechnikov's role in the history of primatology, Robert Yerkes wrote: 'Even before my ideas became "motor", the eminent Russian medical investigator Mechnikov, then located in France and supplied with monkeys and apes by the African colonies of his adopted country, used them for important studies of human disease. By 1910 he had become convinced of their high value as experimental subjects, had employed them extensively himself, and encouraged the development of provisions which should make them readily available for use in laboratories of medical research. Subsequently, about 1923, the Pasteur Institute of Paris, undoubtedly as a result of Mechnikov's reports and recommendations, established in Africa a station for the use of monkeys and chimpanzees and for their collection and shipment to France' (Yerkes, 1943, p. 292).

Figure 2.2 Professor I. I. Mechnikov in his laboratory ('Uzhniy Kray' Newspaper, 1912).

Mechnikov's work and that of his followers established a majority of the trends in biomedical research on primates in the 20th century. We regard Elie Mechnikov as the founder of modern medical primatology (Figure 2.2).

HISTORY OF WORLD SIMIAN NURSERIES AND PRIMATE CENTERS; THE RAPID GROWTH OF PRIMATE STUDIES

Maintaining primates in captivity has always presented complex problems. Their susceptibility to stress during capture and the subsequently long period of transportation, and their particular sensitivity to the certain captive conditions and the change of climate accounted for their initially short lives in zoological gardens, menageries, and other places in which they were kept. This was pointed out by many authors of the 18th, 19th, and 20th centuries. Darwin wrote (1871) that apes could never be brought to maturity in European climates. It was this obstacle that inspired the idea of establishing primate nurseries after medical primatology had come into its own in the first quarter of the 20th century.

The first attempts at maintaining monkeys and apes in captivity, naturally, bore no relation to medical experiments, but were undertaken either in the regions of these animals' immediate distribution, or in places with similar climatic conditions (not in zoological gardens). The first historical precedent for this was a chance 'experiment', which, amazingly, continues to the present time. I refer to the island of St. Kitts, one of the West Indies, where African guenons, green and mona monkeys became acclimatized. These monkeys were imported from Senegal by slave-traders at the end of the 17th century. Over the course of 300 years of absolute isolation, a large population of African monkeys evolved on this island of the New World, which, according to some sources (McGuire, 1974), amounts to several thousand animals. The comparative study of ecology and group interrelations of this population of monkeys is of great interest. It appears that approximately 30,000 crab-eating macaques that inhabit Mauritius today are also immigrants, brought from Java about 300 years ago. The same is true for about 600 monkeys of the same species that today live on the island of Angaur (Micronesia).

In 1763, Duke von Schliffen brought several dozen Barbary apes from Gibraltar to his personal reserve, a park in Windhausen, not far from Kassel, Germany, where he had built special shelters for them. These monkeys were maintained for 20 years, until one was bitten by a rabid dog (Sanderson and Steinbacher, 1957). Books by Huxley, Darwin, and Bram reported that Renger kept broad-nosed monkeys in Paraguay in the 1820s and Bram disposed of a quarantine enclosure in Africa in the middle of the 19th century.

With the beginning of medical studies on primates, some researchers kept monkeys and chimpanzees in their laboratories (e.g., the Parisian syphilidologist, Crishabaire). At the beginning of the 1890s, the Petersburg Institute of Experimental Medicine imported monkeys for its research programs. The animals did not breed in any of these instances, however, so none of the above laboratories could be called a primate nursery.

The first actual primate nursery, as mentioned above, was established at Quinta Palatina, not far from Havana, Cuba in 1906 by Madame Rosalia Abreu. In 1930, when it was abandoned following its owner's death, the nursery contained 150 primates, including 17 chimpanzees, several orangutans and baboons, and over 100 monkeys[3]. In November of the same year, some of the chimpanzees were transported to Robert Yerkes' newly established primate center in Florida, the Yale Laboratories of Primate Biology. The significance of Madame Abreu's nursery was emphasized above. It was here that the first live chimpanzee was born and here it was first demonstrated that primates could live complete lives in captivity, albeit in a subtropical climate.

Beginning in the first decade of this century, a number of comparative psychologists (or psychobiologists, the term Yerkes preferred) became especially interested in the behavioral capacities of apes. A number of German, Russian and American scientists (G. V. Hamilton, W. Köhler, R. M. Yerkes, N. N. Ladigina-Kohts) procured anthropoids, and even set up small ape colonies (Glaser, 1996; Nadler, 1996). One of the most widely known was Köhler's Anthropoid Station on the Canary Island of Tenerife, where this professor of psychology from Berlin carried out his now famous study on the intellectual abilities of chimpanzees (1912–1918) (Köhler, 1973 (1917)).

After the end of World War I, new attempts were undertaken to use primates for experiments in specialized centers. S. A. Voronov, a distinguished doctor of Russian origin and the director of an experimental surgery laboratory at the Paris Physiology Station, began using baboons and chimpanzees for transplantation of reproductive glands (testes) from simians to humans. In 1922, Voronov initiated the construction of an ape nursery in the south of France (near Mentona) which was completed by 1926. The main purpose of this primate center was to obtain testicular tissue from simians for transplantation in men for 'rejuvenation', but endocrinological and oncological experiments were also carried out. It was visited by many scientists. Voronov also provided simians to many Russian researchers. For example, he gave two chimpanzees, Rose and Rafael, to the Russian academician, I. P. Pavlov, which became the subjects of his famous observations of the 1930s. In 1965, the mayor of Mentona informed the author of this book that Voronov's center was destroyed in 1940 during World War II.

In 1920, J. Corner established a small macaque laboratory at the Carnegie Institute (Baltimore, Maryland, USA) to study reproductive biology. After Corner's departure for Rochester, New York, in 1925 (where he also had chimpanzees and monkeys at his disposal), his place was taken by C. Hartman. In October 1925, Hartman established a stable macaque colony which existed until the beginning of the 1940s. In the 1960s, the primate laboratory at the medical school of Johns Hopkins University was restored; it was designed primarily for the study of infectious diseases.

Hartman's colony is seldom mentioned in the scientific literature, but it is of great scientific importance nevertheless. The research orientation of the laboratory was multidisciplinary. It was engaged in anatomical and physiological studies of the reproductive system (later on, work was done in other fields as well). This laboratory was the first to achieve stable rates of breeding rhesus macaques in captivity, namely, in cages. More than 100 births had occurred by 1936. In 1928, Hartman captured on film (with the assistance of R. M. Yerkes) the birth process in a female rhesus monkey. Hartman introduced techniques for establishing the duration of the menstrual cycle and of pregnancy. He became the first investigator to make reliable observations on reproduction in rhesus macaques, including the phenomenon of seasonality, ovulation, pregnancy, anovulatory menstrual cycles, and the role of hormones on reproductive

---

3 A letter of January 6 1930, by Mdm Rosalia Abreu. Archives of IEPT, F. 32, p. 60.

cyclicity. The development of the primate embryo was studied in great detail, including the implantation and formation of the placenta and the growth of the fetus. The studies carried out in Hartman's laboratory are of fundamental significance for understanding the biology of human reproduction, to which I will give special consideration below. Hartman devoted 89 out of his 238 publications to apes and monkeys. The laboratory played a significant role in encouraging scientists to investigate the human reproductive system by means of experimentation on primates[4].

A major role in the history of primatology was played by the 'Pastoria' station of the Pasteur Institute in Paris which obtained its chimpanzees in French Guinea. Establishment of the laboratory was discussed as early as 1913. After the end of World War I, in November 1922, A. Calmette, one of the two inventors of the anti-tuberculosis vaccine (BCG), was then head of the Pasteur Institute. He formulated an agreement with the French colonial authorities to build Pastoria on 35 hectares near the town of Kindia (West Africa). A lack of finances delayed construction for a few years[5]. As early as 1923, however, chimpanzees were brought to Pastoria in large numbers. Thus, its basic function was to obtain and ship chimpanzees to Paris (later on the station also engaged in studies with other primate species). In the 1930s, 2–3 research workers were always at the station and it was intermittently visited by various groups of scientists. In 1959, after Guinea's secession from the French Commonwealth, all research personnel were relieved of their duties and Pastoria was transferred to the jurisdiction of the Ministry of Health of the Republic of Guinea. At that time the station's primates numbered about 2,000, mostly monkeys. During its existence, the station had prepared vaccines against poliomyelitis, tuberculosis, and pox. Pastoria ceased to exist as a research institution in the 1960s.

The significance of Pastoria in the history of medical primatology is enormous. Investigations conducted there made valuable comparisons of primate populations from different climatic zones. I. I. Ivanov, one of the three founders of the Sukhumi Primate Center, who had spent a long time at Pastoria (1927), wrote about the diseases and longevity of chimpanzees at this primate station. It is apparent that the diseases of these apes (tuberculosis, amoebic dysentery, and intestinal worm infestation) and their longevity, especially during the initial period after capture, are entirely similar to those of apes maintained in moderate climates. Other factors of captivity and maintenance, therefore, apart from habitat and climatic conditions, are the most important for primate acclimatization during the most dangerous, initial period following capture[6].

Naturally, it was not only the standards for breeding and maintenance of primates, advanced by the Pastoria researchers, that accounted for the importance of this laboratory. The Pastoria's primates were used in many noteworthy experiments (resulting in the invention of vaccines and serums against tuberculosis, typhus fever, and polio) which became the foundation of modern medicine. Successful medical studies on tropical diseases were also carried out at Pastoria, including those on malaria, piroplasmosis,

---

4 According to the list of publications by C. Hartman, prepared on the event of the 85th birthday of the scientist, and kindly given to the author by the Primate Information Center at the University of Washington (Seattle).

5 A letter by A. Calmette and E. Roux to I. I. Ivanov, 1926. St. Petersburg Department of Archives of the Russian Academy of Sciences. F. 2, op. 1, f. 34, P. 9.

6 Report of Professor I. I. Ivanov about his business trip to Western Africa, 1928. St. Petersburg Department of Archives of the Russian Academy of Sciences. F. 2. op. 1, f. 105.

trypanosomiasis, and others which killed thousands of people. All of the research from 1923 to 1959 done at the Pasteur Institute, one of the greatest benefactors of mankind, was in one way or another connected to Pastoria. The Pastoria station also provided a great stimulus to the development of medical primatology. Many prominent scientists who studied primates visited the station, among them R. M. Yerkes, S. Zuckerman, and H. W. Nissen. Yerkes, in fact, obtained a number of his chimpanzees from Pastoria when he was establishing the Yale Laboratories of Primate Biology in Orange Park, Florida.

Thus, by the mid-1920s, a small number of primatological institutions with a medical orientation had been established. None of them, however, still existed in the 1960s when a large number of new primate centers were built. Nevertheless, scientists were already well aware of the need for medical experiments on primates. During a conference on this issue in 1924, A. Calmette asserted, 'From now on we cannot expect to progress further by using ordinary laboratory animals if we are concerned with leprosy, typhus fever, measles, scarlet fever, yellow fever, the flu, trachoma and a number of other tropical diseases, such as osteitis of the upper jaw, leishmaniasis and trypanosomiasis in man. All of these diseases and many others can be reproduced only in simia; some of the most dangerous for man can only be transmitted to chimpanzees, the simia nearest to our species' (Calmette, 1924, p. 10). But this appeal by the distinguished scientist had no practical consequences in the 1920s.

It is my opinion that the reason for the indifference towards experimental primate laboratories was not only the underestimation by society of the urgency of this issue and the lack of financial means at the disposal of enthusiastic scientists (which, of course, was a fact), but also the extremely negative attitude towards Darwinism, quite common in those days. This was directly pointed out by A. Schultz, one of the classical primatologists; 'In 1925, when William Jennings Bryan and Darrow excited public opinion during the anti-evolutionary incident in Dayton, it was hardly feasible to procure government subsidies for studies of monkeys, who had "tarnished" their reputation by claiming the closest kinship with man' (Schultz, 1966, p. 15). (Schultz was referring to the famous 'Monkey trial' in the United States.) Similar statements regarding this 'dislike' of monkeys at those times were made by other prominent scientists, such as R. M. Yerkes, C. Gregory, and G. H. Bourne. Interestingly, Yerkes named his first two chimpanzees, Billy and Dwina (after W. J. Bryan and C. Darwin, respectively), in 'commemoration' of the anti-Darwinist trial in Dayton. A. Schultz did the same, naming his chimpanzee, Dayton, and thus 'immortalizing' that same event. Few people know that one of the first primate journals, *Folia Primatologica*, which is still published today in three languages, had a large picture of Dayton, the chimpanzee, on its cover in the 1960s and 1970s.

Such anti-evolution sentiments were nonexistent in the Soviet Union at that time, where the famous Sukhumi Primate Center was established in 1927. We can only wonder today how a research center, which for more than 50 years was admired by scientists throughout the world, could have been built in a country with such a weak economy and under conditions of abject poverty and the use of bread coupons. The first primates (chimpanzees and baboons) were brought to Sukhumi on August 24 1927. The Center had been founded by N. A. Semashko, the First People's Commissioner for Health Care in Russia, I. I. Ivanov, Professor of biology, Ya. A. Tobolkin, Professor of endocrinology, G. A. Kozhevnikov, a zoologist, and other Moscow scientists (Figure 2.3). The Center was a branch of the Moscow Institute of Experimental Endocrinology. In 1930,

Figure 2.3 The founders of Sukhumi Primate Center (left to right): Professor I. I. Ivanov, Dr. N. A. Semashko and Dr. Y. A. Tobolkin, 1927.

it became an independent institution for a short time. Then, in 1931, it became a branch of the Leningrad Institute of Experimental Medicine, and over the period 1933 to 1944, its name was changed to the Subtropical Branch of the All-Union Institute of Experimental Medicine (Moscow). At the end of 1944, the Academy of Medical Sciences was established in the Soviet Union and the Subtropical Branch became a Medico-Biological Station of the Academy of Medical Sciences, Soviet Union. It existed in this capacity until 1957, when it was transformed into the Institute of Experimental Pathology and Therapy, Academy of Medical Sciences, Soviet Union. This was the most fruitful period in the history of the Center. In 1977, the Institute received an award from the government to mark the 50th anniversary of the foundation of the Sukhumi Center, and a memorial to the simian was erected on the premises to symbolize the service that experimental primates, and scientists, had rendered humanity by delivering it from many dangerous illnesses.

By the beginning of the 1990s, there were over 7,000 monkeys at the disposal of the Institute and its four affiliations and over 1,000 employees worked there, including almost 300 researchers. In 1992, due to the collapse of the Soviet Union and the war in Abkhazia, a majority of the researchers, led by the academician, B. A. Lapin, Head of the Institute, moved to a former branch institution in the Sochi Region (Adler), where about 2,000 primates were kept at that time. It was here that a new research center was created. Its present name is the Institute of Medical Primatology, Academy of Medical Sciences, Russia. The rest of the scientists from the former Institute of Experimental Pathology and Therapy, and those monkeys in Sukhumi which survived the war (no more than 300 primates did, as reported at the beginning of 1996), were all reorganized into the Abkhazian Primatological Research Center of Experimental Medicine.

Throughout its history, the Sukhumi Center made a unique contribution to the development of medicine and biology. Over 5,000 publications originated there, but no numerical indicators are capable of measuring the great significance of this institution in the history of medical primatology. It was here that the stage of *mass acclimatization of primates in cages and enclosures was carried out, away from their natural habitats, for the purposes of multidisciplinary experimental research*. By 1990, the Center was breeding baboons (*Papio hamadryas*) of the 12th and 13th generations. All in all, there were 20 species of monkeys kept here. Chimpanzees, gorillas, and orangutans were brought in at different times, although they did not breed. In order to resolve the problem of acclimatization, fundamental research in primate biology and pathology was conducted at Sukhumi.

Extensive studies in the fields of central nervous system physiology, pathology, and psychology (initiated under the immediate supervision of academician, I. P. Pavlov) were completed at the Sukhumi Center. The famous neurosis experiments which were conducted on simians are widely known, as well as the first experiments which reproduced neurogenic diseases of the cardiovascular system (hypertension, stenocardia, myocardial infarction). The most important research carried out in the Soviet Union in the field of infectious disease pathology were conducted here as well. These included studies of tetanus, diphtheria, cholera, dysentery, intestinal infections, tick-borne relapsing fever, typhus fever, viral encephalitis, poliomyelitis, measles, mycoplasmas, and L-form bacteria.

Sukhumi achieved world fame with the creation of the first models of simian tumors, initially, it was said, under the supervision of academician, N. N. Petrov, and, later, with experiments on viral hemoblastosis, conducted by Sukhumi oncologists led by B. A. Lapin.

Sukhumi was the main experimental laboratory for medical science on primates in the Soviet Union. Therefore, whenever there was an urgent need to conduct any kind of investigation, verification of a technique, even surgery or a pre-flight astronautical experiment, it was either done at the Sukhumi Center, or used primates from Sukhumi. Penicillin, the first antibiotic, was thus tested in great secrecy during World War II in 1944 (Z. V. Ermolyeva, an aspiring academician of the Academy of Medical Sciences), bilateral experimental ablation of the lobes of the lung was performed by the outstanding surgeon, A. A. Vishnevsky; M. P. Chumakov and his team studied the problem of devising an anti-polio vaccine for over 10 years. It was also at Sukhumi that radiation sickness was studied in primates and the first protective devices against radiation were tested (L. F. Semenov). In this Center, the ostracized team of 'formal geneticists' led by an aspiring academician, N. P. Dubinin, demonstrated the threat of chromosomal aberrations in germinal cells posed by small doses of radiation in primates, including humans. It was this discovery on which the United Nations Committee on Radiation based its recommendations for the Treaty of 1963 banning nuclear tests in three media.

Since the Sukhumi Primate Center always functioned under the authority of the main health organizations of the Soviet Union, and since experimental nonhuman primates always attracted the attention of medical researchers, the most prominent Soviet investigators worked at this center. Many of them directed its research efforts at various times, including P. Zdrodovsky, P. Sergiev, L. Zilber, V. Troitsky, N. Petrov,

Figure 2.4 Director of the Institute of Experimental Pathology and Therapy (Sukhumi Primate Center) Professor Boris A. Lapin (left) tells about the Institute to the visitor, Nobel Prize laureate (literature) Michael A. Sholochov. Behind, at center – the Scientific Secretary of IEPT Dr. Igor T. Dzjeliev, 1973.

M. Chumakov, Z. Ermolyeva, L. Voronin, B. A. Lapin, and many others. B. A. Lapin deserves special mention, in this regard, because he is a unique figure in the history of medical primatology. Although gravely wounded in World War II, he went on to attend one of the best medical schools in Moscow and was taught by the most prominent pathology anatomist in the country, academician I. Davidovsky. He went to Sukhumi in 1952 and within only one year became one of the leaders of the Medico-Biological Station (Figure 2.4). In fact, he became the head of the new Institute of Experimental Pathology and Therapy of the Academy of Medical Sciences when it was established and remained in that position until its demise in 1992 at the beginning of the war in Abkhazia. Headed by Lapin, the Institute of Experimental Pathology and Therapy became a large-scale research institution, recognized internationally as well as within the Soviet Union. He is a world famous scientist who has conducted important studies in the fields of primate pathology, cardiology, and oncovirology (to which I shall return later). Lapin is also a scientist of outstanding organizational capabilities. It was on his initiative that primate research was intensified not only at the Institute, but throughout Russia and in a number of other countries. While the population of monkeys at the Institute amounted to no more than 200 animals during the early 1950s, it had, as mentioned, grown to over 7,000 by 1990. Lapin, moreover, managed to get some of the higher state officials to become personally involved in the Institute's progress. On instructions from the Prime Minister of the Soviet Union, a governmental decree was prepared and released for the expansion of primate research at Sukhumi. A large sum of money was given to the Institute to initiate an ongoing program for the construction of new laboratory buildings. The Institute's affiliated organizations were established in Adler and Tamysh (Abkhazia) and primate preserves were set up in the natural surroundings of the Caucasian woods.

Lapin organized wide-ranging scientific collaborations with other institutions in many parts of the country, encouraged the establishment of other primate nurseries, and invited the most talented scientists to work at the Sukhumi laboratories. This scientist's greatest contribution, however, was the establishment of strong international connections with researchers from many countries and various scientific orientations. At that time, when the state frowned upon any contacts with foreign scientists, especially Westerners, the Institute of Experimental Pathology and Therapy cooperated on a regular basis not only with 'socialist countries', but also with investigators from the United States, France, England, and other countries. Several long-term programs were pursued with American scientists in oncology, primatology, and endocrinology, and in publishing the Current Primate References information bulletin. Sukhumi also sponsored international symposia, conferences, and seminars on different problems in medical primatology, sometimes under a direct initiative by the World Health Organization (WHO).

During the recent period of economic hardship in Russia, Lapin behaved in a manner appropriate to a patriarch of medical primatology. While science in the new Russia found itself on the edge of total collapse and many 'old school' scientists, quite understandably, were psychologically incapable or unwilling to continue their work, Lapin established a new research center in Adler, based on the prior affiliation of the Institute of Experimental Pathology and Therapy. The new center was called the Institute of Medical Primatology, Russian Academy of Medical Sciences. The team of researchers from the Institute was preserved, as were the theoretical foundations, traditions and the precious population of monkeys.

The oldest primatological center of the world, the Sukhumi Center, had stimulated the development of medical primatology in the Soviet Union and in many other countries. When all of the primatological institutions in the Soviet Union were being established (the Koltushi branch of the Academician Pavlov Institute of Physiology near Leningrad, the Institute of Poliomyelitis in the Moscow Region which now bears Chumakov's name, and laboratories in Moscow and Tbilisi), they received both human resources and experimental primates from the Institute. R. M. Yerkes contacted the Sukhumi Center to profit from its experience when he was setting up the first chimpanzee colony in the United States[7]. C. Hartman wrote in 1936: 'We hope to get good advice from the Sukhumi Center and use it in our further work' (Hartman, 1936, p. 632). Before the national primate center program was established in the United States, prominent American scientists twice came to Sukhumi to see the animal facilities. They were interested, in part, in its experience with the large-scale acclimatization of monkeys. The neurogenic hypertension modeling that they saw there, moreover, later provided the impetus for establishing, by an act of the United States Congress, a

---

7 A letter by R. M. Yerkes to the Sukhumi Simian Nursery, May 5 1930. Archives of IEPT, F. 32, p. 54.

whole network of regional primate research centers initially promoted by the National Institute of Cardiology (Whitehair and Gay, 1981).

In June 1930, the Yale Laboratories of Primate Biology became the next primate center to be established. It is the oldest primate center in the United States and it has a more interesting history than most. It was here that systematic breeding of chimpanzees was first achieved. Nowadays, the Center (renamed the Yerkes Regional Primate Research Center) boasts one of the largest hominoid collections in the world, including all the ape species: common and pygmy (bonobo), chimpanzees, gorillas, orangutans, and gibbons. For almost 30 years, from the time he was a graduate student at Harvard University, Yerkes had nurtured the idea of building 'a special research institute for comparative psychobiology' (Yerkes, 1943). The idea was based on his conviction that 'nonhuman primates, especially chimpanzees, would reveal principles of behavioral regulation which could be extrapolated to humans . . .' (Nadler, 1996).

In the 1920s, Yerkes actively supported the expansion of primate research (Yerkes, 1927). He managed to procure grant funds from the Rockefeller Foundation which he used to purchase 200 acres of land not far from Jacksonville, Florida. There, on July 11 1930, 'an institute of primate psychobiology' was opened. Primates were shipped in from different places: 4 chimpanzees from his laboratory in New Haven, Connecticut, among them the pregnant Dwina, 13 from the former Abreu colony, and 16 from Pastoria.

On September 11 1930, the first infant chimpanzee was born at the Laboratories. It was a female, given the name Alpha, which ultimately lived 38 years. Since that time, this particular hominoid species has bred in the colony on a regular basis. By 1942, the Laboratories already had 2 generations of chimpanzees born in captivity. This was a great success, which was explained by carefully elaborated planning of colony organization, specialized care for the animals, and sensible nutrition.

After Yerkes retired in 1942, the chimpanzee colony was managed by the subsequent Directors, Karl S. Lashley (1942–1955), Henry W. Nissen (1955–1958), Arthur J. Riopelle (1959–1962), Geoffrey H. Bourne (1962–1978), and Frederick A. King (1978–1995). T. Insel became Director of the Center in 1995. In 1965, the Yale Laboratories moved to Emory University in Atlanta, Georgia, as the Yerkes Regional Primate Research Center, one of seven regional primate research centers supported by the National Institutes of Health. 'For the first time in the history of the Yerkes Laboratories,' one of the directors wrote, 'their financing is guaranteed for several years in advance' (Bourne, 1965). In 1991, when the author of this book visited this wonderful center, its primate population amounted to 2,600, including about 200 apes. The rest of the primates were monkeys, including *Macaca, Cercocebus, Saimiri,* and *Cercopithecus,* as well as members of other primate genera. At the time, three chimpanzees were older than 50 years, including the female, Gamma, which was 59, a record age for a hominoid in captivity. There were four departments at the Center: Pathology, Behaviour, Reproductive Biology, and Neurobiology. During the administration of F. A. King the Yerkes Center, like other contemporary American primate centers, established collaborations with countries in which apes and monkeys are indigenous. In 1979, for example, King initiated a cooperative program with the Institute of Primate Research in Kenya, where, in the vicinity of the Tana River, a primate research field station was established which carried out AIDS tests on primates obtained in the wild (King and Yarbrough, 1994).

Research at the Laboratories was initially directed toward psychobiology (behaviour, memory, brain function, social life, emotions, etc.). It was not designed to be a medical research facility, although certain experimental issues in medicine (neurosurgery, polio-myelitis) were investigated there. After the Laboratories came under the jurisdiction of the National Institutes of Health as the Yerkes Regional Primate Research Center, and under the authority of Emory University it became a modern and powerful institution of primate biology and experimental medicine.

Robert Mearns Yerkes and his famous Laboratories played an exceptional role in the history of medical primatology (Figure 2.5). Yerkes demonstrated the practicality of breeding chimpanzees systematically in captivity for participation in scientific research. His Laboratories became a dominant influence on chimpanzee research not only in the United States, but also in other countries. Yerkes himself visited Pavlov in Leningrad in 1929. The two scientists agreed to conduct joint experiments in Florida and Pavlov was promised a pregnant female chimpanzee from the Yerkes colony for his studies. It has been established that Yerkes personally produced over 2,000 pages of experi-mental and theoretical material. At least 119 authors of four generations published studies which were influenced by Yerkes' research. This accounts for 486 publications, or four-fifths of the total number of primate studies published before 1960. Yerkes' two classical works (1929, 1943) survived four editions (Rohles, 1969). They have not become out-dated even today. For more detailed information on Yerkes' contributions to science see L. Carmichael's review (1969). Yerkes was the most authoritative primato-logist for many decades. It is noteworthy that N. N. Ladigina-Kohts, the famous Russian scientist, dedicated her first monograph on rhesus macaque behaviour in 1923 to Robert Yerkes.

With the description of the Yerkes Center, we have essentially exhausted the issue of those world primate centers that played a major role in research before 1960. Nevertheless, other studies were conducted, especially after the mid-1930s. Over 41,000 nonhuman primates were used for research in the United States from 1936 to 1938 (Carpenter, 1940). Two extraordinary laboratories functioned in the country at that time, both of them subsidiaries of Yale University. First, there was the colony established in 1935 by Gertrude Van Wagenen, C. Hartman's student (Department of Obstetrics and Gynecology). The second was John Fulton's neurophysiology laboratory, in which about 100 monkeys and chimpanzees were maintained by 1946. A. Schultz maintained chimpanzees for anthropological studies at Johns Hopkins University after 1927. In 1938 he had 6 chimpanzees and 1 orangutan.

In 1938, C. R. Carpenter released 350 rhesus macaques on the wooded island of Cayo Santiago, located east of Puerto Rico. The macaques adapted well, although the colony had no continuous supervisor and there were food supply problems. During World War II some of the animals were recaptured for the purposes of military med-ical experiments. In 1956, the colony was transferred to the authority of the National Institute of Neurological Illnesses and Blindness at Bethesda. By that time there were only 150 monkeys left. In April 1939, Carpenter also attempted to acclimatize gibbons here. Despite the birth of a live infant, the attempt had to be aborted. The infant gibbon was killed by macaques, and the gibbons themselves proved so aggress-ive toward service personnel that Carpenter was compelled to have them recaptured and 'deported' from the island in the spring of 1941. Nevertheless, this experiment demonstrated the general principles of breeding these valuable primates in a semi-free-ranging condition, as well as the need for special ways of handling gibbons in an artificial environment. As far as the macaques are concerned, they have been breeding there without difficulty since 1956; the group grew to 1,000 animals by 1970. The

Figure 2.5 Professor Robert M. Yerkes, the founder of Yerkes Regional Primate Research Center (currently at Emory University, Atlanta).

colony itself now belongs to the Caribbean Primate Center of the University of Puerto Rico (Carpenter, 1972; Kessler, 1989).

Colonies of a commercial nature were also built in the United States in the 1930s, which were designed for public entertainment rather than experimental research. 'Monkey Jungle' in Florida is one example, a rhesus colony in the same state is another, and A. Deniz's chimpanzee group is a third. In the next decade, however, attempts were made to establish primate colonies specifically for the needs of American medical laboratories (mainly for polio studies). During World War II, several hundred monkeys were released on a small island near the Isle of Pines (Cuban territory). The colony was disbanded in the 1950s. A similar fate awaited another colony in Bluffton, South Carolina, a rather larger one. At different times, it maintained from 3,000 to 7,000 primates, including different species of macaques, baboons, and spider monkeys. The colony was established in 1947, but closed by the mid-1950s. The interest in systematic medical research in specialized primate centers had not matured in the United States at that time. At about that same time (1940s), 300 rhesus macaques were released onto an island not far from Rio de Janeiro, Brazil. These monkeys were intended to participate in studies on yellow fever; at the beginning of the 1970s, about 100 of these animals were still preserved here (Hausfater, 1974).

At the end of the 1940s, the military became interested in primate experiments in connection with aerospace research. Special laboratories were established at United States Air Force bases, for example, Wright-Patterson in Ohio, Holloman in New Mexico, and Brooks in Texas. A department of space medicine was established at Brooks in 1949 where a large colony of rhesus macaques was maintained for radiation experiments. The effect of cobalt radiation on arterial blood pressure and brain blood flow was studied here, among other subjects. At different times, there were over 1,000 monkeys kept at the department (Godwin, 1972). It is still in operation. Starting in 1950, chimpanzee research was initiated at the Holloman laboratories. It was from here that Ham and Enos, two famous chimpanzees, were taken to be launched into space in January and November 1961, before the flights of A. Shepard (1961) and J. Glenn (1962) of the United States. The laboratory subsequently became the Primate Research Institute (Hobson et al., 1991).

In the 1950s a primate center was established at the Southwest Foundation for Research and Education in San Antonio, Texas. Pharmacological experiments were carried out here on different species of baboon and many other biomedical issues have also been studied. The organization and affiliation of this center has been changed; today it is the Southwest Foundation for Biomedical Research. Over 3,000 primates are kept here, mainly baboons and chimpanzees, but also macaques and a small number of other monkeys. Research is pursued in the fields of genetics, cardiology, virology, endocrinology, and immunology.

In England, the Birmingham University primate laboratory has existed since the 1930s, headed by the prominent physiologist and primatologist, S. Zuckerman. At the National Institute for Medical Research, research with primates has been conducted since 1949. Primate breeding began in the 1960s, mainly for pharmacological needs.

After 1948, systematic studies on the ecology and behavior of Japanese macaques (M. fuscata) in their natural habitat were initiated. In October 1956, the Japan Monkey Center of Kyoto University was built to the north of Nagoya (Inuyama). Its collection of primate species was one of the richest in the world. By the end of the 1960s, it had already used about 2,000 primates of 80 species, including all the hominoids, many prosimians, and monkeys. This variety of species, unique for captive conditions, still remains and has even been increased in recent times. Today, there are about 90 different species, including 9 hominoids (from the *International Directory of Primatology*, 1994). In 1957, the Center began to publish the professional journal, *Primates*, the first modern research journal in the field. A Primate Institute, consisting of eight research departments, a field station, and a primate breeding facility was established at Inuyama in 1967. The Japanese primatologists became famous for their studies of primate ecology and behavior in their own country, as well as abroad in Africa and Asia. However, not only biological but also biomedical research has been conducted in Japan. In 1978, the prominent Tsukuba Primate Center for Medical Science was established in the vicinity of Ibaraki, as an affiliate of the National Institutes of Health.

Despite the existence of the above primatological institutions, medical research with monkeys and apes was relatively rare in the first half of the 20th century, although, as will be shown below, there were many noteworthy results obtained in this field. In the late 1950s, a new phase was about to begin in the history of medical primatology, stimulated by the growing demands of experimental medicine and biology and backed by the great discoveries in the field of genetics and molecular biology. This was the *stage of extensive primate research in the developed countries, which was characterized by the establishment of specialized primate centers.* It began in the 1960s with the creation of the Regional Primate Research Center Program in the United States, administered by the National Institutes of Health, in cooperation with a number of universities.

In 1960, the Congress of the United States allocated $18.5 million for the construction of seven regional centers (by the end of 1967, more than $51 million had already been spent). The centers began operations between 1960 and 1966 (Goodwin, 1972). They include the Yerkes Regional Primate Research Center (originally founded in 1930), the Oregon Regional Primate Research Center (1960), the Wisconsin Regional Primate Research Center (1960 according to the *International Directory of Primatology*; 1964 according to Goodwin, 1972), the Regional Primate Research Center at the University of Washington (1961), the California (formerly: National) Regional Primate Research Center (1962), the Tulane (formerly: Delta) Regional Primate Research Center (1962), and the New England Regional Primate Research Center (1966). A new, eighth, Regional Primate Research Center has been established (1999) at the South-West Foundation for Biomedical Research in San Antonio (Texas). The regional centers are also associated administratively with university medical schools, but all research programs are mainly coordinated and sponsored by the National Institutes of Health, which has an official office in Bethesda, Maryland, under the auspices of the Animal Resources Program. The term 'regional' has both scientific and geographical implications. Each center is an independent research institute which has: 1) a mission, or specific research objectives; 2) a program of non-staff scientists, as well as; 3) programs similar to those at other centers. The centers are also sponsored by other organizations, private companies, and firms for conducting research activities unrelated to those of the National Institutes of Health.

The present day National Institutes of Health primate centers are highly efficient research institutes which play an important role in the history of science. As of 1994, a total number of 1,100 investigators at the doctoral level had carried out research at these centers. Over 17,000 primates of 32 species have been maintained for this purpose. Through their research programs and primate colonies, the centers are associated with over 400 research institutes around the world; they publish approximately

2,000 scientific papers annually. Over 2,500 young primates, including chimpanzees, are born at the centers in open-air cages and vivariums each year. The remarkable Primate Information Center, which was founded by T. Ruch, has been in operation for over 30 years, presently headed by J. Pritchard. It collects primate data and articles from all over the world and delivers them to scientists world-wide (Washington Primate Center, Seattle). In 1994, the *International Directory of Primatology* (editors L. Jacobsen and R. Hamel) was compiled at the Wisconsin Regional Primate Research Center, crowning many years of hard work on this world-wide encyclopedia of primate research.

The centers are concerned with the most urgent issues of modern medicine, pharmacology, and biology of primates, including humans. Those issues include AIDS, Alzheimer's disease, leprosy, glaucoma, organ transplantation, reproductive biotechnologies, vaccines against various infections, oncological research, and mutation induction (Hearn, 1994; Vaitukaitis, 1994). The influence that the United States regional primate research centers exert on modern experimental science extends far beyond the borders of the country.

After the 1960s, the establishment of primate centers, institutes, and primate colonies became a characteristic feature in the development of medical primatology. This phase was first dominated by the developed countries of the world and then by the countries in which the primates originate. The latter countries often became bases of cooperation with the research centers of the former countries. The *International Directory of Primatology* has a total of 272 primatological institutions registered world-wide. Not all of these are up to a standard sufficient to be called research institutes, which is warranted in approximately 100 of them. The largest number of registered centers (128 of the 272) is accounted for by the United States. There are also large primate centers in Japan, Russia, the Netherlands, Germany, and China. Similar centers may be found in many South American countries, while native primate colonies exist, of course, in countries where apes and monkeys have their natural habitats.

What is the particular value of specialized primatological institutions and, primarily, centers? The centers regularly supply researchers with primates for experimental use and with data on their genealogical, genetic, and pathological peculiarities. They offer the possibility of obtaining pregnant females and newborn infants which are sometimes required for experimental purposes. The centers provide healthy primates, free from infection, virus, and parasite pathogens, which are characteristic of primates that are obtained from the natural habitat. The centers maintain secure quarantine facilities, precluding the hazard of epidemics, not only for the primates but also for the humans that work with them. Also important is the presence of trained personnel with the necessary skills for the difficult task of maintaining and breeding monkeys and apes. Finally, primate centers and their colonies indisputably help to preserve a most valuable fauna, the nearest relatives to humans.

The centers have been and always will be powerful incentives for the development of medical primatology, a fact which is easily verified by the history of the world's oldest primatological centers. It was in the 1960s, when a majority of the primate centers were established, that there was a dramatic increase in research on nonhuman primates. This is clear from the growth of professional publications in the field of medical primatology.

There was a group of data analysts concerned with issues in medical primatology who worked under the authority of Sukhumi Primate Information Center. The Center, which was established in 1966, operated in conjunction with an analogous center at the University of Washington. The group analyzed the publication rate of research in

Table 2.1 Annual publications on nonhuman primates in the area of medical primatology during the period from 1967 to 1976, and for 1985.

1967	1968	1969	1970	1971	1972	1973	1974	1975	1976	1985
2500	3116	3572	4236	4400	5302	4854	4802	5331	4700	6000

medical primatology world-wide and published the results in Russian and in English on virtually an annual basis (1970–1988). Table 2.1 contains the data obtained by the analysts on the annual number of publications for the 10-year period from 1967 to 1976 inclusive, and for 1985.

In a symposium at Sukhumi, T. Ruch reported that a total of only 3,000 articles were published during the 1950s (Ruch, 1966), while 1,000 articles were published in 1964 alone. Starting in 1968, moreover, the number of articles published each year was considerably greater than the total number published within the entire 10-year period of the 1950s (Table 2.1). The number of publications per year more than doubled between 1964 and 1967, a period of only three years. The annual number of publications in the 1970s, moreover, exceeded the total volume of the primate literature throughout history, from Aristotle to World War II! We should not overestimate, of course, the statistical significance of these figures. It is noteworthy, however, that even in the dynamic area of nuclear physics during the 1960s, a doubling in the rate of professional publications took 10 years (Sorokin, 1970). As far as medical science is concerned, where a doubling of professional publications has required from 12 to 15 years, the substitution of primates for other experimental animals was associated with a sharp increase in publications on medical primatology. The annual increase of 16% in the first 5 years was extraordinary (the analogous average increase in other fields of science was 7%, while the maximum increase in some was only 10%). This increase was indicative of a turning point in the development of medical primatology that reached a peak in the 1960s and 1970s. This is shown in Figure 2.1, based on T. Ruch's data, and in Figure 2.6, based on my own investigations.

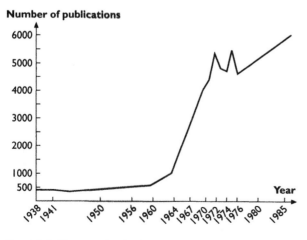

Figure 2.6 The growth of publications on nonhuman primates in the area of medical primatology between 1938 and 1985 (based on personal observations and other sources).

The number of new publications is directly related to the increasing number of new primatological centers. I draw the conclusion, therefore, that these centers were a characteristic feature of the period of increased primate research. Among other influences, the centers also encouraged the use of primates in other, non-primatological institutions. The growing achievements of medical primatology gave impetus, in particular, to more intensive research on monkeys and apes in biomedical laboratories. The number of biomedical laboratories increased as well. During the period from 1968 to 1976, the Sukhumi Primate Informational Center registered 2,584 non-primatological research institutes that published research results on primates (Fridman, 1977a,b).

After 1975, there was a reduction in the rate of growth of medical primatology, although the high annual volume of publications continued. This was accounted for primarily by three factors. Mathematically, there was a certain degree of 'satiation'; exponential curves cannot continue to increase indefinitely. Politically, there was an embargo on the exportation of primates by a majority of supplier countries in the second half of the 1970s. The embargo was initiated by India's example to protest experimentation on monkeys for military purposes. Economically, there was a considerable increase in the cost of nonhuman primates due to the decreasing number of these animals in their natural habitats. This situation was improved by measures which I discuss below, such that by the mid-1980s, the number of publications on primates increased to as many as 6,000 per year (Fridman and Popova, 1988). In 1998 the figure was 7,157 publications annually according to Current Primate References, and in 1999, 6,601. Considering the fact that data from other sources must be added to those contained in Current Primate References, and making several other allowances for determining the number of publications per year, I conclude that the annual rate of new publications on primates today amounts to as much as 7,500–8,000. A majority of these publications originate, naturally, in the United States (57%–62%), the rest are distributed, in descending order, between Great Britain, the Soviet Union, Japan, France, Germany, Canada, the Netherlands, Switzerland, the Republic of South Africa, and other countries.

The total number of primate species used in experimentation of all types, including non-medical experiments, was 57–59 in the period from 1967 to 1971, and increased in 1976 to 98. A majority of experiments (75%, on an average, in the first 5 years in question, and 89% in 1971), however, were performed with only 5–6 species: *Macaca mulatta* (up to 40% of all registered publications); *Saimiri sciureus*; *Pan troglodytes*; *M. fascicularis*; and *Cercopithecus aethiops*. Other species were discussed in only a very few studies. Afterwards, the situation remained very much the same, with the exception that *M. fascicularis* became the second most frequently used species. On the whole, the genus *Macaca* has been the most widely used primate in research, accounting for about 50% of all the primates used (Fridman and Popova, 1983).

The total number of nonhuman primates that were removed from their natural habitats in the 1960s and 1970s reached rather large figures, according to some sources, about 250,000 animals annually (Dukelow, 1972). This rate of exportation could not continue indefinitely. As noted above, the number of animals in the order Primates is decreasing (with the exception of humans, of course). Over 80 species in this order have been entered in the Red Book of threatened and endangered species and some of those that have not been entered might well have been if hunting for them had continued at the previous rate. Primate researchers themselves were the first to sound the alarm (Goldsmith, 1968; WHO, 1971). The participants of the Fourth Congress of the

International Primatological Society formed a Group for Primate Protection. An international convention was signed which substantially limited trade in all primates. The most decisive measures, however, were taken by 'monkey exporting' countries. During the period from 1973 to 1976, they first limited primate exports and then some banned the practice altogether. India's embargo was the most distressing one. The United States, which had only recently been buying 100,000 monkeys annually, received only 30,000 from India in 1973, and only 15,000 in 1974 (Primate Supply, 1974). The cost of monkeys sky-rocketed. Rhesus macaques, which had cost $10–$15 in 1972, were sold for $400 in 1976, $1,000 in 1980 and $3,000 in 1990. (According to D. Bowden and C. Johnson-Delaney (1996) in the early 1990s the cost was '$1200 or more' per monkey).

Nevertheless, the developed countries were relatively prepared to deal with the changed situation in medical primatology, primarily due to the existence of the primate centers. In the United States, where the Interagency Primate Steering Committee was already in operation, the National Primate Plan was initiated (1978), which was later amended on several occasions. Breeding programs for macaques, chimpanzees, and other species were introduced (Hobson et al., 1991). In 1977, the European Economic Community also formed a work group in Brussels concerned with the use of primates (today the European Primate Resources Network operates using the European Primate Information System of the European Union). The Soviet Union formed the Primatological Commission at the Academy of Medical Sciences Presidium. WHO, which also made its contribution, called for 'activating the development and execution of primate breeding programs' (WHO, 29th session, 1976).

Among other steps taken by scientists and officials to resolve the crisis was the more prudent use of primates in experiments. More stringent criteria for proposals to use nonhuman primates in an experiment were introduced. These measures proved so effective that they brought about a significant scientific 'paradox'. The number of imported primates plummeted, while the number of publications in medical primatology in the United States, Japan, and some other countries continued to grow (Fridman et al., 1990). At the beginning of the 1980s there were 2,500 rhesus macaques imported into the United States in comparison to tens to hundreds of thousands in the 1960s. The Primate Supply Information Clearinghouse was established and is still operating today under the auspices of the Primate Information Center at the University of Washington in Seattle. The clearinghouse advises researchers about animals that can be used recurrently after being released from other experiments. Primate tissue cultures began to be used more intensively. Attention was focused on increasing the natural populations of nonhuman primates on special reservations. But the most remarkable step was the establishment of breeding colonies of certain primate species in the United States as well as in the Soviet Union, Japan, China, and European countries.

The efforts made by American researchers to obtain nonhuman primates for research can be said to have no precedent in the history of science. In a relatively short timespan dozens of new colonies designed specifically for breeding were established. Millions of dollars in grants were offered to all the Regional Primate Research Centers and to other institutions to this end. New breeding colonies were established on four islands near Florida, on an island near Puerto Rico, on islands off the East Coast of South Carolina, in Louisiana, and in south Texas. The United States Air Force colony at the Holloman Air Force Base became a Food and Drug Administration colony for chimpanzee breeding. By 1984, United States researchers were supplied with about 8,000 colony-born primate infants per year, including 5,000 rhesus macaques (Johnsen, 1995). This number could not completely satisfy the needs of science, however, since some of these animals were taken for vaccine safety experiments (vaccines against measles, mumps, rubella, polio, hepatitis). In conjunction with thousands of nonhuman primates that were imported from those countries that had not reduced their supply (e.g., the Philippines and Indonesia), however, the domestically bred primates were a substantial benefit to the United States laboratories. Starting in the mid-1980s, a new round of support for breeding primates in colonies was initiated in connection with the spread of AIDS, a problem impossible to resolve without the use of primates. According to institutional reports, research in the USA used about 50,000 primates in 1992 (48,051 in 1985). Half of them were bred in the USA. 'Despite the impact of a rapid succession of negative social and economic factors of historic magnitude, primate research in the United States has continued to grow during the past decade' (Bowden and Johnson-Delaney, 1996, p. 56).

We must also recognize the role of the animal rights activists who increased their efforts in response to the expansion of research on primates and other animals. Everyone, especially the researchers, acknowledges the urgency of preserving the world's precious fauna. It is necessary, however, that we prepare a carefully reasoned and reasonable compromise to this controversy. We cannot promote one noble cause at the expense of another, namely, we cannot restrict the use of animals in medical research and condemn humans to new diseases and premature death. As noted above, it was the medical primatologists themselves who were the first to call for the protection of primates. In November 1981, the WHO Workshop concluded: 'According to rather understated accounts, almost 5 million children die every year and almost 5 million become handicapped as a result of diseases which could have been prevented by immunization. Many adults also die or become disabled due to the same diseases. The United Nations charter states that it is WHO's objective to attain the highest possible health standard for all nations. Therefore, WHO believes it expedient to use tens of thousands of primates annually in order to save approximately 5 million lives and prevent the same number of children from becoming invalids' (A footnote clarified that 40,000 or fewer primates are used for these purposes every year) (Press Release WHO 29, 1981). Regarding the measures for protecting primates and the participation of various organizations, see Johnsen (1995).

Programs for the preservation of the order Primates, as well as for individual species, have been introduced, for the most part, by research primatologists who recognize the value of these animals to science. For many years, energetic efforts have been made to protect the wild population of gorillas; there is a periodical concerned with this issue and an extensive international program. There are also special projects for preserving pygmy chimpanzees, orangutans, gibbons, tamarins, and many other primate species.

A significant contribution to the development of medical primatology and to the cause of primate preservation was made by WHO. An organization of great authority, it possesses substantial financial and scientific capabilities. It began its work with primates in the mid-1960s. Since that time, as indicated by the Sukhumi Primate Information Center files, WHO publications on primates have appeared virtually every year, including professional papers, expert recommendations and accounts, and press releases. WHO, however, was concerned not only with 'the primate issue' when considering the problems of malaria, the hepatitis virus, hemorrhagic fever, filariasis, schistosomiasis, pregnancy toxicology, vaccine development, drug teratogenicity and the standardization of biological preparations, and especially, the AIDS issue. All of these health problems and many others were dealt with by WHO experts from 1965 to 1990, problems that cannot be solved without using experimental primates. What WHO was directly concerned with were the issues of primate use and primate biology, as well as primate preservation. WHO was the host of several international primate symposia (WHO, Use of Nonhuman Primates . . . , 1971; WHO, Sense of Nonhuman Primates . . . , 1988, and others). WHO expenditures on primate issues were estimated to be $100,000 per year (DRAFT:DBD:22/10/1980). WHO established collaborative centers concerned with different areas of medical primatology, including primate virology (San Antonio, USA), hematology (New York, USA), primate zoonoses (Atlanta, USA), reproductive endocrinology (Sukhumi, USSR), and others.

As indicated above, the growth in primate research was accompanied by a corresponding increase in the number of research publications on primates. Primate studies were published mainly in professional journals concerned with biology and medicine. In the 1960s, the practice of publishing specialized studies in medical primatology was established, including the proceedings of congresses and symposia, compilations and serial articles on biomedical experiments in primates, and studies on individual primate species. Monographs in fundamental biomedical science, namely, biology, physiology, endocrinology, hematology, pharmacology, etc., contained an increasing number of primate studies and citations. New scientific journals emerged in the 1960s, devoted specifically to primates. The journal *Primates*, mentioned above, was reorganized and has been published regularly since 1962, and *Folia Primatologica*, since 1963. In 1972, the prominent scientists of medicine and primatology, E. Goldsmith and J. Moor-Jankowski began publishing the essentially international *Journal of Medical Primatology*. Moor-Jankowski is also widely known as the founder (in 1965) and long-time Director of the Laboratory for Experimental Medicine and Surgery in Primates of New York University, which contains a WHO Hematology Center. At about the same time the Publishers, S. Karger (Basel, Switzerland), issued a series of primate publications, including *Primates in Medicine, Medical Primatology*, and others.

The term 'medical primatology', the meaning of which was discussed at the beginning of the chapter, was first used by us in print in Russian and in Latvian (Fridman, 1963). In 1967, the anniversary compilation of studies by the Sukhumi Institute of Experimental Pathology and Therapy was published under the title of *Medical Primatology*, edited by B. A. Lapin (Lapin, 1967). In 1965, the term came into use in the United States, at the same time that Moor-Jankowski's Laboratory was established (Muchmore *et al.*, 1971). As mentioned above, the term is included in the name of a research institute of the Academy of Medical Sciences, Russia (Adler, Sochi) and it has been entered in the *Comprehensive Medical Encyclopedia* (Moscow).

At the end of the 1970s and the beginning of the 1980s, two other primatology journals began publication, the *International Journal of Primatology* and the *American Journal of Primatology*. In the 1990s *Asian Primates, Neotropical Primates*, and *African Primates* were added to the primatological literature. In addition to the standard research journals, there are about 50 bulletins published world-wide on various primate issues, separate species, and specifically, on medical primatology issues, thus reflecting the ongoing dynamic development of research in this field. As a rule, such bulletins are published by all primatological centers of the world.

MAJOR ACHIEVEMENTS OF MEDICAL PRIMATOLOGY IN THE 20TH CENTURY

Research in *virology* warrants a special status among the medical sciences. As is well-known, all the basic human virus groups were experimentally studied in primates. The achievements in the extensive field of virology, moreover, as a rule are associated with experiments on nonhuman primates (Kalter, 1969; Gibbs and Gajdusek, 1976; Sabin, 1985, 1993). The history of this area of research began with poliomyelitis. The prevention of this disease by means of vaccines is an outstanding achievement of modern medicine which saved the lives of millions of people who probably never fully appreciated its significance.

In the context of the present monograph, the solution of the polio problem is associated with two powerful factors in the development of medical primatology. This disease has been the strongest stimulus for conducting biomedical research on primates since the 1940s and 1950s. It was the basis for one of the other great successes of modern science which was crowned with the Nobel Prize, namely, the technique of virus replication in primate tissue culture.

It is known that poliomyelitis is impossible to reproduce in any animal other than a nonhuman primate, at least with respect to two virus types, one of which is the most dangerous for humans. The absence of an experimental model not only rendered medicine powerless in the fight against this dangerous disease, but also precluded the prospect of studying it experimentally, which was vital. As was mentioned above, Carl Landsteiner and Erwin Popper first reproduced polio in two rhesus macaques in 1908. The next year they also established the viral etiology of this disease using rhesus macaques and hamadryas baboons. A relatively active program of virological and immunological research was initiated. Attempts were made to immunize humans, but they were aborted in the 1930s due to the threat presented by the preparations tested. The end of the 1940s became a turning point when a group of researchers learned of the possibility of cultivating the virus in proliferating primate tissue cells (Enders *et al.*, 1949). The subsequent history of the fight against polio is now widely known; the discoveries in the United States of the 'killed vaccine' by J. Salk and then the 'live vaccine' by A. Sabin. The latter vaccine is most commonly used nowadays for the prevention of polio.

An important role in making the Sabin vaccine available for immunizing people was played by the Russian virologists M. P. Chumakov and A. A. Smorodintsev, whose own children and grand-children were the first humans to be inoculated with the new vaccine after its trials on monkeys. Chumakov, whose name would later be given to a virological institute in the Moscow region, conducted polio experiments on primates in Sukhumi for 11 years. The Sabin vaccine was introduced and used initially in Soviet medicine at the end of the 1950s. The virus was grown in cultures of primate kidney tissue. Safety testing of the vaccine for neural virulence was also done on primates. Interestingly, spinal neurons of chimpanzees (and, as extrapolated, of humans) proved more resistant to the polio virus than they were in monkeys, while gastrointestinal observations yielded contrary results (Sabin, 1985).

The live peroral vaccine against polio became the panacea for humanity. It facilitated the complete elimination of polio epidemics in the Soviet Union. By 1967, the number of people afflicted with this disease, according to the Minister of Health, was reduced by a factor of 54. When the Sabin vaccine was accepted in the United States and applied in health care, the number of polio patients was reduced from 100 to 28 per million people within the time-frame of the first 5 years. After 20 years approximately 5 million cases of poliomyelitis were prevented, primarily in countries of moderate climate with a total population of almost 2 billion people (Sabin, 1985). A number of statisticians attempted to calculate the dollar amount that was saved by solving the polio problem. Whereas the amount of money spent on polio research was about $65 million annually, the saving was about $6 billion per year (according to Prof. Zinader in the mass media). No amount of money, however, could be equated with the lives of those saved from poliomyelitis, considering that by 1956 there were 300,000 invalids with paralytic polio conditions in the United States alone, not even counting those who died of the disease (Margath and Reeve, 1993; Chanock, 1996). Why was this brilliant victory of science not crowned with a Nobel Prize?

Success in fighting measles is also associated with research on primates, the only animals that could provide an adequate model of this disease for human testing. Early attempts to obtain a model in chimpanzees at the beginning of the century had failed. In 1911, several groups of American and French researchers initially showed the general susceptibility of monkeys to measles (Anderson and Goldberger, 1911; Nicolle and Conseil, 1991). Further attempts to fight the illness, however, were delayed for many years. Measles research began at the Sukhumi simian nursery as early as 1940, but it was interrupted by World War II. After the War, a multitude of measles experiments were initiated there, led by P. G. Sergiyev, on etiology, pathogenesis, immunology, and vaccine testing. Similar research was successful in the United States and a number of other countries. Isolation of the measles virus in primate tissue culture by Enders changed the state of affairs in this field, and soon the first live anti-measles vaccine was formulated. At the beginning of the 1960s the first vaccination procedures were available. The rates of this sickness in the Soviet Union decreased 4- or 5-fold. Mass immunization by the vaccine protected 90% of the children vaccinated, securing immunity for over 4 years. Anti-measles vaccines are constantly modernized as new and more effective versions are created. The old scourge of humanity, which for hundreds of years had been afflicting children, was now weakened. It is noteworthy that testing of the vaccine's safety and effectiveness was carried out on monkeys.

Also of exceptional significance in the area of medical primatology was the yellow fever research conducted on monkeys. Mankind paid a dear price for the victory over this disease; not only did hundreds of thousands of patients die, but also some of the scientists who had been seeking the means to fight it. At the beginning of the century a number of American scientists led by W. Reed showed that the yellow fever pathogen was a 'filterable' virus. The study was conducted on humans. In 1907, the English scientist H. Thomas determined that yellow fever in chimpanzees could cause the illness in humans. Aborigines on the island of Trinidad were able to predict yellow fever epidemics by a growing death rate among the local population of howler monkeys (Alouatta caraya). The role of monkeys in the etiology of yellow fever was established definitively during the epidemic of 1928 in Rio de Janeiro. At about the same time, the experimental species for studying this pathogen was found. The Englishman Adrian Stokes reproduced yellow fever in rhesus macaques, thus providing an alternative to studying the disease in humans. At the end of 1928, Stokes died when he contracted yellow fever himself, but the experimental model had been found (Stokes et al., 1928). The Azibi virus strain was isolated and was later successfully used to create a vaccine. The South African microbiologist Max Theiler weakened this strain to absolute harmlessness for monkeys by inoculating white mice. The newly invented 17D vaccine was tested on primates and found to provide macaques with strong immunity. In 1931, the vaccine was first tested on people.

Naturally, further primate experiments were conducted to improve the vaccine and eliminate its negative side-effects. Mass vaccination of people against yellow fever prevented urban epidemics which had raged for 400 years since the 16th century. Studies of yellow fever had been going on for a long time, but the above success in this field became one of the greatest scientific achievements of medical primatology (Galindo, 1973). Theiler was awarded the Nobel Prize in 1951 for developing the vaccine and for his other outstanding contributions to the fight against yellow fever.

Primate research also proved that epidemic parotitis was caused by a virus (Gordan, 1914). Experimental studies of parotitis were possible only with nonhuman primates. As a result of these experiments, a mixed anti-parotitis and anti-measles vaccine and a live parotitis vaccine were prepared (Vinogradov-Volzhinski and Shargorodskaya, 1976). Research in this field continues. The chicken-pox virus was also discovered and studied by means of primate research (Rivers, 1927). Similarly, nonhuman primates played a most important role in the study of dengue and chikungunya. The etiology of a majority of viral encephalitides was determined by primate research, including St. Louis, Scottish, and Russian spring-summer encephalitides (Lapin *et al.*, 1987). The virus of lymphocytic choriomeningitis was discovered, moreover, in association with the study of St. Louis encephalitis (Armstrong and Lillie, 1934).

In 1923, it was established that nonhuman primates were the most suitable animals for smallpox virus studies (Blaxall), a discovery which was soon endorsed by various other scientists. Smallpox has been successfully studied in monkeys since that time. The eradication of smallpox is another major achievement of medical primatology (Arita, 1979). Other successes of primate research include a number of vaccination techniques, including those against herpes simplex, rubella, pseudorabies, cytomegalovirus, infectious mononucleosis, venereal lymphogranuloma, and other viruses.

Studies of 'slow virus infections' in primates have also been very fruitful. The study of amyotrophic lateral sclerosis went on for many years in Sukhumi, led by L. Zilber and N. Konovalov (Gardashyan *et al.*, 1970). D. C. Gajdusek (Bethesda, USA) and his colleagues initiated research on the problem of kuru, a mysterious disease of unknown etiology which almost wiped out a whole tribe in New Guinea. The Gajdusek group managed to obtain a model of kuru in chimpanzees and then in other nonhuman primates. The viral character of kuru was determined and successful measures were taken to eradicate this dangerous disease (Gajdusek, 1977). Gajdusek's experiments gave new impetus to the study of a number of other 'slow' infections of the central nervous system whose etiology remains unknown. Carleton Gajdusek was awarded the Nobel Prize in 1976 for his discovery of a new virus class and for his studies of kuru. At present nonhuman primates are also used to study other dangerous diseases, such as Alzheimer's, Creutzfeldt-Jakob disease, and Parkinson's.

There is another chapter in the history of modern experimental science which is quite impossible to imagine without the influence of primates. I refer to the studies of viral hepatitis. There are no animals other than nonhuman primates that are suitable for obtaining a sound model of these human pathologies, one of which (hepatitis B) is related to the etiology of human fibrolamellar liver cell carcinoma. As is well-known, this field of science has also been marked with the Nobel Prize, in this case for the discovery of the hepatitis B virus antigen by Baruch S. Blumberg in the 1970s. All of the research on etiology, pathogenesis, immunology, and vaccination of hepatitis A, B, C ('non-A, non-B'), D, and E, including vaccine quality and safety tests, have been conducted on various species of primates.

Few people are aware that the discovery of interferon, the remarkable anti-virus drug, is also associated with primate research. In 1935, while studying different strains of the yellow fever virus, I. Haskins discovered the interaction of the yellow fever virus with the Rift Valley fever virus, a phenomenon he called virus interference (Findllay and McCallum, 1937). Further research in this field led to a major discovery which is presently being effectively verified on primates. Great prospects are promised by the new interferon biotechnology (Isaacs and Lindeman, 1957; Koch, 1987).

No new leaf has yet been turned in the dramatic history of AIDS research, which began in the early 1980s. The only satisfactory model of this disease, as is widely known today, is established in a few nonhuman primates. The fight against humanity's new enemy began in primate centers, primarily in the United States, but also in Sukhumi and elsewhere. Most experimental AIDS issues, moreover, are tackled with the help of monkeys and apes.

The use of primates in trachoma research, for which there is no substitute as experimental animal, also has a long history. Not very long ago there were 400 million people suffering from this disease, including 6 million who lost their sight. The trachoma model was first obtained in orangutans as early as the beginning of the century (Halberstaedter and von Prowazek, 1907). In the 1960s, trachoma studies were intensified and extended. It is now known that trachoma is induced by chlamydia, which are related to both viruses and rickettsiae and which cause other dangerous illnesses as well. As in earlier examples, no laboratory animals are suitable for chlamydiosis research other than nonhuman primates, which are required to find the cure for trachoma and similar infections and to prepare modern vaccines (Campos et al., 1995).

Monkeys are absolutely indispensable for studying many other diseases of an *infectious* character. Primates were used in studies of plague, nagana, sleeping disease, cholera, and particularly widely, tuberculosis. It was only through the use of chimpanzees and bonnet macaques (*M. radiata*) that C. Nicolle and his associates were able to draw the final conclusion that the spotted fever pathogen is transmitted by the clothes louse (*Pediculus vestimenti*) (Nicolle et al., 1909). This discovery opened up a variety of research prospects and eventually led to the victory over this centuries-old serious infection. This victory was awarded the Nobel Prize, as already mentioned. Baboons were the only species in which a tetanus model could be established and in which an immunization technique could be developed with subsequent revaccination against tetanus and diphtheria (Zdrodovski, 1961). This discovery saved thousands of lives in the Soviet Union, including World War II soldiers, who were saved from infections resulting from wounds. A similar solution was provided at the front by penicillin, which was tested on monkeys by Z. W. Ermolyeva at Sukhumi, as were all antibiotics afterwards.

A number of human diseases which have not yet been eradicated are being kept under control to some extent as a result of experimental models in primates (which are unattainable in any other animals) and the continuing intensive research they make possible. The diseases in question are dysentery, gonorrhea, leprosy and some others, reproduced only in primates and widely studied today. Malaria is also a member of these problematic diseases. It has been studied for some time with a certain degree of success, but it still has not been defeated. A Nobel Prize was awarded for malaria research, however, at the beginning of the century (D. Ross, in 1902). Experimental modeling of malaria is feasible only in primates (Young, 1973). Exacerbation of the problem in the 1960s with the 'new' dangerous form of malaria caused by *Plasmodium falciparum*, failed to discourage the scientists, who were relatively quick to invent a unique model of this disease in owl monkeys (*Aotus trivirgatus*). This model facilitated the search for drugs and vaccines, including recombinant ones. Fighting different variants of this ancient disease still continues (Perera *et al.*, 1998). On the whole many parasitic diseases that are not reproducible in any other laboratory animals are widely studied in primates, as we shall see.

In conclusion to this brief historical account of infectious disease research on primates, it should be noted that probably any common human infection can be reproduced in some nonhuman primate species. The services rendered by medical primatology to humanity in delivering it from the most dangerous illnesses which have preyed on mankind for centuries and millennia cannot be overestimated.

The most outstanding results in medical primatology were obtained in studies of primate *blood* and related problems. It was already mentioned that investigations of immunological blood reactions in different animal species and in humans were shown to be of special importance at the beginning of the 20th century. G. Friedenthal, who carried out the first blood transfusion from man to chimpanzee, noted the great similarity between the blood of humans and apes (Friedenthal, 1902). The similarity of blood in different species, including primates, was shown by G. Nuttall in his classical work (1902). Carl Landsteiner discovered the human blood groups and devoted many years of hard work studying the same issue in nonhuman primates. In 1925, experiments on 21 apes, including 14 chimpanzees, showed that a blood sample from any one of these primates could be related to one of the four AB0 human blood groups (Landsteiner and Miller, 1925). Gorillas were not used in these experiments.

These studies were later expanded and resulted in the discovery of the blood types of lower simians (Moor-Jankowski and Wiener, 1971; Socha and Ruffie, 1983; Socha, 1986; Blancher and Socha, 1997), establishment of the most important phylogenetic links in the order Primates, extrapolation of the data to biomedical research (in organ and tissue transplantation studies, which preceded clinical applications), and even the direct use of nonhuman primates for curing human illnesses (e.g., xenogeneic liver perfusion in hepatic coma patients) (Fischer, 1979).

An epoch-making achievement of primate blood research dates back to 1937 when C. Landsteiner and A. Wiener discovered the so-called Rhesus factor in the blood of *M. rhesus*, which they reported after only 3 years of experimentation (Landsteiner and Wiener, 1940). This discovery and its further application played an exceptional role in the history of medicine, saved thousands of lives, and preserved the health of many more people. The next discovery in this area was that of the Rh subfactors and various primate blood type antigens, which also made a major contribution to the development of modern medicine and biology (Socha and Moor-Jankowski, 1980), particularly in the essential field of immunohematology.

Primates are also effectively used in human hemopoiesis studies and in various marrow manipulations, including prophylactic ones. The first experiments in marrow transplantation were initiated by E. Donnall Thomas, who used baboons from the University of Washington Primate Center. After all the necessary procedures were perfected in the baboons, the data were extrapolated to the human clinic where marrow transplants are still used to treat leukemia and other blood-related genetic disturbances in humans. The study won the Nobel Prize in 1990.

Extensive research in the field of *endocrinology* has been performed on monkeys and apes since the 1960s, although fundamental studies of a similar orientation had been conducted earlier. It is appropriate, however, to point out the enormous significance for medicine as a whole, and for the development of medical primatology in particular, of the research by Li Choh Hao, also a Nobel Prize winner. This scientist investigated growth hormone and determined that only in humans and simians were somatotropic hormones mutually bioactive, those from other animals producing no specific effect on primates (Li Choh Hao and Papkoff, 1956). There are now a variety of endocrine studies conducted with primates, but primates proved to be of particular importance in studies of the human reproductive system, which is uniquely similar in humans and simians.

In fact, primate research helped establish the fundamental principles of female reproductive function in C. Hartman's laboratory as early as the 1930s. The mechanism of menstruation, the role of the hypophyseal hormones in regulation of the menstrual cycle, and the physiology of the neck of the uterus; all of these have been successfully studied in macaques and described in the classical works by C. Hartman, J. Streeter, R. Dickinson, and others (The Carnegie Monkey Colony, 1925–1959). It was at that time that J. Markee (1940) used primates for his famous experiments on transplanting endometrial tissue to the anterior chamber of the eye in order to clarify the mechanism of menstruation. This researcher spent 9 years working on the problem (Christiaens, 1982). The discovery of luteinizing hormone and its role in the regulation of the human and simian menstrual cycles was rewarded by the Nobel Prize in 1977 (to P. Gimmelin, E. Shelley, and R. Yaloe). This experimental research could only be conducted with primates.

Among the other types of primate research, *radiobiology* deserves at least a brief description. Apparently, the main achievements in this field were kept secret, but what was published openly is sufficient to make clear that primates are indispensable in studies of radiation sickness, in the development of radioprotectants, and in the creation of other drugs for the prevention and treatment of this contemporary disease and its far-reaching consequences (Semenov, 1967). The remarkable cytogenetic study by N. Dubinin and colleagues was already mentioned. It contained experimental data on the 'redoubling dose' of radiation obtained from experiments on primates, rather than on mice, rats, or other animals. This discovery became the basis for recommendations by the United Nations Scientific Committee on Radiation to the General Assembly regarding the harmful effects produced by minor doses of radiation on chromosomes of the human reproductive system. Of considerable importance, the primate research showed that the data based on rodents were inapplicable to humans. Thus, 'monkeys saved mankind', in a sense, when the treaty on banning nuclear tests in three media was signed in 1963. These events represent golden pages in the history of medical primatology (Dubinin and Gubarev, 1966).

Oncological experiments on primates, although rather meager in number, had been conducted since the end of the 19th century. Malignant tumors were first induced in monkeys by chemical and radioactive carcinogens only in 1948 (Petrov *et al.*, 1951). Subsequently, primate experimental oncology became best known for its virological emphasis or oncovirology. Of great significance for extensive use of primates in this field was the discovery of viruses in New World primates that were carcinogenic in other primate species. This discovery was made by the oncologists of the New England Regional Primate Research Center in the United States (Melendez *et al.*, 1971).

Experimental proof of the viral etiology of Burkitt's lymphoma and the isolation of the Epstein-Barr oncogenic virus in England in the 1960s were also very important (Epstein, 1974).

In the second half of the 1960s a model of viral leukosis, transmitted by the blood of sick humans, was obtained in the Sukhumi Primate Center under the leadership of B. Lapin. Many years of research began, during which the researchers managed to establish horizontal transmission of lymphoma in monkeys (Lapin and Yakovleva, 1970) and to elucidate many oncovirological issues in primates. The research in this area continues.

Publications on primates used in the field of *pharmacology* have increased in volume considerably since the 1960s. The practical value of simians in this field, however, seems to exceed the scientific value of the publications. Nonhuman primates were the experimental subjects in the testing of many medical preparations, very often where other laboratory animals were not suitable. The research showed that testing drugs on other animals might prove unreliable or even lead to tragic results, as happened with the tranquilizer thalidomide in the 1960s. As noted, primates were used in testing vaccines and antibiotics, as well as hypotensive drugs. Of special importance were experimental monkeys in neuropharmacological and psychopharmacological studies. The tragedy of thalidomide provided a stimulus not only for pharmacological studies in primates, but for the development of medical primatology as a whole. Following the case of thalidomide, WHO experts insistently recommended testing all neuropharmacological preparations on primates prior to their adoption (Reynolds, 1969).

It should also be noted that primate research made it possible to model all types of alcohol-induced damage to the human liver, a finding which could not be reproduced to full value in any other animals (Rubin and Lieber, 1973).

Experimental reproduction of human diseases in nonhuman primates could not have been accomplished without scientific knowledge of the spontaneous or normal *pathology* of these animals, at least as determined in captivity. Extraordinary results were obtained in this field by T. Ruch (1959), who collected data from the literature for 25 years before presenting his brilliant summary of them, as well as by R. Finnes (1967). The same may be said for the outstanding morphological pathologists B. Lapin and L. Yakovleva. These scientists studied similar problems for more than 45 years, resulting in data on tens of thousands of autopsies. In one of their latest publications on primate diseases they discussed the similarity of primate nosological profiles. The data were obtained not only under captive conditions, but also in the natural settings of Vietnam and Nigeria; they call the nonhuman primate a 'laboratory double' for humans (Lapin and Yakovleva, 1994).

When discussing the history of primate research in the field of *neurophysiology* and *behavior*, it should be emphasized that this aspect of medical primatology is the most extensive one. According to my data, this field accounted for 20% (1967) to 30% (1985) of all publications on primates (1,836 of 6,000). C. Darwin pointed out the great potential of physiological research on primates and published the first celebrated book on their expression of sensations and emotions (1872), drawing special attention to monkeys and apes. The simian did not become a significant species in psychobiological research, however, until the 1960s, in spite of a number of remarkable publications in this field.

There are different opinions in the literature as to who was the first to conduct studies in comparative psychology with primates. The first experiments on hominoids were conducted by L. Hobhouse, N. N. Ladigina-Kohts, and E. Thorndike. It can be added that R. Garner studied the language of simians in the 1890s using an incredibly large number of primate species (*Cebus*, *Macaca*, *Papio*, *Ateles*, and *Pan*), and A. Kinnaman in the early 1900s studied the mentality of rhesus monkeys in captivity. There is also evidence that W. Wundt established the first laboratory of experimental psychology in Germany in 1879, where a chimpanzee was set the task of obtaining a banana (Azimov, 1967). I should note that the first animal psychologists employed anthropomorphism in evaluating their primate data, probably due to their dispute with anti-Darwinists. Though regrettable, this is not surprising, considering the considerable intelligence they encountered in these creatures; the chimpanzee is still capable of arousing admiration even today. The anthropomorphic interpretations, however, discredited the idea of behavioral similarities between apes and humans. Even after carefully conducted studies somewhat later revealed homologous intelligence in apes and humans due to their strong biological resemblance, many scientists retained their negative feelings in response to the earlier anthropomorphism. I believe that Soviet physiologists assumed a similar attitude. In concordance with the Soviet tradition, they focused their attention on criticizing the 'tendentious' (biased) interpretations by 'bourgeois' scientists, rather than focusing on the results of the behavioral studies *per se*.

In the first half of the 20th century three names were most notable in the field of primate behavioral research: N. N. Ladigina-Kohts, W. Köhler, and R. M. Yerkes. Nadezhda Nikolayevna Ladigina-Kohts (1888–1963) was an outstanding researcher; a pioneer in great ape research. She conducted experiments on chimpanzees in Moscow beginning in 1913. Using her famous method of 'sample selection' and other original techniques, she made a great contribution to the study of primate cognitive capabilities, the mentality of chimpanzees, and the theory of instrumental and constructive activities of hominoids. She determined the ability of chimpanzees to distinguish chromatic and achromatic colors, object size and shape, and therefore the ability to produce elementary mental images of an abstract nature in object perception. Her comparisons of the behavioral peculiarities of a human baby and those of a young chimpanzee became classics in comparative psychology (Ladigina-Kohts, 1935). Her research proves that she was well aware of the tremendous potential of chimpanzee intellectual capabilities. This compels us to conclude that her statements regarding qualitatively different emotions and early indications of abstraction capabilities in humans and apes, widely referred to by her followers, were to an extent the price she paid to the strict ideology which reigned in the Soviet Union at the time and which became particularly aggressive after the 'Pavlov session' of the Academy of Sciences. Categorical protests against an 'exaggerated' identification of chimpanzee and human behavior by Köhler, and especially against their resemblance shortly after birth, were objectively dismissed in experiments by L. Firsov, A. and B. Gardner, D. Premack, D. Rumbaugh, S. Savage-Rumbaugh, as well as other modern scientists. Nevertheless, the experiments by N. N. Ladigina-Kohts remain a glorious milestone in the history of primate research. The high appraisal of her work given by Köhler, Yerkes, Dembovsky, and many other foreign and Soviet scientists will always remain valid.

A notable mark in the history of science was made by Wolfgang Köhler (1887–1965), the prominent German animal psychologist who became one of the founders of Gestalt psychology. Köhler's experiments on 8 chimpanzees, carried out from 1912 to 1918 on the island of Tenerife, continue to incite admiration because of their faultless character. Ignoring the relevance of primate natural history, and on the basis of experimental results alone, Köhler was the first to reveal the high level of chimpanzee intelligence and to propose the salient idea of individual differences in chimpanzee mentality. This scientist arrived at the revolutionary conclusion (which had been muted in the past), that the chimpanzee stands out from the rest of the animal kingdom and not only approaches humans in its morphology and physiological properties, but also 'reveals that behavioral manner, which is specifically human' (Köhler, 1973 (1917), p. 203). Attempting to prove chimpanzee intelligence, he maintained that this primate, as well as the human, is capable of having insights by means of a sudden revelation, and by grasping the Gestalt structure and the integral field. This was considered an 'idealistic' interpretation of the experimental data in the Soviet Union, one that feeds enormous speculation. B. Beck (1977) proposed that Köhler contributed a great deal more to science than is commonly believed and a great deal more than he himself thought. Köhler described almost all forms of instrumental activities in chimpanzees, which were demonstrated only much later in other studies. Köhler must also be given credit for demonstrating the key role played by experiments conducted under controlled conditions in captivity, different but no less important than observations carried out in natural settings.

Practically the same conclusion about chimpanzee intelligence was drawn by Robert M. Yerkes (1876–1956), a proponent of behaviorism. I have already discussed the outstanding contribution to science made by this scientist. Even today there are essays and papers being written on different aspects of Yerkes' scientific heritage (e.g., Nadler, 1996). His works are inexhaustible. In the context of the present monograph, Yerkes' most important accomplishment was the establishment of the highly developed intellect in chimpanzees and the other apes. He discovered indications of all human psychological functions in chimpanzees. He asserted that any differences in the psychological processes of chimpanzees from those of humans are quantitative, rather than qualitative (Yerkes, 1943), as Darwin also believed. Thus, the mentality of a hominoid that exists in an environment considerably different from human society still approaches the mentality of humans. I have already mentioned the remarkable influence that Yerkes' research exerted on other scientists. Now I would like to emphasize that Yerkes' research also influenced such a distinguished pundit of physiology as the academician I. P. Pavlov. Pavlov held Yerkes in high esteem, was in correspondence with him beginning in 1909, and intended to work with him in Florida in 1936 (the year Pavlov died). We can assume that Pavlov began his observations on chimpanzees in Koltushi, the Soviet Union, influenced to a large degree by Yerkes. He sent his colleagues to Sukhumi to conduct further research with chimpanzees, which led him to such far-reaching conclusions that not even his students could accept them and they were only openly discussed 30 years later (I will return to this subject later).

Of enduring importance in the area of medical primatology are the neurophysiological studies on simians by one of the classicists of modern physiology, Charles Sherrington (1857–1952). In his great book, *Integrative Activity of the Nervous System* (1906) he wrote: 'I have mainly worked on simians' (Sherrington, 1969, p. 267). By 1917, when his review on the anthropoid sensory cortex was published (Leyton and Sherrington, 1971), Sherrington and his colleagues had used 22 chimpanzees, 3 gorillas, 3 orangutans, and dozens of monkeys in their experiments at the University of Liverpool.

As was mentioned above, Sherrington's first primate study was published at the end of the previous century (Sherrington, 1889). Sherrington also introduced into the scientific literature the presently common term 'genital skin' (Langley and Sherrington, 1891), to refer to the specialized female primate tissues that become enlarged during estrus. The main issue on which he worked, however, was the anthropoid cortex, which he studied from 1901 to 1917. We learn about the methods Sherrington used in his research on nonhuman primates from a letter by G. Cushing. Within one week three apes were studied; an orangutan on Wednesday, a chimpanzee on Friday, and a gorilla on Saturday (Cushing, 1969). This was a rather fascinating example of the use of the comparative method in research on three great apes, certainly a rarity in the history of medical primatology.

Sherrington applied surgical and electrophysiological methods to describe the topography of localized functions in the cortex of great apes. This research was fundamental to conceptualizing similar functions in humans. He used primates to study the functions of the motor cortex, visual cortex, reflex mechanisms, the mechanisms of spinal shock, the integrative activity of the brain, tetanus pathophysiology, and even the effects of an anti-tetanus serum. He demonstrated the strong resemblance between the spinal cords of humans and other hominoids which he had studied for many years. Sherrington produced the most valuable data on the comparison between the cortices of humans and great apes, a finding which is still highly significant. On the other hand, when he compared monkeys and apes, he found (as did T. Huxley) a far greater similarity in the former case. He also presented data on the comparative relationship among the brains of the various great apes. Remarkably, Sherrington also appreciated the psychological approach in studying the brains of the great apes. He was awarded the Nobel Prize in 1932.

As was mentioned, after the 1930s the volume of studies in primate physiology and behavior gradually increased and became the largest division of medical primatology. Studies of special interest include the experiments conducted in the Sukhumi Primate Center, initially led by I. P. Pavlov, and then by S. Kaminsky, A. Slonim, L. Voronin, and D. Miminoshvili; the research in comparative psychology by N. Voitonis, N. Tikh, and G. Roginsky; Pavlov's own observations in Koltushi, and then the observations by his followers (Denisov, Shtodin, Vatsuro); the research by N. N. Ladigina-Kohts at the Moscow Zoo; and studies by V. Protopopov's school in the Ukraine, by L. Firsov in Koltushi, and by A. Kats in Sukhumi and Tbilisi.

That was also the period of several successful experiments by American psychologists, e.g., R. M. Yerkes' followers, H. W. Nissen and K. S. Lashley, and some others who conducted notable physiological studies. These latter scientists, more often than not, were also influenced by Yerkes or associated with his research. The 1930s was also the time that J. Fulton's school (Yale University) was conducting highly successful research. According to S. Zuckerman, Yerkes' and Fulton's laboratories alone yielded more data on primate behavior than all the other Western laboratories combined. It was rather significant in the historical sense that Sherrington's student, the distinguished physiologist John Fulton, who founded the Yale Medical School, came to the conclusion that nonhuman primates are 'more than just laboratory animals', they are 'research material of special quality'. This conclusion had important implications for the history of medical primatology. According to Fulton, his laboratory adopted the following motto in the 1930s: 'If not man, then simia' (Fulton, 1941, 1962). Fulton predicted that medicine of the future, the way it is taught and its research perspectives, would be closely related to primate biology. His prediction came true, at least in the United States; virtually every medical school in the country has a primate center or a laboratory. It should be noted that Fulton was in active correspondence with researchers in the Sukhumi Center, as evidenced by this institution's archives.

According to my calculations, of the 408 publications by Fulton that were published during the period from 1920 to 1955, 57 are devoted to experiments on nonhuman primates. Of great interest are the chimpanzee experiments conducted in his laboratory (Fulton, 1937–1939). After 1933, Jacobsen, Fulton's colleague, began to study the effects of brain lesions in two chimpanzees aimed at repressing aggressive reactions. The experiments took place at Yerkes' Laboratories and culminated with the experimental development of leukotomy, a procedure which was first performed on sick humans by the Portuguese neurosurgeon Antonio Egas Moniz. In K. S. Lashley's opinion (quoted by G. H. Bourne, 1971), the development of this operation covered all the expenses connected with the Yerkes' Laboratories since the day it was founded. Moniz was awarded the Nobel Prize in 1949 for the discovery of the therapeutic application of leukotomy in some human psychoses.

Among other studies of the 1930s, a very special role was played by the primate research carried out by I. P. Pavlov and his students. The experiments were conducted by Pavlov's own students in Sukhumi (Fursikov, Podkopayev, Voskresensky, Frolov, Dolin, Mayorov, Galperin, and others) and by Pavlov himself in collaboration with P. Denisov in Koltushi near Leningrad in 1933–1935, using the chimpanzees Rafael and Rose. A concise outline of Pavlov's contributions to experimental primatology, in my opinion, is as follows: 1) Pavlov and his students were the first to conduct neurophysiological research on primates, successfully applying the classical method of conditioned reflexes introduced by this distinguished scientist; 2) Pavlov demonstrated the peculiarities of primate brain physiology, including the speed of performing conditioned responses, high lability of basic nervous processes (excitement and inhibition), characteristic properties of the inhibitory process, prevalence of excitement over inhibition, a high level of orientational and exploratory activity, and rudimentary rationality 'differing from ours only in the poverty of associations' (Pavlovskie sredi, v. 3, pp. 431–432). Pavlov gave direct instructions for the experimental study of neuroses and other human psychopathological problems on Sukhumi primates, which were highly successful.

This account does not exhaust the role of primates in Pavlov's research, however. I should point out one negative occurrence in the history of brain research and primate psychology. Almost all prominent physiologists and psychologists of the West (Yerkes, Köhler, Giliom, Bingham, and many others, i.e., almost everyone who conducted experimental studies with chimpanzees) noted the absence of qualitative differences in the advanced behavioral activities of humans and great apes. Soviet physiologists, however, on the pretext of promoting scientific materialism, constantly emphasized the contrary. Pavlov also participated in the tough debate with 'gestaltists', initially focusing on the differences in human and nonhuman primate intelligence. Being a true scientist, however, Pavlov never disregarded scientific facts. During the 'Pavlovskie sredi' (Pavlov's Wednesdays) meetings, for example, where the results of primate experiments were discussed, Pavlov increasingly often pointed out the similarity between human and nonhuman primate mentalities.

Unfortunately, the enormous amount of data which Pavlov obtained on chimpanzees was not summarized. The great scientist did not live to participate in the International Congress of Psychologists in Madrid (1936), where he intended to present a review of his experiments on chimpanzees. Subsequently, some of Pavlov's comments and assertions, sometimes controversial, were misinterpreted. On the one hand, there was the identification of advanced behavioral activity in simians and dogs in the 'Pavlovskie sredi', with an emphasis on its conditioned reflex basis. On the other hand, there was Pavlov's obvious fascination with the extraordinary intelligence of chimpanzees ('I am admiring them now') and its similarity to that of humans. In analyzing the intellectual activities of primates, Pavlov avoided the term 'conditioned reflex' and used the concept of 'associations'. Ultimately he declared that it was not appropriate to explain some instances of chimpanzee behavior in terms of conditioned reflexes. An analysis of the chimpanzee's behavior involved in constructing a pyramid to obtain a piece of fruit led Pavlov to assert that it 'cannot be called a conditioned reflex', but rather 'it is a case of knowledge formation, grasping the natural relations between objects' (v. 3, pp. 262–263). At the next meeting, moreover, he reiterated this idea and regretted that the audience of his students had not ascribed sufficient importance to it. Naturally, it was not easy for his students to hear the founder of conditioned reflex theory say that there were behavioral instances which could not be explained by his theory. The use of such wording could entail strict punishment as a consequence. Pavlov's words, therefore, could only be expressed 30 years later (Asratyan, 1970; Bassin, 1971). I can only conclude, in the context of the history of medical primatology, that primate research, as a special area of biological research, essentially caused a great theory to be amended by its own founder. Pavlov's chimpanzee studies were most fruitfully continued by L. A. Firsov in Koltushi, who noted the special character of behavioral regulation in the chimpanzee. He believed that hominoids possess a third mechanism in addition to the congenital and acquired behavioral mechanisms, namely, imitative activity associated with memory. He maintained that a chimpanzee was capable of forming ideas at a preverbal level and that a chimpanzee's mind was not comparable with that of any mammal other than the human (Firsov, 1972, 1982).

Among some of the most outstanding achievements of medical primatology are the studies of experimental neurosis and psychosomatic pathology in baboons and macaques, also conducted at Sukhumi. They were initiated by I. P. Pavlov in the 1930s, as was already mentioned (S. Kaminsky, L. Bam). In the 1950s, after many failures, D. Miminoshvili managed to create a model of neurosis in monkeys by making natural (gregarious) inclinations with different biological implications conflict with each other. Following this demonstration, neuroses were experimentally produced in monkeys by means of manipulating the natural 24 hour periodicity (G. Cherkovich). These experiments yielded direct experimental support for the hypothesis that some cardiovascular pathologies, e.g., hypertension, stenocardia, and myocardial infarction, may have a neurophysiological basis (G. Magakyan, G. Kokaya). Other pathological disorders induced by neurosis were also determined, e.g., disturbances of neurophysiological and vegetative functions (N. Lagutina, L. Norkina), amenorrhea (L. Alekseeva), achylia and prediabetic conditions (V. Starzev), and hysterical paralyses and hyperkineses (T. Urmancheeva). The mechanism of nervous systemic damage in neurosis was demonstrated (Starzev, 1971) and many important reviews of this research were published (Cherkovich and Lapin, 1973).

The modeling of neurogenic human pathology achieved great fame world-wide. These experiments, as was mentioned, attracted the attention of American cardiologists and were one of the arguments for establishing the National Regional Primate Research Center program in the United States (Whitehair and Gay, 1981). These experiments, moreover, demonstrated the relevance of primates for the study of human cardiovascular diseases and provided the impetus for cardiology research on primates, which is still conducted nowadays, especially in the field of atherosclerosis.

For over 20 years notable psychological studies were carried out on rhesus macaques at Wisconsin University by professor Harry F. Harlow, who, together with his spouse Margaret K. Harlow, conducted research on the effects of separating a young primate from its mother. Removal of the young from their mothers, peers, and other members of the group for varying periods of time produced physically healthy mature animals which were nevertheless 'emotional invalids'. Such maternally separated animals were unable to live a normal, socially competent life and were characterized by severe degeneration of their 'personalities'. Harlow showed by his experiments that social communication (mother-infant, infant-peer) preceded sexual relations, and that normal communication of the infant with the mother was one of the necessary conditions for development of a social primate. Similar symptoms were discovered in children who had been brought up in orphanages, and in adolescents and adults who had been placed in psychiatric clinics and deprived of social communication (Harlow and Harlow, 1962). This research had a substantial impact on Freud's interpretation of human sexual pathology and his idea of 'inner compulsion'. Harlow drew provocative conclusions about the grave consequences of the infant's removal for the mother herself and about the meaning of the infant's love for the mother's psychological and sexual health (Harlow, 1972). It is difficult to enumerate the many fields of biomedicine and certainly, psychology, where Harlow's discoveries found application. The idea of awarding Harlow the Nobel Prize was expressed in print (Benjamin, 1968) and was supported by the strongest arguments.

Primates also played a significant role in brain mapping studies. As we have seen, the tradition goes back to C. Sherrington. In the 1930s and 1940s, data on cortical localization were obtained mainly from neural surgery on humans. After the 1960s, however, experimental work in this field was done on primates for the most part (Falk, 1982). Mapping the visual fields of simians became a major field of concentration. Such experiments were mostly performed on monkeys at the New England Regional Primate Research Center of Harvard University and led to a brilliant scientific success. D. Hubel and T. Wiesel (together with R. Sperry) were awarded the Nobel Prize in 1981 for their discoveries in the field of visual regulation by the central nervous system. This research opened up new prospects not only in our theoretical knowledge, but also in practical ophthalmology. In particular, it compelled doctors to reevaluate the principles of early medicinal visual deprivation.

The 20th century became the era of describing the ecology and social life and behavior of anthropoid primates, primarily chimpanzees, gorillas, orangutans, and many species of monkeys. Outstanding research on great ape behavior in the laboratories, described above, was followed by the field research of J. Goodall, T. Nishida, G. B. Schaller, D. Fossey, B. Galdikas, and others. Their studies have made a great contribution to the understanding of human biology and anthropology in general. Influenced by primate research, the latter science has gone through major changes in the last 40 years, transforming the ideas regarding the origin and evolution of humans (this will be discussed in detail below). An avalanche of studies in biochemistry, molecular biology, genetics, and paleontology has brought the last common progenitor of man and the great apes 15–20 million years closer to us.

I should also mention again the 'quiet' accomplishments of medical primatology, when primate use in biomedical research, for various reasons, was not reflected in announcements, awards, or other signs of general recognition. Yet these accomplishments did occur in such fields of knowledge as radiobiology, aerospace research (including flight preparation for the first woman cosmonaut), organ transplantation, toxicology, and neuropharmacology.

The century-long history of medical primatology, which played a tremendous role in the fate of humankind in delivering it from illnesses and improving living standards, was written under the influence of strong stimuli. Firstly, there were the great scientists whom I have already named, including Mechnikov, Yerkes, Landsteiner, Sherrington, Calmette, Fulton, and Lapin, to mention just a few. Secondly, there were the research institutions, such as the Pasteur Institute in Paris, the Sukhumi Primate Center, the Yale Laboratories of Primate Biology (Yerkes Primate Center), and the American Regional Primate Research Centers. Finally, there were the scientific problems, if you will, including polio, the mysteries of intelligence, nuclear energy, radiation, cancer, thalidomide, hepatitis, DNA, and AIDS.

In 1960, the monograph *Reproduction of Human Diseases in Experiment* by two prominent scientists, D. Sarkisov and P. Remizov, was published in Moscow. This voluminous book, numbering 780 pages, may be considered a historic transition in experimental science from one period to another. Primates are mentioned only in a very few instances, almost exclusively in the field of infectious disease pathology. At that time, scientists always wondered at the infrequency of primate use in experimental research (Tappen, 1960). The situation changed dramatically in the second half of the 20th century. As we have seen, many acute medical problems of the century were solved in experiments on simians. A memorial to the primates is well-deserved.

*   *   *

Part II

Biological foundations of medical primatology

The biological foundations of medical primatology include:

a) taxonomic and evolutionary-phylogenetic relationships between human and other primates;
b) phenotypic signs of the affinity of human to other primates;
c) possibilities of modeling human pathological and biological processes in primates on the basis of this affinity.

Undoubtedly, spontaneous pathology of monkeys and apes could also be included in the list, if such diseases were considered to be a *biological* foundation for research. The existence of thorough reviews on this issue, made by highly reputed experts, makes it sufficient only to make reference to these works where appropriate.

Chapter 3

Taxonomy and a concise evolutionary-phylogenetic description of the order Primates

Which species belong to the primates,[1] and what is characteristic of these animals as a group? A general description of the order was given as early as 1873 by the English scholar, St. George Mivart, in his book on prosimians: '. . . ungiculate claviculate placental mammal with orbits encircled by bone; three kinds of teeth at least at one time of life; brain always with a posterior lobe and calcarine fissure; the innermost digits of at least one pair of extremities opposable; hallux with a flat nail or none; a well-developed caecum; penis pendulous; testes scrotal; always two pectoral mammae.'

This definition, despite being over a hundred years old and somewhat rough, is still popular among scholars, who quote it even nowadays (Whitney, 1995). However, new evidence has accumulated in the meantime to complement the old data, a century being a rather long period for a dynamically evolving science. W. E. Le Gros Clark (1958), true to his concept of 'general morphological character', called attention to the combination of characters that prove the common evolutionary trends in primates. These characters include progressive growth of the skull, eye orbit rotation to the forward-facing position, diminution of the zygomatic arch, downward, rather than forward, protrusion of the face, reduction of the nasal ossa, and progressive rotation of the foramen magnum. Napier and Napier (1967) supplemented the list by adding high finger mobility, progressive evolution of the torso towards bipedalism, and a prolonged postnatal period. There also exists a more precise contemporary definition, made on the basis of the previous authors' studies, which takes into account the retention of primitive morphological characters and development of the progressive characters which were acquired in the course of evolution (Thorington and Anderson, 1984). It is apparent that all definitions of primates, as well as those of other organisms, have a morphological basis, which became common in systematics as early as in J. Ray's time (17th century). Yet, much water has passed under the bridge since the even more recent date when E. Haekel (1866) proposed the method of 'triple parallelism' in systematics, i.e., the combined analysis of data from paleontology, comparative anatomy, and embryology. Even though it may not have changed in its entirety, biology has acquired enormous amounts of new information since that time which must not be dismissed from consideration.

Returning to the characteristics of the order Primates, we should probably add such unique features as the ovarian menstrual cycle in females (Tarsioidea, Anthropoidea), stereoscopic 3-colour vision, a number of 'new' properties of the brain (e.g., the presence of frontal lobes, a large area of associative cortex, and features responsible for cognitive

---

1 'Primatis' – 'one of the first,' the genitive case of the Latin *primas*.

function), and, in particular, characteristic karyotype features, DNA and RNA homology, and amino acid substitution limits in protein sequences. The number of such characters is permanently increasing. A pseudogene – the unique marker in the genome found only in the order Primates – has recently been discovered (Devor, 1998).

As an introduction to primate taxonomy, it should be noted that this group, naturally related to anthropogenesis, is most often described by one of the three basic trends of classification (phylogenetic, evolutionary, and phenetic [Mayr, 1965]), namely, by the phylogenetic, or cladogenetic trend. More precisely, the most wide-spread description is the combination of phylogenetic and evolutionary classifications. This is in accordance with the precepts of medical primatology, where affinity characters must be homologous (although in some experimental instances the utility of convergent resemblance must not be ruled out either). Since Darwin's time, common origin has been considered the only possible cause of close affinity and the only truly scientific ('natural') foundation for biological classification. 'To be related biologically and genealogically means to have a common progenitor; to be related more closely means to have a closer progenitor. Thus the branched tree, which has been drawn up on the basis of the cladogram, is the only objective foundation of the taxonomic model' (Groves, 1986, p. 187). David Pilbeam (1986, p. 295) called phylogenetic systematics or cladism 'one of the most important biological revolutions of the last two decades.'

Despite certain faults present in the phylogenetic classification (e.g., related to the estimation of the 'weight' of each character or to the establishment of 'side branches' (Habgood, 1988), many authors believe it describes phylogeny well and is appropriate in primate systematics (Delson, 1977; Groves, 1989; de Queiroz and Gauthier, 1992; Goodman, 1996; Shoshani et al., 1996; Goodman et al., 1998).

As far as modern systematics and its methods are concerned, and primate systematics in particular, it should be kept in mind that this important division of primatology (and biology in general) began to grow rapidly in the 1960s. In my opinion, there is a landmark book of that time (from the 1960s onward) by George Simpson (1962b), a leading figure in biology and systematics of the 20th century. This book encourages systematicians to rely on anatomical data, or even 'primarily' on museum items and rarities, thus preserving the morphological criterion as a fundamental principle of systematics. Although Simpson acknowledged the possible benefits offered by non-morphological methods (his speech at one symposium was preceded by a discussion of immunochemistry and DNA in application to primate phylogenetics), especially by physiology, he nevertheless expressed doubt as to the applicability of experimental results obtained in biochemistry, molecular biology, and in behavioral studies, to estimation of the phylogeny of animals in general and primates in particular. Interestingly, one of the reasons for his rejection of new methods was a communication by M. Goodman about the uneven tempo of biological evolution. In addition, he went on to question the utility of sophisticated electronics in primatological (as well as zoological) investigations and observed that some publications reminded one of a statement saying that since there are mountains, there must be mountain climbers.

Thus, the morphological approach was predominant at that time in the ongoing development of systematics and in the definition of phylogenetic affinities. In terms of primate classification as well as classification of other animals, this approach was applied in 'horizontal systematics' (among living primates, or 'neontology' primates) as well as in 'vertical systematics' (fossils, or paleontology primates). The explanation of this 'monopoly' is simple and not devoid of logic; it concerns the traditions and the needs of paleontology. In paleontology there is only tangible material at the investigator's disposal; teeth, skull, and postcranial skeletal fragments. It should be noted, moreover, that the development of morphology and paleontology did not stop after the 1950s. Both were enriched by new concepts, methods, and discoveries (e.g., the use of radioactive isotopes, termo-luminescence, and other new means of assessing the age of fossils, the significance of plesiomorphic (underived) and apomorphic (derived) characters, variations in morphometric analyses used for the estimation of anatomical data, discoveries which led to a considerable aging of the genus *Homo*, discovery and identification of 'new' australopithecines, and reevaluation of the status of ramapithecines in the history of the hominoids, etc.). All of the foregoing fits into the framework of the morphological tradition, which continues today (Swindler and Erwin, 1986; Shoshani *et al.*, 1996).

It can be said that 1963 'officially' became the year for introducing new methods into primate phylogeny because it was then that three publications appeared simultaneously to mark the occasion; those of M. Goodman (1963) on serum proteins, E. Zuckerkandl (1963) on haemoglobin, and H. Klinger *et al.* (1963) on chromosomes. All three were based on the most recent achievements in biology and were devoted to the affinity of humans and other higher primates. In fact, Goodman revived the old approach to evaluating organism affinity, originally proposed by Nuttall and Friedenthal, which was mentioned in Chapter 1. These investigators based their conclusions upon the immunological affinity of proteins in precipitation reactions. Goodman, however, had an advantage in his studies, having improved the technique of diffusion in agar using electrophoresis, and having advanced methods of data-processing which could accommodate tremendous numbers of protein comparisons. Data on albumin, transferrin, ceruloplasmin, tireo-globulin, and gamma-globulin were analyzed in this manner and generally showed identical results. Goodman's group then performed an extensive study of amino acid sequences of various proteins, obtaining reliable results on a multitude of DNA nucleotide comparisons (Goodman, 1992; Goodman *et al.*, 1970, 1994). The main achievement of these studies was not only the confirmation of earlier immunochemical evidence on the profound affinity of humans and great apes, but also establishment of the fact that the African hominoids (chimpanzee and gorilla) were biologically closest to humans. The orangutan is somewhat more distant from the other three, and still further distant is the gibbon (there had been some knowledge of this in the past).

Immunological studies of primate affinities received new support. V. Sarich and A. Wilson (1967) and V. Sarich and J. Cronin (1976) conducted these studies using the somewhat different method of microcomplement quantitative binding. On the whole, the outcome was approximately the same as with immunodiffusion. Moreover, in studying antigen reactions of serum albumin and transferrin, the above researchers established the 'immunological distance' between organisms and the indexes of their similarity and dissimilarity, which allow the determination of their common ancestor's age (note that the age of the common ancestor of humans and chimpanzees proved to be substantially nearer to us than it was according to the paleontologists; 2–5 million years rather than 20–30 million, or even 15 million years). The concept of 'molecular clocks', which had been proposed earlier, was said to have been confirmed (Zuckerhandl and Pauling, 1965). Although this idea *per se* proved to be wrong, due to unequal rates of evolution at different sites of the genome and in different phylogenetic groups (probably owing to the varying longevity of generations), it nevertheless played a positive role in the development of new methods of systematics, especially with respect to use of the idea of 'local molecular clocks' (Goodman, 1996).

Let us reiterate that simultaneously to the strictly biochemical investigations of primate phylogeny by the above and many other scientists, studies in molecular biology with similar objectives began to increase during the 1960s. Starting with the study conducted by M. Martin and B. Hoyer (1967), research on primate DNA and RNA affinities, along with studies of proteins and enzymes, became a favorite concern for many scientists who were interested in studying the evolution and phylogenetic relations of the animals nearest to humans. The variety of methods and approaches for evaluating the relevant data was incredibly large. Morris Goodman (1991) quoted several authors with regard to some of these varied methods for the analysis of immunological distances in constructing tree algorithms (cladograms) by two models of evolution: parsimony analysis of fibrinopeptide, globin, and carbon anhydrase amino acid sequences; data on amino acid sequences of α-crystalline A chains; analysis of distances based on DNA-DNA hybridization data (several versions, including the well-known studies by C. Sibley and J. Ahlquist [1984, 1987]); data analysis of nuclear DNA and mitochondrial DNA nucleotide sequences; analysis of the DNA of various pseudogenes of β-globin and other noncoding DNA; and construction of phylogenetic trees on the basis of DNA data according to the above-mentioned principle of parsimony – the neighbor-joining method and the maximum likelihood method, which allow the creation of stochastic models of mediated phylogeny patterns and phylogenetic trees. One should also add the methods of nucleotide site sequencing and ribosome gene restrictive enzymatic mapping (Wilson et al., 1989), RNA data comparisons, use of recombinant DNA techniques in Southern blot hybridization of newly cloned DNA samples in order to identify closely related primate species (Fukui et al., 1994), and other immunological methods for determining the affinity between different species (e.g., indirect immuno-fluorescence of T lymphocytes [Letvin et al., 1983]). A review of these techniques may be found also in articles by M. Ruvolo and T. Smith (1986), C. Oxnard (1983), R. Stanyon (1989), and in other publications.

Dynamic developments in primate molecular biology, naturally, were conducive to the development of new concepts on the genetics of humans and their near relatives. However, primate karyology alone contributed substantially to primate systematics in the 1960s–1990s, especially in those cases where other 'competing' techniques gave rise to controversial or inconsistent results. Without claiming to give a comprehensive summary of all the new methods in cytogenetics, let us only mention some of those which have played an important role in obtaining new data on the affinity of humans with other primates. Besides routine karyotype morphology and physiological studies, we should include heterochromatin fluorescence methods, striated chromosome methods, which include a variety of high-resolution dying techniques, comparative gene mapping, and determination of related binding groups. All of these approaches and techniques, designed for different purposes, have objectively revealed the degree of homology of genetic structures in primates, particularly, in humans and other hominoids, and led to the determination of the phylogenetic relations present in the order Primates (Grouchy et al., 1972; Chiarelli, 1974; Stanyon, 1989).

Unfortunately, other sciences have not come up with as many proposals to evaluate or reevaluate the disputed issues in systematics and primate evolution. To an extent, neurophysiology and, more substantially, behavior may be considered exceptions. There are comparisons of cortical organization, for example, which permit differentiation of the large primate taxa, prosimians, platyrrhines, and catarrhines (Kaas, 1984). Also of interest are the neurophysiological criteria by which species are differentiated, even within one genus (Balzamo, 1981). Attempts have even been made to determine phylogenetic inter-relations from the properties of neurons in different brain subsystems (Pettigrew, 1991). Especially noteworthy are the discoveries of certain learning capabilities in apes and the recently discovered cognitive capacities of living hominoids which essentially revised the interpretation of anthropogenetic studies on the pre-*Homo* stage (discussed later).

There have also been attempts, often quite expedient, to use the data from various sciences which had no traditional connection to systematics, e.g., pharmacology, where the term 'pharmacozootaxonomy' was coined (Williams, 1976), immunoserology (Moor-Jankowski *et al.*, 1973), and parasitology (Ardito, 1980), but they were relatively few.

In using the above variety of methods and approaches to solve the complicated problems of primate systematics, discrepancies and disagreements inevitably arose, as usually happens. Perfect agreement of all experimental results and conclusions by two different groups of researchers, even when the data are obtained by similar techniques, is virtually impossible. Naturally, such agreement is even more unlikely in experiments of basically different types which yield significantly discrepant conclusions, to be sure! Heated controversy with regard to primate taxonomy and phylogeny arose between morphologists and molecular biologists, the latter being allied with biochemists and geneticists. The central issue at the 3rd International Congress on Systematics and Evolutionary Biology (1985) was the relationship between molecular and morphological evolution, especially with regard to higher vertebrate phylogeny, including certainly that of humans and their nearest relatives. It became clear, however, that classifications based on morphological and molecular characteristics more often than not, though not invariably, proved congruent.

It also became clear by the 1970s that traditional paleoanthropological models of primate phylogeny quite often proved controversial (Hoffstetter, 1979). Cross-immunology, molecular biology, and new cytogenetic methods inspired a certain hope, of course, but they were met with wariness, at least by some of the traditional morphologists. Soon it was revealed, however, that cladograms constructed on the basis of minimal distances with multivariate statistical analysis of quantitative (morphometric) data were in full accord with the conclusions of biochemists and molecular biologists. This was shown by dental morphometry (Mahaney and Sciulli, 1984), as well as by numerous anatomical measurements made earlier by A. Schultz and then analyzed morphometrically by C. Oxnard (1983, 1984). Traditional morphologists and paleonto-logists, however, were generally favorable toward the new data introduced by biochemists at one or another level of classification and were supportive in calling for consistent cooperation between themselves and those who obtained other types of data. The con-clusions of morphologists do not always correlate exactly with the data obtained by molecular biologists in all taxonomic categories of the order up to the species level. The phylogeny, evolutionary chronology, and the models advocated by morphologists and molecular biologists, however, are in close agreement, with little or no exception.

Thus, L. Martin (1986) maintained a skeptical attitude towards the hypotheses of molecular biology and disagreed with the notion of a special affinity between *Pan* and *Homo*, which was implied by these hypotheses, as well as with other issues in biochemical phylogeny. He acknowledged, however, that the closest affinity existed between humans and the African apes and agreed with the inclusion of all hominoids, living as well as extinct, in one family with *Homo*. The new data were quite sympathetically and con-tentiously supported by such morphologists and paleontologists as P. Gingerich (1985), C. Groves (1986, 1989), D. Pilbeam (1986), and others. Among consistent opponents of the molecular systematicians, one should mention J. Schwartz (1986, 1988), who considered the new methods rather uncertain and whose morphological investigations imply that it is the contemporary orangutan (*Pongo*) that is the nearest to humans phylogenetically. According to Schwartz, *Pan* and *Gorilla*, which have a common ancestor, are but a sister group in relation to the preceding clade. It is apparent, therefore, that the opposition between morphology and molecular biology in primate phylogeny remains topical even today (Shoshani *et al.*, 1996; Goodman *et al.*, 1998). Nevertheless, now it has a positive perspective thanks to the support of most authoritative morphologists.

In conclusion to this brief review of the methods and principles of modern primate systematics, it is clear that highly intensive studies are carried out in this field according to the most contemporary scientific standards. The heterogeneity of methods and the frequently discrepant results notwithstanding, I should note the fundamental agreement in whole among the conclusions and the correspondence of those conclusions to the traditional primate taxonomy which has been shaped over the past one hundred years. The theoretical and essential background of evolutionary and phylogenetic studies has been considerably enriched, significantly facilitating the determination of all taxa of the order and, especially, the explanation of historical events in the group Hominoidea.

I should also emphasize the special value of the new methods in systematics and the theoretical foundation they provide for the development of medical primatology, regardless of discrepant conclusions by representatives of different schools and trends. Along with the basics of anatomy and physiology, it is DNA, genetics, biochemistry, and immunosystematics, in a broad sense, that determine the affinity of immunological reactions and pathology in humans with those in other primates, which is of primary importance in medical and biological modeling. On the other hand, experimental pathology, which I discuss below, can make additional contributions to resolving controversial issues in primate classification.

Biomedical studies are carried out on living primates, every species of which is a terminal product of millions of years of divergent evolution. These living forms, as was mentioned, are subjects of horizontal systematics (Simpson, 1945). Horizontal systematics, however, is a 'section' of a horizontal plane through the upper branches of the primate phylogenetic tree. This cluster of branches goes back to prehistoric times, becoming more closely affiliated towards the top of the tree at some point near the end of the Mesozoic, in the Upper Cretaceous (or possibly the Paleocene, in the Cenozoic, according to Gingerich and Uhen, 1994). On the surface of this plane, there remain discrete and, it seems at first, morphologically isolated taxa among the presently living species. In the context of this monograph, it is useful to describe the similarity of traits in different primates, especially between those of humans and other primates. This requires proof of their affinity, which as has already been noted, can only be confirmed by common origin. Hence, my interest in primate evolution, of which I am able to give here only a fragmented presentation. In places where I touch upon 'vertical systematics', therefore, I do so not to follow the structural canons of classical systematics, but solely to demonstrate one of the foundations of medical primatology.

As stated in Chapter 1, confusion in primate classification was one of the long-standing problems of primatology. I consider C. Linnaeus' systematics (1758) the beginning of scientific primate taxonomy, though we must keep in mind that until C. Darwin's time it remained artificial and quite chaotic, despite the many attempts to organize it. Moreover, the efforts by a significant number of scientists notwithstanding (including T. Huxley, E. Haekel, St. Mivart, G. Forbes, M. Weber, D. Elliot, W. Gregory, and others), this situation lingered on at least until the middle of the 20th century. In fact, it was G. G. Simpson who first brought order to primate classification. His treatise, *The Principles of Classification and a Classification of Mammals* (1945), was a fundamental study that summarized 200 years in the development of mammalian classification. It showed that primate systematics was in a pitiable condition. Especially confused was the nomenclature, where, as the author justly pointed out, one could not name two scientists who used the same terminology. Simpson quoted 26 'equivalent' Latin names for the genus *Macaca*, noting that even that was not the total; macaques were also associated with other forms (*Papio, Cercocebus*, etc.) amounting to 25 names more, even discounting those which were spelled differently.

At that time, controversy arose even in regard to the inclusion of large groups like prosimians in the order Primates, to the identification of suborders, and to the determination of many other taxonomic strata. Thus, chaos ruled in nomenclature. Authors followed their predecessors in dismissing the laws of zoological nomenclature (the laws themselves were imperfect; a standard *International Code of Zoological Nomenclature* came into force later). Also at that time, the principles and theoretical foundation of modern systematics were only beginning to take shape (Simpson, 1961; Mayr, 1971). Simpson (1945) brilliantly applied explicit foundations for the theoretical classification of an enormous amount of literature to substantiate a taxonomic model. His monograph confirmed the differentiation of two suborders in the order Primates, introduced three infraorders of prosimians (first proposed by W. Gregory, 1915), included tree shrews in the order, and divided Hominoidea into two families, Pongidae and Hominidae. The nomenclature was presented in strict correspondence to the rule of precedence, backed by scrupulous historical and bibliographical analysis. From the viewpoint of the present time, the inclusion of gibbons in the Pongidae is unacceptable since the constitution of this family appears to be different. Anthropoidea must be divided into Platyrrhini and Catarrhini, and the status of Tupaiidae, together with the problem of distinguishing between Strepsirhini and Haplorhini, requires additional consideration. The position of the group Hominoidea is given special treatment below. Though the author himself considered it out-dated later on, Simpson's classification became the basis for further improvement in modern primate taxonomy. In combination with new data, Simpson's classification was used in the publication of a handbook and a monograph (Napier and Napier, 1967, 1985), as well as the constructive labor of F. Szalay and E. Delson (1979) and C. Groves (1986, 1993), which profoundly substantiated the taxonomy of living primates.

A great number of distinguished scientists of various biological specialties have participated in the foundation of modern group-based primate classification. Taxon phylogeny studies at different levels have become especially wide-spread in recent years; studies of genera, families, suborders, and, naturally, of the order itself. For many years these studies have been carried out by scientists who, regardless of the schools and trends to which they belong, the conclusions to which they come or even the scientific goals they establish, have been making tremendous contributions (of which they are sometimes unaware) to the biological foundations of modern medical primatology and to the discovery of more affinities and similarities between humans and other primates. They have every right, therefore, to claim a connection to the achievements of this specific division of experimental science, whose direct purpose is the preservation of people's health. Included among these scientists are P. J. Andrews, R. L. Ciochon,

E. Delson, J. G. Fleagle, M. Goodman, C. P. Groves, M. Hasegawa, S. Horai, B. F. Koop, L. B. Martin, C. E. Oxnard, J. D. Pettigrew, D. R. Pilbeam, V. M. Sarich, J. H. Schwartz, C. G. Sibley, E. L. Simons, F. S. Szalay, R. H. Tuttle, S. L. Washburn, A. Wilson, and, of course, many others.

In the 1990s alone, knowledge of primate phylogeny had been enriched by numerous monographs and compilations, including *Primate Origins and Evolution: A Phylogenetic Reconstruction* (Martin, 1990), *Primate Evolution* (Conroy, 1990), *Primates and Their Relatives in Phylogenetic Perspective* (MacPhee, ed., 1993), *A Theory of Human and Primate Evolution* (Groves, 1993) and *Anthropoid Origins* (Fleagle and Kay, eds., 1994). As far as phylogeny studies of separate groups of the order Primates are concerned (genera, families, and others), their number and heterogeneity are so great at present that it may soon become difficult to account for them all without special computer-aided calculations and analysis. Especially important are the previously mentioned publications of J. Shoshani, C. Groves, E. Simons, and G. Gunnell (1996), M. Goodman (1996), and M. Goodman, C. Porter, J. Czelusniak, S. Page, H. Scheider, J. Shoshani, G. Gunnell and C. Groves (1998), to which special attention will be given below.

I shall now make a transition to the consideration of primate systematics. We should recall that G. G. Simpson observed more than 40 years ago (1962a) that there are no discoveries of new species of living primates made nowadays; only taxonomic reevaluations can be made. With little exception, this classical scientist has been proved correct in his judgment.*

COMPOSITION OF THE ORDER PRIMATES

Two large groups of living primates are assigned to the order, Strepsirhini and Haplorhini. This distinction, which had long ago been established by anatomical data (Hill, 1953), was not always taken into account taxonomically. Today however, considering the new research data, this distinction must be represented. Each of these two groups has related forms among fossils. The first includes the extinct Adapidae, Archaeolemuridae, and Palaeopropithecidae which, together with living primates, make up the group of prosimians (strepsirhines). The second includes the extinct Omomyidae, Microchoeridae, Parapithecidae, Oreopithecidae, and Pliopithecidae; together with the living primates of this kind (tarsiers, monkeys, apes and *Homo*), they make up a distinct group of haplorhines ('plain-nosed'). These two groups are sometimes supplemented with another completely extinct group of Plesiadapiformes (Thorington and Anderson, 1984), although some scientists disagree with this addition. On the other hand, haplorhine phylogeny was recently presented in a different light from what was stated above, but it did not affect the essentially closer affinity of Tarsiiformes to Anthropoidea, than to common prosimians (Kay and Williams, 1994). Yet the nomenclature of the Anthropoidea proved rather complex and controversial (Wyss and Flynn, 1995), as is the whole of paleoprimate systematics, as well.

There is a debate regarding a relatively precise date for the beginning of primate evolution. Since early in the 20th century, it was assumed to have begun in the Upper Cretaceous, this corresponding to the oldest discoveries, which have been aged at approximately 65 million years (Ma). At present, as a consequence of the mathematically determined age of 80 Ma (Martin, 1993), the date of the origin of the primates is assumed to be between 63 and 55 Ma (Gingerich and Uhen, 1994). M. Lynch (1991) also approached

---

* This statement now (2002) requires clarification. A. Rylands has placed on the internet (www.primate.wisc.edu/pin/ newspecies.html) a list of 36 forms of primates discovered and published from 1990 to 2000. Among these there are: one Old World species (and 10 subspecies); New World, 9 (and 2); Prosimians, 11 (and 3). It is to be assumed that of these, some forms have merely been given a new status.

this date by means of a mathematical algorithm which set the time at the beginning of primate evolution as approximately 60 Ma. As can be seen, the last three dates are in close agreement with earlier paleontological data. E. Simons (1992) accepts 90–65 Ma, whereas M. Goodman *et al.* (1998) propose a date no earlier than 63 Ma.

Which animals, then, were the predecessors and relatives of the primates? The order Primates is one of almost twenty orders of the class Mammalia, subclass Eutheria. Its evolution occurred at the time of expansion of this class, but some time after the advent of the order Insectivora (contemporary moles, hedgehogs, and shrews). It was from the ancient insectivores, in fact, that the divergence of primates began. But which animals were the *direct* predecessors of the primates? Were they the archaic Paleocenic-Eocenic Plesiadapiformes? Should *Purgatorius* from the Upper Cretaceous of North America be considered the first primate? Which was the first haplorhine; *Shoshonius* from Wyoming or *Catopithecus* from Egypt? These and many other questions in paleoprimatology, which deals with 60–50 million-year-old material, cannot be answered with certainty.

As early as the beginning of this century the superorder Archonta was proposed, which included the order Primates and other relatives of the mammalian order, e.g., tree shrews (Scandentia), flying lemurs (Dermoptera), bats (Chiroptera), and primates (McKenna, 1975). Evidently these are the nearest relatives of primates, a conclusion which is drawn not only on the basis of macroanatomical data, but also from properties of brain subsystems of neurons (Pettigrew, 1991). Selection of the monophyly of Archonta, on the other hand, has been both rejected (Kay *et al.*, 1990) as well as accepted (Greenwald, 1991) by morphologists. In such cases (and not infrequently) the data from the 'new' sciences, i.e., those sciences that have recently begun to influence systematics and phylogeny issues, may prove to be of great value. According to these data (on α-crystalline, myoglobin, and hemoglobin amino acid sequences [Chnishi, 1991]), the nearest relatives of primates are tree shrews (Tupaiidae), rodents (Rodentia), hares (Lagomorpha), and armadillos (Dasypodidae), which, together with insectivores, in the opinion of the author, form a monophyletic group. At the same time, there is evidence of a profound immunological distinction of insectivores and rodents from lemurs and lorises, i.e., from the lowest primates (Tsutsumi *et al.*, 1988). As far as the superorder Archonta is concerned, a review of molecular data on DNA sequences (Goodman *et al.*, 1994) confirms the closest affinity of primates to Scandentia and Dermoptera, and excludes Chiroptera from the superorder. The former animals, therefore, are the predecessors and relatives of the order Primates.

As noted above, horizontal systematics is the final product of many divergent evolutionary trends. As Le Gros Clark observed (1959), however, classification does not give evidence of linear evolution, nor is evolution itself drastically linear. Nevertheless, large primate groups (near the taxonomic suborder) must have evolved gradually from protoprimates (Plesiadapiformes?) to Strepsirhini and Haplorhini in the Eocene. The former yielded Lemuriformes and Lorisiformes (i.e., contemporary Prosimii), and the latter, about 45 (?) Ma became divided into contemporary Tarsiiformes (tarsiers) and Anthropoidea (all monkeys, apes, and humans). In the Oliogocene or at its very beginning (40–35 Ma), the latter Anthropoidea diverged into Platyrrhini (flat-nosed or New World monkeys) and Catarrhini (Old World monkeys, apes, and humans). This evolutionary schedule, or one closely similar to it, is accepted by both paleontological morphologists (Delson and Rosenberger, 1984; Niemitz, 1985; Shoshani *et al.*, 1996) and biochemists (protein amino acid sequences [Barnabas *et al.*, 1987]; review of data on DNA and protein sequences [Goodman, 1991]). Note, as well, the ages of the last common ancestors in Goodman *et al.* (1998), in which the classification of primates is based on DNA data complemented by fossil evidence and theoretically, on the 'age-related classification concept.' The authors are well-known scientists (morphologists and biochemists). They offer a 'provisional phylogenetic classification of Primates, in which the taxa represent actual clades and in which ages of the clades determine the ranks of the taxa' (Ibid., p. 586). This concept is not widely accepted, nevertheless, I shall refer to this interesting article repeatedly in this book.

In terms of medical primatology, it is of interest to determine the (quantitative) degree to which these primate groups are related to each other; of primary importance, the biological 'distance' between humans and other animals. Undoubtedly, this kind of data will be more precise and have a wider scope in the future. Some of the sciences are already making use of relatively precise data. M. Goodman (1991) cited the following figures: rapidly evolving noncoding DNA sequences in primates differed from those in other placental mammals by 45%; sequences in Haplorhini differed from those in Strepsirhini by 35%; differences between Anthropoidea and Tarsioidea, and Catarrhini and Platyrrhini were 32% and 12.3%, respectively. These numbers may not be completely accurate and they must be improved upon using other indicators, but they do provide us with some idea about the essence of the problem and they are in sufficient agreement with the primate genealogy scheme discussed above.

When considering the composition of the order Primates, it is imperative to touch upon two other issues, the status of tree shrews and tarsiers. Tupaiiformes (commonly called tree shrews) have presented a lingering puzzle for systematicians for over 100 years now. These are 'the most primate-like insectivores or the most insectivorous primates' (Simpson, 1945). Initially they were referred to as insectivores (Diard – in 1820). Later on, however, their anatomical similarity to primates was discovered (Doran, 1878; Parker, 1886; and others). At the beginning of the century, W. Gregory (1910) and others assigned these animals to the primates, a proposal which has been debated throughout the 20th century. Indeed, in terms of their reproductive system, embryology, brain, and paleo-phylogenetic characteristics, tree shrews are not primates. Yet, a number of other anatomical characteristics (arterial system, dentition, nasopharyngeal tract, muscles, and limbs) and principles of thermoregulation in tree shrews, testify to their inclusion among the primates. It is noteworthy, moreover, that tree shrews are occasionally (more often than prosimians) used to advantage in experimental medicine. G. G. Simpson (1945) included them in the order Primates at the level of the superfamily, having artificially combined them with Lemuriformes, as did Napier and Napier (1967).

Paleontologists and morphologists refused to consider tree shrews as primates more frequently than others did (Thorington and Anderson, 1984; Martin, 1985), but similar protests also came from neurophysiologists (Balzamo, 1981). Earlier studies notwithstanding, similar conclusions were drawn by biochemists (Goodman *et al.*, 1982, 1994; Matsuda, 1985), ethologists (Deriagina and Efremora, 1998), and other scientists. A significant contribution to the solution of Tupaiidae taxonomy problems was made by a compendium on tree shrews published in 1980 (W. P. Luckett, ed.). At present, systematicians exclude this group from the order, isolating them as a separate mammalian order, Scadentia (Napier and Napier, 1985). The present monograph also distinguishes tree shrews from primates, although it maintains that these albuminoid animals of Southeast and South Asia are closely related to primates. As such, if a captive breeding program were established for them, they could become useful experimental subjects in biomedical research.

Since medical research on Tupaiidae is indeed carried out, let us give a concise description of their taxonomy. The family Tupaiidae is divided into two subfamilies, Tupaiinae and Ptilocercinae. The former comprises 4 genera (16–17 species, the most common and the best known of which is *Tupaia glis*, or common tree shrew) and the latter 1 genus (with the single species *Ptilocercus lowii*, or feather-tailed tree shrew). Body length is usually 14–24 cm without the tail (in *T. tana*, up to 25 cm); the tail approximately equals the body in length. The weight is 140–155 g on average (the Phillipine male tree shrew, *Urogale everetti*, may weigh as much as 355 g). There are 38 teeth and the chromosome diploid number is 44–68 (which varies in different species). Puberty occurs on the 90th–100th day of life and pregnancy, according to different authors, continues for 41–48 days (in *U. everetti*, for 50–56 days). Maximum longevity in captivity is 12.5 years (*T. glis*).

The systematics of Tarsiiformes, whose classification also causes much confusion, is somewhat different. Their inclusion in the order Primates has never been questioned. As was stated above, they definitely belong to one monophyletic group with all higher primates, including humans, namely, to the Haplorhini. The evolutionary and phylogenetic foundation upon which this inclusion is based is especially solid. For the 'technical' convenience of taxonomy, however, they were often referred to as prosimians (Simpson, 1945; Napier and Napier, 1967; Fridman, 1979b). Yet, as has been shown by numerous morphological studies of phylogeny as well as in biochemistry (with respect to the structure of proteins, as well as the DNA data) (Groves, 1989; Goodman *et al.*, 1994; Joffe and Dunbar, 1998), Tarsiiformes are not prosimians and they require a separate status. They were recently considered to be a transitional link between lower and higher primates, but their phylogeny and, accordingly, classification is debatable even today (Shoshani *et al.*, 1996). Napier and Napier (1985) reasonably returned to one of the early models and separated tarsiers from prosimians as a self-standing suborder Tarsioidea. The scheme presented below retains this division, but in deference to earlier authors I emphasize the affinity between tarsiers and Anthropoidea, doing so with a certain degree of taxonomic imprecision; strict systematicians attempt to avoid a fourth category between the two main ones (in this case, 'order – family'). The fundamental status of all primate suborders is retained, therefore, as accepted by the authors cited above, as well as by others.

The scheme of modern taxonomy for higher taxa classification of Primates is shown below.

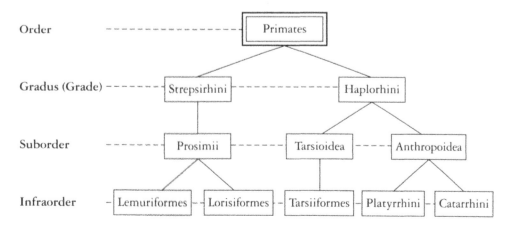

The piquant aspect of this paradigm (at which we have arrived after decades of controversy and hundreds or even thousands of studies in the traditional and newer sciences, which have contributed to a great deal of polemics) is that, in constructing a general model of the order Primates (and ignoring the classification of categories lower than infraorder and taking into account the exception of the tree shrews), we have, in fact, returned to the taxonomic scheme of the earlier primatologists (Nesturch, 1960). We are clearly at a new stage in biological science.

Strepsirhini

Suborder Prosimii

The names of two ancient divergent primate groups are associated with the construction of the rhinarium. Primates whose nostrils are shaped like upturned commas ('strepsis' meaning 'rotation' and 'rhinos' meaning 'nose' in Greek) are referred to as *strepsirhines* and primates with round nostrils ('haplo' meaning 'plain') as *haplorhines*. Strepsirhini are also distinguished by a rigid and hairless upper lip, peculiarities in placental structure and embryo development, and other anatomical characters which I discuss below. In Haplorhini, on the contrary, a hairy flexible upper lip is present together with other anatomical traits which make them distinct from strepsirhines. Of all living primates, only Prosimii are referred to as strepsirhines, which I touch upon below. Prosimians are divided into two sections, or infraorders, Lemuriformes and Lorisiformes.

Prosimians are lower primates; in many respects they are positioned taxonomically on the borderline between simians and other mammals. They are called half-monkeys, early, or pre-monkeys in some languages. Their scientific Latin name of Prosimii may be translated similarly. They retain many of the primitive traits of other animals that occupy 'lower steps' (than primates) on the evolutionary ladder. The brain of these primates is small and possesses few fissures and convolutions; along with nails they may possess typical claws, and they have special scent glands for marking territory (they may also mark with urine). The uterus in these animals is bicornuate; instead of two mammae, as in humans and simians, they have more. The facial portion of the skull projects significantly forward in many species, as does the nose (although there are species with a weakly developed facial cranium). The face and the eyebrows are supplied with sensitive whiskers. Large eyes are designed for nocturnal vision. The ears, which are large and flexible, may cover the auditory canal in some cases. The number of teeth is 36 in many species (30 in Indriidae). Some of the lower incisors in a majority of prosimians are frontally directed, rather than growing upward, and thus form a 'dental comb' which is used for scraping gum off trees and cleaning fur. In the indriids, a fold of skin on the forelimbs is reminiscent of a flying web. A majority of prosimian species are small animals, although they may be of medium size (in the sense of a medium-sized dog). All of them have tails, usually long or medium in length, although some may be very short (as in indriids, and in slender and slow lorises).

A special sensitivity to olfactory signals, the presence of whiskers, large flexible ears, and a specific eye structure (with a light-reflecting tapetum) are all indications of the adaptation of prosimians to nocturnal life and the more important role of hearing and smell relative to eye-sight, in comparison with the higher primates. In conjunction with paleontological and anatomical data, the data on DNA nucleotide sequences, particularly the 5 loci of β-globin (-∈, -γ, -η, -σ, -β) which correspond to the data on DNA cross-hybridization and its 'melting' temperature (delta-T50H), as well as the order of amino acid sequences of α-crystalline and the immunological properties of blood serum proteins, essentially confirm the monophyly (affinity due to descent from a common ancestor) of strepsirhine primates in relation to the rest of the order (Porter *et al.*, 1995). At the same time, prosimians (strepsirhines) remain true primates, related to higher primates not only in their origin. Among 40 Eocene fossil prosimian genera I may mention *Teilhardina* and *Cantus*, which were similar to mice or rats and lived 56 Ma (Shoshani *et al.*, 1996). The monophyly of strepsirhines is indicated by their common descent from adapids as well. Fluorescence studies of chromosomes reveal karyotype homology of galago strepsirhines with such haplorhines as tarsiers and humans (Healy, 1995). Prosimii are relatively seldom used in experimental biomedical research, which is the reason why the present monograph gives but a brief description of their taxonomy.

* * *

All numerical characteristics of primate biology presented below are found in earlier books (Fridman, 1979b, 1991) which gave an account of studies by other authors (Nesturch, 1960; Napier and Napier, 1967, 1985; and others), various reviews (maximum primate longevities, Jones, 1980; biometry, Robinson and Ramirez, 1982), and in other publications which present more comprehensive data. The space accorded to the description of the taxa will vary with their relevance to the objectives of this book.

Infraorder Lemuriformes

This infraorder includes two superfamilies, Lemuroidea (lemurs, indriids, cheirogaleines or dwarf lemurs) and Daubentonioidea (aye-aye). The former is divided into two families, Lemuridae (lemurs proper, subfamily Lemurinae, and dwarf lemurs, subfamily Cheirogaleinae) and Indriidae (3 genera: *Indri* proper, *Lichanotus*, and *Propithecus* [sifakas]). Lemurids have 36 teeth (lepilemurs 32) and indriids have 30. By virtue of a number of acquired characters in cheirogaleines, which are similar to those in lorisids, some morphologists combine these two groups, a form of classification which is challenged by others, in particular, R. Thorington and S. Anderson (1984). It may be assumed that the matter is somewhat clarified by the biochemists, who, on the basis of DNA and protein data, assign dwarf lemurs only to Lemuriformes (Goodman *et al.*, 1994), and not to Lorisiformes. Nevertheless, these and other authors recognize cheirogaleines as a separate family, although traditional systematicians hold an opposite view (Napier and Napier, 1985).

All living species of this infraorder inhabit Madagascar and the Comoro Islands. Ancestral forms of contemporary lemurs had evidently crossed the Mozambique Channel on floating logs at the beginning of the Tertiary, coming from the shores of eastern Africa. Possibly these were nocturnal primates from the group of the Eocene adapids and notarctids (Shoshani *et al.*, 1996). Monophyletic descent of lemurimorphic primates, long since determined by anatomists, is also confirmed biochemically and cytogenetically (Ishak *et al.*, 1988).

Figure 3.1 Lemur catta (See colour plate section)

FAMILY LEMURIDAE

Subfamily Lemurinae   This subfamily includes five genera: true lemurs (*Lemur* with 1 species; *Eulemur* – 4 species [Figure 3.1]); ruffed lemur or vari (*Varecia*, 1 species); gentle lemurs (*Hapalemur*, 3 species); and weasel or sportive lemur (*Lepilemur*, 1 species). These are small primates; the head and body are 30–45 cm (ruffed lemur, 60 cm and up; *Hapalemur*, up to 90 cm), the tail is 30–60 cm, and the body weight is about 2 kg. They have nails on all their fingers and toes (or on only one or sometimes two toes), and a protruding face with 4–5 clusters of whiskers. They have a chromosome diploid number in different species of 44–60 (in *Hapalemur* subspecies, 22–38). Gestation lasts for 120–145 days (*L. catta* [Figure 3.2] and *E. macaco*) and there is a great diversity of coat colours. Data on species affinity are rather ambiguous. In its morphology and cytogenetics, *L. catta*, while being distant from the rest of the lemur species, approaches the hapalemurs (Simons and Rumpler, 1988). *Lepilemur* is nearer to the indriids than it is to the true lemurs (Groves and Eaglen, 1988). Its hemoglobin and cytogenetics put *L. catta* next to the ruffed lemur or give it an intermediate position between *Lemur* and *Hapalemur* (Rumpler and Dutrillaux, 1978). The primary structure of the hemoglobin β-chain in *L. catta* differs from that in humans in 26 positions (for the sake of comparison, I note that this indicator in *Macaca mulatta* is different in 8 positions only) (Nute and Mahoney, 1980; Coppenhaver *et al.*, 1983). The *Homo/Eulemur* difference in

Figure 3.2 Eulemur mongoz (See colour plate section)

noncoding DNA of the β-globin cluster is 22.6% (rhesus monkey: 7%, chimpanzee: 1.7%). The figure for the delta-T50H difference is 22.3% (rhesus: 6.9%–7.3%, chimpanzee: 1.6%) (Miyamoto and Goodman, 1990). The record for longevity of *E. fulvus* (brown lemur) in captivity is over 35.5 years*[2]. We know of infanticide among *E. macaco* (Andrews, 1998).

A sensation was caused by B. Meier's discovery in the 1980s of the 'golden lemur' (*Hapalemur aureus*), which became the third hapalemur species (Meier *et al.*, 1987).

Subfamily Cheirogalinae   The Cheirogalinae are another subfamily of lemurids. These are among the smallest primates on Earth. They are called dwarf or mouse lemurs, although, as emphasized above, a number of scientists have proposed that they be placed in a different infraorder by virtue of their anatomical characters. These primates are also classified into 4 genera: *Cheirogaleus* (2 species), *Microcebus* (3 or 4 species), *Phaner* (1 species), and *Allocebus* (1 species). Of these tiny animals the best known, allegedly, is *M. murinus*, ordinarily called the 'mouse lemur'. Its head and body measure about 13 cm, the tail 17 cm, and it weighs 60 g on average (sometimes as little as 40 g). It is omnivorous. The onset of puberty occurs at about 7–10 months, there is a seasonal

2 An asterisk marking a life-span indicates that the animal was still alive at the time of registration (Jones, 1980).

reproductive cycle, and gestation lasts for 59–62 days. There are two births within one season, usually twins, and a newborn weighs from 3 g to 5 g. The chromosome diploid number is 66. Other cheirogaleines, although they may be larger in size, are still rather small (head and body length does not exceed 25 cm–28 cm). The record for longevity in captivity is 15 years (*M. coquereli*). The species *M. ravalobensis* was discovered recently (Zimmerman *et al.*, 1997).

It should be noted that the common name 'lemur' is incorrectly used, not only for the family Lemuridae described above, but also for all prosimians. This mistake originated in the earliest taxonomy by C. Linnaeus when this great natural scientist introduced the 'genus' *Lemur*, to mean, in fact, all prosimians. Since Linnaeus' time, however, primate systematics has undergone considerable change. To call a galago a 'lemur' today would be the same as calling a fox a 'dog', since they both belong to the family of canines. Even more controversial is the title of a 1995 paper, where a '. . . lemur (*Loris tardigradus*)' is mentioned. It is clear that only the animals of the family described above may correctly be called lemurs, and even then with certain reservations.

FAMILY INDRIIDAE

The second family of the infraorder in question is Indriidae (indriid prosimians). This family comprises the largest of the prosimians, although even some of these species may be of relatively small size. Head and body length is 30 cm–70 cm and the tail ranges from 3 cm–55 cm. The digits have nails, but the second toe has a claw. Both the number and the size of the whiskers are less than those in lemurids. The face is shortened and covered with hair. The lower incisors form a dental comb, but with only 4 teeth instead of 6 as in the previous family. As noted, they have a total of 30 teeth.

As stated above, the family includes 3 contemporary genera: indri proper (*Indri* with 1 species); lichanotus or hairy indri, or long-haired avahi (*Lichanotus* is sometimes incorrectly designated by the term *Avahi*; 1 species) and sifaka (*Propithecus*, 3 species). The latter had its third species identified recently by E. Simons and named *Propithecus tattersalli* in honor of the prominent contemporary primatologist, Ian Tattersall (USA).

Indri proper (*Indri indri*) is the largest species in the family. Head and body length is about 70 cm–90 cm (45 cm–55 cm in *Propithecus*), and its tail length is unique for the prosimian suborder, measuring only 3 cm (it equals the body in length or is even longer in *Propithecus*). Reproduction is seasonal; the gestation length is 2 months in indri and 5 months in sifaka. One young is usually produced in both genera. Puberty occurs at the age of 2.5 years in a sifaka female. Chromosome diploid number is 44 in indri and 48 in sifaka. Maximum longevity in captivity is 18 years (*P. verreauxi* [Figure 3.3]).

The superfamily Daubentonioidea includes the only family of living prosimians, Daubentoniidae, with the only genus and species of *Daubentonia madagaskariensis*, or aye-aye, a rather rare primate. This animal is justly included in the infraorder Lemuriformes. According to the latest biochemical and DNA data (Porter *et al.*, 1995), as well as the long-established anatomical data, the aye-aye is a sister group to the other Malagasy strepsirhines, separate from the Lorisiformes. The aye-aye is about the size of a domestic cat (about 40 cm in length), with a long and thick, though rigid, fox-like tail (up to

Figure 3.3 Propithecus verreauxi (See colour plate section)

60 cm long) and 18 rodent incisors. This is the only strepsirhine lacking the dental comb. The face is short, drastically widening at its upper part, with large eyes and ears, the latter rather prominent, flexible, and thin. The ears, hands, and feet are black. The fur is coarse and scant, with a dark brown or black undercoat. All fingers and toes, except the hallux, have claws. The hallux has a flat nail which is opposable to the others. The claw on the third finger is notably thin, sharp, and long; an efficient device for making incisions in tree bark and extracting insects. Reproduction is seasonal; one young at a time is produced. The chromosome diploid number is 30. The record for longevity in captivity is 23 years.

I should now mention the status of the present-day primates as representatives of the Earth's diminishing fauna. By the mid-1980s, about half of living primates were under threat of extinction and were entered in the Red Book. This sad list is growing longer, according to the Convention on International Trade in Endangered Species of Wild Fauna and Flora (CITES), which has two appendices: Appendix I (Endangered) and Appendix II (Threatened). The Convention was enacted in the mid-1970s and has been regularly revised since then. All of the Earth's primates were originally entered in the 'Threatened' appendix and most were entered in the 'Endangered' appendix by 1990. To the other primates threatened with extinction, the latter appendix recently added all the prosimian species of Madagascar and nearby islands, i.e., all of the Lemuriformes (Lemuridae, Indriidae, Daubentoniidae).

Infraorder Lorisiformes

FAMILY LORISIDAE

The second infraorder of living strepsirhines, of the suborder Prosimii, is more homogeneous than the first one. It has only one family, *Lorisidae*. It comprises relatively small primates (body length of 12 to 40 cm) with tails that are medium-sized, short, or completely absent. The dental comb is made up of 6 teeth; the total number of teeth is 36. Each hand has five fingers and each foot has five toes. All fingers and toes bear nails, except the second toe, which has a claw. The head is round with a short facial section. Close-set eyes are large and often have dark rings around them. There are whiskers on the face, though fewer than in lemurs. In some galagos, whiskers may grow on the forearms. The ears in galagines, in contrast to lorisines, are large and prominent, and capable of independent motion (lorisines have relatively small ears). Representatives of both the galagines and lorisines have 2–3 pairs of mammae. As a rule, these are nocturnal arboreal primates. They may be insectivorous, herbivorous, or even carnivorous. The forebears of the galagines and lorisines evidently lived in Africa. Among the fossils of Fayum from the Eocene was found a loris-like primate (Simons, 1992b).

The family Lorisidae is divided in two subfamilies: Lorisinae and Galaginae. The former comprises only 4 genera of slow-moving primates: *Loris* (1 species), *Nycticebus* (2 species [Figure 3.4]), *Arctocebus* (1 species), and *Perodicticus* (1 species, potto). The former two genera, the slender loris and slow loris, inhabit Asia; the latter two, *Arctocebus* and *Perodicticus*, live in Africa. The subfamily Galaginae comprises 1 or 4 genera, depending upon the taxonomy that is applied. The number of genera and species of Galaginae is debatable, in fact, for anatomical reasons. Certain subspecies may be considered species and some subgenera may be considered genera. Equally unstable is the taxonomic status of the subfamily itself, which is sometimes regarded as a family. I shall abide by the opinion of Napier and Napier (1985) and other authoritative scholars who

Figure 3.4 Nycticebus coucang

combine all fast-moving Galaginae in one subfamily with 4 genera: *Galago* (3 species of galago proper, or bushbabies), *Otolemur* (1 species), *Euoticus* (1 species), and *Galagoides* (probably 4 species; see Rylands and Caram, 1998, about the discovery of a new species). Let us reiterate that this species classification is far from being recognized as valid by all experts; some name up to 9 or even 11 species (Nash *et al.*, 1989).

Contemporary galagines inhabit Africa. Body size ranges from that of a mouse (*Galagoides demidovii*, or Demidoff's galago) to that of a rabbit (*Otolemur crassicaudatus*, or thick-tailed bushbaby). They are sometimes used in experimental medicine, for which they are bred in captive colonies. The percent difference in galago β-globin noncoding DNA nucleotide sequences from those in humans is 28.9% and the DNA hybridization delta-T50H difference is 28.0% (Miyamoto and Goodman, 1990).

The slender loris (*L. tardigradus*) is an interesting representative of the Lorisinae. The head and body measure 18 cm–26 cm in length and body weight is 300 g. The tail is undetectable. The fur is thick, soft, yellow and gray or dark brown in color. The duration of gestation is 5–6 months and one or two young are normally produced. The chromosome diploid number is 62. It inhabits the forests of Sri Lanka and South India. The record for longevity in captivity is 12 years.

The Senegal galago or lesser bushbaby (*Galago senegalensis*) may be considered a typical representative of the subfamily. The body and tail measure 16 and 23 cm, respectively, and body weight is 1200 g. The fur is soft, long, and orange-brown in color. A female gives birth twice per year to 1–2 young; the newborn weighs 8 g–13 g. The chromosome diploid number is 36–38. It inhabits the forests and underbrush of equatorial Africa. It is known that infanticide occurs in at least 5 species of galago (Tartabini, 1991). In captivity galagos are afflicted with many diseases which are characteristic of humans, among others, atherosclerosis, myocarditis, glomerulonephritis, fatty liver, etc. The record for longevity in captivity is 16.5 years*.

Haplorhini

Suborder Tarsioidea

In order to understand the general characteristics of haplorhine primates we must examine the primary characteristics of strepsirhines (structural peculiarities of the nostrils, lips, placenta, etc.) and point out the differences in each case. As was stated above, there is a group which is distinctly noteworthy among the living haplorhines at the suborder level, that is, the tarsiers. These primates have been known to science since the beginning of the 18th century (Niemitz, 1984). The name given to these animals in primatology is an unusually precise reflection of a particular anatomical characteristic in these primates; each foot of the rather long hind limbs ends in an elongated heel section, called a 'tarsius', which accounts for both the scientific and the common name of this fascinating creature. The animal's biological status is of great interest to scientists, since it has the features of both prosimians and anthropoids. These features include their skull and limb anatomy, the number and position of the mammae (1 pair on the chest and 1 pair on the stomach, although there may also be 3 pairs), huge yellow eyes (in comparison to the small body; the largest in size among all mammals), a body length of 8.5 cm–16.0 cm, a hairless tassel tail of 13 cm–27 cm, large flexible ears, a uterus of bicornuate structure, claws on the feet (two on each foot, not one as in Prosimii), a frog-like locomotor pattern, and nocturnal habits; all of these characterize tarsiers as something less than monkeys. Moreover, the tibia and fibula in the hind limbs are fused in the lower third as in a rabbit. The fingers and toes are bony, thin, and long, surmounted by circular digital pads designed specifically for tree-climbing.

Yet tarsiers also have many anthropoid characters, e.g., they lack a naked rhinarium and dental comb and their lower incisors grow vertically (there is a total of 34 teeth). The head is round and is positioned relatively vertically on the spine. The face is very flexible and capable of grimacing due to its well-developed muscles. As in anthropoids, the skull has a postorbital partition which isolates the orbit from the temporal fossa, although it is less well-developed in tarsiers. The orbits are more frontally directed than in the prosimians. The retina is characterized by a central fossa, the tapetum lucidum is lacking, and the visual cortex is considerably enlarged. These latter traits place tarsiers closer to anthropoids.

Many similarities may be found between tarsiers and anthropoids in terms of their reproductive systems. They have, for example, a monodiscoidal, hemochorial placenta, scrotal testicles, and the females are polycyclic. It is with tarsiers, moreover, that we first encounter a physiological phenomenon of the reproductive system which is unusual in the animal world. This is the menstrual cycle of females, which is characteristic of human beings and other primates (with some exceptions). In at least two species (*Tarsius bancanus* and *T. spectrum*) the length of the cycle is 24 days and the menstrual period lasts for 3–4 days (Permadi *et al.*, 1994). Gestation lasts 6 months and a single young is usually produced. Tarsiers are capable of reproduction throughout the year. A newborn tarsier usually weighs 25 g–27 g, has sight when it is born, and possesses a strong grasping reflex.

As was noted above, the evolutionary basis for assigning tarsiers to the order Primates is rather solid. Once the opinion was expressed that contemporary tarsiers were 'living fossils', i.e., they were very close to their ancestors. Whether or not this is true, today we know up to 46 genera of fossil tarsiers and only one is now extanct (Shoshani *et al.*, 1996). Most probably, tarsiers descended from anaptomorfines of the family Omomyidae, which evolved in the Lower Eocene and the Upper Oligocene of North America, Europe, and possibly Asia. The possibility that tarsiid omomyids were the ancestors of anthropoids must not be ruled out either (Szalay, 1988), although there are opinions that this role is played by adapis (Ibid.). Supposedly, Omomyinae-Tarsiiformes gave rise to two branches of primates which emerged approximately 45 Ma. One of them eventually led to contemporary New World monkeys (Cebidae, Atelidae, and Callitrichidae), and the other one to contemporary Old World catarrhines, and in the long run, to the advent of humans (Niemitz, 1985). There are, in fact, other paradigms for representing the evolution of the higher primates (Simons, 1995). Let us, however, return to tarsiers.

Apparently, tarsiers are the least similar to other primates in terms of karyotypes. The chromosome diploid number of 80 is unique among primates to tarsiers. A study of somatic karyotype in the Phillipine tarsier (*T. syrichta*), by means of the differential chromosome dying technique (C-, G-, and R-type bands), revealed a rather close affinity in chromosome morphology with another related species of the same genus (*T. bancanus* [Figure 3.5]). There were significant evolutionary structural differences, however, in comparison with other primates, prosimians as well as anthropoids. Scientists went so far as to define this peculiarity taxonomically at the order level (Tarsiiformes) (Dutrillaux and Rumpler, 1989). We find less categorical conclusions in another cytogenetic study,

Figure 3.5 Tarsius bancanus

the author of which maintained that, since tarsiers cannot be placed either in the suborder Prosimii, nor with Anthropoidea, they should therefore be identified as a separate suborder, Tarsioidea (Mai, 1985). This approach, in fact, has been adopted by systematicians and is also accepted in the present monograph.

Nevertheless, in many other ways, as was stated, tarsiers are very closely related to monkeys, apes, and humans. I have already emphasized the substantially closer affinity of tarsiers with anthropoids, rather than prosimians, based on molecular biology studies in general. Delta- and beta-globin DNA data also substantiate the placement of tarsiers with anthropoids (Koop *et al.*, 1989), as do statistical methods of constructing phylogenetic trees based on nuclear and mitochondrial sequences of nucleotides (Hasegawa, 1990). Finally, the same conclusion is derived from studies of hemoglobin amino acid sequences and lens α-crystalline in the tarsier eye (Beard and Goodman, 1976; de Jong and Goodman, 1988). Position 146 is taken by isoleucine, as in anthropoids, including humans, while in prosimians this position is taken by valine. This change, in relation to other mammals, is a newly acquired character of the haplorhines' common ancestor, and therefore, testifies to their closer affinity to humans.

This hypothesis is supported by vitamin C biosynthesis, a remarkable phylogenetic indicator in primate biochemistry. It is a well-known fact that this vital vitamin is not produced in primates, including humans, due to the absence of L-gulono-1, 4-lactonoxidase enzyme which catalyzes the final phase of ascorbic acid synthesis. A study of 15 prosimian and tarsier species revealed a low concentration of the above enzyme in the kidneys of all animals and its complete absence in the liver of tarsiers as well as of anthropoids, another finding which is indicative of the closer affinity of tarsiers with the higher primates (Pollock and Mullin, 1987). As far as percentage differences (and similarities) in data on DNA hybridization and duplex melting temperatures are concerned, the figures for humans and tarsiers are 24.6 and 25.4, respectively (Miyamoto and Goodman, 1990). Data on the reproductive system and the brain were mentioned above.

The classification of the living tarsiers, therefore, can now be described as follows. Suborder Tarsioidea includes one *family Tarsiidae* with a single genus (*Tarsius*). All representatives of the genus inhabit South East Asia (Sumatra, Borneo, Sulawesi) and the Philippines. Until recently, 3 species were traditionally assigned to this genus; the Phillipine tarsier (*T. syrichta*), the Bankan or western tarsier (*T. bancanus*), and the spectral tarsier (*T. spectrum*). In the 1980s, Carsten Niemitz proved the species identification of *T. pumilus* (from the Celebes) which had been described in the early 1920s, but was later classified as a subspecies of *T. spectrum*. Other authors agreed with him (Musser, 1987). However, Niemitz and his colleagues also proposed to separate another tarsier species from the same group and to name it *T. dianae* after Dian Fossey, the prominent primatologist who strived to save East African mountain gorillas from extinction and who was killed in Africa (Niemitz *et al.*, 1991). Over 10 tarsier subspecies are known.

Let us point out that the tarsier, which weighs 120 g–130 g, is one of the smallest primates on Earth. Tarsiers may live alone, in pairs, or in small groups. They eat insects, small lizards, and eggs of birds. A tarsier carries food to its mouth by standing upright on its feet and leaning back on its tail. It laps water, however, as monkeys do. The head is capable of 180 degree rotation. The coat is primarily gray with shades of red and brown (*T. syrichta*) or with gold and brown speckles (*T. bancanus*, *T. spectrum*; the latter is distinguished by a long tuft of fur on the tail and white spots behind the ears). Locomotion is effected by leaping (1 m and further). Tarsiers thrust back their hind limbs in a manner similar to that of frogs or grasshoppers, using the tail for balance. The oldest tarsier in captivity lived 13.5 years (*T. syrichta*).

Species of the suborder Tarsioidea are threatened with extinction and therefore require protection. All Tarsiiformes, in fact, have been classified as endangered.

Suborder Anthropoidea

Of all primates (of all animals on Earth, in fact), this suborder has a special status; it includes all monkeys, apes, and humans. It is not by chance that the name of this group of primates means 'anthropomorphous' in translation from Latin. It contains the closest relatives of humans. The nonhuman representatives of this suborder have played an exceptional role in studying human diseases and pharmacology. They are the experimental foundation of medical primatology.

Despite the great diversity of species included in the suborder (with head and body length ranging from 10 cm–13 cm [*Cebuella*] to 180 cm–200 cm [*Gorilla*, *Homo*]), morphologists, paleontologists, geneticists, biochemists, and systematicians all recognize the monophyly of this suborder. Summarizing the morphological data, J. H. Schwartz (1986) described the monophylic features in Anthropoidea which I use herein to describe the suborder, along with other peculiarities of this group. These features include a round head and short face (excluding baboons and related species), a large brain with a great number of fissures and convolutions (except in some platyrrhines), early fusion of metopic and mandibular symphyses, a reduced number of premolars (2–3) with a total of 32 teeth (36 in many platyrrhines), frontally directed orbits, an orbital cavity separated from the temporal fossa by a ring of bone, a retina with central fossa, indicating that the eyes are designed for color, stereoscopic and diurnal vision (except in *Aotus* [owl monkeys, nocturnal]), small and, as a rule, immobile ears, flat nails on all fingers and toes (except in marmosets and tamarins), a reduced number of whiskers which may be found only in lower simians and exclusively on the face (2–3 tufts on the chin, above the upper lip, and above the eyes), tactile skin patterns developed on the palms of the hands, fingers, and soles of the feet (in all monkeys, apes and humans), a plain uterus, a discoidal, hemochorial placenta, one pair of mammae, an ovarian menstrual cycle, and a uvula (with the exception of *Ateles*, in which it is absent, as in prosimians [Szostakiewicz-Sawicka *et al.*, 1981]). Vision in anthropoids is the predominant analytic faculty, as opposed to smell and hearing in prosimians. The thumb and hallux are opposable in most species, except in some platyrrhines in which they are not always so on the forelimbs. In a number of species the thumb and hallux are reduced or completely absent (Atelinae, Colobinae). All species, except humans, retain a coat with varying hair length among species, but no undercoat. Fur is frequently brightly colored.

No tail is present in the hominoids, while other species of the suborder may bear short, medium-sized or tails which are longer than their bodies, and which may be prehensile, semi-prehensile or not prehensile. Among the anatomical arguments and characteristics that prove the monophyly of the suborder are a number of other peculiarities of skeletal and vascular structure (Schwartz, 1986), and also of DNA and protein affinity (Goodman, 1991; Goodman *et al.*, 1994). The similarity between Old World and New World primates is indicated by abundant evidence from other sciences, such as genetics and immunology, and such 'indirect' evidence as the similarity of ocular dominance in *Cebus* and *Macaca* (Rosa *et al.*, 1988) and of parasites in platyrrhines and catarrhines in contrast to prosimians (Ardito, 1980). It should be noted that while the percentage difference of noncoding DNA for Strepsirhini and Haplorhini is 35, it is only 12.3 for Platyrrhini and Catarrhini (Goodman, 1991).

As stated previously, the branch which subsequently developed into Anthropoidea apparently began to evolve in the Upper Eocene – Lower Oligocene. Possibly, the ancestors of these primates were ancient forms close to the ones discovered recently in Fayum, Egypt (Simons, 1990, 1993). Among the oldest and most primitive anthropoids are such early Fayum forms as *Aegiptopithecus*, *Oligopithecus*, *Parapithecus*, *Catopithecus*, *Propliopithecus* and others. The oldest are *Catopithecus* and *Catrania*.

The divergence of anthropoids began in the Oligocene (40 Ma–35 Ma?) (and even earlier for prosimians) and led to contemporary platyrrhines and catarrhines. Considering the fact that the ankle bone in the Argentinean *Dolichocebus* of the Lower Oligocene is similar to that in the extant platyrrhines, the tarsal bones in *Aegiptopithecus* to those in the future hominoids of the Miocene, and calcaneus in *Apidium* to that in the lower catarrhines (Gebo and Simons, 1987), one might assume that the platyrrhines and catarrhines existed as early as the Lower Oligocene. It cannot be said with certainty that their common ancestor was an anthropoid, since it has been hypothesized that this division had already begun with prosimians, and that thereafter the two modern infraorders evolved independently (Hershkovitz, 1981).

At present, as was shown above, there exists an undeniable division of the suborder Anthropoidea in two infraorders, Platyrrhini and Catarrhini. The meaning of these terms makes it obvious that the division is based on the difference in structure of the rhinarium. This Latin nomenclature has Greek roots which are anatomically accurate. In New World monkeys, the septum is broad and the nostrils point sideways, while in Old World primates, including humans, the septum is narrow and the nostrils point downwards. Clearly, this is not the only difference between the two infraorders. I discuss this in more detail below. Many representatives of the suborder in question have been entered in the Red Book and are banned from being bought and sold, owing to the serious danger of extinction they are facing in the wild.

Infraorder Platyrrhini

This infraorder includes primates of the New World or 'broad-nosed' monkeys, which are classified as simians, or higher primates (as opposed to the lower primates described above), but they are nevertheless lower simians (as are catarrhine monkeys) in relation to higher simians, or apes. Naturally, there is another catarrhine primate species which belongs to the higher primates together with monkeys and apes; the human.

Dissimilarities between catarrhines and platyrrhines are, naturally enough, revealed through descriptions of the two groups. A number of general characters of Platyrrhini are considered primitive (Thorington and Anderson, 1984), including: the absence of auditory canal osteal tube; 3 premolars (Cebidae, Atelidae and *Callimico* have 3 molars in each quadrant and a total of 36 teeth each, while in Callitrichinae the figures are 2 and 32, respectively); cheek pouches, which are characteristic of many catarrhines and absent in platyrrhines; and the lack of ischial callosities in platyrrhines. Platyrrhines also possess general acquired characters which are concerned mostly with the shape of the teeth (e.g., the absence of hypoconulid on the third molar), certain ossa, plus reduced noseblade cartilages, placental structure, and particulars of embryogenesis. They possess a bidiscoidal placenta (monodiscoidal in *Alouatta*), no definite appendix, a large intestine with no sigmoidal curve, and thumb muscles that are either degenerated (in arachnoids) or weak. In arachnoids the first finger on the forelimbs is absent, in others it is opposable, but to varying degrees; it is perfectly opposable on the hind limbs. The nails are flat in most species, but claw-like and sharp in callitrichids, which still bear flat nails on the hallux. Platyrrhines are characterized by a relatively heavy brain (relative to body weight), especially in the Cebinae in which it is 1:18 or 1:16. This is comparable to or even exceeds the respective ratio in humans.

New World monkeys are arboreal primates of small and medium size, from 10 cm–13 cm (*Cebuella*) to 90 cm (*Alouatta*). The tail is long (one-and-one-half times the body length in *Ateles*) or of medium length (less than half the body length in *Cacajao*), and it is prehensile in many instances. The larger the primate, the more prehensile is the tail (it is especially strong in *Ateles* and *Alouatta*). Hind limbs are normally longer than forelimbs. The fur is thicker than in the catarrhines, usually a single color in larger species (black, gray, brown, red), but with contrasting spots of white or black on the face and head. Callitrichines are still more brightly colored black, brown, gray and red, and golden, while some specimens are only partially colored, or completely white (tamarin, and a subspecies of another genus – uakari). Sexual dichromatism is characteristic of some howler monkeys and sakis. The Callitrichines have crests on top of the head, above the ears, and also 'sideburns' and whiskers. Some monkeys are bearded (howler, saki) or even bald (uakari).

Gestation in Callitrichinae lasts 140–170 days and results in 1–3 young, normally 2. In other platyrrhines, gestation continues for approximately 6 months and only 1 young is born. Social habits range from monogamy in *Callicebus* to bonded pairs assembled within large groups (Callitrichinae), and promiscuity in large (Alouttinae) and small (Atelinae) groups.

The small-sized species are primarily insectivorous or consume plant juice and resin. Larger platyrrhines are herbivorous. Howler monkeys usually eat leaves rather than fruit. Platyrrhines in captivity are quite sensitive to vitamin D2 and D3 deficiency, which causes so-called cage paralysis, a rachitic affliction of the ossa. Almost all New World monkeys possess a very specific means of vocal communication. The cries of howler monkeys (*Alouatta*) and titis (*Callicebus*) can be heard a great distance away. An important role is played by olfactory communication; platyrrhines mark their territory, partners, and young by means of various secretions and are quite sensitive to smell.

Modern platyrrhines account for up to one-third of all genera of the order Primates (16 genera). New World primates inhabit South and Central America from southern Mexico to southern Argentina. All platyrrhine species may be successfully maintained in captivity, although under very specific conditions. It is of medical interest that some species of these primates, e.g., marmosets, tamarins, squirrel and spider monkeys, are normally carriers of various strains of herpesvirus which are absolutely harmless to themselves, but lethal to other monkey species. In part for this reason, many platyrrhines are successfully used for the purposes of medical primatology.

As stated above, platyrrhines began to evolve in the Oligocene (40 Ma–35 Ma). The suggestion was made (by T. Huxley) in the previous century that this group, which represents an exceptional anatomical variety of species, may be polyphyletic. As noted by J. Schwartz (1986), platyrrhine monophyly cannot be easily proven, and it is still more difficult to assert the 'naturalness' of modern New World monkeys. Outstanding works have been written, however, which indeed advocate platyrrhine monophyly (Ford, 1986; Rosenberger *et al.*, 1990). The problem of monophyly is associated with the controversial idea of the African origin of the ancestors of the American monkeys.

The statements made above, in conjunction with the great diversity of the extant platyrrhines, accounts for the somewhat problematic and unstable classification of the New World primates. There are few if any variations in platyrrhine systematics (at least above the genus level) that have not been proposed for grouping these monkeys taxonomically. In my opinion, however, no matter how extensive the variety of modern platyrrhine taxonomies, the common origin and present day affinity of this primate group is clearly indicated. So far, no agreement has been reached on this issue in classification, due to the different approaches and criteria which have been used. Nonetheless, exclusively on the basis of morphological criteria (i.e., discounting biochemical and related approaches), this affinity allows us to combine all New World monkeys in one family, Cebidae (Thorington and Anderson, 1984; Shoshani *et al.*, 1996), divide them into two families, Cebidae and Atelidae (Rosenberger, 1990), or into three families, Cebidae, Callitrichidae, and Atelidae (Ford, 1986). Goodman *et al.* (1998) provided some other data. In fact, many other schemes have been proposed by other systematicians and non-morphological researchers. Below additional suggestions are cited in order to show certain instances of affinity within this group of primates which is indispensable to medical primatology.

The South American genus *Branisella* from the Oligocene should be considered the oldest representative of the Platyrrhini (Delson and Rosenberger, 1984). Besides *Branisella*, other extinct forms are known, including *Dolichocebus* and *Tremacebus* from the Oligocene, and *Homunculus*, *Neosaimiri*, *Cebupithecia*, and *Stirtonia* from the Miocene (Szalay and Delson, 1979). Morphological systematicians (Hayasaka *et al.*, 1992) have recently established a connection between the descent of modern forms and the evolutionary process in the Middle Miocene (15 Ma–12 Ma). Thus, phyletic affinities are assumed between modern *Saimiri* and extinct *Neosaimiri* (which, in its turn, is related to the above-mentioned *Dolichocebus*), between the pitheciines and *Cebupithecia*, owl monkeys and *Aotus dindensis*, and howler monkeys and *Stirtonia*. The living Callitrichinae are phyletically related to *Mahanamico* and *Micodon* from the Miocene. I should also note the work of M. Goodman's group, which used two methods to delineate phylogenetic trees on the basis of $\psi\eta$-globin, and found that spider and night monkeys are considerably closer to each other than to capuchins (neighbor-joining method). Yet a different strategy (maximum parsimony method) proved the contrary, i.e., it is capuchins and night monkeys that form two closely entwined branches (I regard discrepancies of this type as signs of the close affinity of all three branches). The biochemists believe that these three primate groups diverged from one ancestor within a relatively short span of time about 22 Ma–16 Ma (Hayasaka *et al.*, 1992).

In studying the nucleotides of other genes (γ-globin, ϵ-globin), Goodman's group (with some membership modifications) obtained different, but basically correlated results on platyrrhine phylogeny. These studies suggested that there were two families, Cebidae and Atelidae, that diverged about 20 Ma and their subfamilies 19 Ma–16 Ma, assuming 35 Ma as the point of divergence between platyrrhines and catarrhines (Schneider *et al.*, 1993), or three families, in the case where the Pithecidae are classified as a separate family (Meireles *et al.*, 1995). In one of his concluding publications, M. Goodman (1996) suggested the times of the last common ancestors using calculations based on the 'local molecular clocks' procedure; for families Cebidae 16.5 Ma and Atelidae 18.2 Ma; correspondingly, for subfamilies Cebinae 15.7 Ma and Atelinae 12.9 Ma (see also Goodman *et al.*, 1998).

At this point I should discuss platyrrhine clades, regardless of the number of families recognized. Clades certainly reflect phylogenetic affinity, but their taxonomy may be described quite differently by different scientists. Assuming an African origin for the common ancestor of New World monkeys, J. Cronin and Sarich (1975) concluded from their immunological studies of primate albumin and transferrin that divergence of the platyrrhines began in the Lower Miocene and led to the formation of seven branches: 1) Aotus; 2) Callicebus; 3) Cebus; 4) Saimiri; 5) Ateles-Lagothrix-Alouatta; 6) Pithecia-Cacajao; and 7) Callimico-Callithrix-Cebuella. This classification appeared adequate; it was also in line with what morphologists believed at the time (the only qualification being that tamarins [*Saguinus*] should have been included in the 7th group).

On the basis of morphological cladogenesis, which assigns the seven clades of Platyrrhini to the period of 23 Ma–17 Ma, Goodman's group of biochemists also proposed in the 1990s to classify the evolution of these seven clades into three contemporary families (common names are used): family Atelidae, 1 clade (howler monkey, spider monkey, wooly monkey, wooly spider monkey); family Pitheciidae, 2 clades (1 for titi monkey and 1 for saki and uakari monkeys); and family Cebidae, the most heterogeneous family comprising 4 clades (night monkey, capuchin monkey, squirrel monkey and marmoset). The marmoset group also includes tamarins (*Saguinus, Leontopithecus*) and Goeldi's monkeys (*Callimico goeldii*).

I should note again that the different models of primate evolution and classification are given here for the purpose of discussing their affinity and similarity as foundations of contemporary medical primatology. Naturally, these models are taken into account in delineating primate taxonomy, not as something inflexible and ultimately definitive, but rather with reference to historical background. Thus, major problems arise in connection with the status of the genus *Callimico*, whose taxonomic history resembles the problematic history of *Tarsius*. The single species in this genus has been intermittently assigned by authoritative scientists to the family Callitrichidae and to Cebidae, thus making them close either to tamarins and marmosets, or to capuchin-like monkeys. Sometimes they are even defined as an independent family, Callimiconidae. This uncertainty is more than justified. Important taxonomic characters, such as skull structure and shape, dental formula (3 molars, 36 teeth total), peculiarities in dental morphology, and single births (as in most platyrrhines), rather than twin or triplet births (as in marmosets and tamarins), all put *Callimico* close to Cebidae. On the other hand, *Callimico* also possesses traits characteristic of callitrichines, for example, small body size, claw-like nails, imperfectly opposable first finger on the forelimbs, and a shortened first toe which also bears a nail.

In recent decades, a multitude of data have been collected which favor the affinity between *Callimico* and callitrichines, e.g., anatomical data. Dental ontogenesis and cusp patterns in *Callimico* resemble those in tamarins (Byrd, 1981). Femur structure in both proves their primitiveness, common origin, and significant difference from the more advanced Cebidae (Ciochon and Gorruccini, 1975). The same is revealed by forelimb musculature studies (Dunlap *et al.*, 1985). Results of chromosome evolution research unequivocally confirm that *Callimico* ancestors were definitely closer to the line of marmoset and tamarin ancestors, than to the ancestors of other platyrrhines (Dutrillaux *et al.*, 1988). Considering these and previous recommendations by morphologists, as well as multilateral investigations of DNA and proteins, the present monograph, in contradiction to previous publications (Fridman, 1979b, 1991), includes *Callimico* in the group of marmosets and tamarins as a subfamily, Callimiconiinae, of the family Callitrichidae. Most probably, this group emerged no later than the Middle Miocene.

Another problem of great concern in the taxonomy of the platyrrhines is the situation in the until recently indivisible family Cebidae. As already mentioned, it was sometimes considered to comprise all the primates of the New World, but more often it was separated into one or two families. It appeared to be a rather polyphyletic group with widely varying anatomies, karyotypes, external features, and evolutionary origins. Paleontological data of recent years indicate a clear atelid-cebid dichotomy of 20 Ma and subsequent independent divergence of each of these two branches. As a result, we now recognize two platyrrhine families, Cebidae and Atelidae, each comprising two subfamilies (Rosenberger *et al.*, 1990). All callitrichids are assigned to the cebids. This decision is confirmed by globin DNA investigations (Meireles *et al.*, 1995), which, in turn, require also the separation of another family, Pitheciidae.

Taking into account the above discussion on the status of Callitrichidae and also the possibility of a higher position of pithecies, my version of the platyrrhine taxonomy scheme is shown below (I note that the scheme presented is, on the whole, in fundamental agreement with recent phylogenetic analyses, although it may be discrepant in some details. In particular, there is a certain deviation from cladogenesis based on a very interesting $\in$-globin DNA analysis [Schneider *et al.*, 1993]).

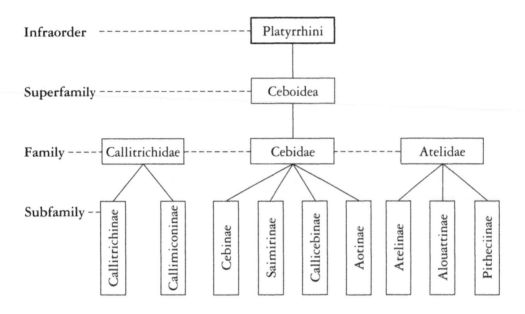

It must be said that this scheme does not pretend to be the final word, especially for families.

As it happens, it is the last category in my scheme, the Subfamily, which is the most important for understanding the systematics of the living New World primates (in the context of the present monograph). The nine subfamilies presented above are contemporary groups of living platyrrhines, regardless of the way in which they are classified.

FAMILY CALLITRICHIDAE

The general characteristics of this family were described above. These are the smallest monkeys in the suborder (comparable in size to a mouse, a rat, or a squirrel). They are also the most primitive. Females may often be larger than males. The tail is long, but not prehensile. Long and soft fur is characterized by peculiar coloration. The face is almost completely hairless (except in *Cebuella*), but is adorned with sideburns, often ear tufts (marmosets) and moustaches, and some species have manes. The ears are relatively large and the eyes are small and blue, relatively rare characteristics in primates. They are highly emotional and possess a variety of communicatory abilities, including facial expressions, ear and hair movements, various vocal signals, and scent secretions in saliva, urine, and feces to mark territory. It may be assumed from paleontological data that this group was one of the first to diverge from the common ancestor of platyrrhines over 25 Ma (Rosenberger, 1984).

As can be seen from the scheme above, the family includes two subfamilies: Callitrichinae and Callimiconinae. The former includes 4 genera: Callithrix (marmosets; the number of species is debated by different authors, but the total of 8–10 taxa may be classified into 3 species [Hershkovitz, 1977], 7 species [Coimbra-Filho, 1981], and so on), Cebuella (*C. pygmaea*, pygmy marmoset; 1 species), Saguinus (tamarins; about 20 taxa, the number of species varies according to different sources, e.g., 10 [Natori, 1988], 11 or 16 in 2 previous references), and Leontopithecus (golden lion tamarins; 4 species).

Figure 3.6 Callithrix geoffroyi (See colour plate section)

The dentition in Callitrichinae is characterized by 2 incisors, 1 canine, 3 premolars, 2 molars, and a total of 32 teeth, as mentioned above. The first two genera, *Callithrix* and *Cebuella*, are the most primitive primates in the infraorder. Their faces make them easily distinguishable from the other two genera (*Saguinus* and *Leontopithecus*). The former two genera are characterized by a V-shaped mandible and relatively large incisors which are not smaller than the canines. The latter two genera have a U-shaped mandible and the lower canines are notably longer than the incisors and project beyond them. The former two genera are marmosets (including pygmy marmosets), and the latter two are tamarins (including the golden lion tamarins). Tamarins are somewhat larger than marmosets. While an adult pygmy marmoset weighs only 110 g–150 g, the largest tamarins may weigh 600 g or more (a golden lion tamarin weighs about 750 g). The average weight of marmosets is 250 g–500 g, whereas that of *Callimico* is 470 g.

In medical primatology, very extensive use is made of Callithrix monkeys ('callos' = 'beautiful', 'thrix' = 'hair') (Figure 3.6). These primates are comparable in size to a squirrel, with the head and body measuring 15 cm–25 cm in length and the tail measuring 25 cm–40 cm. Coat color varies in different species and may be gray, silver, brown, or almost black. The ear tufts are white, black, or yellow and there are light and dark rings on the tail. They inhabit the equatorial rain forests of South America and are most abundant in Brazil, Peru, and Ecuador.

The common, or cotton-eared marmoset (*C. jacchus*) is arboreal, but often comes down to the ground (east Brazil). Locomotion consists primarily of quadrupedal leaps. Their pattern of aggressive behavior is curious (although these monkeys are rather peaceful in their family environment). They lay back their ears, raise their fur on end, and bring their eyebrows together. Dominance is indicated by an arching of the body. Whereas turning the back on another and exhibiting the genitalia with an upturned tail means submission and resignation in many species of Old World monkey, in marmosets it is a threatening sign.

Marmosets breed throughout the year; they have no seasonal reproduction. Gestation in *C. jacchus* lasts 144–148 days on average and parturition occurs at night. Two young are usually produced, sometimes 3–4, or, more rarely, only 1. A newborn infant weighs 25 g. Although the female begins nursing the young immediately after birth, a considerable amount of postnatal care is provided by the male (as in a majority of platyrrhines). The chromosome diploid number is 46. In the natural habitat, it has an omnivorous diet consisting of insects, fruit, berries, small lizards, rodents, and nestlings. It adapts well to captivity, where the registered longevity is 12 years. On the other hand, mass dysentery and measles epidemics have also been known to afflict these monkeys. In captivity they can display infanticide (Kirpatrick *et al.*, 1998).

The marmoset is of tremendous value to experimental medicine. Its representatives have been successfully used in studies of hepatitis, leukosis, immunology, organ transplantation, and other types of research. According to my own unpublished data, marmosets and tamarins were the sixth most widely used experimental group of primates in the 1970s–1980s. Summing up the facts on common marmosets in toxicological investigations conducted by the Institute of Toxicology and Embryopharmacology (Free University Berlin) over the last 15 years, R. Stahlmann and R. Neubert (1995) emphasized the indispensability of these primates, their considerable advantage over rodents, and the applicability of experimental results to human biology and medicine.

It is imperative, however, not to forget the fact that certain marmoset species (or subspecies, depending on the classification used) are on the brink of extinction. Some of the endangered species are *C. argentata*, *C. aurita*, *C. flaviceps*, and *C. humeralifer*. The government of Brazil, which has banned primate exports completely, permits no sale of marmosets either.

Tamarins, also effectively used in medical research, present a profound taxonomic diversity. The almost 20 species of the *genus Saguinus* may be classified into three groups, regardless of species classification (these three groups being designated as subgenera). These groups are *Saguinus* proper, which are black- or white-faced tamarins with hairy faces, cotton-top tamarins (*Oedipomidas*), and bare-faced tamarins (*Marikina*). The species designation may often be accounted for by coat color. They differ from the marmosets in that they have no ear tufts or colored rings on their tails. They inhabit the forests of South America from Panama and Costa Rica to south-eastern Brazil.

The cotton-top tamarin (*Saguinus oedipus* [Figure 3.7]) is one of the most valuable experimental animals used in medical research. The head and body measure 22 cm–27 cm and the tail is 36 cm–38 cm in length. Sweeping back from the forehead is a white mane of hair (which is raised on end in agitation, along with backward-forward movements of the ears and exhibition of impressive canines). The face is almost square, dark, and has very little hair. The fur is dark brown on the back and light on the stomach. Communication is by vocal signals and territory- and partner-marking. They inhabit Panama and Columbia. This species is distinguished by seasonality in births, in contrast to other tamarins, in which reproductive cycles continue uninterrupted throughout the year.

Figure 3.7 Saguinus oedipus. Reproduced with permission from Napier and Napier (1967).

The cotton-top tamarin adapts well in captivity, but is fastidious in its diet, preferring bananas, apples, and other fruit, as well as insects, mice, eggs, and boiled meat. It is effectively bred in captivity, with a gestation period of 140–165 days in *S. oedipus* and 165–170 days in *S. geoffroyi*. Twin births are common. The young are born with naked stomachs and chests and are therefore in need of warmth. The chromosome diploid number in many tamarins is 46. These useful experimental animals have been employed in investigations of viral leukoses, hepatitis A, colorectal cancer, colitis, and in immunological and pharmacological experiments. Several species are defined as endangered, including *S. oedipus*, *S. bicolor* (pied or bare-faced tamarin), and *S. leucopus* (white-footed tamarin). Exports of some other tamarins have also been restricted. Tamarins must be bred in special colonies, which has been done successfully in support of biomedical research. The record for longevity in captivity is 16.3 years* (*S. imperator*), but a tamarin's life-span normally does not exceed 7–9 years.

The final representatives of the tamarin species (and of the whole subfamily Callitrichinae) are lion tamarins (*Leontopithecus*). These are the largest animals in the

Figure 3.8 Leontopithecus chrisomelas (See colour plate section)

family, with a head and body length of 23 cm–37 cm and a tail of comparable or even greater length. A luxuriant lion's mane accounts for the common name. These primates are quite similar to other tamarins, although the lower portion of the face is not so angular and the feet and hands are elongated with long digits. Another peculiarity is seen in the skin membranes between the second and third toes and between the second, third, and fourth fingers. The eyes, moreover, are bright blue.

Apparently, a majority of experts agree to the division of tamarins into several species, although there also exists a different classification which recognizes only one species (Thorington and Anderson, 1984). The external features of lion tamarins, among other things, enables them to be differentiated on the basis of coat color. For example, some are golden all over the body (*L. rosalia*), some golden only on the head (*L. chrisomelas* [Figure 3.8]), and some golden on the back (*L. chrysopygus*). Quite recently the world's scientists learned of a sensational primatological discovery; a golden-backed, black-faced tamarin was found not far from Sao Paolo (lion tamarins are found only in Brazil), which was designated *L. caissara* by Lorini and Person (Rylands, 1994a,b). All lion tamarins have seasonal reproductive cycles and usually produce twins following an average gestation of 128 days. Monogamy is characteristic of this monkey, which lives in family groups. The chromosome diploid number is 4 and the record for longevity in captivity is 14.5 years (*L. rosalia*).

Lion tamarins are not used in medical research and all forms of golden tamarins have been entered in the Red Book as endangered. In nature, there are no more than

560 *L. rosalia* and 260 *L. caissara* (Ibid.). Special programs are underway for breeding golden tamarins in zoological gardens and research centers. One behavioral habit, rare among primates, should be noted with regard to tamarins; they share food with their conspecifics.

The final genus of this family is the already mentioned *Callimico* ('mico' from Greek means 'tiny'). It comprises the single species *C. goeldii*, or Goeldi's marmoset, which is an incorrect name since this species does not belong to the marmosets. The present monograph, as stated above, puts *Callimico* at the subfamily level (Callimiconinae). It is a small, dark brown monkey with shaggy, mane-like hair and brown eyes. The head and body length is 19 cm–22 cm, the tail 25 cm–33 cm. The low position of the bridge of the nose makes these primates look snub-nosed. They inhabit the forests of the Upper Amazon (Brazil, Peru) and live in the lower and middle canopy. On the ground, they travel by rapid leaps, rather than by running, as in the tamarins. They are adept at hunting snakes and frogs. Groups usually consist of 20–30 specimens. A variety of facial expressions and vocalizations are used for communication.

Breeding of *Collimico* in captivity is possible, but rare. The length of gestation is 149–152 days (Cologne Zoo). The chromosome diploid number is 48. The diet in captivity consists of locusts, worms, newborn mice, lizards, butterflies, crickets, shredded meat, but also fruit, vegetables, biscuits, and bread. It is a rare and endangered animal which is entered in the Red Book and is not used in biomedical research. The record for longevity in captivity is 9.3 years[*].

FAMILY CEBIDAE

A general description of this family was given above. I here present the conditional characteristics of the taxonomy of this family which was earlier assumed to comprise either all platyrrhines, or all platyrrhines excluding callitrichides (Napier and Napier, 1967, 1985; Fridman, 1979b; Whitney, 1995). The classification of three subfamilies as a separate family, Atelidae, appears reasonable on morphological and biochemical grounds (see above), but may be inappropriate in other respects. For instance, while all five genera of the prehensile-tailed monkeys were previously included in one family (*Cebus*, *Alouatta*, *Ateles*, *Lagothrix*, *Brachyteles*), they are now considered to belong in different families. Another disputed issue in this classification is the status of Pitheciinae, which, as with Aotinae, are considerably distant from capuchins, although owl monkeys are still assigned to the family Cebidae.

Prehensile tails in the above genera are an important diagnostic character and play an important role in the lives of these primates. The undersurface of the tail is bare and has tactile combs, i.e., sensitive skin formations (with the exception of capuchins, in which the undersurface of the tail is not bare). This type of tail possesses a grasping capability and serves as an additional limb. The digits on both forelimbs and hind limbs have flat nails rather than claws, as in callitrichids. The first finger is shortened, while hallux is well-developed and opposable to the other toes. The dentition is similar to that in *Callimico*; due to the presence of a third molar the total number of teeth is 36. These genera have large eyes with well-developed eyelids. Reproduction is nonseasonal. The family includes medium-sized monkeys, excluding *Saimiri*, whose size does not exceed that of a squirrel, and *Aotus* which is not much larger. They are diurnal, except *Aotus* (owl monkey) which is nocturnal. Locomotion is both arboreal and terrestrial and they live in large groups.

Figure 3.9 Cebus apella

Subfamily Cebinae   This subfamily, which gave its name to the whole family, includes, according to my taxonomy model, all capuchins (the genus name *Cebus* is derived from the Greek 'kebos' and means 'tailed simia'). The group comprises approximately 35 capuchin taxa (subspecies) which are classified into 4 living species. The head and body length is 32 cm–57 cm and the prehensile tail is 34 cm–56 cm. Body weight of adult specimens may be up to 3 kg–5 kg. Unlike other platyrrhines, capuchins have functionally opposable thumbs and are capable of skillful manipulations. They have a round head, a relatively short face, and prominent jaws with sabre-like canines. Capuchins have a comparatively well-developed brain, are distinguished by their intellect, and can acquire sophisticated skills and mimic human actions. In the Sukhumi vivarium, black-capped capuchins (*C. apella* [Figure 3.9]) 'did the laundry', 'hammered nails', and 'mopped the floor'. These animals have also been known to use objects as tools in the wild, for instance, sticks to kill snakes. Their behavior is comparable to that of chimpanzees (Visalberghi and McGrew, 1997).

In their natural habitat (vast areas from Honduras to Paraguay to southern Brazil), capuchin groups may consist of up to 30–40 individuals. The young account for about half of the group; the ratio of males to females is usually 1:1. Males dominate females and adults dominate the young. Communication is carried out by vocalizations, facial expressions, bodily gestures, and the hygienic and social performance of grooming. Infanticide has been observed in the natural habitat.

Capuchins mostly eat fruit and leaves, but may also be carnivorous, with a large insectivorous element. The diet also includes small squirrels, frogs, and other animals. They perform an act of 'self-anointing' or rubbing various odoriferous plants on their skin or, in captivity, such things as orange juice, onions, other plants, and even squashed

Figure 3.10 Saimiri sciureus (See colour plate section)

insects. Puberty occurs at the age of 3 in females and somewhat later in males. Gestation lasts for approximately 180 days (*C. apella*) and occurs biennially. They adapt well in captivity and breed effectively. The chromosome diploid number in all representatives of the genus is 54. The record for longevity in a vivarium is 53 years (*C. capucinus*, or white-throated capuchin, an endangered species).

Capuchins are of great value to researchers, especially in research on brain physiology and behavior. They may also be used for investigations of human diseases (*C. albifrons*, the brown and white, or white-fronted capuchin and *C. nigrivittatus*, the weeper capuchin). In spite of different chromosome numbers in species of capuchin and human, significant homology was found between these species (Clemente *et al.*, 1987).

Subfamily Saimirinae   This subfamily contains the single *genus Saimiri* with only two species, *S. sciureus* (squirrel monkey [Figure 3.10]) and *S. oerstedii* (red-backed squirrel monkey, an endangered species). The former comprises 6–7 subspecies and the latter 2 subspecies. The species are sometimes joined together; the whole genus was until recently assumed to belong to the same subfamily as capuchins. In any case, we may speak of the phylogenetic affinity between these two genera as that between sister groups (Harada *et al.*, 1995). There are other proposed classifications, e.g., one which designates 4 species, taking into account Roman and Gothic squirrel monkeys (which are distinguished by round and pointed patches of white above the eyes, respectively) (Hershkovitz, 1984). A great variety of external features and coat colors are noted, dependent on habitat (from Costa Rica to Paraguay and from the Atlantic to the Andes, in Bolivia, Brazil, Peru, Panama, and Columbia).

The Saimirinae are the smallest primates in the family. Body length ranges from 25 cm–31 cm, tail length from 37 cm–47 cm, and body weight from 0.5 kg–1.2 kg. The eyes are close-set, the ears large, and the tail is nonprehensile. There is a diversity of coat colorations in these animals, including a black or olive-gray head, a green and blue or red back, red and yellow limbs, and as in most primates, a light stomach. Especially distinctive is the facial region which is characterized by a dark ring encircling the mouth and that contrasts with the otherwise white background (this type of coloration accounts for the common name of 'dead head' which is given to *Saimiri*). The brain is not fissured (the skull is dolichocephalic), and it is small (the average weight is 24 g), although its relative weight is close to that of the human brain.

These animals inhabit tropical rain forests. The size of *Saimiri* groups is a function of the density of the forest and its area. In Panama and Columbia, the number of specimens in a group varied from 10 to 35, while in thick rain forests of the Amazon basin (Brazil) group size varied from 120 to 300 and up. There was evidence that behavior differed between relatively small and large groups. Adult males dominate the group, although no single leaders are distinguished. A number of specific behavioral traits characterize *Saimiri*. For example, exhibition of the genitalia is a sign of victory or dominance. Leaders roll on the ground to emphasize their dominant status and at times of rest the tail is wrapped around the animal's back. Rubbing urine on the soles of the feet and on the hands is associated with thermoregulation, but may also have some social significance. Curiously, when two *Saimiri* meet, they 'kiss' each other on the mouth (i.e., they make mouth-to-mouth contact), as do some other primate species. *Saimiri* is characterized by extensive vocal communication, including screeching, squealing, ululating, and even clucking. Their diet in the wild consists of fruit, flowers, berries, nuts, leaf buds, insects, and small animals, and they are sensitive to vitamin D3 deficiency in captivity. Birth is clearly seasonal and changes in captivity with geographic location. Males undergo dramatic seasonal changes in appearance; they gain weight and change their coats during the breeding season. This condition, termed 'fatting', is regulated by the male sexual hormone (testosterone). The age of puberty is 3 years for females and 4–6 years for males. The length of gestation is 150–180 days and the newborn weighs 65 g–70 g. The chromosome diploid number is 44 (within this range, chromosome morphology in different *Saimiri* species may vary substantially).

Due to the significant similarity of these monkeys to humans in terms of physiological, biochemical, and immunological processes, the ease with which they are bred in captivity, and their relatively high abundance in the natural habitat, *Saimiri* is used extensively in biomedical research. In the 1960s–1980s, *S. sciureus* was alternately the second and third most widely mentioned species among all the primates described in scientific publications. *Saimiri* has been used to study brain physiology and pathology, illnesses of hearing and sight, atherosclerosis, intrauterine growth retardation, malaria, helminthiasis, and other maladies. It should be noted, however, that one species of this genus is on the brink of extinction, and that the abundance of less endangered forms in the natural habitat is decreasing steadily. It is important, therefore, that these primates be bred conscientiously in captivity to ensure their availability for research. The record for longevity in captivity is 21 years (*S. sciureus*).

Figure 3.11 Aotus trivirgatus (See colour plate section)

Subfamily Aotinae   This subfamily also has only one genus, *Aotus*, the night or owl monkey ('aotus' means 'earless'; the ears in these monkeys are almost completely covered with fur). *Aotus* is the only nocturnal genus of all the Anthropoidea. The best known species is *A. trivirgatus* or 'douroucouli' (the local name given to these animals because of the three-segment pattern on the forehead formed by two patches of white above the eyes) (Figure 3.11). Nevertheless, whether or not these monkeys should be included in one species has always aroused controversy, which was recently reinforced by new scholarly arguments. The number of different species that have been proposed varies considerably from 1 or 3 to as high as 15. At present, the predominant point of view is very likely that of P. Hershkovitz (1983), a most distinguished authority on platyrrhines, who identifies 9 separate species of night monkeys. These primates are a good example of the confusion that occurs as a result of the relation between genotype and phenotype. According to their bodily form, night monkeys may be classified into 1 or 3 species. However, their genotypes are so numerically and morphologically different that it appears impossible to designate them as one species (this probably also accounts for the fact that until recently these monkeys did not breed well in captivity). Genetic data, coat color and structure, and characteristic vocal signals are in perfect agreement with the differences in serum proteins and immunological reactions to infestation with malaria parasites, for studies of which night monkeys are indispensable.

Thus, Hershkovitz classified night monkeys into two groups, four gray-necked species (more primitive), *A. trivirgatus*, *A. lemurinus* (two subspecies), *A. vociferans*, and *A. brumbacki*, and five red-necked species, *A. miconax*, *A. nigriceps*, *A. infulatus*, *A. azarae* (two subspecies), and *A. nancymai* (a relatively new species). The evolutionary origin of this group was already discussed above. I should add that the most ancient ancestor of this group may be related to *Tremacebus* of the Lower Oligocene (Delson and Rosenberger, 1984).

Night monkeys are not at all flat-nosed, representing an exception in the infraorder due to their narrow septums. They are rather small, with head and body lengths of about 24 cm–47 cm and a tail of about the same magnitude (semi-prehensile only in young specimens). Body weight ranges from 800 g–1,250 g, with the female somewhat larger than the male. They have a short facial skeleton with a protruding lower portion and huge eyes that take up as much as half of the muzzle. They have thick fur that may be gray or brown in color. Another peculiarity of night owls is a throat pouch, which, in conjunction with an enlarged trachea, enables these small frenzied monkeys to produce loud and threatening sounds. Night monkeys use up to 50 different means of communication, e.g., they can squeal and bark like dogs or meow like cats and jaguars. They also scent-mark their territories and rub urine on their hands and feet. They inhabit Central and South America, living within the thick rain forests of the Amazon and Orinoco basins.

The night monkey is a cautious nocturnal primate, noiselessly and adeptly moving through the canopy and underbrush. In the daytime it rests in the hollows of tree trunks. It eats insects, bats, small birds, eggs, frogs, snails, fruit, and tree leaves. Night monkeys live in relatively small groups. Breeding is nonseasonal, although there are birth peaks (e.g., June–August in Panama). Females become sexually mature at the age of about 3 years and gestation lasts for about 133 days on average. By approximately the tenth day of postnatal life, the father takes over support of the newborn. The chromosome diploid number varies in different species and can be 50, 52, 54, or 58.

Night monkeys are not easily maintained in captivity because they are highly fearful of humans, react negatively to unusual noises, and are easily afflicted with illnesses. As was mentioned above, WHO has taken special measures to maintain and breed night monkeys for laboratory use because of their extraordinary experimental value. In fact, these are the only animals which may be employed in falciparum malaria research. They are also used in investigations of eye diseases, blood cancer, viral infections, and immunology. For several decades they have been in the 'top ten' of the most valuable primate species for biomedical research in the world. The maximum longevity (in a zoological garden) is about 25 years (*A. trivirgatus*).

Subfamily Callicebinae   This is a separate primitive branch of the family Cebidae. It contains 1 genus, *Callicebus* or leapers or titi monkey. Since there are about 15 subspecies of these monkeys, they are highly diverse in their external and other features (chromosomes, muscles, dermatoglyphics), although the exact number of species is debatable. Let us note that several different classifications of the titi group exist, e.g., biochemists assign titis to the family or subfamily of pithecins (Schneider *et al.*, 1993). I distinguish 3 species, although quite recently (1990) Hershkovitz suggested that these primates should be classified into 13 species, making up 4 groups. Since these monkeys are not used in medical primatology, I shall not discuss this classification in detail.

The 3 species are *C. personatus* (masked titi), *C. moloch* (dusked titi), and *C. torquatus* (white-handed or widow monkey). These are relatively small monkeys with a body length of 29 cm–39 cm, tail length of up to 50 cm (the tail is nonprehensile), and body weight of 1,200 g–1,270 g (*C. moloch*). They have a flat facial region and small ears that are almost fully concealed by thick silky fur. The coat is red and gray, dark brown, brown, or completely white. They inhabit the forests of Bolivia, Brazil, Paraguay, Peru, and Ecuador. *Callicebus* occupy relatively small areas which they defend aggressively and loudly. They are adept tree-climbers and leapers, using their tails for balance. They live as couples within family groups. Couples may often be found sitting next to each other, entwining their lowered tails. Titis are omnivorous in the wild and share food. In captivity they must be given large doses of vitamin D3. Births are non-seasonal. Gestation lasts about 167 days on average (dusked titi), producing a single young. Shortly after birth, the female gives up the infant to the male for most of its care. The chromosome diploid number varies according to species; 20, 46, and 50. The record for longevity in captivity is 12 years (dusked titi). The masked titi has been assigned to the list of endangered species.

FAMILY ATELIDAE

This family, with its 3 subfamilies, is the final platyrrhine family in my model. All the general characters of atelids, including anatomy, evolution, and phylogeny have been discussed above. The conditional character and problem with their classification should be kept in mind. I proceed in accordance with my method of first describing the genera which give the names to families.

Subfamily Atelinae   This subfamily includes 3 genera: *Ateles*, or spider monkey, or coata (4 species), *Brachyteles*, or woolly spider monkey, or muriqui (1 species), and *Lagothrix*, or woolly monkeys (1 species). There are molecular data that the latter two genera are sister-groups (Meireles *et al.*, 1999).

Four species of the *genus Ateles* encompass 16 subspecies which inhabit large areas of land from southern Mexico to southern Brazil and Bolivia. Spider monkeys had been so named for their long and thin limbs (the forelimbs are longer than the hind limbs) and their long prehensile tails (60 cm–90 cm or 1.5 times longer than body) which indeed make them look like spiders. They are relatively large primates with a body weight of 8–9 kg. They have elongated hands which lack a thumb (or it is considerably reduced). The head is small with protruding jaws and the nostrils are set wide apart. In a majority of coatas, the facial region is black with white or pink rings. The fur is coarse, scant, and primarily black (*A. paniscus*), the head may be brown (*A. fusciceps*), and the hands may be black with a golden brown body (*A. geoffroyi*, black-handed coata). There is also a long-haired coata (*A. belzebuth*).

Ateles is a typical brachiator which may sometimes descend to the ground. Females are usually larger than males. These animals live in groups with no marked dominance, although there are more females than males. Their diet is primarily herbivorous. They are considerate in their social relations with each other and they exhibit a relatively rare example of altruism in captivity; they share their food. Breeding is nonseasonal. The female becomes sexually mature at 4 years of age, the male one year later. Gestation lasts 215–230 days (*A. geoffroyi*). The young is born black, but its coat color becomes light at about the age of 6 months. The chromosome diploid number is 34 (*A. fusciceps*, *A. geoffroyi*). *Ateles* adapts well in captivity and may be easily tamed. The record for longevity in captivity is 33 years (*A. geoffroyi*). The abundance of spider monkeys in the natural habitat is falling rapidly. In some places these animals are used as food. Several subspecies of Geoffroy's coata have been assigned to the list of endangered species (Nicaragua, Costa Rica, Panama). These animals are rarely used in experiments. The percent difference in *Homo/Ateles* noncoding DNA sequences (β-globin) is 10.8 (cf: the human/rhesus is 7.0) and the delta T50 H, DNA-hybridization is 11.2 (cf: the human/rhesus is 6.9–7.3) (Miyamoto and Goodman, 1990).

The *genus Brachyteles*, as stated above, is represented by one species, *B. arachnoides*, the woolly spider monkey, or muriqui. It is basically similar to the previously described species, but somewhat larger and more thickset than the coatas. The head and body measure 65 cm in length, the tail 65 cm–80 cm, and the body weight of an adult specimen ranges from 12 kg–15 kg. The skull is globular with a bare facial region of reddish color. All of the limbs are long and there are no thumbs on the forelimbs. The fur is thick and yellow and gray or different shades of brown. It is found in the Brazilian rain forests. The chromosome diploid number is 62. It is an exceptionally rare primate and highly endangered, but it is not used in biomedical research. The record for longevity in captivity is 8.3 years.

The *genus Lagothrix* (woolly monkeys) comprises 2 species, *L. lagothricha* (Humboldt's woolly monkey [Figure 3.12]) and *L. flavicauda* (the yellow-tailed woolly monkey). The tail (55 cm–73 cm) is longer than the body (40 cm–58 cm) and body weight is about 10 kg. The forelimbs are shorter than the hind limbs and the tail is prehensile.

Figure 3.12 Lagothrix lagothricha (See colour plate section)

The digits are short and thick, the thumb is well-developed, and the nails are pointed. The head is large and globular and the facial region is almost black and hairless. The canines are rather long. The fur is short, thick, and soft. Coat color is gray, blue and gray, or dark brown, and is darker on the head and fairer on the stomach. Four sub-species of these monkeys are known.

Humboldt's woolly monkey lives in the basin of the Upper Amazon (Columbia, Ecuador, Peru, and Brazil) in groups of 20 to 70 animals (usually 30–40). Locomotion is quadrupedal, but these animals are also capable of walking a short distance on their two hind limbs. Groups of these primates contain more males than do those of other platyrrhines. In captivity, the group is dominated by a male leader. Intra-group relations are peaceful and benevolent and grooming is common. In the wild these primates may live in groups mixed with other species (such as capuchins or saimiri). They eat a herbivorous diet of flowers and tree bark.

Breeding occurs throughout the year, but data on the duration of gestation in these primates are contradictory; the most reliable sources give it as about 230 days. The chromosome diploid number is 62. It is known that homology in the karyotypes (banding studies) of this monkey and human is significant (Clemente et al., 1987). L. lagothricha are amusing, peaceful, and clever primates, but also somewhat obese and slow, which makes them easy prey for hunters who use their meat as food. Their abundance in the natural habitat is decreasing, with the exception of those areas where they have been provided with special protection. They are very sensitive to vitamin D3 deficiency and rarely breed in captivity. They are especially useful in experimental studies of behavior, but they have also been used in cancer and virology research. The record for longevity in captivity is 24.7 years.

L. flavicauda was placed on the endangered species list after it had been recently rediscovered in the mountains of Peru (it was previously assumed to be extinct).

Subfamily Alouattinae   This subfamily consists of 1 genus, Alouatta, or howler monkeys, which includes 5 or 6 species (dependent on classification) and over 20 subspecies. These monkeys have been known since the 16th century. Their distinctive vocalization accounts for the name that was given to the genus. Adult males produce loud, menacing roars by means of a special anatomical resonating mechanism (a laryngeal pouch and enlarged tongue bone shaped like an air-bladder). These are the largest New World primates (and the most 'narrow-nosed' among the 'broad-nosed' platyrrhines). Head and body length varies from 40 cm to 90 cm, according to different sources, and body weight is 6 kg–9 kg. The tail is approximately the same length as the head and body or somewhat longer, and it is naked on the lower third where it has sensitive skin combs. The tail is highly prehensile and enables these primates to cling to a tree branch and eat in an upside-down position, freely manipulating all four limbs. The forelimbs and hind limbs are equal in length and each has 5 digits (the thumbs and halluces are shorter than the other digits).

The upper part of the body is especially large, including the head which has a bare facial region and prominent ears. There is a thick beard and a mantle-like mane sweeping back from the top of the otherwise short-haired head. The lower part of the body is covered with short fur which is especially short on the limbs. The jaws project frontally and the canines are large. Coat color varies in different species and can be yellow and brown, red, black, and black and brown. It varies not only according to species, but also by age and sex. Sexual dimorphism is conspicuous, with females substantially smaller than males.

The distribution of these primates covers large territories from Veracruz in Mexico to southern Argentine and Brazil, including offshore islands where they inhabit deciduous rain forests. The black howler (*A. caraya*) has a vegetarian diet. In the region of the Panama Canal, tree leaves account for about 50% of all the food it consumes, the rest being fruit, flowers, and buds. Howler monkeys live in groups of 10–40 (*A. fusca*, the brown howler of Argentina, Bolivia, and Brazil). There is no distinct leadership within the group, although the brunt of defending a group's territory is carried out primarily by mature males (in case the roars alone fail to frighten off intruders). Howlers may be found in the same groups as spider monkeys. There is no evidence of seasonal reproductive cycles, but birth peaks are present, e.g., during the dry season (December–May) in red howlers (*A. seniculus* of Brazil, Venezuela, and Guyana). Puberty occurs at 3–4 years of age and gestation lasts about 190 days. One young is produced, although twin births may also occur. Infanticide has been known to occur in red howler groups. The chromosome diploid number is 44 (red howler) and 52 (black howler).

In their natural habitat, howler monkeys avoid the vicinity of human settlements and retreat into the heart of the forests. They are hunted for their meat, although it is not very valuable (*A. belzebul*, the red-handed, or black-and-brown howler, and other species). Guatemalan (*A. vilossa*) and Columbian or mantled howlers (*A. palliata*), as well as black (*A. caraya*) and brown (*A. fusca*) howlers have been placed under protection and are registered as endangered. They generally adapt poorly in captivity, where they have been known to live for up to 12 years (the Columbian howler), but they are rarely used in biomedical research.

Subfamily Pitheciinae This subfamily consists of 3 genera of unusual medium-sized monkeys. Present day pithecins descend from one of the early platyrrhine groups. They undoubtedly had a fossil relative (*Cebupithecia*) in the Miocene of Columbia (Kay, 1990). The whole subfamily is sometimes incorrectly called 'Saki', after the name of one of the genera. Other trivial names of these monkeys are associated with mysticism, such as 'devils' (this nickname was once used in the title of a scientific article), 'satan's monkeys', 'satans', etc. These names are accounted for by the extraordinary external features of these primates. All 3 genera are distinguished from other New World monkeys by a broad chest, protruding upper and lower incisors which are separated from the canines with a wide diastema, the broadest septum in all the platyrrhines, and, finally, unusual coat coloration.

The *genus Pithecia* (common sakis, or sakiwinkis) has 2 species, the white-headed saki (*P. pithecia* [Figure 3.13]) and the monk saki (*P. monachus*). Head and body length in both species is 30 cm–50 cm and the nonprehensile tail of similar length is thick and fluffy. Body weight is about 2 kg. The white-headed male saki has a face covered with white (cream-colored) hair which grows radially from the middle of the forehead and which contrasts with the flat black nose and dark coat over the rest of the body. The female may be mistaken for a different species of primate, so much does sexual dimorphism affect coat color; the muzzle of the female is less hairy, dark, and has only two light stripes stretching from the noseblades to the corners of the mouth (these stripes are absent in immature females). The fur is thick, long, and coarse. The monk saki has a hairless face, but there is long hair on the top of the head which covers the forehead, ears, and shoulders (like a monk's cloak). The fur is gray on the upper

Figure 3.13 Pithecia pithecia (See colour plate section)

part of the body and pale yellow or red on the lower part. There is no sexual dimorphism in the color pattern of this species.

The natural habitat of the common saki encompasses the Amazon basin up to
Orinoco in the north and from Guyana in the east to the foot of the Andes in the west.
These animals adapt poorly and rarely breed in captivity. Gestation lasts from 163 to
177 days (*P. pithecia*). The chromosome diploid number is 46. The record for longevity
in a zoo is about 14 years (*P. pithecia*). They are not used in medical research.

The *genus Chiropotes* (chiropotes or red-backed sakis) also comprises 2 species, the
black or bearded saki (*C. satanas* [Figure 3.14]) and the white-nosed saki (*C. albinasus*).
Sometimes one of the subspecies is considered to be an additional species of the same
genus. *Chiropotes* have thick-set bodies, measuring 40 cm–46 cm, and thick tails which
are 35 cm–38 cm in length. Body weight is about 3 kg. The face is bare, there are two
tufts of black hair on the forehead and the hair on the head is long and black. The coat
color on the back and on the limbs is a lighter shade of brown. The white-nosed saki
has a white and pink patch stretching from the eyes down to the upper lip.

White-nosed sakis live in small groups in the dense rain forests of Brazil, southern
Venezuela, and Guyana. Their diet consists of fruit, leaves, and small animals. They
drink water by the handful rather than by lapping. The screeching vocalization of
the *Chiropotes* sounds like a whistle that switches from one key to another in a unique
manner. In conjunction with these animals' unusual appearance, this sound demonstrates
the basis of their 'satanic nature'. Gestation lasts about 4 months, but they adapt poorly

Figure 3.14 Chiropotes satanas

and rarely breed in captivity. The chromosome diploid number is 54 (*C. satanas*). The white-nosed saki has been entered in the Red Book of endangered species and is not used in biomedical research. The record for longevity in captivity is 15 years (*C. satanas*).

The *genus Cacajao* (the uakaris) contains 3 species, the black-headed uakari (*C. melanocephalus*), the bald uakari (*C. calvus* [Figure 3.15]) and the red uakari (*C. rubicundus*) (the latter may contain completely white subspecies). These monkeys have the shortest tails among the New World primates (the tail is one-third the length of the body, which measures 45 cm–50 cm), and are also the most distinctive. They have shaggy, long hair which blows in the wind and almost no hair on the face or on top of the large head. Their ears, which resemble those of humans, are also naked. Some species may have long whiskers. In the bald and red uakari, the facial region is crimson and in the black-headed uakari, it is black. The lack of sunlight in captivity results in a pale complexion. At puberty the face gains color and in a state of excitement it turns almost red. As the animal gets older, its coat color becomes darker and its hair recedes on the head. In the bald uakari, which has no hair on its head, the rest of the body is covered with long silvery fur. The black-headed male uakari acquires a black mantle in adolescence which, together with black hands and feet, contrasts sharply with a background of long chestnut-brown fur that covers the body. The hands and feet of the uakari are disproportionately large.

The distribution of the uakari extends from western Brazil to eastern Peru (the forests of the Amazon basin). These primates live in small family groups. Their diet includes fruit, leaves, insects, and small vertebrates. They adapt poorly and rarely breed in captivity, although there have been some births when special care was provided.

Figure 3.15 Cacajao calvus (See colour plate section)

One young is normally produced. The chromosome diploid number is 46 (the red uakari). All three species are endangered and have been entered in the Red Book. They are not used in experimental research.

Infraorder Catarrhini

The infraorder of catarrhines contains the lower simians of the Old World (monkeys), the higher simians (apes, or hominoids), and humans (which are also hominoids). That this infraorder is monophyletic, having descended from a single branch, is no longer questioned by any investigators, neither morphologists nor molecular biologists. There is another remarkable feature of this infraorder, with respect to the present monograph; since this taxon includes humans and their nearest relatives, and since the demographic survival of its species was not threatened in the past, it is the simians of the infraorder Catarrhini (the genera *Macaca, Papio, Cercopithecus,* and *Pan*) that have been the subjects of most experimental research – no less than 70% – in the field of medical primatology.

The characteristics listed above in the general descriptions of Primates, Haplorhini and Anthropoidea, naturally enough apply to the catarrhines of these various taxa. They differ from platyrrhines by possessing a number of characteristics regarded as hereditary acquired (apomorphous). These include a narrow cartilaginous nasal septum between the nostrils and shorter nasal wings of cartilage; a decrease in the number of premolars in each quadrant of the jaw to two (i.e., each member of the infraorder, including human, has 32 teeth) and a specific molar morphology; the presence of an ectotympanic septum and a relatively flat bulla with a long external auditory meatus; and a completely developed postorbital septum (Thorington and Anderson, 1984; Schwartz, 1986).

Other distinctive features worth mentioning are ischial callosities (in lower catarrhines, less commonly in hominoids) covered with a cornified thickening that is insensitive to temperature gradients – a visible adaptation to the sitting position; cheek pouches (mainly in cercopithecines); a dorsoventrally compressed chest (only in hominoids); a well-developed clavicle; in some species, longer forelimbs than hind limbs, the longest digit is the third, followed in length by the fourth, second, fifth, and first, the latter being opposable to the rest (one exception are the colobines), and all the digits have flat nails. Specialization of the cubital articulation and of the well-developed clavicle and wrist, together with the highly mobile digits of the hand, convert the forelimb into a true arm suitable for very refined manipulations.

The tail is not prehensile. The hominoids have no tails, while other members of the infraorder have tails of varied length, from quite short up to 110 cm. The hair coat is sparser than in all the other primates and is absent from the face, which in some forms has whiskers and/or a beard; some species have a crest on the head. The ischial callosities and the palms and soles of the hands and feet are hairless. In some species, the upper part of the trunk is covered with a mantle. In general, the hair cover in the Catarrhini is less dense than it is in the Platyrrhini – about one-half less dense on the back and one-third on the abdomen in cercopithecines; the higher simians (apes) are even less hairy, the least being the human.

The olfactory system is reduced in comparison to all the other primates, but the visual system is more advanced. The macular area of the retina, the central fovea, and the cones are well-developed and make possible adequate trichromatic vision. Approximately one-half of the fibers of the optic nerve do not cross over in the optic chiasm; this, together with the more progressive development of the corresponding cerebral structures, provide for adequate stereoscopic vision which other animals lack. The facial musculature is well-developed. The brain is large, with large numbers of sulci and gyri. The facial area of the roundish skull is flattened, except in a few species (baboons and the related mandrills). The cecum is usually shorter than it is in platyrrhines; an appendix vermiformis is rarely found in cercopithecines, but is present in apes and humans. The genital tissues of females swell to varying degrees prior to ovulation (under stimulation of estrogen), except in orangutans and humans. An ovarian/menstrual cycle is characteristic of all primates of this infraorder. The clitoris is smaller than in the cebids, the penis is pendulous, and the testes are in a scrotum. The placenta is hemochorial (bidiscoidal in cercopithecines, monodiscoidal in hominoids), and all forms have only one pair of nipples. Females usually give birth to one young; twins are rare, being produced at approximately the same rate as in humans.

The stomach is simple. In colombines it is bag-shaped and composite and is adapted, like the stomach of ruminants, for processing plant food. In the higher catarrhines, the colon has a sigmoid flexure. Most species are omnivorous, but some are herbivorous.

Catarrhines occur, as already noted, in the Old World, or more precisely, in Africa and Asia (except for some which are found in North Africa and the southernmost part of the Pyrenees, Gibraltar, and of course, the globally distributed *Homo sapiens*), inhabiting a wide diversity of places including forests, mountains, savannas, river banks, marshes, rocks, and even deserts. In the Himalayas, they are found at altitudes as high as 4,000 m above sea level. Catarrhine primates range in body weight from 3 kg to 200–250 kg. They live in large and small groups, as well as singly. Sexual relations include polygamy, monogamy, and promiscuity.

As already pointed out, this infraorder is certainly monophyletic, as are all of its constituent groups and subgroups. The origins of the catarrhines date back to the early Oligocene of Africa (Fleagle *et al.*, 1986), if not to a still earlier period, the Eocene (de Bonis *et al.*, 1988; Simons, 1995). One must appreciate the above-mentioned difficulties encountered in the area of paleoprimatology, which is concerned with a period of time spanning tens of millions of years, to be patient in evaluating uniquely uninterpretable schemes of an objectively proceeding evolution. In any case, the aforementioned datings of 40 Ma–34 Ma as the time of divergence from the oldest anthropoids of their two lineages, including the catarrhines, are not difficult to accept today.

Of primary importance for the study of catarrhine evolution are the aforementioned artifacts which were discovered in the Faiyum depression (Egypt), findings which have been carefully studied since the beginning of the 20th century. The parapithecines and pliopithecines from the early Oligocene of Egypt, the most ancient of catarrhines, bore little resemblance to present day Old World primates. It is only during the early Miocene that macroadaptational traits such as large body size, leaf-eating, terrestrial locomotion, and other morphologic and ecologic features appeared to create the appropriate conditions for further divergence to the taxa of modern primates (Fleagle, 1986). Over the past 35 years or so, at least 10 genera of fossil anthropoids have been discovered in Faiyum, much credit for which is due to the American anthropologist, Elwyn L. Simons. So far, 19 primate species belonging to 6 families have been identified from Faiyum (Simons, 1995). After reappraising many structural features of the skulls, jaws, and teeth of the Faiyum primates, Simons and Rasmusson (1991) proposed to classify these fossils within the suborder Anthropoidea, including many forms which, if not the ancestors of catarrhines, are at least the oldest members of this infraorder.

According to this classification, the first superfamily, Parapithecoidea, is represented by one family (Parapithecidae), which includes 3 genera from Faiyum – *Quatrania*, *Apidium*, and *Parapithecus* (each genus includes 2 species). The second superfamily of catarrhines, Cercopithecoidea, whose recent (neotologic) species still exist, is also composed of one family of extinct higher primates, Propliopithecidae (but with two subfamilies); it is taken to include 5 genera – *Proteopithecus* (1 species), *Catopithecus* (1 species), *Oligopithecus* (1 species), *Propliopithecus* (4 species), and *Aegyptopithecus* (1 species). The third superfamily, Hominoidea, will be discussed below.

Although one may disagree with this classification and construct other putative schemes (for example, see Shoshani *et al.*, 1996), it is beyond doubt that the species listed above do include the oldest forms of catarrhines and not only lower forms, but higher ones as well. Recently, Simons himself (with his co-author) assigned one extinct form, *Prohylobates tandyi* from Egypt, to the superfamily Cercopithecoidea, as the first among the lower catarrhines (Miller and Simons, 1996).

A highly important problem from the perspective of this monograph is that of the cladogenesis of the catarrhines at the Cercopithecoidea-Hominoidea *junction*, i.e., how and when the lower simians and hominoids diverged. A variety of dates and schemes have been proposed, and the results of molecular and morphologic studies have been increasingly concordant in recent years. The most likely time of divergence is the Upper Oligocene or early Miocene. Paleontologic findings strongly indicate that lower catarrhines definitely existed in Africa during the intervals between 22 Ma and 17 Ma (Harrison, 1988), as did higher ones (Simons, 1987). It is probable that a special role was played by *Victoriapithecus* in the origin of catarrhines (Benefit, 1999). Of interest are the studies of proconsuls from the early Miocene of Kenya, in particular, *P. nyanzae*. Specimens of this primate are quite similar in the anatomy of the femur and its proportions and articulation to the extant *Cercopithecus* monkeys on the one hand, and on the other they are no less similar to the pygmy chimpanzee or bonobo. The *P. nyanzae* specimens examined were dated to about 18 Ma. In general, the genus *Proconsul* (early and middle Miocene of Africa, 23 Ma–15 Ma) combined primitive, advanced, and unique traits which, in the opinion of a group of scientists including E. Simons, suggests its divergence to relatively large monkeys and to certain apes. The *Afropithecus*, which is related to the proconsuls, had a distinct connection to the earlier *Egyptopithecus* (Shoshani *et al.*, 1996). Three genera of small apes from the Miocene of Africa – *Micropithecus*, *Limnopithecus*, and *Dendropithecus* – could well be either close relatives of the recent Hylobatidae or simply primitive catarrhines; their well-defined sexual dimorphism is not shared by the present day gibbons (Ibid.).

According to molecular studies, these two lineages diverged between 26 Ma and 22 Ma (Bailey *et al.*, 1992), which is generally consistent with the modern datings of paleontologists. Other dates have been obtained, however, by other approaches. For example, geneticist R. Stanyon (1989) dates the separation of higher and lower catarrhines at 33 Ma to 37 Ma by comparison of karyotypes based on the DNA/ DNA hybridization data of C. Sibley and J. Ahlquist. Currently we have data that the divergence of Asian and African anthropoid primates occurred near 36 Ma (Jaeger *et al.*, 1998).

While phylogeny of the hominoids is specifically dealt with below, I will consider here the evolutionary history of cercopithecoids or, more precisely, those clades which led to the living lower catarrhines, Cercopithecidae. The major events in this history evolved during the Miocene, although they extend to the Pliocene and even to the glacial Pleistocene. An attractive phylogenetic scheme is the one resulting from an immunologic collation of albumin and transferrin molecules with fixation of microcomplement comparisons by electrophoretic techniques. Blood proteins were taken from close to 40 species from all major cercopithecide groups. Although the study is based on the principles of a molecular clock, the results were consistent with paleontologic findings at the generic and higher levels (Cronin, 1989).

This study dates the origin of the family Cercopithecidae to 20 Ma (+/− 2 Ma); the separation of the subfamilies Cercopithecinae/Colombinae to 10 Ma (+/− 2 Ma); that of the tribes Papionini/Cercopothecini to 6–7 Ma; that of the subtribes Macacina/ Papionina to 5 Ma; and that of the subtribes Colobina/Presbytina to 4–5 Ma. It also dates the emergence of *Mandrillus* to 5 Ma; *Cercocebus* to 4 Ma; *Lophocebus* to 3–4 Ma; *Therohithecus* to 3–4 Ma; *Papio* to 3–4 Ma; *P. papio/P. hamadryas* to 1 Ma; Asia *Macaca* to 2 Ma; and *Erythrocebus patas* to 1–2 Ma. These datings coincide almost completely, with the minor exception of the lower taxa, with those of paleoprimatologists (Delson and Rosenberger, 1984; and others).

We can see that the phylogeny of cercopithecides is thus roughly coincident in the time of evolutionary development with that of platyrrhinian anthropoids. As discussed below, moreover, similar temporal ranges are characteristic of the hominoid dichotomies (as well as of the mammalian class as a whole [Romer, 1971; Goodman, 1996; Shoshani *et al.*, 1996]).

There are innumerable examples of phenotypic similarities combining with specific differences in primate taxa of various levels and many of them will be described in the next chapter. In this chapter, too, some examples will be given in discussing individual taxonomic forms of catarrhines as the animals that are most extensively used in medical primatology and that are closest to humans. The overall scheme of catarrhinian affinities now rests on a powerful foundation of data obtained in anatomy, physiology, and, over the past few decades, in biochemistry, molecular biology, genetics, immunology, and even experimental pathology.

I here present just two relatively recent examples. When sera from different primates were compared with antisera to human apolipoproteins, it was found that the human apolipoproteins were completely identical immunologically to those of the chimpanzee and gorilla; the reactions with orangutan sera were weaker. Only plasma from all the anthropoids reacted with the anti-LP-D fraction and only plasma from catarrhines, and not platyrrhines, reacted with the anti-II and anti-III fractions (McConathy and Weech, 1981).

In another study, the DNA fragment p82H from the alphoid family of repeated sequences located in the centromeric regions of all human chromosomes, following hybridization of nucleotides with homologous regions from different animals, demonstrated identical profiles of blot hybridization for humans and all the apes. After *in situ hybridization*, the p82H probe also yielded heteroduplexes with centromeric regions of the cercopithecide lion-tailed macaque (*M. silenus*), but in this case the hybrid heteroduplexes were unstable. Hybridization with the corresponding DNA of the platyrrhine owl monkey (*A. trivirgatus*) failed to occur at all. There was, of course, no hybridization with DNA of the house mouse (Miller *et al.*, 1988). A host of similar examples exist, and some, as already stated, will be described below.

In line with the evolutionary tendencies and phenotypic characters, and also considering both the historic practice and recent publications in the field of systematics, I accept the following taxonomic scheme for living Catarrhini (excluding the Hominoidea, discussed below):

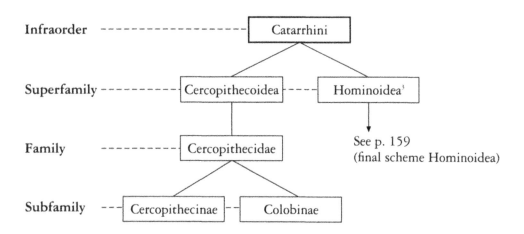

See p. 159
(final scheme Hominoidea)

---

3 Goodman *et al.* (1998) lowered the group of hominoids to the level of the family (Hominidae) and included the latter in Cercopithecoidea. This action is not accepted in this book.

Superfamily Cercopithecoidea

FAMILY CERCOPITHECIDAE

This is the only family of living lower catarrhine simians. They are small (32–50 cm) to medium-sized (55–100 cm) primates. The forelimbs are equal to or somewhat shorter than the hind limbs. Males are invariably larger than females, they are often more brightly colored, and sometimes they have a mane (mantle). The feet are longer than the hands and the upper canines are longer than the lower. They are diurnal animals and live in groups of various sizes. The groups are usually dominated by a single male or several males headed by a leader or *alpha male*.

The ancestors of the present day Cercopithecidae family may be considered to have diverged into the two main evolutionary lineages of lower catarrhines, which are currently classified into two subfamilies, Cercopithecinae and Colobinae. The latter are placed by some in a separate family, a placement not accepted by this monograph. I believe, as most other authors do, that the division of the catarrhines into two sub-families is quite satisfactory (which is not so at the generic level – see below).

Subfamily Cercopithecinae   The Cercopithecinae subfamily is customarily subdivided into three generic groups: 1) the long-tailed monkeys of Africa, guenons and closely related forms, including patas; 2) the mangabeys; and 3) the group of macaques, baboons, and forms closely related to the latter (mandrills, and geladas). Taxonomically, this comprises seven genera (if the *Miopithecus*, talapoins, and *Allenopithecus* [Allen's guenons] are held to be subgenera of the genus *Cercopithecus*, rather than taken to be separate genera). As in other taxa, the classification of cercopithecines has a morphologic 'precedence', but in recent decades the systematics of this group has received strong support from fundamental molecular genetic research. I may note briefly that the affinity of guenons, macaques, and baboons has been well-validated by studies of chromosomes (Giusto and Margulis, 1981). Macaques are closer to baboons than to guenons, according to the results of hybridizing unique DNA sequences (Hoyer and Van de Velde, 1975), and mandrills and geladas, which traditionally were regarded as being close to baboons, have been shown to derive from a single clade and are assigned to one tribe.

As regards the biological affinity of Cercopithecinae primates to humans, this group may be considered to be nearer to *H. sapiens* than any other lower anthropoids and, indeed, than any other primates except apes, i.e., Hominoidea. As shown by my analysis (extending over many years) of the results reported by various investigators, the overall data on immunologic affinities by DNA and RNA hybridization, amino acid sequences of proteins, chromosome homologies, and other indicators of similarities between humans and all other primates give a percentage resemblance to humans ranging from 50% for *Cercopithecus* to 80% for *Papio* (Fridman and Popova, 1988; Fridman, 1991). There may be marginal variations, of course, in some indicators.

The *genus Cercopithecus* has the largest number of species (23–25), is one of the largest genera in terms of subspecies (over 70), and is probably the most extensive primate genus in demographic terms. Its members are the smallest (and the most primitive) Old World monkeys, with a head-trunk length ranging from 32 cm to 52 cm and a tail length from 35 cm to 110 cm. Females weigh 3–4 kg and males 4–6 kg (green monkeys). They have cheek pouches and small ischial callosities. The hair coat is dense, soft, and highly variegated in different species; green (*C. aethiops*), blue (*C. mitis*), olive

(*C. mona*), almost black with yellow stripes (*C. nigroviridis*, subgenus *Allenpothecus*), yellowish-gray with a black tail tip (*C. talapoin*, subgenus *Miopithecus*, or dwarf guenon), and red (*C. erythrotis*). There are red-tailed monkeys (*C. ascanius*) and white-nosed ones (*C. nictitans*, *C. ascanius*, *C. petaurista*), as well as bearded (*C. diana*, *C. neglectus*, *C. l'hoesti*), crowned (*C. pogonias*), moustached (*C. cephus*), and several species with white or yellowish whiskers. Facial skin may be brown, dark or light flesh-colored, blue, or violet. The scrotum is blue in some species. In addition, there may be prominent light spots near the eyes, on the chest, around the throat, or on the sacrum.

The classification of *Cercopithecus*, though not uniquely interpretable, seems to be quite satisfactory in general. The traditional morphologic evidence has received good support from recent studies of this genus. Thus immunologic and electrophoretic studies on the variability of loci coding for serum proteins and erythrocyte enzymes have shown a high degree of affinity between the different species of guenons and provided further arguments for placing them in a single genus, including Allen's guenons and talapoins (and even *Erythrocebus*, which I do not accept for various reasons – see below) (Dugoujon *et al.*, 1989). Some uncertainty remains because the group of green monkeys is sometimes assigned to a single species. These monkeys, however, have been generally regarded as compromising three separate species – *C. aethiops* (grivet) (Figure 3.16), *C. pygerythrus* (vervet), and *C. sabaeus* (tantalus) on the basis of morphologic data (Dandelot, 1959). It is of particular importance to define the systematics of this group accurately because of the wide use of these species of guenons in biology and medicine.

Figure 3.16 Cercopithecus aethiops

In recent decades, the scientific community on several occasions has been excited by the discovery of new *Cercopithecus* species, for instance, *C. solatus* (Harrison, 1988), *C. sclateri*, and *C. salongo*. Actually, such discoveries not infrequently prove to be nothing more than reappraisals of the taxonomic level, as in the case of *C. salongo* (Colyn *et al.*, 1991).

Guenons live in Africa, their habitats ranging from sub-Sahara to the southern tip of the continent. As already noted, the present day occurrence of green monkeys and monas in the West Indies (on the Islands of S. Kitts and Nevus) is the result of their accidental importation in the 17th century. Most guenons are arboreal animals, but some species are found outside forests, in brushwood, and along river banks (e.g., green monkeys). They live in groups that vary widely in size (e.g., from 10 to 150 individuals) and in which the dominance hierarchy is not very rigid. A group is headed by a physically strong and resolute male or a group of males. Guenons are adroit and mobile, but cautious, for a guenon may be torn to pieces for a meal, not only by a leopard, but even by a hungry chimpanzee. Until the last few decades, guenons were considered to be herbivorous although, like nearly all primates, they also eat the eggs of birds. Green guenons in southern Senegal ('mangrove monkeys'), moreover, were seen catching and eating crabs with great dexterity from the Salout River (from shallow places, although they can swim) (Galat and Galat-Luong, 1976). It is also now known that monkeys of this species hunt birds, hares, and lizards, and that blue monkeys hunt galagos and other small animals (as probably do guenons of other species). Moreover, infanticide has been recorded among blue monkeys (Macleod, 1996).

Vocal communication is largely limited to twittering and loud barking and active grooming is widespread. They may threaten their enemies by thrusting out their chests and displaying their sexual organs.

Guenons reproduce in all seasons, but there is a birth peak. The female becomes sexually mature by the age of 3 years and has a menstrual cycle typically lasting for 30 days. The male is capable of reproduction by the age of 4 years (green monkeys). Gestation lasts for an average of 165 days in green monkeys and 140 days in talapoins. The newborn weighs 200 g–250 g. The diploid chromosome number varies in different species from 54 (*C. talapoin*) to 72 (*C. mitis*); in green monkeys it is 60. The adaptation to captivity is not bad. The record life-span is 28 years (*C. aethiops*), 31 (*C. diana*), and 33 (*C. campbelli*). Guenons are exterminated in some areas because of the harm they do to agriculture. The number of guenons in the natural habitat, even those belonging to 'protected' species (green monkeys), is diminishing. The species considered endangered include *C. diana*, *C. erythrogaster* (red-bellied guenon), *C. erythrotis*, *C. l'hoesti*, and *C. preussi*.

Monkeys of this genus, primarily green monkeys, are widely used in experimental laboratories. This is justified not only by the results obtained in studies using them as models of human disease, but also by fundamental principles of biology. Indeed, as was shown as early as the 1960s, the immunologic distance from humans for albumin ('dissimilarity index') is 1.14 in chimpanzees, 2.23 in macaques, and 2.59 in guenons, as compared to 35 in swine (if this index is taken as 1 in humans) (Sarich, 1968). It is in guenons that a pericentromeric, highly repeated, alphoid DNA was discovered, which is regarded as specific for primates; the homology of tandem repeats in guenon and human DNA reaches 65% (Manuelidis and Wu, 1978). Other investigators found the DNA to be even more similar (Sharples *et al.*, 1987). This affinity corresponds to the data of cytogenetics: comparable studies of gene disposition on the accordance chromosomes of more than 15 enzymes detected not only morphological similarity, but also identity of group genes in green monkeys and humans (Creau-Goldberg *et al.*, 1984). When the α- and β-chains of hemoglobin from green monkeys and humans were compared, these chains proved to differ by only 5 and 6 amino acid substitutions, respectively (Maita *et al.*, 1979).

The kidneys of green monkeys have been a particularly convenient medium for culturing viruses over the decades, including those of poliomyelitis, and for preparing vaccines against this disease, as already noted. Guenons have also been successfully used to study human diseases such as measles, viral hepatitis, leprosy, hemorrhagic fever, and hemoblastosis, as well as many pharmacologic agents. Guenons, however, may be sources of dangerous diseases, as exemplified by the notorious Marburg disease among laboratory workers in West Germany and Yugoslavia who were exposed to green monkeys. Of about 30 persons who fell ill, 9 died. Other fatal cases were also described later (Johnson *et al.*, 1996; IPN, 2000).

According to our data, the guenons were among the three or four primate genera most often used for research purposes in the 1960s. By the mid-1980s, they accounted for 4.2% of all primates used for these purposes (Fridman and Popova, 1987).

The *genus Erythrocebus* consists of one species, *E. patas* (Figure 3.17), or red monkeys (also known as 'hussar monkeys'). They are related to guenons, but differ from them not only anatomically (in having an elongated skull, long slender limbs adapted for rapid running on the ground, a large body, and in the structure and locomotor function of

Figure 3.17 *Erythrocebus patas*

the spinal column [Hurov, 1986]) and genetically (in karyotype morphology), but also ecologically. Patas monkeys are terrestrial primates living in open spaces, predominantly in grassy and semidesert areas of Africa (in Cameroon, Nigeria, Sudan, and Tanzania). The *Patas* species is subdivided into four species on the basis of habitat. The trunk and head measure 58–75 cm and the tail is approximately of the same length. The body weight of males (8–13 kg) is twice that of females (4–7 kg). The upper canines are large. There are tufts of hair on the cheeks, light moustaches, and a black line of hair above the eyes which widens as it approaches the ears. The hair coat is red except on the abdomen and chest where it is light-colored. The hair coat of patas living in captivity in the absence of sunlight grows pale, but completely regains its color when exposed to the sun. The ischial callosities are small.

The dominance hierarchy of the group is more rigorous than in guenons. The group (20 or more monkeys) is always headed by a leader (alpha male), although several adult males in a group may act cooperatively. One may observe the leader rising up on its hind legs from time to time, perhaps with some object in its hands, to look around the territory vigilantly, with tail abutting the ground to help maintain the erect position. When excited, a patas may bend its back and jump like a cat. The feeding habits of patas do not differ from those of guenons.

Females begin menstruating at the age of about 2.5 years, whereas males become sexually mature 1–2 years later. The gestational period has been variously reported to last from 170 to 215 days. One reproductive feature of patas needs to be emphasized, namely, that females, unlike other primates, tend to deliver in the daytime rather than at night (which is associated with the unique habit of this species of sleeping individually, without grouping together as many other monkeys [Chism *et al.*, 1978]). In the Sukhumi Primate Center, patas were never observed to mate in the daytime. The diploid chromosome number is 54. The longest life-span recorded in captivity is 21 years. Patas are valuable in experimental research, but they are used sparingly because of their small numbers.

Monkeys of the *genus Cercocebus*, or mangabeys, combine the anatomical features of guenons and macaques, and even baboons. They are medium-sized monkeys, with the body ranging in length from 45 to 60 cm and the tail from 60 to 95 cm. The face is pointed anteriorly, with noticeable supraorbital ridges (as in macaques). The ischial callosities are large and in females swell at about the time of ovulation (as in baboons). At the same time, their genital area reddens. Mangabeys are arboreal animals and in this respect resemble guenons, but they are larger. The second and third digits are almost fused, while the third and fourth are approximately half-fused. Their habitat is in the middle of Africa, ranging from Guinea and Nigeria in the west to Kenya in the east.

This genus consists of 4 or 5 species, comprising about 10 subspecies, which are usually divided into two groups of species. The first group includes the uncrested mangabeys *C. torquatus* (white-collared mangabey) and *C. galeritus* (Tana River mangabey), to which *C. atys* (sooty mangabey) is sometimes added by those who consider it to be a subspecies of the white-collared mangabeys. All patas are now assigned to the endangered category. The second group comprises *C. albigena* (white-cheeked mangabey) and *C. aterrimus* (black mangabey). This pair of species differs from the first group genetically and biochemically. They are so similar to each other that it was recommended, with lukewarm support by morphologists, that they be placed in a separate genus, *Lophocebus*. This proposal was not supported by the systematists (e.g., Napier and Napier, 1985). It should be appreciated, however, that as far as characteristics important for medical primatology are concerned (immunologic, hematologic, and karyologic characteristics), *Lophocebus* (subgenus?), in the opinion of geneticists, is closer to *Macaca* and *Papio* than to *Cercocebus* (Dutrillaux *et al.*, 1979).

Although they have short fur (of gray-green, golden brown, black, or smokey color), mangabeys bear luxurious long whiskers and are also distinguishable from other monkeys by white stripes over the upper eyelids. Most of them live in trees, but some (white-collared mangabey) fare rather well on the ground.

Depending on the environmental conditions and species, groups of mangabeys (8 to 20) may be headed by one or more males. The means of communication are relatively diverse. These monkeys cackle, twitter, bark, and grunt. They smack their lips and shake their heads in an amusing way in response to a perceived threat, for example, an approaching animal or human being. They actively groom each other. Their main food is plant material and there is no seasonality in reproduction. The menstrual cycle is 30–32 days and the duration of gestation is 174–179 days (white-cheeked mangabey). The diploid chromosome number is 42 in all species. Mangabeys adapt well to captivity where the longest recorded life-span was nearly 33 years (white-cheeked mangabey).

Mangabeys are used infrequently in biomedical research, but have become well-known in the last 15 years because of their use as a unique model of leprosy and for the development of vaccines against this disease. They are of additional interest as symptomless carriers of the simian immunodeficiency virus (SIV) and therefore, are also used for developing vaccines against AIDS. The experimental potential of *Cercocebus* does not appear to have been fully tapped. Since the species of this genus are disappearing from the wild, however, and are close to becoming extinct, they should be specially bred for the purposes of medical primatology.

The *genus Macaca* (the trivial name – macaques) is the most widely distributed primate genus, both geographically (extending from Gibraltar and Morocco through Japan and China to Vietnam, Indonesia, Micronesia and the Philippines) and demographically (the number of rhesus and Java macaques is probably larger than that of any other primates). It is also the most popular and widely used genus in medical primatology. Macaques are the subject of scientific volumes and make up over 50% of all primates used in the more dynamically developing lines of experimental research. All this is indicative, among other things, of the unusual ecologic adaptability of these monkeys, which become habituated to various temperate zones of the Earth and readily adapt to captivity.

Macaques are medium-sized primates. In *M. mulatta*, the trunk and head are 40–75 cm long and the body weight is 4–10 kg in females and 5–11 kg in males. There are also species of smaller (*M. fascicularis*) and larger (*M. nemestrina, M. sylvanus,* and others) size. Depending on the species, tails are short like 'stumps' (*M. arctoides, M. tonkeana,* and other macaques found on the Island of Sulawesi), half the length of the trunk (as in *M. mulatta* [Figure 3.18]), or up to 50 cm long (*M. fascicularis, M. radiata,* and others); there are also tailless macaques (*M. sylvanus*). Sexual dimorphism is evident; males have a larger trunk and canines than females. The head is round, with moderate protrusion of the face. The forelimbs are shorter than (or equal to) the hind limbs. There are cheek pouches. The hair coat is of medium length, but *M. silenus* has a lion's mane and light-colored beard and whiskers. Macaques living under harsh climatic conditions (*M. assamensis, M. fuscata*) have longer and denser fur. The face in

Figure 3.18 Macaca mulatta

most species is hairless, with various coloring of the skin. The hair coat is also of different colors – yellow or yellowish-green (*M. mulatta*), brown (*M. arctoides*), dark brown (*M. nemestrina*), olive (*M. silenus*), or black (*M. nigra*).

The predecessors of macaques appear to have already diverged by the Miocene from the common evolutionary flow of cercopithecines, but the major events in the differentiation of species in this genus occurred later, in the Pliocene and Pleistocene. The earliest finds of macaques come from North Africa and are 5 million years old (Delson and Rosenberger, 1984). They had already inhabited Europe by the middle Pleistocene and spread throughout the continent up to the British Isles in the late Pleistocene (Fladere, 1991). Ancient forms of macaques had reached East Asia by 2–3 Ma (Jablonski and Pan, 1988). Approximately at that time, speciation of the four main extant macaque groups had begun – *sylvanus, sinica, fascicularis* and *arctoides*. About 1 Ma, the differentiation of seven present day Celebes macaques began. This phylogenetic scheme is well-documented, both morphologically (Fooden, 1988) and biochemically (Hayasaka *et al.*, 1988; Melnick *et al.*, 1989).

Macaques are undoubtedly closely allied to other cercopithecines, including baboons (*Papio*). These genera are particularly close in karyotypes (Finaz *et al.*, 1979; and others) and blood proteins (Takenaka, 1985). On the other hand, serologic tests (electrophoresis, immunodiffusion) point to differences in transferrin fractions and complement factors (Biedermann, 1982). Obvious differences also exist in the system of blood groups from erythrocyte antigens, although the two genera have been shown to have a common ancestral antigen (Socha *et al.*, 1983). Differences are revealed, moreover, by neurophysiological methods and by the differentiation of behavioral characteristics (Balzamo, 1981). Finally, intergeneric distinctions are, of course, evident from morphologic data which were the first to be relied upon in systematics. At the same time there exist well-defined interspecific differences within the *Macaca* genus itself, not only the traditional ones, but also those revealed by experimental studies. Thus, the first models of human leukemia were obtained with *M. arctoides*, whereas attempts to obtain them with *M. mulatta* failed (Lapin *et al.*, 1975). Also, the very closely related *M. mulatta* and *M. fascicularis* were each found to have distinct metabolic pathways (Wilson *et al.*, 1978) and, in the case of *M. nemestrina*, not only the species, but even a subspecies, was an important consideration when responses to alcohol were evaluated (Kamback, 1972).

In studying how closely *Macaca* species are related, one may encounter different variants of interspecific proximity, but this would by no means indicate that the conclusions are of low validity. The macaques, I emphasize, constitute a closely allied group of monkeys. Not infrequently, however, the use of one set of criteria permits identification of various signs of similarity where no similarity is apparent by other criteria. A good example of the high concordance between the different approaches to systematics is the classification of the genus *Macaca* by the well-known American primatologist, Jack Fooden. This investigator divided the macaques into four groups comprising 19 species on the basis of anatomical traits of the reproductive system (1976, 1980). In his study of these primates, initiated in the 1960s and continuing afterwards, Fooden worked out the nomenclature and taxonomy for *M. arctoides* (Fooden, 1986). He eliminated the confusion that existed in the systematics of the Celebes macaques, having identified 7 species which previously had not even been regarded as macaques, and demonstrated their clear-cut taxonomy (Fooden, 1969). He also developed the systematics of *M. fascicularis* (Fooden, 1991), established the status of *M. thibetana* as a species, and published several monographs devoted to the *sinica* group of macaques which had previously been confused with other species. The validity of Fooden's systematics has been confirmed by specialists in various departments of biology.

The *silenus* group (lion-tailed macaques) consists of *M. sylvanus* (barbary macaque), *M. nemestrina* (pig-tailed macaque) and the 7 species of Celebes macaques; *M. tonkeana* (Tonkean macaque), *M. maura* (moor macaque), *M. ochreata* (booted macaque), *M. brunnescens* (Muna-Butung macaque), *M. hecki* (Heck's macaque), *M. nigrescens* (Gorontalo macaque) and *M. nigra* (Celebes black ape).

The *sinica* group includes *M. sinica* (toque macaque), *M. radiata* (bonnet macaque), *M. assamensis* (Assamese macaque) and *M. thibetana* (Père David's stumptailed macaque, or Tibetan macaque).

The *fascicularis* group is comprised of *M. fascicularis* (Java, crab-eating, or cynomolgus macaque), *M. cyclopis* (Taiwan or Formosan rock macaque), *M. mulatta* (rhesus macaque), and *M. fuscata* (Japanese macaque).

The arctoides group includes only one species, *M. arctoides* (red-faced, or stumptailed, or bear macaque [Fooden, 1980]).

To the first of these groups, one more species is sometimes added, *M. pagensis* (Mentaweian macaque), which is recognized here as a subspecies of *M. nemestrina*. In all, there are about 50 macaque subspecies.

M. mulatta, the famous rhesus monkey, is the 'classical' laboratory primate. It entered the history of medicine, not only by virtue of being the animal in which the *rhesus factor* was discovered, but one in which many diseases have been investigated. These macaques are distributed in Asian countries, including Afghanistan, China, India, Vietnam, and Thailand. They live predominantly in trees, but often forage for food on the ground and may visit human settlements. They live in groups that may be large, consisting of more than 200 animals, or in smaller groups. Members of the familial clans tend to concentrate around a leader or alpha male. As among monkeys of other species, there are hermits which live in solitude, either individually (male) or in small groups, sometimes joining conventional groups. Rhesus macaques are rather aggressive monkeys, but the frequency of conflicts within groups often depends on environmental conditions, particularly the availability of food. In captivity, they may bite human beings when threatened.

The types of communication are fairly rich. They screech, coo (young), squeak, and growl, and are capable of distinctive facial expressions (especially for frightening others). They frequently groom each other (socially significant), use gestures, and express their heightened emotions by postures and by positioning of the tail. They diversify their diet of plant food (leaves, sprouts, grains, vegetables, fruits) with worms, insects, and small animals. Their characteristic feature is the annual mating season, which is associated with hormonal changes, and the seasonality of reproduction, which depends on environmental conditions (reproduction is not seasonal in the Java macaque and inapparently seasonal in the stumptailed macaque). Macaques, at least the stumptailed variety, are capable of experiencing physiologic orgasm (de Waal, 1995). Infanticide has been recorded in captivity among pig-tailed macaques.

The rhesus female begins to menstruate at the age of 3 years, and the menstrual cycle lasts 28 days on average (31 days in the Java and stumptailed macaques). The duration of gestation is, on average, 165 days (178 days in the stumptailed macaque, 165 in Java macaque). The male becomes sexually mature by the age of 4–4.5 years. The weight of a newborn rhesus is 450 g on average (350 g in the Java macaque, 475 g in the stumptailed macaque). The diploid chromosome number in all macaques is 42 (as it is in mangabeys, baboons, mandrills, and geladas). The longest life-spans recorded in captivity are 38 years for *M. fascicularis*, 30 for *M. fuscata*, 30 for *M. arctoides* and 39 for *M. mulatta*. The latter figure is recorded from a recent analysis of autopsies at the Wisconsin Primate Center (Kemnitz *et al.*, 1996). Although macaques of some species occur in the wild in relatively large numbers, several other species are threatened with extinction. The latter include primarily the picturesque lion-tailed macaque (whose number in the wild is estimated at about 2,000–3,500 [Kaumanns *et al.*, 2000]), Japanese, toque, Taiwan, and stumptailed macaques.

As already noted, the macaques are well-adapted to captive conditions (although they are prone to contact infectious diseases during the adaptation period, primarily intestinal infections and tuberculosis. This condition is probably linked to the severe stress they experience and their compromised immunity (immunity is inhibited in approximately 50% of imported monkeys). Successful adaptation and acclimatization are among the prerequisites for the use of monkeys in biomedical experiments. My analyses of publications describing studies on primates (mainly for the period 1965–1985) revealed that monkeys of the genus *Macaca* were used in 42% of them. The latter include studies of a purely biological nature which were undertaken in areas where prior research with macaques had reached its peak long before. Considering only experimental studies in the biomedical sciences, the relative proportions of macaques used in individual sciences were as follows: 61% in endocrinology (43% for M. *mulatta* relative to all other primates); 52% in dentistry (32% for M. *mulatta*); 52% in cardiology (30%); 74% in physiology of the nervous system (41%); 57% in pharmacology (45%); 63% in radiobiology (55%); 55% in biochemistry (41%); 50% in immunology (35%); 63% in ophthalmology (51%); and 53% in space biology and medicine (27%) (Fridman and Popova, 1983, 1987). It is on the macaques that major advances were made in the area of medical primatology, primarily in the study of infections. In this area, the proportion of studies concerned with communicable diseases in which macaques were used has decreased from 31% to 24% over the past 30 years. This decline in the use of macaques is due, on the one hand, to the successful solution of certain medical problems and, on the other hand, to the use of primates from other genera.

Does a biological basis exist for the use of this genus in research on such a phenomenally large scale (and with such spectacular results)? A detailed answer to this question came only after the spectacular experimental results themselves.

The very fact that *Macaca* and *Homo* shared the same infraorder since the second half of the 19th century (mainly on an anatomical basis) was the first fundamental justification for using these monkeys to study human diseases. But it was only in the 20th century that subtleties of the similarities between monkeys and humans were explained in detail. Some similarities between macaques and humans were already mentioned above, in particular, in discussing the infraorder Catarrhini and the sub-family Cercopithecinae. I here present some additional examples of data which were generated in the greatest amounts during the time when medical primatology was rapidly developing.

It had become evident in the 1960s that rhesus and human DNA hybridize at levels reaching 66%–74% (Goodman, 1967; Barnicot, 1969). This remarkable fact is particularly striking considering that the corresponding percentages were 10%–12% for birds and 19%–20% for such conventional laboratory animals as rodents (mice). These findings were later expended and confirmed in various comparative studies of DNA and other *intimate* biological characteristics. For example, the primary structures of two variants of rhesus and human alphoid DNA were shown to be 70% homologous (Pike *et al.*, 1986). Similar results were obtained in studies on mRNA (Martin *et al.*, 1981; and many other investigators). This corresponds quite well to the homology of chromosomes, which was detected in macaques to human (Seth *et al.*, 1983; Huang *et al.*, 1993). The intensity of precipitates formed by human serum proteins against those of macaques was 71% (as demonstrated by double diffusion in agar) (Annenkov, 1974; Bauer, 1974). Hematological and biochemical characteristics of human and macaque were shown to be very similar (Bourne *et al.*, 1973). Analogues of the ABO blood groups and other genetic systems in these two species were also demonstrated (Wiener *et al.*, 1972), as were similarities (or even virtual interchangeability) between strictly species-specific growth hormones (Li Choh Hao and Parkoff, 1956) and between the structure and radiosensitivity of their karyotypes (Chiarelli, 1967; Jemilev, 1970). Furthermore, macaques and humans were shown to be strikingly similar as regards the processes of ovogenesis and spermatogenesis (while widely differing in these respects from other animals [Baker, 1972]), fetoplacental physiology (Solomon and Leung, 1972), embryogenesis, and gestation in general (Knorre, 1969; Hendrickx, 1972a). That macaques and humans have similar α- and β-chains of hemoglobin has already been mentioned, but their fetal hemoglobins are also similar. The γ-chain of fetal hemoglobin in Japanese and rhesus macaques differs from its human counterpart by only 3 amino acid residues (Takenaka *et al.*, 1986). When ribosomal DNA sequences were tested in 12 assays for differences between *H. sapiens* and other primates, a mean difference of only 14.1% was obtained for *M. fuscata* (Suzuki *et al.*, 1994). Human and rhesus placental gonadotropin releasing hormone is identical (Dello and Boyle, 1996).

Although, as already stated and as can be inferred from the figures given above, *M. mulatta* was the primate species most widely used in biomedical research, crab-eating macaques (cynomolgus monkeys) were substituted for rhesus after their exportation of the latter was banned. The cynomolgus monkeys proved to be readily available for importation from Malaysia and the Philippines (and, incidentally, are less aggressive than rhesus). Stumptailed and pig-tailed macaques as well as some other *Macaca* species are also used in research, on a moderate scale.

The *genus Papio* comprises baboons, which are sometimes called 'apes' because they were long considered to be at the boundary between lower and higher simians (in the aforementioned medieval book of K. Gesner, the baboon is depicted in an upright posture with a stick in its hand). Possibly, this clever monkey is indeed superior to all other monkeys as far as the similarity of their biological parameters to those of human beings is concerned.

Baboons are larger than any other lower simians. The head and body measure up to 100 cm in male anubis and chacma baboons, which may weigh up to 35 kg. Hamadryas, or sacred baboons, are smaller (up to 85 cm and 30–32 kg) and Guinea baboons are the smallest (up to 60 cm). The tail in males is long (42–60 cm). Females are approximately half the size of males (anubis females weigh 18–22 kg and hamadryas females, 14–16 kg). Baboons have an elongated, dog-like muzzle and are therefore also called dog-headed monkeys. The nostrils at the tip of the muzzle appear to be cut vertically. They have powerful canines, conspicuous superciliary ridges, and large ischial callosities (which swell severely in females during ovulation) of pink, red, or (in anubis baboons) brown. The hair coat is coarse and of a smoky gray color (in the hamadryas), yellow (in the yellow baboon, *P. cynocephalus*), reddish (in the Guinea baboon), green-olive (in the anubis baboon), and dark brown to black (in the chacma baboon). The facial skin ranges in color from pinkish to dark brown and is hairless. There are capacious cheek pouches. A silvery mantle-like mane appears at puberty in the hamadryas male, so that monkeys of this species are sometimes called mantled baboons.

Precursors of the Papionini tribe (which, in addition to baboons, includes geladas and mandrills) appear to have separated in the Miocene, at the end of which and during the early Pliocene a divergence was occurring to the ancestors of the contemporary genera, *Papio*, *Mandrillus*, and *Theropithecus*. The latter genus seems to be particularly close to *Papio* (Cronin and Meikle, 1979; Delson and Dean, 1991). Baboons and geladas are thought to have first appeared about 4 Ma (Eck and Jablonski, 1984). At the close of the Pliocene of Africa (2.5 Ma) there lived species, now detectable as fossils, that were very close to the extant forms of baboons whose speciation continued in the Pleistocene (2–1 Ma).

The affinities of *Papio* with the two other genera of the tribe, on the one hand, and those among the *Papio* species themselves, on the other, has generated, as not uncommonly happens in systematics, two interrelated problems. Should all three allied species be placed in one genus or should the five currently existing baboon species be reduced to one or two species (*P. hamadryas*, which appreciably differs in morphology from the other four species, as one species, and the others, representing the 'savanna baboons' or '*P. cytocephalus* group'). Arguments in favor of both decisions exist, based not only on morphological but also genetic and biochemical data. The debate that arose was kindled by the detection, in areas inhabited by baboons, of zones where hybridization occurred between baboons that had been considered as belonging to distinct species – namely, hybridization of hamadryas with anubis baboons (Nagel, 1971), of anubis with yellow baboons (Maples, 1972), and of yellow with chacma baboons (Hayes *et al.*, 1990). If the principle of 'reproductive isolation' (Mayr) was followed, this effectively converted the 5 baboon forms into subspecies.

This principle, however, is not universally applicable; it does not apply, for example, in the classification of certain species of fish and birds. Moreover, a large body of evidence indicates that the extant forms of baboons are highly differentiated. In our paper on the *Papio* species in experimental research, we presented arguments, based on evidence from medical primatology, that demonstrate the generic and specific separation of the baboons (Fridman and Popova, 1988, Part I). A helpful decision in bringing order to the classification of baboons was reached by the International Nomenclature Committee in preserving the terms *Papio* and *Mandrillus* for designating, respectively, the baboons and the mandrills (Delson, 1982). At present, although attempts are still made to reduce the baboons to one or two species, and to place the mandrills and geladas in the genus *Papio*, the principle of separating these monkeys is maintained when their taxonomy is described (Napier and Napier, 1985; Whitney, 1995).

It follows, therefore, that I recognize 5 species in the genus *Papio* – *P. hamadryas* (hamadryas or sacred baboon) (Figure 3.19), *P. cynocephalus* (yellow baboon), *P. papio*

Figure 3.19 Papio hamadryas

(Guinea baboon), *P. anubis* (anubis or olive baboon), and *P. ursinus* (chacma baboon). These species comprise 10 subspecies. All baboons are found in sub-Saharan Africa where their range extends to the southwestern tip of the continent. The habitat of *P. hamadryas* is confined to Ethiopia, the Sudan, Somalia, and Arabia, and that of *P. papio* to Guinea, Cameroon, and Senegal. The other three species inhabit vast territories, with *P. anubis* ranging from east to west Africa, north of the equatorial forest, *P. ursinus* from Kenya to Angola and South Africa, and *P. cynocephalus* from east to west in Central Africa.

Group relations, in general, rest on the principle of a dominance hierarchy; it is less rigid and has greater parity between sexes in the savanna baboons than in the terrestrial hamadryas baboons. The latter congregate exclusively around a single adult male leader – usually a group of females with their young. Not uncommonly, several such groups join together to form a community that may number several hundred individuals which interact in defending one another, in raiding plantations, and in other situations. The hierarchy is linear, in that all members are subordinate to the leader, the sexually mature females dominate the other females, and the stronger and more active individuals rule over the weaker and less active ones (juveniles and aged). If one member of the group, whatever rank, is threatened by some danger, the group comes to his defense, with the sexually mature males entering the conflict first and in concert (these males are capable of starting bloody fights between their own clans in normal life). Savanna baboons gather in groups of 30 to 80 animals, but this number can sometimes reach 200. Anubis prefer to gather in the trees for the night. Although they are ground-dwelling animals, hamadryas are also good tree climbers. They sleep in groups in a sitting posture.

Communication between hamadryas baboons takes a very rich variety of forms – vocal signals, various gestures, several sorts of glances and grooming, which is used widely. Baboons are very bright monkeys which have been taught to throw objects at a target and to exchange toys; and they are capable of activity with tools. One characteristic behavioral phenomenon of hamadryas (as found in other monkeys) is curious; a 'guilty' member of the group will approach the leader backwards with a penetrating shriek, 'subordinating himself' and thus 'apologizing', while demonstrating his loyalty. If the leader accepts the 'apology', he touches the subordinate with his hand. Infanticide has been noted among the chacma baboons (and among the hamadryas in captivity).

Baboons feed on fruits, cereals, and plants and may also be predatory, especially the yellow baboon, which hunts not only insects, frogs, and birds, but also hares, gazelles, and guenons.

Reproduction among baboons proceeds all year round. The menstrual cycle begins at the age of 3.5–4 years and lasts for 28–32 days. Sexual maturity in the male starts at the age of 4.5–6 years (hamadryas). Pregnancy takes an average of 175 days for the hamadryas, 165 for the yellow baboon, and 185 for anubis. Births are usually single (at Sukhumi, only one hamadryas birth in 50 yielded twins). The young are black, their color starting to lighten at three months. A newborn weighs an average of 760 g. As mentioned above, the diploid number of chromosomes is 42. Maximum life-span in captivity is 37 years for hamadryas and 45 years for chacma baboons. We must underline that P. hamadryas is a species with a high reproductive-rate in captivity (data of Sukhumi Primate Center and Zoo of Cologne). A hamadryas male leader named Murrey, which lived in the Sukhumi Primate Center (Nursery) for more than 30 years, was the father of 348 sons and daughters. A monument erected in Sukhumi in honor of laboratory monkeys is simply called 'Murrey' by the workers of the nursery.

Baboons adapt well to life in captivity, especially if kept in spacious, open-air cages. They have been used in experimental biomedical research for many decades and comprised 10.3% of all the primates used in medical primatology in 1965–1985. This is a large proportion in view of the relative scarcity of the genus *Papio*, in terms of both the number of species and the number of individuals. The figures are much higher when their use in certain individual disciplines is considered. According to our estimates, baboons accounted for 38.4% of all primates used in surgery, 21.6% in hematology, 21% in cardiology, and 16.3% in endocrinology, over the period indicated (Fridman and Popova, 1987). Baboons have been used in important studies on neurogenic cardiovascular diseases, organ transplantation, reproduction, drug testing, and in other areas. Since I place members of the genus *Papio* among those lower simians that are particularly close to humans, a more detailed discussion of their affinity with humans is warranted.

Despite the enormous differences that exist between baboons and humans in general anatomy, there are morphological similarities in various organs, as revealed, for example, by computerized tomography of the cranium (Balzamo, 1980) and by studies of spleen structure (Nute and Mahoney, 1979), thymus and lymph node morphology (Charin, 1979a, 1979b), groin anatomy (Vagtborg, 1967), and the structure of the cardiovascular system, in particular, the coronary arteries (Hitchcock, 1969; van Zyl, 1971).

As mentioned above, the similarity of baboons to humans has been estimated to be as high as 80% in a number of characteristics. These primarily include humoral, biochemical, molecular biological, and possibly karyological factors for which homologies can be expressed quantitatively. This permits comparisons with data from other animals which demonstrate that they stand farther from humans than simians do.

Using a variety of staining techniques (for C-band, Q-band, R-band, T-band, G-11.6 band, and staining after incorporation of BrdU), analogous chromosome banding patterns were established between *P. papio* and humans. Although baboons have fewer chromosomes, the banding pattern of the *P. papio* karyotype is quite close to that of the orangutan, whose karyotype is in turn similar to that of the human (Dutrillaux *et al.*, 1979). As shown by a comparison of highly repetitive nucleotide sequences in human DNA with those in the DNA of five other primate species, rats, and rabbits, the human, chimpanzee, and yellow baboon DNA are much more similar to each other than to the DNA of other species. This clearly indicates that the genomes of the primates are closely allied (Dandiey and Lucotte, 1984). Genetic linkage map for loci of the bone physiology in baboons are homologous for 4 or 5 loci in human chromosomes 18, 19, and 22 (Rogers *et al.*, 1996), and these investigators believe that similarities of this kind can be found for other systems as well. More recently comparative genome analysis has been carried out in *P. hamadryas* and humans (Rogers *et al.*, 2000).

Baboons are also similar to humans in a number of hematological parameters, including hemoglobin, platelets, packed cell volume, cell counts, and the capacity of prostaglandin to inhibit platelet aggregation (Karim and Adaikan, 1979). The forementioned intensity of precipitates produced by blood proteins from baboons with antiserum against human blood proteins with double diffusion in agar reached 80%, as compared to 36% and 11% in the case of blood proteins from prosimia and swine, respectively (Annenkov, 1974). Most components of baboon and human blood sera are similar (McGraw and Sim, 1972). Analogous ABO blood groups were also established between baboons and humans. The O group is relatively uncommon in baboons, but all four groups are found at different frequencies in various species of this genus (Wiener *et al.*, 1974). Chacma baboons and humans share several characteristics of immunity, both cell-mediated and humoral immunity (cross immunoreactivity of IgG and IgM, immunoglobulin concentrations, T- and B-lymphocyte levels, and mutagen sensitivity [Mendelow *et al.*, 1980]). In the order of their decreasing similarity to humans in genes of the major histocompatibility complex, the animal species examined were arranged as follows: chimpanzee, baboon, marmoset, *Microcebus*, and rat (Chausse *et al.*, 1984). Baboons and macaques are very close to humans in the amino acid structure of hemoglobin, there being only 11 and 8 differences in the α- and β-chains, respectively (Mahoney and Nute, 1980). In the yellow baboon, the primary structure of the fetal hemoglobin γ-chain differs from that of humans by only 3 substitutions (Takenaka *et al.*, 1986). The processed P117 gene found in an intergene region of primate alpha globins proved to be more similar in baboons to its human counterpart than to that of macaques (Takenaka and Takenaka, 1996). The erythrocyte enzyme phosphate-dehydrogenase of baboons and humans are as similar as a different version of each others (Verjee and Damji, 1974). The activities of several other enzymes in baboons (as well as chimpanzees) are closer to those of corresponding human enzymes than they are in gibbons, geladas, macaques, and prosimians (Shapiro and Cohen, 1980).

The formation of blood lipids and the characteristics of cholesterol, triglycerides, phospholipids, α- and β-lipoproteins, and fatty acids in baboons and humans are so similar as to make these monkeys eminently suited models for the study of atherosclerosis (Blaton *et al.*, 1972; Howard *et al.*, 1972). Chorionic gonadotropin is excreted throughout gestation in baboons, humans, and other higher primates, whereas it disappears from the blood and urine (but not from the placenta) of rhesus females by day 40 of gestation (Hobson, 1970a). In terms of several characteristics of hormone circulation during pregnancy, the anubis baboon may be defined as an organism intermediate between humans, on the one hand, and macaques and marmosets, on the other (Hodges *et al.*, 1984). The overall pattern of urinary gonadotropin excretion in a pregnant yellow baboon is identical to that of human, chimpanzee, and gorilla females (Hobson, 1970b). A high level of homology between baboons and humans was recorded in studies examining the balance and metabolism of the steroid hormones (Goncharov *et al.*, 1980), hormonal aspects of the fetoplacental system (Solomon and Leung, 1970), thyroid-stimulating growth, and other hormones, and prostaglandins (Webster *et al.*, 1970; Parker *et al.*, 1972; and others). The baboon's endometrium is microscopically indistinguishable from its human counterpart (Kramer *et al.*, 1977), as is the placenta and amnion of the baboon (Lee and Yeh, 1983). The baboon is, therefore, an excellent model for studying human reproduction and reproductive endocrinology, as well as contraceptive vaccines (Castracane, 1996; Stevens, 1996). A. Hendrickx and others at the University of California Primate Center in Davis have used primates in embryological and teratological research for many years. They have shown that baboons are better suited than macaques for studying the reproductive system, for assaying relevant gene-engineered preparations, and for testing antinauseants, corticosteroids, and sex steroids (Peterson *et al.*, 1996). A reliable model of the human menopause is provided by the aging baboon, but not by the macaque female (Karey and Rice, 1996; Johnson and Kapsalis, 1996). Studies similar to those listed above have been conducted with hamadryas baboons for over 30 years at the Sukhumi Primate Center.

There are also considerable affinities between baboons and humans in the chemistry and biochemistry of fluids and tissues. The formation and chemical composition of gallstones are so similar as to render these monkeys an excellent model of the human pathology (Gleen and McSherry, 1970). Folic acid deficiency induces chemical and hematological changes in baboons which are identical to those in humans (Siddons, 1974). Owing to similarities in the composition and activity of drug-metabolizing microsomal enzymes and in other factors, e.g., the physiology of vascular receptors, pharmacological preparations may produce the same effects in baboons as they do in human patients – unlike that in other animals, in particular dogs. This was demonstrated in the case of carbazeran (Kaye *et al.*, 1985) and with isosorbide dinitrate, whose testing in dogs might well have resulted in a tragedy if it were not also tested on anubis baboons. The effects of these drugs on the coronary vessels of dogs was opposite to those found in monkeys and humans (Hitchcock, 1969; Sato *et al.*, 1984). Baboons are a good model to study heart disease (Magakyan, 1977; Hixon, 1999).

The microbial, viral, mycologic, and parasitic flora of a chacma baboon newly caught from the wild proved to resemble those of humans (Brede and Murphy, 1972). Baboons appear to be more advanced than other primates (with the exception of apes) as far as the brain and behavior is concerned. The cytoarchitecture of the frontal cortex in hamadryas baboons is more intricate and more similar to that of humans than it is in rhesus macaques. In baboons, but not in macaques, area 46 of the cortex, considered to be specific for man, is detected. There are other examples (such as sulcus asymmetry, increased cell size, cortical thickness, and area volume) indicative of more progressive brain development in baboons. Moreover, the brain develops at a lower rate in baboons than in macaques, which correlates with data on the physiology of conditioned reflexes and illustrated the advantage of baboons in experiments (Orjechovskaya, 1983). Curiously, the anubis baboon is the only species in the *Papio* genus – and probably among the monkeys in general (let alone other animals besides primates) – which shares with humans, for the most part, characteristics of sleep (Balzamo, 1980), one of the most ancient functions of the nervous system. It is evident from the foregoing brief account that the biological basis for the use of baboons in medical primatology is solid indeed.

All the other primates described below, other than the chimpanzee, are seldom used in biomedical research, if at all.

The *genus Mandrillus* consists of two species, *M. sphinx* (Figure 3.20), or the mandrill, and *M. leucophaeus*, or the drill. (It may be noted in passing that the genes for red cells and serum proteins differ in these two species by about 30% [Lucotte and Jouventin, 1980], a difference which is much larger than those among species in the other genera of cercopithecines or than that between chimpanzees and humans – see below.) Adult mandrill males are 80 cm long or more (male drills are slightly shorter), have a short tail of 5–7 cm length (the drill's tail is about 12 cm), and weigh 32–35 kg, whereas females are 1/3 to 1/4 that size (their body weight is only 8–10 kg). Their colors are surprisingly variegated (Charles Darwin considered the mandrill to be the most brightly colored mammal). When the male has reached puberty (at age 6–7 years), the buccal outgrowths on its horse-like muzzle are light blue and a fiery red color similar to that of the nostrils. The beard is yellow, the ischial callosities are a delicate violet, the bare skin on the tail root is red-violet, and the fur is olive on the back and silvery on the abdomen. Drills are less colorful, with an olive-green fur, black muzzle, and white beard.

Both species live in forested areas of West Africa, on the trees and ground. Groups of 5–10 adult females and a similar number of juvenile females concentrate around

Figure 3.20 Mandrillus sphinx (See colour plate section)

one or more males. These groups may unite to form a community of up to 200 or more individuals. Mandrills are very aggressive, and the group hierarchy is rigid. Communication is limited and characteristically includes a low pitch trill, not unlike the auditory solo of a frog. They mark trees with the secretion of mammary glands. Their diet consists of plants, with the same composition as that of baboons. Some seasonality of reproduction is discernible. Gestation lasts 220–270 days. The newborn infant is colored dark gray, but has a red muzzle. These monkeys adapt to captivity quite well. The records of longest life-span are 31 (mandrill) and 28 years (drill). Both species are considered as endangered. Less than 3,000 drills are reported to remain in the wild (Gadsby and Jenkins, 1996). Although they are rarely used in biomedical research, both species are among the best models of human filariasis (Wahl and Georges, 1995).

The *genus Theropithecus* comprises one contemporary species, *T. gelada*, with two subspecies. Geladas measure 50–75 cm, males being larger than females, and have a long (40–50 cm) and thick tail with a dark tuft on the end. The face protrudes to a lesser extent than in baboons or mandrills and is concave in profile. The nose is somewhat upturned, the nostrils are directed upward and sideways, and there is a depression in the area of cheeks. The first and second digits of the hands and feet are shorter than the third and fourth, with sharp nails adapted for digging out roots and seeds from the soil. The ischial callosities are rather small and of a dark gray color. The facial skin is brown, but light above the eyes. The fur is long and of a dark brown color. The adult male, like the adult hamadryas baboon, has a luxuriant mantle-like mane, but of a chocolate color. The neck and chest have hairless areas which fill with blood and turn red when the male is excited. The same areas in the female turn red during menstruation when these areas are fringed with a necklace of light-colored outgrowths.

Geladas are inhabitants of high mountain pastures in Ethiopia. Their social structure is characterized by groups with one male, several males, or units of many males. Groups of these terrestrial animals can number several hundred individuals, but are composed of smaller autonomous harem groups. Their diet consists of grass, roots, and sprouts. They sleep in the sitting position. A common means of communication is social grooming. The male responds to a threat by opening its mouth and turning out the upper lip to demonstrate powerful canines. Infanticide has been observed in the wild. Reproduction is nonseasonal. The menstrual cycle is comparable in duration to that of baboons. Gestation lasts 170–180 days. They adapt to captivity with difficulty. In the Sukhumi Primate Center they produced offspring when crossed with baboons. The longest life-span on record is 19.2 years. Geladas are not used in biomedical research.

Subfamily Colobinae   In the Russian language, colobines have long been referred to as 'slender-bodied' monkeys, which seems reasonable enough in view of their elegant external appearance. In English, they are often called leaf-eaters, which is also an appropriate term, reflecting in a way their biology and ecology. Colobines are very attractive in color; their beauty brought some species to the brink of extermination because their fur was used to adorn the interior of buildings and women's dresses.

The biological characteristics of this group, mentioned above, are worth repeating here. Most colobines do not have cheek pouches (though some have small ones), but a majority have ischial callosities (though they are relatively small). One of their distinguishing features is a multichambered stomach for digesting leaves. They have sharpened molars for cutting leaves. The first digit of the hand, if present at all, is a hardly-noticeable knob. This subfamily diverged in the Miocene, as already noted; in the Pliocene, the colobuses separated from the other colobines and still later (in the Pleistocene), evolved into the contemporary forms.

The taxonomy and nomenclature of Colobinae are complex and unstable. The scientific name of the subfamily was endorsed by the International Commission relatively recently (Delson, 1982), but some still claim that this group should be considered a separate family, Colobidae, rather than a subfamily, Colobinae. The classification of genera, subgenera and species, which number dozens of subspecies, is also complicated. Since many of the Colobinae are very rare, have been entered on the list of endangered species, and are seldom used in medical research, details of the classification problems involved are not discussed in this book, nor will detailed descriptions be given of these undoubtedly highly interesting monkeys. It should be noted, however, that a large contribution to the systematics of Colobinae has been made in recent decades by C. P. Groves, J. P. Napier, and P. H. Napier, as well as by a group of Chinese scientists, including, among others, Y. Peng, Z. Ye, Y. Zhang, and R. Pan, which has clarified the position of *Rhinopithecus*.

In my view (Fridman, 1979b), some of the seven contemporary Colobinae genera listed below (Napier and Napier, 1985), could be regarded as subgenera. All monkeys of the genus *Colobus* live in Africa and the rest of the species of Colobinae in Asia.

Figure 3.21 Colobus guereza. Reproduced with permission from Napier and Napier (1967).

The *genus Colobus* includes two groups: 1) a black and white group consisting of 4 or 5 species (*C. guereza, C. polycomos, C. satanas, C. vellerosus,* and *C. angolensis,* the latter added by some authors); and 2) a red colobus group of 2 species (*C. kirkii* and *C. badius*). Because colobus monkeys are highly diverse (mainly in color), numbering about 40 subspecies, it is difficult to define the boundaries for individual species and even in general. Hence, there are differences of opinion among different authors with regard to the number of these taxa. The average size of colobus monkeys is 40–75 cm, and they have a fluffy tail of 55–90 cm length. The nose protrudes slightly, but is bent down. Some colobus are almost black (*C. satanas*) and some are totally white (one subspecies of *C. polycomos* in the Kenyan mountains). In the beautiful guereza, which is also called the Abyssinian colobus, the white and black colors of its silky fur, combined with white 'feathers' of long hair along the trunk, make the monkey extremely attractive (Figure 3.21). This does not inhibit the local population from selling their fur or even from using their meat for food. Colobus monkeys inhabit the forests from Senegal to Ethiopia and, in the south, from Angola to Tanzania. Guerezas are the only monkeys which, like prosimians, use the mouth to carry offspring. The lead male emits a frightening roar to rally support from other males – hence, the famous chorals of colobus monkeys in the morning, apparently designed to protect their territory. Colobus adapt poorly to, and rarely produce offspring in, captivity. Pregnancy lasts 178 days

(*C. polycomos*, one birth recorded) and the newborn is white. The diploid chromosome number is 44. Colobus rarely live beyond 10 years in zoos, the longest recorded lifespan being 26 years (the 'king colobus', *C. polycomos*). The black colobus, *C. satanas*, and Kirk's colobus, *C. kirkii*, have been cited as endangered, as have been the subspecies of red colobus (*C. badius*), one of the most endangered primates in Africa.

The *genus Procolobus*, comprising only one species (*P. verus*, or olive colobus), is often placed in the preceding genus, and is an endangered primate.

The *genus Presbytis*, Asia leaf-monkeys, or langurs, consists of 14–16 species, including at least 85 subspecies. These are medium-sized monkeys, with the head and trunk measuring 40–80 cm and the tail 50–110 cm. The head is round, but somewhat posteriorly elongated at the top. The nose does not protrude and is flattened. The limbs are long and thin, the forelimbs being 15–25% shorter than the hind. The thumb is short, but well apposable to the other four. There is a laryngeal pouch in the absence of cheek pouches. The fur is long and gray, brown, dark brown, black, or golden depending on the species, and its color changes with age. The head in some species has a tuft of parted hair forming a crown. Langurs are natives of India, Pakistan, Sri Lanka, Indo-China, the Malay Archipelago, and Malacca. They predominantly live in forests. It has been proposed that some langur species be given the generic name, *Trachypithecus* (Weitzel and Groves, 1985).

The most famous, most abundant, and best studied langur is the hanuman (*P. entellus* [Figure 3.22]), which is a sacred monkey in India, Sri Lanka, and some other countries.

Figure 3.22 Langur monkey (*Presbytis entellus*) wounded by an adult male. Courtesy of Dr S. B. Hrdy. (See colour plate section)

Hanuman langurs are yellowish-gray animals with black limbs and with forward growing superciliary hair. They are adroit and agile monkeys, rapidly moving in the trees
and on the ground. Group relations do not appear to be rigid and sexual dimorphism
in size is relatively slight. Infanticide may be practiced in groups (which may number
7–80 animals) by newly established male leaders. The means used for communication are rich and include vocalizations (distress cries, roaring, 'yoop' calls, and barking
threats), grooming, and touching (one monkey may touch another with its hand or the
tip of its tail). The phenomenon of using urine for washing has been observed among
hanuman and dusky langurs (*P. obscurus*) and may occur in other langurs. The male
is sexually mature at the age of 5–6 years and the female a year to a year and a half
earlier. The menstrual cycle lasts 24–30 days and gestation 180–210 days. The birth
peak in India is from December to April. The infant is born with a black hair coat and
skin of a pale pink color. The diploid chromosome number is 44 (in the hanuman,
dusky, and purple-faced [*P. senex*] langurs). Several species have been recognized as
endangered, including the hanuman, golden (*P. geei*) (the total population of this monkey
in India is no more than 1,200 [Srivastava *et al.*, 1998]), capped (*P. pileatus*), and
Mentawai (*P. potenziana*) langurs; Francois' leaf-monkey (*P. francoisi*) from Northern
Vietnam hardly numbers 160 individuals (Le, 1996). It is difficult to maintain langurs
in captivity, where they usually live for no more than 3–5 years, although the record
for longevity (in the San Diego Zoo) is 31 years (*P. cristatus*, or silvery langur). Langurs
are very rarely used in biomedical research.

The *genus Pygathrix* is represented by one species (*P. nemaeus* [Figure 3.23]), in-
accurately called the douc langur or, more accurately, the douc langur of Laos and
Vietnam. In German, it is called *Klaider-affe*, which is the most accurate name for
this attractive monkey. It has olive-gray fur on the back and chest, reddish fur from
the knees to heels, almost black 'gloves' and 'socks', a red 'necklace' around the neck,
a yellowish or bluish face framed with white whiskers and a beard, and a dark 'cap'
overhanging the forehead (the colors differ somewhat between the two subspecies).
The skull is noticeably wider in its upper part than in the area of the chin, and the
eyes are widely set. The fore- and hind limbs are of equal size. The ischial callosities
are small. The head and trunk measure 55–80 cm and the tail is of similar length.
Females are somewhat smaller than males. In captivity these monkeys may be seen
sharing food with their conspecifics, but they do not adapt readily to captive conditions. Gestation lasts 180–190 days. The diploid chromosome number is 44. This is
an endangered genus.

The *genus Rhinopithecus* is often regarded as a subgenus of the langurs and these
monkeys are therefore called snub-nosed langurs, which is an inaccurate name. Their
snubby nose is similar to that of geladas. The skull is wide in the forehead region
and the eyes are widely set. Cheek pouches are absent in this genus and ischial
callosities are small and dark. The head and trunk are 55–85 cm long and the tail is
in approximately the same range in *R. roxellanae*, longer in *R. brelichi* (to 100 cm), and
nearly twice the body length in the Tonkin rhinopithec (*R. bieti*). The fur is long and
dense, and of gray, brown-gray, or dark gray color. The dark colors of the tail alternate
with light rings. These monkeys inhabit the forests of northern Vietnam and southern
China. Information is limited, but in recent years monkeys of this genus have been
actively studied by Chinese scientists. On the basis of anatomical data, these investigators consider the three *Rhinopithecus* species to be intermediate between the langurs

Figure 3.23 *Pygathrix nemaeus* (See colour plate section)

and the anthropomorphic gibbons (Peng *et al.*, 1989). They are seldom kept in captivity. The three species have been recognized as endangered; some forms (e.g., *R. avunculus*) number only between 150 and 200 specimens in the wild (Ren, 1996).

The *genus Nasalis* includes 2 species, *N. concolor* (snub-nosed or pig-tailed leaf-monkey, or simias) and *N. larvatus* (proboscis monkey). C. Groves (1970a), with good reason in my view, placed them in a single genus, for anatomically they have much in common. The skull is rounder than in the *Pygathrix* and the interorbital space is narrower, and the two species have equally narrow and long nasal bones and similar hair coats, with tufts on the sides of the body (see also Shoshani *et al.*, 1996). In some taxonomic publications, however, two genera are recognized (Napier and Napier, 1985; Whitney, 1995). Simias, monkeys found on the Mentawai Islands, live in boggy forests. The head and trunk measure 45–55 cm and the tails are short (13–19 cm) and hairless, except for a few hairs on the tip. The fur is brown to black on the back and there are cream-colored spots on the chest and abdomen. The face is black and the hands and feet are dark colored. The ischial callosities are well-developed. They live in family groups of 3–5 individuals (parents with offspring). Simias are rare and need protection.

Figure 3.24 Nasalis larvatus (See colour plate section)

The second species, *N. larvatus* or proboscis monkey (Figure 3.24), is remarkable in that it is one of the strangest looking creatures on Earth, owing largely to its long nose. The nose increases in length with age up to 8 cm (especially in males) and over-hangs the mouth like an appendage or tail. The head and trunk measure 60–75 cm; females are about half the size of the males. The naked face is framed by long hair which forms a red-brown collar around the neck and shoulders. The head and back are of a similar but darker color. The limbs are of gray color shot with pink and the tail is creamy white. The ischial callosities are large. Proboscis monkeys occur on Kalimantan and Malaysian Borneo, within coastal swampy forests and on plains. They commonly live in trees but prefer moving through the forest on the ground. They can be seen on the banks of rivers and bays where they adroitly dive and swim. They live in groups with several males, but the hierarchy is rigid, with each group being headed by a single male leader. At least 15 acoustic signals have been recorded from these monkeys, including the famous warning 'ka-khau' cry, for which they are sometimes called *kakhau*. They can adapt to captivity, but it is important to remember that they need high ambient temperatures and a diet of leaves and other plant parts. They also eat locusts and worms. They have bred in the San Diego Zoo where pregnancy lasted 166 days. The diploid chromosome number is 48. The longest life-span on record is 14.6 years. Both subspecies of proboscis monkey are endangered.

Superfamily Hominoidea

The last superfamily of the order Primates, suborder Anthropoidea, infraorder Catarrhini, comprises gibbons or lesser apes (*Hylobates*), the great apes (orangutan [*Pongo*], gorilla [*Gorilla*], and the two chimpanzee species [*Pan*]), and humans (*Homo*). Primates of this group are characterized by a round head, large, highly developed brain, a protruding nose, and their dentition, as already mentioned above: 2/2 incisors; 1/1 canines; 2/2 premolars; and 3/3 molars; giving a total of 32 teeth and a cusp pattern on the lower molars. The specific anatomical characteristics include long forelimbs, a well-developed clavicle and wrist, five fingers with flat nails, and a thumb set off opposite to the other fingers. The chest is wide, the scapulae are located on the back in one plain, the number of vertebrae is 29–36, and the number of ribs is 12–14. The form of the pelvis and the S-shaped shift of the vertebral column as well as the modification of the legs for stability favor the tendency for vertical locomotion while conserving the capability for brachial activity in certain species. Stable orthogradity (bipedalism) is inherent only in humans. Gibbons achieve bipedalism relatively easily, while chimpanzees and gorillas are able to assume the vertical position, but travel only short distances in this manner. It is thought that the ability of the representatives of these two families of hominoids to walk erect developed independently on the basis of the principle of parallel evolution (Alexander, 1992) (I note that many monkeys, especially catarrhines, sometimes stand on their hind limbs, carry objects in this position, and sleep in a semi-vertical [sitting] position). The hominoids lack an outer tail and cheek pouches, and most of the specimens do not have ischial callosities. The number of vibrissa buns is reduced to 2 in the apes and they are lacking altogether in humans.

The stomach of the hominoids is simple and there is a caecum with a vermiform appendix. The gestation period is longer than that in the lower simians, the period of helplessness of the newborn is longer, puberty occurs later, and the life-span is longer. They have a rich means of communication which includes posturing, staring, gesticulating, and vocalizing.

EVOLUTIONARY, COMPARATIVE-BIOLOGICAL, AND HISTORICAL BASES OF THE MODERN TAXONOMY OF HOMINOIDEA

Discussion of phylogenetic problems and taxonomy of the superfamily Hominoidea has been ongoing and especially intense for 35–40 years. It not only touches upon the main subjects of present day biology and anthropology, but also influences the world outlook of human beings. It is likely that, starting in the 1960s, there has not been a single scientific meeting on anthropology, evolution, taxonomy, and other closely related areas where these problems were not discussed. Hundreds of articles and books are devoted to this theme, which has become something of a scientific sensation. It appears that by the middle of the 1990s, however, this discussion had reached its logical conclusion.

There are two hypotheses regarding hominoid evolution and taxonomy of apes and humans; one suggests an early divergence (15 Ma–30 Ma or even 50 Ma), whereas the other a later divergence (4 Ma–8 Ma) of the pongid-hominoid branch. If we accept the first hypothesis, humans (taking into account various indications of similarity) are taxonomically distinct from all other modern hominoids and are classified in a separate family, Hominidae, as before. The three great apes (*Pongo*, *Pan*, *Gorilla*) are also united in a separate family, Pongidae. If we accept the second concept, then humans and all other hominoids are biologically closer and are classified as a single family, Hominidae (or as two families with a separate family for *Hylobates*). The first hypothesis is based exclusively on the data of morphology and paleontology. The second hypothesis takes the traditional grounds into account, but is based primarily on the recent data of genetics, biochemistry (primarily immunology and the sequences of amino acids of proteins), molecular biology, and behavior.

E. Mayr, one of the most prominent taxonomists of the 20th century, discussed the evolutionary data, especially for the taxonomy of hominoids; 'Relationships between the blood of hominids and apes and their expression within a certain range of taxa are a new illustration of the importance of the evolutionary approach. The study of chromosomes, numerous biological characteristics, and parasites reveal great similarity between humans and chimpanzees, such that some authors suggest that they should be classified as one family. Yet humans have occupied such a unique and sharply defined adaptive zone that J. S. Huxley (I am not speaking about Thomas H. Huxley, but about his grandson Julian – E. F.) even proposed to place them in a special group, Psychozoa. This seems superfluous, but the evolutionary isolation of humans undoubtedly justifies their assignment to a special family' (Mayr, 1971, p. 268, from the Russian translation) (besides his conviction of the strict isolation of *Homo*, we see from this citation that Mayr left no room for the possible influence of genetic and biochemical data on the 'evolutionary approach'!).

The research of G. Simpson (1945) on taxonomy of mammals, mentioned earlier, is the basis of the modern taxonomy of the hominoids. This taxonomy is based entirely on morphological and paleontological data. From an ideological point of view, this research led T. Huxley (1863) to conclude that there is a greater anatomical proximity between humans and apes than between the latter and other simians. G. Simpson's taxonomy of primates, corrected somewhat in detail, was consolidated in an early (1967) and later publication (1985) by J. and P. Napier. According to their scheme, Hominoidea includes three modern families; Hylobatidae (*Hylobates* and *Symphalangus*), Pongidae (*Pongo*, *Pan*, *Gorilla*) and Hominidae (*Homo*). I have already mentioned the great scientific value of Simpson's work, which significantly advanced the taxonomy of primates. Nevertheless, in accordance with an early evolutionary divergence, humans are taxonomically placed in a distant position from the other hominoids at a relatively high level in the family. This separation satisfied few morphologists, and it became more doubtful when confronted by the intrusion of the new data from genetics, molecular biology, and behavior. I note that G. Simpson was criticized mainly for the inaccuracy of his concept of the monophyletic origin of primates, and in particular, of the evolutionary origin of hominoids, rather than a failure to consider the data of the molecular biologists, who at the time were only initiating their activities in this field (Reed, 1963).

The usual difficulties encountered in validating taxonomic propositions on the evolutionary basis for classifying the hominoids were aggravated by poor paleontological data. No one at the beginning of the 1960s could imagine that rapidly developing branches of science, such as genetics, biochemistry, and molecular biology, burgeoning since the mid-1950s and inspired by Nobel Prizes, would be able to resolve this problem. Nevertheless, this is precisely what happened between 1960 and 1990, with the support of paleontology and morphology.

Let me relate the key events of this scientific progress. Starting in 1962 and 1963, as mentioned above, there were significant scientific publications from the laboratory of M. Goodman, the scientist who revived and improved upon the early methods of precipitation (Nuttal, 1902, 1904), using a method of two-dimensional starch-gel electrophoresis. This research demonstrated a special similarity between the proteins of human blood and that of the higher African simians (chimpanzee and gorilla), which was essentially unknown at the beginning of the century. Deceleration in the evolutionary rate of hominoids compared to other primates was suggested, which this investigator attributed to favorable conditions of feeding and oxygen supply to the fetus through the hemochorial placenta peculiar to higher primates. He further proposed that other functions provided immunological resistance to the fetus against harmful antigens, and accounted for the unusual growth of the cerebral cortex and the general prolongation of generation in hominids. Deceleration, in the investigator's opinion, was advantageous because it favored natural selection in the evolutionary flow of the primates in the direction towards the hominids (Goodman, 1962, 1963).

Since that time, intensive research has been carried out in Goodman's laboratory to find links of affinity between humans and other primates and between primates and other animals. The transition to phylogenetic foundations of taxonomy, and in particular, to molecular cladistics, was another milestone on this path. Different principles, methods, and even new fields of science are being used which require the inclusion of humans and all other living hominoids in one taxonomic family. Within this family, there is a close affinity between humans, chimpanzees, and gorillas (at the subfamily level) and an unique affinity between humans and chimpanzees (at a subfamily, tribe, or even the same genus). Asian apes (orangutans and the more distant gibbons) are placed at a somewhat greater distance from humans, although they are classified in the same Hominidae family at the level of different subfamilies (Goodman, 1996).

The next landmark was the earlier mentioned presentation by V. Sarich and A. Wilson in 1967. These investigators used a different immunological approach (microprecipitation) which also demonstrated the closest and equal affinity between albumin in the blood of humans and the African apes. Using the concept of 'molecular clocks' for the first time to construct a scheme of hominoid evolution, they demonstrated the existence of a quite recent common ancestor in the evolution of humans, chimpanzees, and gorillas whose lineages diverged somewhere around 5 Ma ± 1.5 Ma. Shortly thereafter (1975), M.-K. King and A. Wilson published results demonstrating a general affinity between humans and chimpanzees in terms of various proteins. They concluded that amino acid sequences and immunological and electrophoretic comparisons of proteins indicate a relationship in mean polypeptide of human and his 'double-chimpanzee' which exceeds 99% (1975, p. 115). Concurrently, methods for constructing evolutionary schemes from chromosomal data were developed and improved. New methods were also developed for determining homologies in the very *contents* of chromosomes, i.e., in 'the substance of heredity,' DNA and RNA. Phylogenetic trees of primates and other animals were calculated on this fundamental basis. Since the 1960s, surprising discoveries have also been made regarding the behavior of higher simians, both under natural conditions and in laboratory studies.

At the end of the 1950s and the beginning of the 1960s, discoveries of fossil primates (by L. and M. Leakey) that were given the name *Homo habilis*, extended the genealogy of our genus (*Homo*) by 2 million years. In 1974, the discovery of 'Lucy' (D. Johanson with collaborators) made it possible to place the previously problematic australopithecines (included in the family of humans after a heated debate) in the history of human origins. Later, *Potwar sivapithecus* was discovered (D. Pilbeam and collaborators), which appeared to be close to the orangutan evolutionary branch. This also determined the place of the other group of hominoids, the ramapithecines, in the history of the primates. All these paleontological discoveries made their contribution, although indirect, to the highly debated problem of the time of divergence of the lines of evolution of humans and apes and their respective positions in taxonomy.

By the end of the 1970s, morphologists more and more often supported the bio-chemists in accepting the unity of the family of humans and great apes. In the middle of the 1980s, this was reflected in its most expressive form in the important work of Colin P. Groves (1986) and later, in a joint paper by a group of reputed specialists, J. Shoshani, C. P. Groves, E. L. Simons, and G. F. Gunnel (1996). These investigators conducted a computer analysis of 264 morphological characters distinguishing the synapomorphies and the taxonomic range of the living hominoids and other primates. The evidence convincingly demonstrated the validity of including *Homo*, *Pongo*, *Gorilla*, and *Pan* in one family, Hominidae, thereby accepting the special affinity of humans and chimpanzees, and confirming the validity of the long-term research of M. Goodman and other biochemists, molecular biologists and geneticists. I consider this in more detail below.

We see that the debate between supporters and opponents of bridging the gap between humans and the higher simians has been ongoing for many years and continues today (Jaeger *et al.*, 1998). Sometimes these arguments even develop between the advocates within the same 'camp.' There is, for example, the controversy between biochemists about the problem of deceleration in the evolution of hominoids and the specific times of divergence, and that between morphologists about the maximum affinity of humans and each genus of ape, etc. At present, however, we can confidently accept that living humans are much closer to other primates than was thought at the middle of this century. I emphasize that the great affinity established between humans and simians in the fundamentals of biology strengthened the positions of theoreticians and provided the scientific grounds and stimulus for the development of medical primatology, which in itself supported the ideas of theoreticians.

As stated above, in the 1960s it was thought that the divergence of the pongid-hominid lineages occurred 30 Ma or even 50 Ma (Nesturch, 1970; Washburn, 1982). In the middle of the 1970s, paleontologists dated the time of this divergence as 14 Ma–15 Ma (Walker, 1976). By the middle of the 1990s, however, few would question that the figures were closer to 4 Ma–8 Ma. Consequently, the common ancestor of humans and other higher primates was gradually approaching present time. Let us consider this issue in more detail.

Cytogenetic evidence

In the 1960s, the Turin group of cytogeneticists headed by B. Chiarelli, using the traditional methods of karyotype research, showed a significant morphological affinity between the human genome and those of the catarrhine simians, especially with the highest ones. This occurred despite the fact that the diploid set of chromosomes is greater by 2 in the great apes (48) than in humans (46) (Chiarelli, 1962, 1967). This result was confirmed repeatedly both by these researchers themselves, using various methods, and by other geneticists. In England, the group of P. Pearson (1973), using the method of acrichine fluorescence, discovered a surprising phenomenon. Of all 27 species under study, characteristic fluorescence of the sexual Y-chromosome was found only in humans, gorillas, and chimpanzees. At the same time, the French geneticists, using the method of chromosome differentiation by denaturation with various stains ('disc method', 'banding'), were able to identify chromosomal reconstruction with an accuracy previously unavailable. They found homology in the majority of human and other hominoid chromosomes. The topography of dark and light stripes was identical in 17–18 pairs of human chromosomes and those of the three great apes, and maximal in humans and chimpanzees (Turleau *et al.*, 1973). It is intriguing that in the next decade, when the methods of chromosome staining were significantly improved and cytogenetic methods were available for calculating the common ancestor, it was found that 18 of the 23 pairs of chromosomes in the common ancestor were identical and the 5 other pairs had only insignificant differences. The investigators concluded that initially only the orangutan line branched from this common ancestor due to mutations in 6 pairs of chromosomes. Later the line of the gorilla genus branched from the common ancestor of gorilla, chimpanzee, and human as a result of changes in 9 pairs. Finally, 3 chromosome rearrangements of the common ancestor of human and chimpanzee led to the divergence of these two genera (Yunis and Prakash, 1982).

Twice in the 1970s (1971, 1975), the Paris conference on genetics favored the table of chromosome homologies for humans and the three great apes (Mitchell and Gosden, 1978, p. 285). According to this table almost all 46 human chromosomes have essentially equivalent homology with 48 chromosomes in chimpanzee, gorilla, and orangutan, and with maximal consistency between the chromosomes in human and chimpanzee. It was becoming clear that with the combined homology of karyotypes for the entire order Primates, that the chromosomes of humans and the three great apes are surprisingly similar (98% homology) and vary mainly in a small number of pericentric inversions (Dutrillaux, 1975; de Grouchy, 1979). We see from the aforementioned table that the second chromosome in humans, which is absent from the karyotype of the three great apes, is present as the 12th and 13th chromosomes in chimpanzee and gorilla and as the 11th and 12th in orangutan. This explains the basis of the lesser number of chromosomes in humans, which was formed due to Robertson fusion (de Grouchy, 1987). The characteristic banding of its shoulders, moreover, is surprisingly identical to the ones for the 12th and 13th chromosomes of chimpanzee. In fact, the chromosomal homologies of these shoulders appear in almost all primates of the Catarrhini infraorder. Genetic mapping also shows that the genes of this chromosome in humans (N2) are not only found on the chromosomes of chimpanzee, but they are also found on the homologous chromosomes of the lower simians of the Catarrhini (Chiarelli, 1985). New data concerning the second chromosome of humans were recently presented by T. Haaf and P. Bray-Ward (1996).

The chromatin in humans and other higher primates generally has a similar differential banding (not taking heterochromatin into account), although specific chromosomal segments in certain species are combined in different ways. The sections with equivalent banding, however, contain the same genes (de Grouchy, 1987). In the previously mentioned research of Chiarelli, the time of divergence of the human and African ape line is defined as 8 Ma–6 Ma. On the other hand, only a low degree of chromosome homology with humans was found in the 'small' hominoids, gibbons and siamangs. Numerous differences in the synthenia were discovered which indicates a comparatively low level of cytogenetic affinity of gibbons with humans; they diverged earlier from the evolutionary stem of the hominoids (Guisto and Margulis, 1981; Turleau *et al.*, 1983). As we see, among geneticists the specific karyotypic affinity of human, chimpanzee, and gorilla is beyond question. This is reflected taxonomically by the union of the last three genera in one family. But which species among the two hominoids presently living in Africa is closer to humans? The answer to this question was not a simple one for cytogeneticists.

Until recently, it was thought that the karyotypes of these two genera had equivalent affinity with the human one. Consequently, the divergence of the lines of these three genera either occurred simultaneously (trichotomy), or the divergence of humans and the common ancestor of chimpanzee and gorilla occurred first, and was then followed by a dichotomy of the latter (Ayala, 1980; Nishio *et al.*, 1995). In an astute review about the origin of humans, R. Stanyon (1989) conducted a detailed analysis of various studies on chromosomal banding by different investigators. The results indicate that depending on the method, the three closest hominoids may be grouped according to two schemes, human-chimpanzee and chimpanzee-gorilla. The author concluded 'Karyotypic data cannot define with confidence which of these two phylogenetic schemes is correct' (p. 41). There was some slight reason, however, to favor the *Homo-Pan* pair (28 mutations against 29 in the *Gorilla-Pan* pair). He also suggested that the time of divergence of humans from the two apes was 5 Ma–8 Ma and noticed a slowing down of evolutionary time of the human karyotype compared to the ape. There was some evidence, moreover, which suggested that the gorilla was the closest genetic relative of humans (Miller, 1977). If we take into account the limits on precision of scientific instruments and the variability in individual studies, the different results may be described merely as 'noise' in a system which combines homologous chromosomal material from three hominoids.

Nevertheless, considerable research on cytogenetics has for some time considered the chimpanzee to be the nearest relative of humans (Chiarelli, 1973; Turleau *et al.*, 1973). An article in *Science* on the affinity of chimpanzee and human chromosomes described 'The striking resemblance of . . .' these species. The analysis was conducted by G-staining with high resolution (1,200 lines) at the late prophase, prometaphase and early metaphase (Yunis *et al.*, 1980). They wrote, 'The affinity (of the chromosomes of humans and chimpanzees) is so great that it hampers the explanation of the phenotypic differences between them' (p. 1145). An analysis of the haploid karyotype was conducted by B. Dutrillaux (1985) with a new improved level of differential staining. The results showed that similar to his previous research (mentioned by Yunis *et al.*, 1980), *Homo* and *Pan* differ in 9 pericentric inversions, one translocation ('the end to end') and one addition of heterochromatin in chromosomes #1 in humans and #13 in chimpanzees. The phylogenetic model built on the basis of these data places the two species of chimpanzee closest to the human, followed by the gorilla and at a greater distance the orangutan (Dutrillaux and Coutrier, 1986). The same conclusion follows from the analysis of acrocentric chromosomes and from the study of gene localization by ribosomic RNA (Guisto and Margulis, 1981), which suggested that the great apes should be classified in one family with humans. M. Nei (1978) considered the chimpanzee and human to be closest on the basis of the analysis of genetic distance. He dated the divergence of their evolutionary lines as 4 Ma–6 Ma.

I shall consider the similarity of the specific phenotypical characters of humans and simians in the next chapter, but here I give an example of the cytogenetic affinity of humans and chimpanzees in terms of the major histocompatibility complex. In particular, this system is characterized by genes that have unusually great variability in alleles, with up to 100 allele variants that differ significantly in nucleotide composition. Some genes may have 90 nucleotide substitutions which correspond respectively to 20–30 amino acid versions of the protein. Comparison of the alleles of the human HLA-DRBI gene with the homologous gene in chimpanzee (ChLA-DRBI) showed that part of the human alleles are structurally closer to those of chimpanzee than to the other alleles of their own gene! That is to say that individual alleles of the major histocompatibility complex in humans and chimpanzees are more homologous than different alleles at the same locus within the same species. One can explain this striking identity only by the fact that the two species are descendants of a common ancestor. In the opinion of some authors, the divergence of the majority of the genes occurred before the branching of the lines of humans and chimpanzees (Lawlor et al., 1988; Klein et al., 1990). I note that a significant genetic affinity was found in the major histocompatibility complex of humans and monkeys as well. The DOB gene in chimpanzee appeared to be 99% homologous with the human gene by nucleotide sequences of cDNA and the polypeptides coded by the gene were 98.5% homologous (Kazahara et al., 1989).

The remarkable cytogenetic affinity of the living primates, especially that of humans and chimpanzees, besides having fundamental evolutionary importance, strengthened the foundations for further development of medical primatology, including of course research models for the most severe hereditary diseases of humans.

DNA data

The various areas of molecular biology studies are so intertwined in our subject matter that it is difficult to establish the sequence in which to discuss them. I follow the logic of science, namely, I first ask, 'What does this affinity of chromosomes and genes represent?' For our next consideration, I choose the main substrate of heredity which resides in the DNA of chromosomes and RNA. Of course, immunogenetics and the rest of biochemistry could also claim a certain priority, especially if the historical primacy mentioned above is taken into account. M. Goodman, who played no small role in this research, wrote that he quit immunodiffusion in the 1970s when he understood that '. . . my immunodiffusion distance data were not amenable to . . . cladistic analysis' (1996, p. 273) (but he continued studies on amino acid sequencing of proteins).

In the second half of the 1960s, study was initiated on the affinity between humans and other primates, in terms of nucleic acids. This research has continued on the basis of DNA hybridization, thermostability of hybrid duplexes, analysis of nucleotide sequences, sequencing DNA (restrictional analysis: digestion with restriction enzymes; Southern blotting), and study of repeated sequences of nucleotides. The study of nuclear DNA and mitochondrial DNA (where the evolutionary rate is much higher than that of nuclear DNA) became important. Similar methods are also used for comparing RNA.

The first experiments on hybridization of DNA, conducted in the laboratory of B. Hoyer, though imperfect, showed a high level of homology between nonhuman primates and humans. The homology, expressed in percent with respect to human DNA, exceeded 95% in chimpanzees, 80% in macaques (rhesus), 50% in lemurs, not even 20% in mice, and 10% in hens (Martin and Hoyer, 1967). Despite the flaws in the first experiments, these figures were actually close to the values obtained using cross immunological reactions of blood proteins, amino acid sequences of proteins, and cytogenetics. A short time later, phylogenetic trees were constructed using the dissociative temperature of hybrid duplexes. In these studies, chimpanzees appeared to be the nearest relative of humans and orangutans were found to be the most distant among the large hominoids (Kohne et al., 1972). Again, there is agreement with other molecular data.

It is worth recalling the fundamental progress which occurred in the theory and methods of taxonomy in the beginning of the 1970s. M. Goodman (1996) reached the conclusion that a pure Darwinian genealogical homology consists of the proximity of organisms to their mutual ancestor, which is sealed in the genotype. Consequently, we have the key to constructing an actual phylogenetic tree of the kingdom of nature. I emphasize that these facts are known to everyone who studied Darwinism. According to Goodman, in reality, metaphysical constructions dominated taxonomy (arbitrarily selected morphological signs, grades, and the so-called 'adaptive zones'). An obvious reflection of this was the anthropocentric view of nature and primarily, the isolation of Homo from its nearest hominoid relatives in taxonomy at the level of different families. Goodman turned to purely monophyletic grouping of the taxa, which in the environment of a successfully progressing new science, resulted in the establishment of molecular cladistics. He performed this work on a computer with the aid of G. Moore, a mathematician (Goodman and Moore, 1971). The programs, written by W. Fitch, were used later to study the amino acid sequences of proteins and the sequences of nucleotides. A new approach, the method of maximum parsimony, which is presently very popular, was accepted on the basis of Fitch's program. The algorithm takes into account the origin of modern sequences to determine the topology of the phylogenetic tree.

The study of phylogeny of the highest primates became entrapped in a new problem, as mentioned above, the degree of relationship between different species of great apes and humans, and consequently, the model of their evolution. By the end of the 1970s, the main concept in this discussion, the inclusion of all great apes into one family with humans, was receiving greater and greater support among the specialists, including morphologists (Szalay and Delson, 1979). As a result, the problem of the special proximity between humans and chimpanzees, as the closest pair of genera in evolution, which was supported by M. Goodman and his followers, was actively opposed by those who adhered to the scheme of trichotomy, that is, the simultaneous splitting of the branches of humans, chimpanzees, and gorillas (Sarich and Cronin, 1976). The latter authors used the microcomplement fixation method and proceeded from the assumption of the 'evolutionary clocks.' They denied the deceleration of evolution, but proved the later ancestral separation of humans and both of the African apes as 5 Ma, and later as 4 Ma. This was very important for further progress in knowledge about the origin of humans and also for strengthening the scientific foundations of medical primatology.

In the 1970s, biochemical studies of various proteins dominated the field of hominoid evolution, which I shall consider specifically below. Even on their own 'field of battle', they were actively strengthening the concept about the close phylogenetic relationship of humans, chimpanzees, and gorillas. It is the opinion of M. Goodman (1996) that the turning point in the study of phylogeny with DNA occurred when R. Britten and D. Kohne (1968) discovered that 40–50% of the mammal's genome comprises repeated sequences of DNA (later it was found that there are significantly more of them in certain cases). Their functions are unknown, but due to their independence from the influence of natural selection, the pseudogenes seemed to be more representative in these studies than coding unique parts of DNA. Nevertheless, until the middle of the 1980s, only a few papers were published on hybridization of the hominoid's nuclear DNA. In these publications one can find conclusions postulating both trichotomy of the lines of *Homo*, *Pan*, *Gorilla* with a divergence time of 6.7 Ma (Cronin and Meikle, 1982) or 5 Ma (Lowenstein and Zihlman, 1984), and the dichotomy of *Homo* and *Pan*. It was found that the DNA of these two species is closer than those of the zebra and horse, and the time of their divergence was estimated as 4.5 Ma (Dorozynski, 1982). Using restrictional analysis of repeated DNA with 13 endonucleases, investigators reported a significant similarity between the DNA of humans and chimpanzees (as well as baboons) (Dandieu *et al.*, 1984). New indications of unique similarity in the DNA of the sexual chromosomes were found primarily in humans and chimpanzees, where homology of the DNA of Y-chromosomes was found with digestion by 8 different restrictases. Later it was shown that the differences in DNA of the Y-chromosomes in these species did not exceed 1.7% (Daiger *et al.*, 1987).

Starting at the end of the 1970s, the number of papers on mitochondrial DNA (mtDNA) increased. The mitochondrial gene appeared to be a very convenient subject for phylogenetic investigation. Its size is small, it consists of approximately 16 thousand pairs of nucleotides in mammals, and it has a very economic structure without noncoding sequences. The first experiments using restrictional endonucleases were conducted in the laboratory of Allan Wilson (Brown *et al.*, 1980; Ferris *et al.*, 1981). The results showed the efficiency of this approach in the study of the evolutionary affinity of primates, supported the general scheme of their molecular taxonomy, and suggested a special affinity of humans and African apes, with noticeable differences in the orangutan and even greater differences in the gibbon. Nevertheless, the first analyses of mtDNA argued against a trichotomy during the evolution of humans, chimpanzees, and gorillas, and an even earlier branching of humans from the line of the two apes, whose mtDNA seemed more similar to each other (Barton and Jones, 1983). Different investigators, including those who worked with RNA until the second half of the 1980s, held this point of view (Martin *et al.*, 1981; Smouse and Li, 1987). Although shortly after, when the technology of research with DNA improved, the viewpoints of those who studied mtDNA started to change in favor of a closer affinity between humans and chimpanzees.

We see that the use of nucleic acids to establish the true evolutionary affinity of the primates, the specific similarity of humans and the four hominoids, primarily human to chimpanzee and gorilla, and even the exclusive affinity of humans and chimpanzees, was on the rise. This was a sound approach in the first half of the 1980s, stimulated by the research and analyses carried out in the laboratory of M. Goodman (on other material), which included DNA data. Nevertheless, the paper by C. Sibley and J. Ahlquist 'The Phylogeny of the Hominoid Primates as Indicated by DNA-DNA Hybridisation' (1984) justified these assumptions and stimulated extraordinary interest.

These investigators gained considerable experience in research on the evolution and taxonomy of various birds using hybridization of single-copy nuclear DNA (without repeated sequences). They devoted many publications to this problem, which were backed by thousands of experiments. In this research, based on hybridization of radioactively marked DNA sequences, Sibley and Ahlquist obtained full matrices of values of delta-T5OH (the temperature at which half of the hybrid duplexes dissociate). With this information, they calculated the time of branching of the phylogenetic tree for the catarrhine primates. The rate of DNA evolution, the change in longevity of generation, and difficult problems of calibrating the dates of divergences, were thoroughly taken into account in this research. The results obtained were compared to the literature on primate phylogeny, but the problem of phylogenesis of the Hominoidea was treated separately. The conclusions of these investigators were in complete agreement with the general scheme of the evolution of primates. They were unambiguous in indicating the maximum similarity between humans and chimpanzees among the order Primates, which had a common recent ancestor in this group. The time of divergence of the line of the catarrhines was determined (citing the most recent paper of the authors with a significantly greater number of hybrids). The divergence of the line of Old World monkeys was estimated as 25 Ma–34 Ma, gibbons as 16.4 Ma–23 Ma, orangutans as 12.2 Ma–17 Ma, gorillas as 7.7 Ma–11 Ma, and chimpanzees and humans as 5.5 Ma–7.7 Ma (Sibley and Ahlquist, 1987).

Clearly, such an article could only increase controversy which immediately followed its publication. The criticism concerned the methods of analysis, statistics of data processing, and a very important problem, the authenticity of homology. Are the properties of affinity inherited from a recent common ancestor, were the orthogonous divergences actually obtained, and weren't the misleading parallel divergences obtained instead? Are these not artifacts of the technique of hybridization itself? Don't the authors base their conclusions on primitive retentions and thus not true homology ('not' synapomorphies, indicators of synapomorphy), but only convergent similarity? Not all of these problems were solved by molecular cladistics (as was done, for example, by morphologists). This research was criticized not only by morphological cladists, but by molecular biologists as well (see the discussion by J. Marks and J. Diamond [1988] and Marks [1991]).

Nevertheless, after repeated testing by the investigators themselves (Sibley and Ahlquist, 1987; Sibley et al., 1991) and independently by other researchers, the results remained the same. They demonstrated the affinity of each hominoid with humans and suggested that chimpanzees are the closest relatives of humans because their divergence occurred last (8 Ma–6 Ma) (Felsenstein, 1987; Caccone and Powell, 1989). The studies of M. Goodman and his colleagues, based on the shift in their research to DNA, provided the strongest support for this concept.

A significant length of nucleotide sequence equal to 2.190 bp was obtained in cooperation with Jerry Slighton, who studied the genes of embryonic and adult beta-globulin. It was found that 95% of the cluster of human beta-globulin is comprised of the nucleotides of noncoding DNA, which appeared to be an excellent source of molecular synapomorphies. Further, the accumulations of the η-globulin pseudogene were obtained from recombinant human genomic library clones as well as from the libraries of chimpanzee and gorilla (together with E. Chang). Each of these Ψη-sequences reached a length of approximately 2.160 bp. The sequences of lemur, owl monkey, and goat were used; later B. Koop investigated the sequences of orangutan and rhesus macaque. The comparison of these nucleotides indicated the existence of a human-chimpanzee-gorilla clade, but gave no indication of any correlation within this group. Only when Goodman's colleague, M. Miyamoto, determined that the value of the locus of the pseudogene, Ψη-globulin, was 7.100 bp long (human, chimpanzee, gorilla, orangutan), did the picture become clear. The cladistic analysis of these four enlarged sequences by the method of maximum parsimony clearly indicated, for the first time by DNA, the presence of a human-chimpanzee clade, which was supported by 8 synapomorphies (and later by 10, with a greater number of nucleotides). At the same time, the chimpanzee-gorilla clade was supported by only 3 synapomorphies, and the gorilla-human clade by only 2! (Miyamoto *et al.*, 1987). Humans and the large hominoids differ by only 1.61% (chimpanzee) and 3.52% (orangutan) of these sequences (Miyamoto *et al.*, 1988). Clear statistical confirmation of the *Homo-Pan* clade was also obtained with the study of gamma-globulin genes.

In this research, both functional genes and noncoding parts (flanking and intergenic) of DNA were used. Different variations of the parsimony analyses were tested and more and more precise procedures of statistical analysis were developed. The data on nuclear DNA were compared to data on mitochondrial DNA, to the amino acid sequence of the proteins, and of course, to the new data of comparative anatomy and biochemistry (Williams and Goodman, 1989; Bailey *et al.*, 1991, 1992; Czelusniak and Goodman, 1995). The results generally had a single interpretation. Monophyly of the order Primates was confirmed, as was monophyly of the catarrhine and platyrrhine primates, monophyly of the hominoids, monophyly of humans and all the great apes, monophyly of humans and the two genera of African apes, and finally, the existence of a separate clade of humans and chimpanzees. DNA data on the date of evolution of the catarrhine primates were presented. They suggested that the split of Cercopithecoidea-Hominoidea was 25 Ma, the gibbon 17.9 Ma, orangutan 13.9 Ma, gorilla 7.2 Ma, and the chimpanzee-human split 5.9 Ma. The hominoids are correspondingly classified according to M. Goodman (1996, p. 279), as shown below:

Cladistic Classification of the Extant Hominoid Genera (Goodman, 1996)
Superfamily Hominoidea
    Family Hominidae (17.9 Ma)
        Subfamily Hylobatinae
            Hylobates: gibbon and siamang
        Subfamily Homininae (13.9 Ma)
            Tribe Pongini
                Pongo: orangutan
            Tribe Hominini (7.2 Ma)
                Subtribe Gorillina
                      Gorilla: gorilla
                Subtribe Hominina (5.9 Ma)
                      Pan: common and pygmy chimpanzee
                      Homo: human

All the hominoids reside in one family, all great apes in one subfamily with humans, humans and African apes in one tribe, and humans and chimpanzees in one subtribe. The affinity of humans and chimpanzees by coding DNA is 99.5% and by noncoding DNA is 98.3%.

The utility of the molecular evolutionary clock is distinguished only at short time intervals, that is, the local clock at reliable initial paleontological dating. Slowing of molecular evolution of the primates (with large variability in the speed of different groups) was confirmed (Britten, 1986; Li *et al.*, 1987; and others), especially in the hominoids, with maximum deceleration in humans. Experiments conducted on other mammals and Ceboidea using the same criteria showed that the time of the evolution of taxa of an equal level and their divergence was comparable to that obtained in the Hominoidea superfamily up to genus (Goodman *et al.*, 1994; Goodman, 1996).

The second half of the 1980s and especially the 1980/1990s boundary saw the triumph of molecular DNA investigations into the phylogeny of humans and their closest relatives. Although there is the opinion (though rarely expressed) that there is a greater affinity between chimpanzee and gorilla than between either of them and humans (the coding of epidermal protein of involucrine [Djian and Green, 1989]) and other solitary objections, the majority of studies on DNA and RNA supports the data presented above and the special affinity of human and chimpanzee. This is shown on the alfoid DNA area of the X-chromosome centromere (Jorgensen et al., 1990), the alfoid DNA of individual autosomes (Baldini et al., 1991), and the fragments of Y-chromosome DNA (Lucotte et al., 1990). This is also shown by the homology (99%) of the c-myc gene (Argaut et al., 1910), the homology of the glycophorine gene (Creau-Goldberg et al., 1989), haptoglobin (MaEvoy and Maeda, 1988), alpha and epsilon genes and C-pseudogene at 3 immunoglobulins (Ueda et al., 1989), and many other vital proteins which play cardinal roles in the processes of tolerance, resistance, pathology, and immunity of the organism (in regard to the contradictions associated with involucrine, they are likely caused by the small value of the chain of nucleotides used [Goodman, 1996]).

New data appeared on the homology of mitochondrial DNA. Even those scholars who previously postulated a trichotomy of humans and African apes acknowledged the maximum affinity of *Pan* and *Homo* (Foran et al., 1988). Consistent results were obtained with various versions of phylogenetic trees with different sizes of mtDNA from different genes of various proteins. The closest primate to human is chimpanzee. The other primates are related to each other according to the sequence corresponding to the data on nuclear DNA (see above) (Hayasaka et al., 1988; Horai, 1990; Saitou, 1991a; Ruvolo et al., 1991; Horai et al., 1995).

A group of scientists from Japan, headed by Shintaroh Ueda, played an important role in establishing the phylogenetic hierarchy of the primates on the basis of the DNA data (mainly on the immunoglobulin data). Ueda and his colleagues stated their belief that DNA pseudogenes of E immunoglobulin in gorilla have greater similarity with human DNA than those that human DNA have with chimpanzee DNA. After studying different immunoglobulin genes, however, they gained support for the maximum affinity of humans and chimpanzees, and presented their estimate for the dates of hominoid evolution. They proposed that the separation of the orangutan line was about 14 Ma, the gorilla 5.9 Ma +/− 0.9 Ma, the chimpanzee 4.9 Ma +/− 0.9 Ma (Ueda et al., 1988; Ueda, 1991). I can only mention the outstanding research on the evolutionary analysis of DNA and RNA of the primates carried out by a researcher in Tokyo, Masami Hasegawa, who in cooperation with H. Kishini, T. Yano, and S Horai, was one of the first to determine, on the basis of various statistical methods, the order of branching of the extant and fossil taxa. His results provide the foundation for the present concepts regarding the phylogeny of humans and their closest relatives. According to M. Hasegawa, the closest relative of the human is the chimpanzee; the divergence of these two forms occurred 4 Ma–5 Ma (Hasegawa et al., 1984; Hasegawa, 1991; Hasegawa and Horai, 1991; Adachi and Hasegawa, 1995). Numerous publications by these investigators and other supporters of the close affinity of humans and other primates have now confirmed one of the most significant scientific results of our time.

The research on primate DNA demonstrates both the isolation of the hominoids, even from the Old World simians, and the affinity with them. Individual fragments of the DNA of human chromosomes hybridize not only with the DNA of hominoids, but also with those of baboon and macaque. They do not hybridize, however, with the DNA of mouse or cow (Kaplan and Duncan, 1990).

As one can see, data from the study of nucleic acids completely correspond with the conclusions of the cytogeneticists. What is more, research in this field in the 1980s through 1990s was focused not so much on the problem of the maximum affinity of humans with other hominoids (this became clear from the first experiments with DNA), but mostly on the dichotomy or trichotomy of *Homo* and the African apes. A unique affinity between these genera was shown, especially for *Homo* and *Pan*, which exceeds 99%. However Dr S. Pääbo from Germany showed at a workshop in Tokyo (March, 2001) that while gene expression in the liver and blood of chimpanzees is very similar to that in humans, the brain activity in the latter is three to four times greater (Normile, 2001). '"Among these three tissues, it seems that the brain is really special in that humans have accelerated patterns of gene activity", Pääbo says.' (Normile, 2001). Study of ribosomal DNA using 12 restriction enzymes revealed a difference between *Homo sapiens* and *Pan paniscus* of 2.3%. As we have seen, this is not the highest homology that has been reported. Using the same method, investigators reported a difference between *Hylobates lar* and *H. syndactylus*, which are now considered to be two species of same hominoid genus, of 2.4% (Suziki *et al.*, 1994). In other words, according to the DNA data, humans and chimpanzees are more closely related than the two genera of gibbons.

What do they code? Evolution and classification of hominoids by other data from modern biochemistry

As I have repeatedly stated, comparative immunological studies of primate blood reactions, aimed at determining their affinity, were initiated by G. Nuttal (1902). The revival of this branch of investigation occurred in the 1960s, although at a new level of science, and was associated with the research of M. Goodman in Detroit. This is precisely where deceleration of evolution was discovered, with respect to blood proteins (despite their great genetic variability), especially in hominoids. A close antigenic affinity was also discovered between human, chimpanzee and gorilla. Asian apes are some more distant from these, especially the gibbons. Thanks to the discovery of the amino acid sequences of proteins and the application of this discovery to the study of hominoid phylogeny by comparative immunology, it was possible to make the significant contribution of introducing and elaborating the *new* phylogeny of the hominoids and to validate its use. When M. Goodman proposed for the first time (1962, 1963) to include chimpanzee and gorilla in the family of humans, and even to include them in the same subfamily (Goodman and Moore, 1971), or when V. Sarich and A. Wilson (1967) stated that the lines of *Pan* and *Homo* diverged only 5 Ma, and not 30 Ma, as was accepted earlier, the majority of scientists, and especially the most prominent ones, did not consider this seriously.

I should note that an effective computer program for investigating phylogenesis was written initially for analyzing amino acid sequences (Fitch and Margoliash, 1967) and later for calculating nucleotide sequences (Fitch, 1971). It was used by Goodman's group (Barnabas *et al.*, 1972; Moore *et al.*, 1973) to find the hypothetical 'codons of the ancestor' (by the maximum parsimony method).

In the 1970s, as was mentioned above, traditional investigations, especially those in biochemistry, dominated molecular research on hominoid phylogeny. By that time, there were already indications of an unique similarity in blood proteins of humans and other primates, and even suggestions of the specific affinity of humans and chimpanzees. K. Bauer (1969) divided 20 proteins of human plasma into 3 groups, one of which reacted only with the serum of chimpanzee (!). The second reacted only with the serum of monkeys and the third reacted only with the serum of other mammals. None of the proteins, however, reacted with the serum of fish or birds. These relationships were especially well-established for the immunoglobulins (Wiener *et al.*, 1970). R. Damian and E. Lichter (1972) demonstrated a maximum affinity between humans and chimpanzees by antigenic determinants of proteins. I already mentioned the research of V. Sarich and A. Wilson (1967), which bothered the scientific community by its calculation of the common ancestor of *Homo* and *Pan*. Of course, the research of M. Goodman was also known. Using the method of immunodiffusion in the beginning of the 1970s, his group carried out about 6,000 computer comparisons of proteins obtained from the sera and plasma of 70 species of primates and 48 species of other animals. The similarity of the primate proteins in this study was unusual, especially in the hominoids, given those sequences that are already known (to the reader of this book) and the significant differences of primates from other animals. Data on albumin, transferrin, ceruloplasmin, and thyreo- and gamma-globulins were also analyzed (Goodman *et al.*, 1970).

Later in the 1970s, M. Goodman and colleagues carried out comparisons of amino acid sequences of different chains of fetal and adult hemoglobin. The result is already known to the reader; unusual affinity among hominoids, including humans, and deceleration of the rate of evolution. Prior to the 1980s, this group stated that the two African apes were closest to humans; from the beginning of this decade, they emphasized the maximum affinity of *Homo* and *Pan*, which are closer to each other by amino acid sequences than either of them is to *Gorilla*. The identical chains of 2 species of chimpanzee and the human species (a total number of 287 positions in 2 chains) differ equally from gorilla by 2 amino acid substitutions of hemoglobin: a single substitution in α- and β-chains. They differ from orangutan by 4 substitutions. The three genera form a branch on the phylogenetic tree of hemoglobin, where the gorilla line diverges earlier than the common branch of human and chimpanzee. The common branch of these three separated significantly later than the line of *Pongo* (Goodman, 1982; Goodman *et al.*, 1983). According to these researchers, the rate of molecular evolution is not at all high (6%) for these genera (6 nucleotide substitutions per 100 codons over 10^8 years), whereas in the higher vertebrates, it is 25% in the α-chain and 29% in the β-chain of the same hemoglobin. I note that there is a correspondence of data on DNA, immunological distance, electrophoresis, and amino acid sequences of hemoglobin.

Later, Goodman showed that amino acid sequences (fibrinopeptides, mioglobin, fetal and adult hemoglobins, carbonic anhydrase) vary more slowly than noncoding DNA (99.6% affinity of *Homo* and *Pan* compared to 98.4% affinity for their nucleotides). Among the hominoids, the change in evolutionary rate of fetal hemoglobin provided the most favorable balance of oxygen transport, prolongation of fetal life, and prenatal development of the brain (Goodman, 1992). Consequently, it was the substance of Darwinian natural selection. Goodman used this example to illustrate the close relationship between molecular evolution and adaptive changes in morphology and behavior, which no doubt is correct and important. I shall now consider the dominant role that these factors played in the biological foundation of medical primatology.

Completely identical amino acid sequences were found in both chains of fibrinopeptides in humans and chimpanzees (Doolittle and Mross, 1970), for mioglobin, lysozyme, cytochrome C, apocytochrome B from liver microsomes, and insulin (Atassi *et al.*, 1970; Barnicot and Wade, 1970; Nobrega and Ozol, 1971; Romera-Hererra *et al.*, 1976) (One should bear in mind that almost all of these authors define the degree of similarity of these proteins in humans and other primates as not less than 90% for the other hominoids, 50%–80% for monkeys, 40%–50% for prosimian, and 0%–20% for other animals). When M.C. King and A. Wilson (1975) published their analysis of the amino acid sequence of 12 proteins or their single chains (44 loci including albumin, transferrin, carbonic anhydrase), it appeared that there were only 19 substitutions in chimpanzee, in comparison to human, out of a total of 2,633 sequences, an affinity in the proteins of these species that exceeds 99%. I emphasize that other proteins with similar affinity were already known by that time, and their list has since enlarged. This affinity not only contradicts splitting the phylogenetic tree at the family level, but in other animals it is also characteristic of twin species. The striking discrepancy between the data from molecular biology and morphology, in the opinion of the researchers at that time, could be accounted for only by peculiarities in the evolution of the regulatory genes. Without denying this concept, M. Goodman, on the basis of these data, supported the idea of deceleration of evolution, especially for the highest primates (Goodman, 1976). No one at that time could imagine that the discrepancy between the biochemists and morphologists was not really so striking.

I note that the affinity of humans and African apes appeared to be even greater according to these data. In the chimpanzee, only 5 substitutions out of 1271 sequences of amino acids were found in 9 chains of proteins (6 types of globulins, fibrinopeptides A and B, cytochrome C, carbonic anhydrase), an affinity of 99.6%. The affinity of humans and gorillas on the one hand, and chimpanzees and gorillas on the other, was the same by the same criteria and equaled 99.3% (Goodman *et al.*, 1983). Nevertheless, there were scientists who did not support this specific phylogenetic relationship between *Homo* and *Pan*.

As I mentioned above, V. Sarich was an active supporter of the trichotomy, *Homo-Pan-Gorilla*. His long standing position (1968) has not changed (Sarich, 1993). In their detailed work on electrophoresis of hominoid proteins (23 loci), E. Bruce and F. Ayala (1979) found that the genetic distance of the *Homo-Pan* pair was even less than the one determined by M. C. King and A. Wilson (0.0386 against 0.620). According to the same study, the value for different genera of the same family of rodents was 1.833. These investigators, however, found the same proximity between *Homo* and *Gorilla* as between *Homo* and *Pan*, which suggested a trichotomy in the evolution of these three genera and an equal assignment to one subfamily of the family Hominidae (p. 1054). Electrophoretic distances of erythrocyte enzymes and repeated blocks in Y-chromosomes define a greater evolutionary affinity between humans and gorillas than between humans and chimpanzees (Lucotte and Lefebvre, 1981). B. Haynes and his collaborators (Haynes *et al.*, 1983) suggested a trichotomy on the basis of monoclonal antibodies to T-lymphocytes, although with a recent common ancestor (5 Ma). The same conclusion can be drawn from individual papers on histocompatibility antigens (Gyllensten and Erlich, 1989) and by the characteristics of certain proteins in the membranes of erythrocytes of gorilla and orangutan (Nickells and Atkinson, 1990). There is a study of allotypes of immunoglobulin (inhibition of hemagglutination), moreover, which places hominoids in the traditional classification of three families (Dugoujon and Hasout, 1987).

Nevertheless, studies of hominoid affinity in terms of the properties of proteins continued in the 1980s and, in general, led more and more scientists to include all the great apes in the family of humans, especially the closest cluster of *Homo* and *Pan*.

Screening blood proteins by electrophoresis (32 loci) confirmed the data of the previous investigators regarding the genetic distance between humans and chimpanzees of 0.5706 and a calculated time of divergence of 2.86 Ma (Nozawa *et al.*, 1982). Polypeptides of fibroblasts and later isozymes and other polypeptides, determined by two-dimensional electrophoresis, demonstrated a maximum affinity of this pair and solved the problem of trichotomy, supporting the closest and most recent relationship of the pair *Homo-Pan* (Goodman *et al.*, 1987; Schmitt *et al.*, 1990). This opinion was confirmed with other diverse methods, including the indirect hemagglutination inhibition test, ELISA (Tsutsumi *et al.*, 1985, 1989), and monoclonal antibody methodology (Rearden, 1986; De Moor, 1986).

Data on the close affinity between humans and chimpanzees were obtained from the most diverse components of blood and tissue. In addition to those mentioned above, there were glycophorines, steroid-binding proteins, gonadotropins, low density lipoproteins, and many others. This affinity was confirmed again in blood groups by different criteria (Socha *et al.*, 1987; Redman *et al.*, 1989). Serum proteins of humans and chimpanzees are so similar, that when human serum was repeatedly injected into chimpanzees with Freund's adjuvant, protein antibodies were rarely produced (to only 1 or 2 antigens) (Prince, 1981). This investigator stated that 'biochemical and immunological data indicate that there are no special considerations leading to differentiation of these species' (Laboratory animals . . . , 1981, p. 9).

Concluding the section on the molecular biological foundations of evolution and taxonomy of the hominoids, after the analysis of different methods N. Saitou supposes that the most reliable approach to the problem seems to be use of the DNA-DNA hybridization data (Saitou, 1991b). It is noteworthy that they confirm the generally accepted hierarchy of primate classification from lemurs to humans. They also confirm the requirement of grouping the large hominoids in one family with humans, whose nearest relative is chimpanzee. Both these genera are closer to each other than species in one genus of drosophila, frogs, squirrels, rodents, horned cattle, or even gibbons. It is possible, of course, that *Homo* and *Pan* may be assigned to one genus (Goodman, 1996) (if this conformed to other data).

It is noteworthy that gorilla, as we have already seen, rather frequently demonstrates insignificant differences from the *Homo-Pan* pair (about 0.3%, at least by 9 polypeptides [Goodman *et al.*, 1983]), which hampers clear differentiation of the three. I also note the deceleration of evolution in the primates, especially hominoids, as determined by the molecular data and the important value of these data as the biological foundation of medical primatology.

What do they code? Evolution and classification of hominoids by paleontology and comparative anatomy

After the appearance of convincing molecular data showing that, regardless of how unshakable the morphological data demonstrating the close affinity of chimpanzee and gorilla, the former was nevertheless closer to humans than to gorilla, the journal *Science* presented the view of a scientific observer. If M. Goodman is right and, despite the general adaptation of *Pan* and *Gorilla* (such as for knuckle-walking), the closest relatives are *Pan* and *Homo*, 'it could strike the death knell of comparative anatomy' (Levin, 1987, p. 275). But this appears far from true. Comparative anatomy, the oldest methodology of phylogeny and taxonomy, objectively made an invaluable contribution to the new taxonomy of primates, including hominoids, to say nothing of the fact that it was the original basis for evaluating the latest molecular and genetic data in this field. Goodman, Sarich, Chiarelli, and many other biochemists and geneticists emphasized in their papers that their data agreed with the present day data of paleontology and morphology.

Nevertheless, as already mentioned above, some paleontologists and comparative anatomists argued against the conclusions of the molecular studies. The most reputed specialist, Louis Leakey (1970), said that the paleontological data were completely inconsistent with the conclusion of the later divergence of humans and the African hominoids. Leakey assumed that by 14 Ma–12 Ma, there already existed Hominidae to which he assigned *Kenyapithecus*, *Ramapithecus*, and the Pongidae, represented by the dryopithecines and *Proconsul*. Other morphologists expressed their disagreement with the biochemists in one form or another. The situation was aggravated by the fact that there is still a gap in the paleontology of the African hominoids from 13.5 Ma to 5 Ma (Shoshani *et al.*, 1996). Nevertheless, by the end of the 1970s, there appeared the first indications of agreement (unless we consider several of the *old* morphologists, such as W. Gregory) to include all the apes in the family of humans at a level of three subfamilies (Szalay and Delson, 1979).

I mentioned earlier that an important role in establishing the present day concept of the phylogeny and taxonomy of hominoids was played by the findings in the field of paleontology. Habilises extended the human line to 2 Ma–2.5 Ma, Hadar australopithecus changed the previous view regarding the time and significance of orthogrodity and the time of development of the brain throughout the history of humans, and finally, the new forms of *Ramapithecus* required a revaluation of their place in the evolution of hominoids. *Ramapithecus* (together with *Sivapithecus*, *Kenyapithecus*, and other close forms) appeared to be European-Asian relatives of the present day orangutan, whose ancestors diverged much earlier than the line of humans (possibly 2 times earlier, if we assume that the ancestor of the orangutan branched 12.5 Ma (Kappelman *et al.*, 1991). *Australopithecus*, that lived in Africa from 5 Ma to 1 Ma, is the oldest among the known representatives of the line, which also includes the ancestors of *Homo*. Species such as *H. habilis* likely descended mainly from this line in Africa. Later *H. erectus* appeared, a larger species whose representatives migrated from Africa to Eurasia not later than 1.5 Ma. It is likely that *H. sapiens* descended from *H. erectus* (with the present interpretation as subspecies, *H. sapiens neandertalensis*, that lived 130–135 thousand years ago and apparently already possessed speech, and *H. s. sapiens* that appeared 90–100 thousand years ago in Africa). This is an assumption of modern paleontologists and morphologists (Brauer and Rimback, 1990; Schmidt, 1991; Shoshani *et al.*, 1996).

Let us return to the source of the hominoids. It is supposed that the ancestor of the gibbons was the first to split from the common stem (20 Ma or earlier [Pilbeam, 1985]). The majority of morphologists agree with this position and it is confirmed by the finding of the common ancestor of humans and great apes with an estimate of 17 Ma–18 Ma (Andrews *et al.*, 1987; Walker and Teaford, 1989). The quadrupedal *Proconsul*, long known to paleontologists, is likely the most recent common ancestor of humans and large hominoids.

In the 1970s–1980s, the attention of scientists was attracted by *Lucy, Australopithecus afarensis*, a hominoid adapted to bipedal locomotion but at the same time, capable of climbing trees (Johanson and White, 1979). Its height is estimated as 106 cm–115 cm (although there were massive *Australopithecus* of 127 cm–176 cm tall), but its brain hardly exceeded that of the chimpanzee (404 ml compared to 383 ml cranial capacity)

(Smith and Topkins, 1995). This was a primate that combined the properties of humans (pelvis, femoral bones, foot, iliac bones [Latimer *et al.*, 1990]) and African apes, especially the pygmy chimpanzee (size of the cerebrum, postcranial skeleton [Zihlman *et al.*, 1979, 1990], cranium [Schoenemann, 1989]). I cannot give all the phylogenetic estimates for *Lucy* and the other australopithecines (ancestors of humans and chimpanzees, ancestors of humans and the chimpanzee-gorilla line, descendants of the common ancestor of all three, etc.), but many authors claim that they are very close both to humans and the living African hominoids. This should find its reflection in taxonomy.

In the 1980s, many paleontologists and anatomists agreed with the concept of a later divergence of the human and the ancestor of the African apes (less than 8 Ma) (Wolpoff, 1982; Pilbeam, 1984; R. Leakey, 1985; and others). This was confirmed by a multitude of comparative anatomical studies. R. Ciochon and R. Corruccini (1982), who noticed 'chimpanzee-like' properties in *Australopithecus*, included *Homo*, *Pan*, and *Gorilla* in a common clade (Ciochon, 1983). In conjunction with an assessment of the position of *Ramapithecus*, R. Kay (1982) combined *Australopithecus* with chimpanzee and gorilla, but not with orangutan. The internal structure of the iliac and femoral bones and the sacrum appeared surprisingly similar in humans and *Australopithecus*. The nearest relatives of the latter were modern chimpanzees, to a lesser extent gorillas, and more distantly to orangutans and gibbons (Suzman, 1981). Y. Deloison (1985) considered the plantigrade locomotion of *Lucy* to be intermediate between human and pongid motion. Confirmation of the affinity of locomotion in humans and apes, carried out following A. Schultz (1966), is especially interesting in the light of bipedal primate ancestors of humans, and in connection with the hundred year division of the order Primates into Bimana and Quadrumana, that was corrected by Thomas Huxley (1863) (see Chapter 1 of this book). It was shown that, regardless of some differences, there is much in common between *Homo* and *Pan* (Prost, 1979; Yamazaki and Ishida, 1986), which supports the hypothesis that a brachiating ancestor preceded bipedal locomotion (Yurovskaya, 1983, 1991). Tomographic study of the ulnar joint of *Pan* and *Homo* established morphological differences between these forms only at the level of the two genera (Senut and Le Floch-Pringent, 1982). I already mentioned the specific anatomical affinity of *Lucy* (and humans) with bonobos (McHenry, 1984; Zihlman, 1984).

These data gave D. Pilbeam (1984) grounds to state, 'Previous debates between anthropologists and molecular biologists on the problem of the character and time of the evolution of hominoids are practically settled at present. The majority of paleontologists (and specialists in the field of comparative anatomy) reached an agreement regarding the conclusions made on the basis of analysis of molecular structures indicating a striking genetic affinity of humans with extant African anthropoids and its almost two times greater difference from Asian apes' (p. 40, from the Russian translation). Nevertheless, the pressing problem of anthropology remained. Which of the two species of extant primates is the closest relative of humans, *Pan* or *Gorilla*?

Individual morphological studies that placed humans and chimpanzees together had appeared in the past, but they gained special attention in relation to the conclusions made by molecular biologists. Significant affinity of modern humans and chimpanzees was found in studies on certain dimensions of hominoid skulls (Krukoff, 1971). Humans and chimpanzees are closer to each other by characteristic properties of the skin than each of them is to gorillas (Grant and Hoff, 1975). R. Corrucini (1982) found a special taxonomic relationship between *Homo* and *Pan* in the composition of dentine and enamel in odontogenesis of various species of extant primates. A multivariate cluster analysis of the odontometric data not only demonstrated the necessity of including humans and all apes in one family, but it also showed the maximum affinity of *Homo* and *Pan*, even when compared to gorilla (Mahaney and Sciulli, 1983).

The year 1986 appeared extremely fruitful for obtaining information about the evolution and comparative anatomy of hominoids. A series of prominent studies in this field appeared as a publication of several collected articles. In 1986, the paleontologist David Pilbeam published a series of papers which, although based on molecular data, nevertheless grouped the hominids with African apes, but 'closer to chimpanzee than to gorilla' (Kelley and Pilbeam, 1986, p. 402). In an oral presentation, Pilbeam (1986) sympathetically gave the time of the divergence of *Homo-Pan* as 5 Ma, whereas he accepted the divergence of gorilla as much earlier (8 Ma) (p. 307). He pointed out the necessity of taking into account homoplasy and true homology, differentiating symplesiomorphy from synapomorphy (Pilbeam, 1985).

L. Martin (1986) presented a differential analysis of morphological characteristics of the hominoids (122 cranial, dental, postcranial, and tissue characteristics), inherited from the common ancestor, and the common characteristics of the catarrhines. The characteristics were further analyzed as to whether they were characteristic of the ancestral (*Australopithecus* and *Sivapithecus*) or the extant forms (human, chimpanzee, gorilla, orangutan). The characteristics and their phylogenetic analysis are of great value, even taken alone, as one of the foundations of medical primatology. In addition, the author determined the evolutionary and taxonomic status of humans and their closest extant relatives. As a result, humans and great apes were included in one family, which provided valuable support for the molecular data, but did not find significant grounds, however, for especially combining *Homo* and *Pan*.

The paleontologist D. Gantt (1986) supported the data of the biochemists, which strengthen the inclusion of the two chimpanzees and gorilla in one family with human. Following an analysis of a large amount of paleontological and modern data on tooth enamel, he expressed his opinion against grouping all the great apes together, since the former three genera are apparently closer to each other than they are to the orangutan.

We find rich information on the anatomy, taxonomy, and evolution of the order Primates, and on hominoids, in particular, in the frequently-mentioned paper of J. H. Schwartz (1986). It is difficult to overestimate the data and theoretical research of this work in establishing the biological background for medical primatology. Nevertheless, Schwartz, in his previous publications, maintained the concept of a maximum affinity of humans and orangutans among the hominoids (I emphasize once more that this was done on the basis of a vast data set, such that it would not be difficult to find a similar source of information about the affinity of humans and orangutans in the scientific literature). This contradicts the entire assemblage of modern knowledge, both paleontological and neoontological, although the author sometimes provided interesting and even highly detailed data (for example, on the reproductive system). In addition, I note as already mentioned, that Schwartz argued against the conclusions of the biochemists. Criticism of Schwartz's concept in the scientific press, therefore, seems quite reasonable (Groves, 1987; Susman, 1989).

Colin P. Groves, a reputed specialist in comparative anatomy of the primates, published a fundamental work on systematics of the hominoids 'Systematics of the Great Apes' (1986). In this paper on comparative anatomy of the hominoids, accompanied by more than 10 tables and appendices, the author conducted a cladistic analysis using his own data and data in the literature (considering inherited and acquired characters, presenting taxonomic interpretations, and synonymy of taxa). No primatologist or researcher in this field should fail to read this paper. It seems that this is the first morphological study, scientifically based on the classical style, in which the great apes are not only included in one family with humans, but also in the same subfamily, Homininae. Although *Homo*, *Pan*, and *Gorilla* are equidistantly located within the sub-family, the author repeatedly (pp. 190, 191, 205) refers to the peculiar affinity of human and chimpanzee by comparative morphological data. The *Homo-Pan* pair have a maxi-mum number of acquired and inherited characters, 25 compared to the other hominoids (12 in *Homo-Gorilla*, and 10 in *Homo-Pongo*). Groves corrects his previous assumption (1970b) about one genus status for gorilla and chimpanzee, favoring the greater affinity of humans and chimpanzees, although he does this with certain provisions.

Groves touches upon one of the trickiest problems in comparing the morphology of humans and great apes, the characteristic use of 'knuckle-walking' by the latter. He admitted that this capability could have been lost in the process of evolution of the intermediate human ancestors. He presented data from the literature about osteological characters of the wrist associated with the locomotion of the fossil ancestors of *H. sapiens* on their knuckles. His conclusions are also based on the data of molecular and behavioral research. According to Groves, *Pongo* forms a separate subfamily, Ponginae, and all *Hylobates* are moved to a special family, Hylobatidae (p. 192). The conclusions of this study for the first time essentially coincide with the data of the molecular biologists, and the results of the research became one of the foundations of another prominent work on morphology and systematics of hominoids (Shoshani *et al.*, 1996), to which I refer below.

In his articles and monographs, Colin Groves continued to propose the affinity of humans and African apes (1989) and the inclusion of orangutan in the same family at the level of a subfamily. As we have seen, although he assumed a special relationship, *Homo-Pan*, he did not further define the three separate tribes in the subfamily, Homininae (1991). By the end of the 1980s, he wrote that there were no contradictions in uniting molecular, genetic, and anatomical data. M. Goodman's model, presented in 1963, was confirmed, 'It is likely that the wheel made a full circle' (Groves, 1989, p. 36). This was correct regarding the admission of the recent common ancestor of *Homo-Pan Gorilla* by morphologists. The discussion was not complete *in detail*, however, and it was considered to be one of the most intriguing ones (Diamond, 1988). Professor J. Marks (1988), who argued most vehemently, called the conclusions of molecular biologists 'absurd' and criticized their 'lack of modesty' (p. 656). The morphologists themselves have not yet formulated a clear concept for distinguishing the three genera. There were studies which defined the superiority of *Gorilla* as the closest to human (Sarmiento, 1988). The conflicting nature of the morphological results on this problem was noted even later (Barriel *et al.*, 1993), and it is likely that this was determined by nature itself.

The work of prominent present day specialists in morphology and paleontology, as was mentioned above, significantly clarifies the problem of hominoid phylogeny. In my opinion, this work concludes a 50-year (if not a thousand year) controversy on one of the most burning problems of science, the origin and affinity of humans and their closest extant relatives. The case in point is the publication, 'Primate Phylogeny: Morphological vs. Molecular Results' (1996) by J. Shoshani, C. Groves, E. Simmons, and G. Gunnell. This article was repeatedly cited in my presentation regarding the different groups of primates and their evolution. Now we are interested in the same problem with respect to the hominoids themselves. Regardless of the importance of this study to the biology of the entire order Primates, its main strength lies in the phylogeny of the hominoids.

The authors used two computer programs to carry out a cladistic analysis of 264 morphological characters (osteology, soft tissue, external anatomy) in 18 extant genera of primates, representing all the main groups of the order and 4 external groups (outgroups) of other mammals. Both of the programs were developed in the 1990s, based on the parsimony method and the capacity for assessing not only the phylogenetic weight function of the characteristic, and its level of primitiveness and true inherited properties, but also its quality (morphological, functional, and combined). The programs provide precise calculation of confidence limits in the form of bootstrap values (BSV) in percent. They use true synapomporphies, thereby avoiding primitive characters while making quantitative estimations of evolutionary changes and testing different combinations of forms and outlining the necessary number of extra steps. Classification carried out on the basis of such a powerful technological foundation deserves credit, considering that the input for the system included first class data (of Groves, Shoshani, and 65 characters chosen from the literature).

Following the analysis of recent data on fossil primates and their ancestors, these investigators presented an analysis of the extant groups of primates. They used data for each clade on the number of characters accepted as synapomorphies to analyze the level of authenticity. The total list of 264 characters is presented at the end of the work as Appendix 2. Detailed classification of extinct and extant primates up to the level of the genus is presented with full scientific documentation as Appendix 1.

The authors give a detailed comparison of the morphological and molecular data and they conclude that they found almost complete correspondence of results. Minor discrepancies relate only to the closest relatives of the order Primates, possible polytomy of the Platyrrhini groups; in other words, those subjects for which the data were not completely clear. I also note that unlike the results of M. Goodman, whose data the authors generally use for comparison, gibbons are not included in the Hominidae, but are placed separately in an independent family as was done previously. All of the remaining divisions are practically the same, although with a different level of confidence!

The cladogram based on morphology is similar to the one constructed on the basis of the molecular data. I present it as a contemporary version of relationships among relatives in the order Primates (Figure 3.25).

The grouping of the order Primates was substantiated with a high degree of confidence on the basis of the analysis of synapomorphies (BSV equals 98%). Significant support was found for Haplorhini (BSV = 99%), Anthropoidea (100%) and other taxa, whereas, both for Strepsirhini and for Lemuroidea, there was poor support (45% for each). The Hominoidea clade, as well as the family Hominidae and subfamily Homininae (*Homo*, *Pan*, *Gorilla*), received rather significant support; the BSV was 99%, 87%, and 99%, respectively. The Hominini clade was obtained for the first time on the basis of morphology. In a taxonomic sense, this means that humans and chimpanzees are included in one tribe (!). Although this received comparatively weak support (the BSV is 42%, which is almost as much as Strepsirhini and Lemuroidea), it was justified in another respect. When calculating branch-swapping within the Primates, the pair *Homo-Pan* (604 steps) did not require any extra steps, whereas the trichotomy *Gorilla*, *Pan*, and *Homo* would require 15 extra steps. That would be 'very costly' (p. 121), according to the authors and is thus, unacceptable. The first publication on objective computer analysis of morphological characteristics using the method of maximum parsimony with a very representative data matrix, therefore, clearly supported the molecular data. Those data seemed nonsensical for 35 years, at least those data which refer to the phylogeny

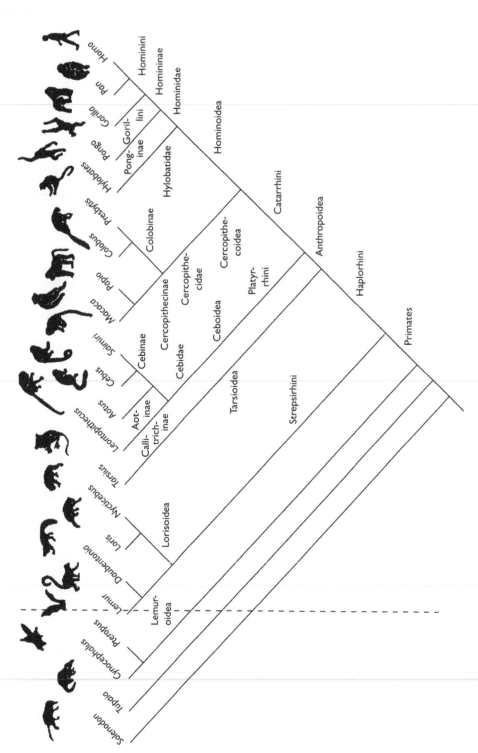

Figure 3.25 Cladogram of the phylogeny of the order Primates based on morphological data (Soshani et al., 1996). Reproduced with permission.

and systematics of the order Primates, including the most mysterious taxon, Hominoidea, which comprises humans and their nearest extant relatives. In general, the other taxa also correspond. The closest to the human is chimpanzee, gorilla is somewhat more distant, and still further distant is orangutan, and finally gibbon. It is my opinion that this is a reasonably acceptable scheme of taxonomy for the extant hominoids (Shoshani *et al.*, 1996).

Superfamily Hominoidea
    Family Hylobatidae –   *Hylobates*
    Family Hominidae
        Subfamily Ponginae –   *Pongo*
        Subfamily Homininae
            Tribe Gorillini – *Gorilla*
            Tribe Hominini – *Pan, Homo*

The authors took a phrase from *Science* as an epigraph to their article, 'If Morris Goodman is correct in his conclusion, we will just have to go back to the anatomical evidence and find out what we've been missing' (Levin, 1987, p. 273). At the end of the article, they emphasize the productivity of combining different approaches to study the same phylogeny, 'It has taken about 100 years, since the days of Nutall (1901, 1904), for morphologists and molecular biologists to generate sufficient bodies of data on hominid relationships for us to realise that these are generally congruent. Disagreements are minor and are usually with regard to the precise relationships and classifications of hominid taxa' (p. 122). The latter likely refers to the gibbon family and subfamily or tribe, rather than the subtribe for *Pan* and *Homo* formulated by M. Goodman.

The simultaneous appearance of these review papers in 1996 is a remarkable illustration of the coincidence of maturing scientific ideas; M. Goodman, J. Shoshani and colleagues, and D. Pilbeam all carried out their research on the same basic problem.

There was no miracle; the gene codes the phene and the DNA determines each material character of the organism. The wheel really has made a complete circle.

New and remarkable finds in Africa (*Ardipithecus ramidus*, *Ardipithecus anamensis*, *Kenianthropus platiops*, *Orrorin tugensis* and others) made by Y. Haile Selassie, M. Pikford, B. Semu and M. Leakey (with colleagues) confirm the genealogical relationships among hominids as described above.

Other characters

This subsection will be the shortest, because the next chapter deals with the analysis of properties which characterize the affinity of humans and other primates. Here, I want to demonstrate only those characters which shed some light on the evolution and systematics of the hominoids.

Let us start with the brain. In the opinion of the primatologists of the old generation, who maintained that 'nothing significant is absent' in the morphology of the brain of the highest simians that is characteristic of the human brain (Weber, 1936, p. 215), I should add specific information about the significant similarity of the neocortex in the highest primates. In 1936, it was found that chimpanzees have *special* human fields 39 and 40 in the inferoparietal area (lower parietal) of the cerebral cortex (Schevtchenko, 1971). Structures that had been considered specifically *human*, moreover, such as supra-temporal field 37 (Blinkov, 1955), fields 44 and 45 of the basal frontal cortex, and evolutionarily new structures of the frontal area, fields 46 and 10 (Kotchetkova, 1962, 1973), were found in chimpanzees. It is also well-known that there are zones homologous with the center of speech in the brain of simians, 'Broca's area' (Lieberman, 1985).

'The modern anatomists state that the brain of gorilla and chimpanzee (at the beginning of 1970 it was thought that gorilla was "closer" to human than chimpanzee – E. F.) is similar to the human, not only in general shape and location of sulci and convolutions, but also in the location of the architectonic systems of the cortex of the large cerebral hemispheres' (Schevtchenko, 1971, p. 168). The human brain is approximately three times as large as the one in chimpanzee, but the calculations indicate that if it was equal to the human by weight, it would have the same relationship between its parts as found in the human (Dagosto, 1984). It is now known that the level of neuronal typology is revealed in the brain inequality amongst the great hominids (including humans), as was demonstrated for cingulate cortex and for other formations (Hof et al., 1998; Nimchinsky et al., 1999).

The unique similarity of brain in the hominoids should be considered when reviewing the striking facts about the behavior and intellectual abilities of chimpanzees and other apes. Starting at the beginning of the 1950s, S. Washburn (1982) stated that there is a specific similarity between chimpanzee and human behavior. Over the past several years, it was found that chimpanzees not only have the capability of using tools (sticks, branches, leaves, sponges, stones), but they also were found to make tools and to remember the locations of those tools (van Lawick-Goodall, 1968; Firsov, 1982; McGrew, 1992; Boesch-Achrmann and Boesch, 1994; Tonooka, 1996). At present, we can validly speak of a 'chimpanzee culture' (Wrangham et al., eds., 1994; Whiten et al., 1999; de Waal, 2001), a concept that was the subject of a special symposium of the XVIth Congress of the International Primatological Society (1996).

A high level of social life has been observed in the great apes, especially in the pygmy chimpanzee, whose group organization could be hardly defined as a group, because it is more like an open society (Itani and Suzuki, 1967; Power, 1991; de Waal, 1995). Comparative sociobiology suggests that not very long ago the two species of chimpanzee and humans had a common ancestor (Ghiglieri, 1987). Chimpanzees have considerable ability for abstraction, self-recognition in a mirror (this ability is typical of the other great apes in one way or another, but it is beyond the abilities of very young children and adults with certain mental diseases), distinguishing groups of objects, categorizing objects of different modalities, and recognizing familiar faces and objects in photographs (Gallup, 1970, 1994; Boysen and Berntson, 1986; Mitchell, 1996). The linguistic competence of chimpanzees, especially the pygmy chimpanzee, and gorilla, was a striking finding in nonhuman creatures (Gardner and Gardner, 1972; Patterson, 1983; Premack, 1984; Savage-Rumbaugh, 1986, Rumbaugh, 1995). If we compare the summarized data from these apes with the human data, we find that the intellect of these hominoids, and primarily that of the pygmy chimpanzee, is equivalent to the intellect of a competent child of 2–3 years (Savage-Rumbaugh and Rumbaugh, 1996).

The social relations of chimpanzees influence the level of neurotransmitters (catecholamines and their derivatives), similarly to the same processes in humans (Kraemer, 1985). The decrease in the level of catecholamines in old simians (which happens in Alzheimer's disease) leads to a decrease in cognitive function similar to the cognitive deficit of aging people (Arnsten and Goldman-Rakic, 1985).

The tendency to bipedalism (ignoring the convergent bipedalism of gibbons), ventroventral (face-to-face) copulation by bonobos, and a series of neurological properties can be related to the characteristics of affinity between humans and chimpanzees (Thompson-Handler *et al.*, 1984; Alexander, 1992). It is interesting to note that the only species for which there were found to be no differences in taste perception of sweetness compared to humans was the chimpanzee (Hellekant and Danilova, 1996). As early as the 1960s, the data of parasitology (malaria, ectoparasites, and helminths) indicated a greater affinity of humans with chimpanzee and gorilla than with the Asian apes (Dunn, 1967).

Other data on the affinity of humans and simians is discussed in the next chapter from the perspective of neurology, hematology, immunology, reproduction, ophthalmology, pathology, pharmacology, etc. We shall see that the results in these areas agree completely with those given above concerning the affinity of *H. sapiens* with other primates. Additionally, the data reinforce the foundations of the modern taxonomy of the order. Medical primatology has shown that there are specific human diseases which are even impossible to reproduce in the more commonly studied simians, but they are adequately simulated in chimpanzee. Such diseases include hepatitis B, gonorrhea, Alzheimer's disease, AIDS, the Herstmann-Straissler syndrome, and others.

Some historical parallels; conclusion to the section

In one generation, there were changes in extremely important paradigms of science. The 'dichotomy' of Aristotle and Descartes (which according to V. & J. Reynolds [1995] held a strong position and was supported ideologically), which postulated an impassable gap between humans and the rest of the animal world, collapsed. One of the main initiators of this event, M. Goodman, thought that the foresight of C. Darwin, regarding objective genetic criteria of taxonomy, had become a reality, and that now we can see without emotion 'how we view ourselves' (1996, p. 282). Biologically, we are very close to those whom we consider animals, but we are no doubt closer to some than to others. The simians and especially the great apes, are our closest relatives. If we calculate this relationship in numerical indices, it reaches a value of 99% by DNA in chimpanzee and even in gorilla. For tens of millions of years we have been in the same evolutionary branch with other primates, 15 million years with large hominoids, but 'only' 5 or 6 million years ago with chimpanzee. In my opinion, the irreconcilable controversy with the 'immoral' biochemists has come to an end. We managed to accomplish this without the passion of inquisition, excommunication, 'monkey trials', or exile to Siberia.

It is quite interesting, that if we carefully analyze the history of science, we can easily see that in the past, before humans were alienated from their relatives in the order Primates, the concept that humans were close to the apes was not offensive. I do not refer only to ignorant and uneducated people who were impressed by chimpanzees that embraced and even kissed each other when they met, or to such trouble-makers with the public grandeur of the French Enlighteners. I refer to actual biologists whom I consider prominent scientists, and 'leniently' forgive them their former ignorance and 'individual mistakes'.

The first scientific work about primates, carried out by a prominent physician of the time, Professor Nicholas Tulp (1641), reported an anatomical similarity of humans with a creature 'who is as high as a three-year-old boy' (Tulp listed the characteristics of this similarity). It can be seen from the name (he called him not only *Satyrus Indicus*, but also *Homo Sylvestris* and *Ourang-Outang*, that Tulp fully understood that *Homo* refers to humans and *Ourang-Outang* means 'man of the forest.' Even the strict scientist, Edward Tyson, M. D. (1699) could not resist naming his 'pygmy chimpanzee' with the Latin term, *Homo Sylvestris.*

We know that the concept of '*genera*,' introduced by C. Linnaeus, united modern large taxa. Nevertheless, there are indications that the prominent classifier of nature, who was afraid of clashing with official dogma (Stankov, 1958; Fridman, 1978), was especially careful in his treatment of the genus *Homo.* He experienced difficulty defining the taxonomy of humans precisely due to their close affinity with relative forms. This is apparent from his 'Laplandia Diary' (the entry of July 11 1732 [Fulton, 1941]) and from his letter to Gmelin on February 14 1747. 'It is not desirable that I locate humans among *Anthropomorpha* . . . If I had called humans as apes or visa versa, I would be attacked by all theologians. Maybe I still have to do this in accordance with science' (Shaparenko, 1935, p. 156). Nevertheless, Linnaeus did not assign *Anthropomorpha* to the genus *Simia.* Rather, he introduced two new species into his system, *Homo sapience* and *Homo troglodytes.* The latter was related to orangutan. The 10th edition of *Systema Naturae* used the terms, *Homo Nocturnus, Homo Sylvestris* orangoutang, and the expression of Linnaeus, 'wild people of the forest' (although I emphasize that Linnaeus did not consider this similarity to be biological affinity).

Thomas Huxley discussed systematics and the affinity of humans and the highest simians. 'From the point of view of taxonomy the cerebral differences between humans and simians have meaning which is no greater than the level of the genus (! – E. F.); the family differences are based primarily on the system of teeth, pelvis, and hind limb structure' (Huxley, 1864, p. 117). When Charles Darwin, convinced that 'true classification has a genealogical tree structure' and 'the natural system has a genealogical tree structure' (Darwin [1859] 1864, pp. 332, 355), was considering this problem following Huxley's statement, he said, as mentioned, that 'humans should comprise only a special family and maybe even only a subfamily' (Darwin, 1871, p. 219). In the 1870s, George Zeidlitz, professor of anatomy of Derpt's University, considered humans and apes to be only different genera, based on the structure of their hind limbs (Bljacher, 1971).

Beginning in the early years of our century, the most prominent primatologists discussed the special affinity of the African apes to humans (Weber, 1904, 1936; Gregory, 1916). Nevertheless, ideological collisions that excited society after the discovery of australopithecus in 1924, attacks by fundamentalists, and other factors had a negative effect on the forming of hominoid classification. The already known relationship of similarity in anatomy, blood chemistry, and diseases of humans and apes required them to be grouped more closely. A prominent anthropologist, Elliot-Smith (1924, p. 125) wrote, 'Negation of the validity of the data about their close affinity is equivalent to and admission of the complete uselessness of the facts of comparative anatomy as an indicator of genetic correlation and return to obscurantism of the Dark Ages.' Within a year, Robert Mearns Yerkes published a book whose name alone revealed the attitude of the author with respect to the position of the chimpanzee, *Almost Human* (1925). Nevertheless, the best that primatologists could accomplish by the end of the 1920s was to assign hominoids, including humans, to one superfamily.

I here present an example of the logic of the specialists in 1930s. G. Simpson was informed that G. Wilder 'fearlessly united' *Homo* and the apes in a common Hominidae family, and that W. Gregory supported this classification (Gregory and Hellman, 1939). Simpson paid high tribute to Gregory's scientific work (in 1971 he devoted an article to his memory) and cited 'his point of view as the most trustworthy among all present day ones.' Simpson wrote that on the basis of their common origin and common features 'such as teeth,' this 'union seems warranted, when characters are analysed objectively' (Simpson, 1945, p. 188). Simpson, however, also objected to this union for two reasons; 1) thinking, which in Simpson's opinion is also a zoological characteristic, should be considered, placing humans a bit more distantly, so as 'not to depreciate our own significance,' and 2) 'there is not a single chance that zoologists and teachers, generally convinced of the close affinity of humans and apes, would agree with the didactic and practical application of such a family' (Ibid.). The second argument has no scientific content; it is related to the ideological and psychological atmosphere in a specific country. As for the first argument, if we agree that *thinking* is a zoological characteristic, then *Homo habilis*, whose thinking is quite inferior to that of *Homo sapiens*, is included in the same genus with the latter. *Australopithecus*, which is not very distant from chimpanzee when judged by the size of the brain, is also placed taxonomically in one subfamily with humans (Shoshani *et al.*, 1996, p. 128). It is clear that Simpson did not know, in 1945, the striking intellectual capabilities of the apes, which were not convincingly demonstrated until the 1960s–1980s. As a result of these discoveries, in the opinion of neuropsychologists, the boundary between the cognitive abilities of human and chimpanzee has been more and more dissolved (Ettlinger, 1984, p. 685).

I have repeatedly emphasized (Fridman, 1989, 1991) that in the past 30–35 years, since the discoveries of L. and M. Leakey, the genealogy of the line *Homo* has *gotten older and older* (from 500 thousand years to 3 million years), and the point of divergence of the lines of humans and apes *more closely approached our time* (from 30 Ma or more to 5–6 Ma). Thus, *graphically*, the distance between them on the line of anthropogenesis has *significantly decreased*. This mathematical calculation implies that, at present, we have indeed much more evolutionary grounds regarding the biological affinity of humans and other primates than we did 35 years ago.

Deceleration in the evolution of DNA and of biochemical properties as a whole, at the stage of the hominoids, and especially, at the stage of *Homo*, indicates exactly the process of transformation of the extinct apes into humans. This deceleration was repeatedly mentioned by M. Goodman, was noted by his teacher, Morris Wilson, in the study of serum albumin (Goodman, 1996), and was confirmed by various other scientists. This deceleration reflects the main strategy of evolution in the 'Life History of the Hominidae – Human Strategy,' 'Live slow, die old' (Smith and Tompkins, 1995, p. 261), and was discussed by moralists, philosophers, and anatomists. Moreover, it is associated with the deceleration in maturation of the brain, which developed in parallel to the development of orthogrady, walking in the upright position. The bones of the lesser pelvis of females drew closer in the course of this process. The fetus, with a large brain and skull, was doomed to remain fetal forever and perish, if it did not accommodate itself to the new conditions. It did accommodate itself; it began to appear with a soft, flexible skull, with un-knitted sutures in the skull, and what is most important, with an *immature* human brain. The time needed for the human brain to mature is approximately 20 years. A relatively long time is precisely what is required to master the vast human 'social program' (Dubinin, 1974), which is most likely not encoded in any genes, and to learn everything *human*, including articulate speech. Only after birth does a human baby gradually acquire personality by non-physical inheritance of the 'social program'.

A series of new concepts regarding the peculiarities of anthropogenesis follows from the previous discussion of the evolutionary aspect. On the basis of some discoveries about the behavior of chimpanzees, it is likely during evolution that labor preceded bipedal walking. Orthogradity appeared earlier than the development of a large brain and articulate speech, but the basics of gestural language were formed very long ago, at a stage close to chimpanzee. In defiance of the critics of Darwin, who stated that there are no intermediate forms between simians and humans among fossilized discoveries (each is related, supposedly, either to apes or *Homo*), we now know that gradual changes occurred in brain size and skull shape, from the highest simians through australopithecines, early skilled humans (*H. habilis*), archanthropes (*H. erectus*), and paleoanthropes (*H. neandertalensis*) to present humans (*H. sapiens*) (Vark *et al.*, 1990). Darwin was correct in supporting cladistic, phylogenetic foundations of taxonomy, in his suggestion that the closest relatives of humans are present day chimpanzee and gorilla, and in his proposal that the main processes of anthropogenesis occurred in Africa. He was also right in his statement that there are few, if any, qualitative differences in the biology and behavior of humans and apes; the differences are mainly only quantitative. The only qualitative difference is articulate speech, that complicated, complex system which is an essential component of human social activity. This activity fundamentally differentiates humans from other present day hominoids. But even in this area, 'the difference between chimpanzees and humans is one of degree' and contemporary data are 'consistent with the Darwinian notion of continuity – that we are all relatives' (Fouts and Fouts, 1999, p. 256), which with 'our fellow apes just happen to be the next of kin in our phylogenetic family' (Ibid.).

As we have seen, the closest relatives of humans are the two species of chimpanzee. It is likely that the nearest relative is the pygmy chimpanzee or bonobo (*P. paniscus*). According to the comparative data of morphology and molecular biology, these two species are not only in one family and one subfamily with humans, but they are also in one tribe (related genera) and even in one subtribe with humans. Gorilla is somewhat more distant from these three species. In my opinion, the controversy regarding its proximity to humans is based only on its great similarity to them and few differences. Unquestionably, gorilla is also a very close extant relative of humans. Taxonomically it represents one genus in a separate tribe in the subfamily of humans. Orangutan, which is more distant in its genealogy from the previous quartet, nevertheless, appears to be very close at the level of a separate subfamily in the same family. Classification of the last present day genus of hominoids, the gibbon and siamang, created some disagreement, but not a major one. Biochemists (and not only they) consider the *Hylobates* to be a subfamily within one family with the others (Goodman *et al.*, 1998), whereas morphologists place them in a separate family. I retain it (temporarily) in a separate family, based on the following three reasons: significantly earlier evolutionary separation from the large hominoids, significant differences in karyotypes, and dissimilarity in anatomical attributes.

At the same time we must remember that one or another hominoid may appear closer to human than chimpanzee on the basis of individual characteristics. Orangutan, for example, is the only ape, in which the female, like women, lacks a cyclic genital swelling associated with the menstrual cycle (G. Gallup [1997] thinks that the intellect of orangutan is closer to that of human than are the other apes). The gestation period of orangutan, on average, is also similar in length to the human (Schwartz, 1986). As was mentioned above, individual biochemical characteristics may indicate a closer relationship between gorilla and human. I attribute these contradictions to the principle of 'mosaic evolution' (Schwartz, 1967), by which specific individual characteristics may become apparent in descendants of different forms in one cluster in the absence of a strict correlation with the main line of phylogenetic development.

To summarize the discussion, my concept of the taxonomy of the present day hominoids is shown diagrammatically below (from Fridman, 1989) (omitting the levels of tribe and subtribe).

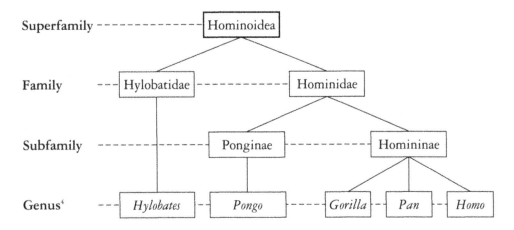

I now continue and finish the taxonomy of the order Primates.

Family Hylobatidae

This family includes gibbons and siamangs. The classification of hylobatides is rather contradictory and depends on the individual scientist, a situation which results from their really 'contradictory' position in biology. On the one hand, these are small primates (40–65 cm long and 5–7 kg and 11–12 kg [for siamang] body weight), which are similar in size to catarrhine monkeys, especially to colobines. Some of the following anatomical and genetic characteristics allow us to relate them to Cercopithecoidea: ischial callosities (sitting pads – though not very large), a small brain volume (100–150 cm^3), similarity in visual areas of the cerebral cortex with those of the lower simians, some skeletal features, and the number and morphology of chromosomes (44 for the majority of species). On the other hand, the absence of an outer tail and cheek pouches, anatomy of the hand and shoulder, sternum and ribs, location of the scapulas, morphology of the molars, antigens of erythrocytes (ABO blood group),

4 In the aforementioned article (Goodman *et al.*, 1998), primate cladograms were presented based on DNA evidence and on fossil osteological evidence. All living *Hylobates*, *Pongo*, *Gorilla*, *Pan*, and *Homo* are assigned to one family Hominidae (which belongs to the superfamily Cercopithecoidea) and to one subfamily Homininae, which includes both the tribe Hominini and the tribe Hylobatini (gibbons and siamang). Hominini is divided into subtribes Pongina for *Pongo* and subtribe Hominina for *Gorilla* and *Homo*. Genus *Homo* itself is divided into two subgenera *H. (Homo)* for human and *H. (Pan)* for chimpanzee and bonobo. This 'provisional age-related phylogenetic classification' is not decided for taxonomy of Hominoidea in the present book. Nevertheless, we should turn our attention especially to the phylogenetically closest of all contemporary catarrhine monkeys and apes with human and of the unique relationship of the sister lineages human and chimpanzee.

structure of proteins and DNA molecules, reveals that gibbons and siamangs are true hominoids. In the opinion of M. Goodman (1996), as mentioned above, it is even possible that they are hominids. These were the reasons that they were classified either as a subfamily of the lower catarrhines, a subfamily of the anthropoids, or as a separate superfamily. As such, their taxonomic position between higher and lower simians does not mean that they serve as a 'bridge' between them. The data on fossil hylobatides are poor. Among their ancient ancestors or relatives, some authors list *Propliopithecus*, which is not indisputable, *Pliopithecus*, and as was already mentioned, *Dendropithecus*, *Micropithecus*, and *Limnopithecus*.

Colin Groves has made a significant contribution to the study and classification of hylobatides during our time (1982, 1984, 1988; Shoshani *et al.*, 1996). It is most expedient to consider these hominoids as one *genus Hylobates* with 9 species (the number of subspecies is about 15–19) (Marshall and Sugardjito, 1986; Mootnick and Haimoff, 1986; Groves, 1988). As such, they may be grouped into 4 subgenera: *Symphalangus* (the siamang) (*Hylobates syndactylus*); *Nomascus*, depending on the subspecies, may be called monotonous, black, or white-cheeked gibbon (*H. concolor*). It is possible that there are three species (Groves, 1988); *Bunopithecus* (hoolock gibbon) (*H. hoolock*); and *Hylobates* proper with 6 species – lar or white-handed gibbon (*H. lar* [Figure 3.26]), agile gibbon (*H. agilis*), Mueller's or gray gibbon (*H. muelleri*), moloch, Javan, or silvery gibbon (*H. moloch*), pileated or capped gibbon (*H. pileatus*), and Kloss or Mentawai gibbon (*H. klossii*). The monotonous gibbon deviates most from the group with its

Figure 3.26 Hylobates lar

pappus crown, 52 chromosomes, gray newborn (which become black during the first year of life), and females that become buff at sexual maturity (puberty) (other species become golden, brown, or gray; sexual and age dichromatism is characteristic of several species). Siamang is the most noticeable. It is a larger animal with two toes grown together down to the nails, with a guttural vocal sac for resonating sounds, with 50 chromosomes, and sometimes with a sixth finger or toe (polydactylism). I should also acknowledge the uniqueness of morphology of the hoolock gibbon and the fact that it has 38 chromosomes (Prouty *et al.*, 1983). It is interesting that the famous singing of gibbons, unique among mammals (territorial?), is a melodiousness song with pure tones which differs between males and females. This stable and species-specific characteristic supports the existence of 9 species (Marshall and Marshall, 1976). The sexual dimorphism of hylobatides is morphologically insignificant.

Gibbons live in the forests of Indochina, southern China, Burma, India, the Malacca peninsula, and the Indonesian islands. They eat bananas, nuts, figs, leaves, and the sprouts of plants. They also eat grasshoppers, ants, nestlings, and bird eggs. Gibbons drink by lowering their hand into water and sucking water from the hair (like howlers). They are classical brachiators; they progress through trees at high speed, jumping distances of up to 15 meters, and they can walk on the ground on their hind limbs, balancing themselves with their arms held aloft and bent at the elbows and wrists.

Gibbons are monogamous animals that usually live as bonded pairs with multiple offspring; sometimes one or two older individuals also live in the family. The social life of gibbons, however, may deviate from this attachment to family life. The pairs are formed after several weeks of courtship by a young male after a maturing female from another group (it is difficult in a zoo to provide gibbons with a choice of partner, which may cause problems of adaptation for gibbons in captivity; these nimble creatures also need relatively large enclosures). Sexual maturity occurs at 6–9 years of age. When a young gibbon is about 10-years-old it leaves its parents. Among *Hylobates*, only the siamang has seasonal reproduction. Gestation lasts 210–255 days (230–238 days in siamang). A young offspring is born every 2–3 years, it has little hair and is helpless until 2-years-old. The record for longevity in captivity is 34 years (*H. pileatus*).

Gibbons have a high level of central nervous system activity and behavior; they can use tools such as sticks and rope. They are very close to humans in their immunological characteristics. They become ill spontaneously with influenza and leukemia, and they are subject to other human diseases. It is possible to model many human diseases in these primates, but they are rarely used in medical primatology because of difficulties in keeping them in captivity. The main reason, however, is that all gibbons and siamangs are on the list of endangered species.

Family Hominidae

These are the large hominoids. The total length of the body and head, excluding humans, is between 70–105 cm and body weight is generally 45–180 kg, but can reach 300 kg. The males are larger than females. The size of the brain is large, it has significant sulci and convolutions on its surface, and the cerebellum is covered by the occipital lobes. The dentition is similar to other catarrhines, but there is a peculiarity in the morphology, for example, of the upper molars (the protocone and methacone are connected slantwise with a crest). The frontal sinus is characteristic of only humans, chimpanzee, and gorilla. The eye sockets are directed forward. There is a well-known regression in the auditory system, associated with immobility of the outer ear. Humans and other hominids have 12–13 pairs of ribs, a short caudal interval and a S-shaped shift of the vertebral column with signs of this characteristic in apes and a full bend in *Homo*. The central arm bone (os centrale) is connected to the navicular (scaphoid) bone (not in orangutan). The sternum is wide and the thorax is not keel-shaped (but deeper in apes). The musculature of the face and body is more similar in humans, chimpanzee, and gorilla. Locomotion is completely vertical in humans and semi-straightened, 'false quadrupedalism', or 'inconvenient' in the apes. The stomach is simple, retort-shaped, and the colon is bent. The uterus is a simple horn. The placenta is monodiscoidal and haemochorial, being expelled. Female chimpanzees have a large genital swelling during the midcycle phase of the menstrual cycle, the female gorilla has a much smaller labial swelling, and orangutan females, as noted above, have no swelling associated with the menstrual cycle, similar to women. The hair integument (cover) is comparatively thin, the hair on the forelimbs extends to the elbow. Significant similarity in embryos and foetuses is observed in all species of the family.

Subfamily Ponginae

The *genus Pongo*, orangutan, comprises 1 species (*P. pygmaeus* [Figure 3.27]) with 2 sub-species living on the islands of Borneo (*P. pygmaeus pygmaeus*) and Sumatra (*P. p. abelii*). These are large primates. The body and head of the Bornean orangutan is on the average 96 cm (male) and 78 cm (female) long, the height of males reaches 158 cm

Figure 3.27 Pongo pygmaeus, a juvenile male (4 years).

(recorded by O. Hill in 1938), and body weight is 189 and 81 kg, respectively. It is well-known that they may grow obese in captivity, and their weight may reach 250 kg. Sexual dimorphism by weight is significant, although it is not very pronounced with respect to height, 137 cm and 115 cm, respectively. The Sumatra orangutans are not as heavy as the Borneo ones (70 kg and 37 kg for males and females, respectively), but they are not shorter than the former (Napier and Napier, 1985).

Body build is awkward; the arms are long and thick with a short first finger and arm-span reaches 2.5 m. The legs are comparatively short (without a nail on the first toe), the ratio of the arm and leg length to the body is 210 and 165, respectively; the abdomen is large. The profile of the face is concave with a prognathous lower mandible and frequently outpouching labium, and the forehead is high. A sexually mature male has elaborate semicircular outgrowths of connective and adipose tissues on its cheeks and a midsagittal bony crest on its head. The ocular ridges are developed. Males have a large guttural vocal sac that expands. Ischial callosities are usually lacking, but sometimes found in small size. The hair cover is thin and long (up to 40 cm), reddish brown, and darker in the Borneo subspecies. Adult males have beards and moustaches (on the sides of the high upper lip). The animals are adapted to live in trees where they make nests for rest and sleep, sometimes with an awning for protection from the sun and rain. On rare occasions when they descend to the ground, they travel predominantly on four limbs, with their weight resting on the middle knuckles of the fingers (and sometimes on the clenched fist) and on the outer edge of the foot. They are nonaggressive, sluggish in their activity, do not aspire to supremacy, and adults rarely associate with each other. A sexually mature male usually leads a solitary life and only interacts with a female when sought out for reproduction. The home range of a mature and dominant male may subsume the home ranges of several adult females with offspring. The young remain with the mother for 4–5 years or longer. Adolescents sometimes join groups of other adolescents and remain in them until they become sexually mature. They eat various fruits, leaves, bark, bird eggs, and insects; meat-eating has been observed, especially gibbons and other primates (Utami, 1994). They drink water that collects in the holes of trees, and they lick rain from the trunks of trees and their own hair. The number of vocalizations is small, but they are able to 'sing' in their own way, and the males utter roars.

Females become sexually mature at about 7-years-old, males 1.5–2 years later. The menstrual cycle lasts 29–32 days, the menstrual period is 3–4 days, and gestation lasts 225–275 days (by different authors). The weight of a newborn is 1,265–1,600 g. Single births are the rule, but twin births occur too. The chromosome diploid number is 48. There has been a catastrophic decline in the number of orangutans living in their natural habitat; in the 1970s there were no more than 4,000 individuals. There are 'rehabilitation centers' for orangutans which house animals that have been confiscated from poachers and 'pet owners' and which attempt to release healthy animals into safe and appropriate locations. Environmental protection programs are operating in both Borneo and Sumatra, but the fate of one of the nearest relatives of humans remains alarming. Orangutans have been kept in captivity for more than two centuries (if sporadic cases are considered, such as the one in the Netherlands at the end of the 18th century), but they began bearing young only at the end of the 1920s. The record of longevity in captivity is 59 years (*P.p. abelii*). Orangutans, as mentioned, are characterized by a high level of intellect and they are capable of using tools. In zoos, they learn to push a reward out of a tube with a stick, chew leaves to make a sponge with which they collect water from a vessel, etc. In experimental studies, they demonstrate self-recognition in a mirror.

Subfamily Homininae

Genus Gorilla is represented by one species (*G. gorilla*) with 3 subspecies; the western lowland gorilla (*G.g. gorilla*), the eastern lowland gorilla (*G.g. graueri*), and the mountain gorilla (*G.g. beringei* [Figure 3.28]). The habitat of each subspecies is Equatorial Africa (excluding the center of the continent). The mountain gorilla lives in the volcanic region (the Virungas) in an inaccessible forest of East Africa (Zaire, Uganda, Rwanda). This is one of the largest primates. The length of the body and head is 80–105 cm and the height is 172 cm (one individual was 196 cm in height). The male weighs 140– 155 kg (one wild specimen weighed 260 kg [Groves, 1986] and one in captivity weighed 300 kg). The body is bulky, barrel-like, the thorax is wide, the abdomen is stout, the neck is short, the shoulders are broad, and the head is large. The arms are longer than the legs; the ratios with respect to the body are 161% and 151%, respectively. The hand is wide with a short, but opposable thumb (as is the large toe), and the toes are frequently webbed. There is a large midsagittal bony crest on the male's skull. The eyes are set wide apart and the superciliary ridge protrudes conspicuously. The ears are small, the upper lip is short, and the nostrils are large and fleshy. The center of the face lacks hair, as do the ears, hands, and bottoms of the feet.

Figure 3.28 Gorilla gorilla beringei (See colour plate section)

Males are larger than females; an adult female weighs about half of the male's weight. The eastern lowland gorilla is larger than the western. The mountain subspecies has a longer body, shorter arms, longer feet, and the midsagittal crest of a male seems larger because of greater hair cover. The hair of gorillas is rough, stiff, and short, the color almost black, but they are born with fair hair. With age, light gray patches and brown dappling appear on the head and back. When males become sexually mature (starting at 9–11 years), gray hair appears on the back; hence the term 'silverback' for mature males.

Gorillas lead a primarily terrestrial life, although they are genetically fit for climbing trees. Their nest-building behavior is a consequence of arboreal life, but they construct their nests predominantly on the ground. They move on the ground using all four limbs with their weight supported on the middle knuckles of the hand (which are calloused). Sometimes they stand erect to free their hands for manipulation, eating, and performing other manual activities. The legendary fierceness attributed to gorillas has been refuted. When threatened, however, the males perform intimidating chest-beating displays, charge, and attack other male gorilla adversaries as well as people (infrequently), especially those who flee. In general, relations within groups of gorillas (from 5 to 30 individuals) are rather quiet and peaceful. Only females emigrate from their natal groups at about puberty to join other groups; males either remain in their natal groups or if they leave, live alone, or join all male groups. The individuals from neighboring groups quite often know each other. These primates are polygynous, the family units include an adult male, the leader, and several females with young. There are also groups with multiple adult males, especially among the mountain gorillas with a specific system of mating, where the dominant male has an advantage in comparison with subordinate males (Robbins, 1996).

Gorillas are phytophagous; they eat leaves, sprouts, and the pith of stems (approximately 80% of their diet), and fruits, though the percentage of the constituents varies with location. They communicate by vocalizations, gestures, and specific expressions. There have been cases in captivity when they begged for food and even offered their favorite foods to people they knew. Grooming is most commonly directed toward youngsters that are nursing, not only by the mother, but also the male and other members of the group. Infanticide has been recorded among mountain gorillas.

They hardly bear captivity, but can in good conditions adjust themselves and can live to be 50-years-old or more. Gorillas have been kept in zoos since the middle of the 19th century (Figure 3.29). The first birth in captivity took place in 1956 at the Columbus Zoo (USA). The gestation period was 257–259 days and the female infant weighed 2.262 kg. The pubertal growth spurt of the male occurs at 8 years of age (Wickings, 1996) and females become sexually mature at 7–8 years of age (younger in captivity). There is no seasonality in reproduction. The menstrual cycle of a female is 33 days on the average. In different zoos the duration of the gestation was recorded as 245 to 289 days. Some cases of twins birth are known. The diploid number of chromosomes is 48. The number of gorillas still living in the natural habitat has declined sharply in recent years and each of the subspecies is considered endangered. The state of the mountain gorillas is the most dangerous; there are only 630 specimens in two of the main pockets of their habitat (Virunga, Bwindi) (Steklis et al., 1996). Political instability in central Africa aggravates this danger. An outstanding achievement of the modern science, the birth of a gorilla following in vitro fertilization and transplantation of embryo into the uterus of a second adult female, aroused considerable interest (Pope

Figure 3.29 Man and ape: with the adult female gorilla Meta (Rostov-on-Don Zoo).

et al., 1996). This work, as well as similar advancements in other species of primates including chimpanzee and even humans, is extremely important and was discussed in a symposium at the 16th Congress of the International Primatological Society.

Gorillas possess a high level of intelligence. They are capable of performing difficult tasks, recognize themselves in mirrors, learn gestural language, and even understand the speech of people (Patterson and Holts, 1989; Patterson and Cohn, 1994).

The *genus Pan* includes two extant species, the common chimpanzee (*P. troglodytes* [Figure 3.30]) and the bonobo, or pygmy chimpanzee (*P. paniscus*). This hominoid is

Figure 3.30 Professor Leonid A. Firsov with a common chimpanzee, *Pan troglodytes*, named Boy (St. Petersburg).

not large, comparatively; it is smaller than the orangutan or gorilla. The length of the body and head is 70–95 cm (*P. troglodytes*), the height does not exceed 150 cm, the weight is usually 45–60 kg, but sometimes it reaches 80 kg. The ratio of the arms and legs to the body length is 168% and 158%, respectively. The females are slightly smaller than the males. The fingers are long and the thumb is shorter and set opposite to the others. The hallux is longer than the thumb, it is noticeably set aside, and there is some webbing between the other toes. The head is round, the jaws are protruding, the nose is not big, the superciliary ridge is developed, the labrum is high, and the ears are comparatively large and upright, very much like the ears of humans. The forehead is slanting, the canines are not very big (bonobo's canines are smaller than those of the common chimpanzee), and the guttural vocal sac is medium in size. Ischial callosities occur in approximately one-third of the specimens. The outward appearance of the pygmy chimpanzee differs from the common one in a somewhat smaller size and weight, although not very much; the lower limits of the common chimpanzee size correspond to the mean characteristics of bonobo (de Waal, 1995). The bonobo looks more elegant. They have longer limbs, narrow shoulders, and a smaller thoracic volume, their ears are comparatively short, the forehead is high, and the hair on the head appears as if it was combed with a part in the middle. The whiskers and the hair on the head are moderately long. Bonobos differs from the common chimpanzee in that there is lesser sexual dimorphism, a greater tendency to bipedalism, and they have red lips, which is a unique morphological peculiarity in nonhuman primates. There are significant distinctions in the ABO blood group and in karyotype banding. It is suggested that the genus diverged to the two present day species 2.9 Ma (Bailey *et al.*, 1992). The greatest difference between the bonobo and the common chimpanzee is in their behavior.

Unlike chimpanzees, in which the foundation of their social behavior is the association of adult males in the group (which helps them to survive in battles and succeed in hunting), the more peaceful bonobos base their relations on the alliance of adult females. The principle of an open group characterizes the community of both species, the so-called 'fission-fusion society' (de Waal, 1995, p. 62). The members of individual groups, usually females, migrate freely from one group to another. Sex is the primary motive that influences power struggles in the social relations among bonobos. As I mentioned above, bonobos are the only primates other than humans in which ventro-ventral copulation occurs frequently. Anatomical and physiological distinctions are characteristic of the specific level of the bonobo.

The hair of chimpanzees is sparse and colored black or dark brown (a population of white headed chimpanzees live in Equatorial Guinea). The face, ears, hands, and feet are not covered with hair and they are colored pink (black in bonobo); in some forms they are pigmented to variable degrees. There are 3 subspecies of common chimpanzee: *P.t. troglodytes*, *P.t. schweinfurtii* and *P.t. verus*, but the distinctions between them are not so obvious as those in subspecies of other hominoids. Common chimpanzees live over the entire equatorial zone of Africa with adjacent regions of forests and savanna, whereas, bonobos live only in the tropical rain forests of Zaire. When moving on the ground they support their weight on the feet and middle phalanx of the fingers. On the ground they often walk erect with their arms free; so too when they move from one tree or bush to another carrying fruit. They are good at climbing trees, where they make nests for night. Similar to gorillas, however, they are not true brachiators. Chimpanzees migrate even when food is in abundance, traveling up to 50 km a day, and they frequently return to their starting place.

Their social life (and even their communication) differs somewhat from one region to another and from one type of terrain to another. The societies of common chimpanzees, as well as bonobos, are constant neither by number nor by age and sex. In some places there are no dominant males, in the others the leadership is not clearly expressed and is temporary (Goodall, 1968, 1971). Nevertheless, a group leader and dominance hierarchy is always present. A male can reach the *alpha* rank (highest) by 15 years of age (Nishida, 1968, 1996). There is evidence of clearly isolated social units of chimpanzees which have their own territory. The societies comprise 30 to 80 individuals which typically consist of small migrating subgroups. The relations within a group are generally quiet and peaceful, although leadership and subordination are clearly observed. Examples of cooperation are found, especially among relatives, and there are cooperative associations among adult males. Conflicts arise among rival males of equal rank, when a female defends its offspring, and when there is competition for food in conditions when food is insufficient.

Chimpanzees greet each other when they meet, they may hug a friend, tap a shoulder, and touch with hands or lips. Fighting occurs within groups, after which the opponents may remain enemies for a long time, or they may quickly reconcile; in this case reconciliation is frequently accompanied by hugs and mutual mouthing. Combat, and even deadly battles, are possible when they meet an unfamiliar group of chimpanzees. Afterwards, the winners may eat the bodies of the dead. When chimpanzees encounter gorillas, there may be mutual aggression (these apes are sympatric in some areas).

The relations within groups of bonobos are different. As was mentioned above, sex plays a special role here. During the sexual cycle, the period of heightened female sexual responsiveness (proceptivity) is accompanied by swelling of the genital area and a willingness to copulate with a male; it is two times longer than that in common chimpanzees. In bonobos (unlike chimpanzees), sex even takes place during the period of lactation. Sex and reproduction is separated to some degree, in the process of evolution of bonobos and other higher primates, including humans. Besides the function of procreation, sex for bonobos acquired a key role in social life. It is a powerful factor in reconciliation following conflict within the society, it plays a significant role in the regulation of feeding, and finally, it leads to a congenial coalitions of adult females. This type of 'female bonding' prevents the unlimited domination by males, which is characteristic of so many species of primates, especially baboons, common chimpanzees, gorillas, and others.

Besides the previously mentioned peculiarity of ventro-ventral copulation, similar to the sexual behavior of humans, this ape engages in a surprising variety of erotic behaviour which is not limited to heterosexual pairs, and which includes lesbianism and other forms of homosexual relations, masturbation of a partner, intense tongue-kissing, and oral sex to orgasm (de Waal, 1995). Sexual interactions are random and governed by the principle of promiscuity. It is likely that this is the reason why social relations in bonobo societies are rather peaceful. They are characterized by sexual relations with equal rights by the male and female and dominance relations that are not based on sexual dimorphism, but depend instead on specific circumstances and in which a female may have the advantage in certain cases.

For a long time, it was thought that chimpanzees were completely herbivorous. From the end of the 1950s, following the observations of J. Goodall and other scientists, it became clear that these hominoids eat not only honey and termites, but they also eat fish and purposely catch birds and mammals, including many species of monkeys and not excluding, as mentioned, cannibalism. A successful hunter (usually an alpha male [Hosaka, 1996]) may share its meat with other members of the group, which are usually old males, females with sexual swellings, or relatives. Infanticide is known to occur in chimpanzees, apparently when the father belongs to another group (Nishida, 1985).

Nevertheless, plants comprise 90% of the chimpanzee's diet (several hundred plant species). It is likely that bonobos are more phytophagous than common chimpanzees, and they are not known to eat birds. There are differences in nutrition (and, no doubt, differences in behavior) characteristic of different populations of chimpanzee, living in various habitats. The means of communication between these primates are apparently the richest among the entire order, excluding humans, of course. There are numerous vocalizations, such as hooting, barking, screaming, humming, and puffing associated with various expressive gestures, expressions, positions, and staring. In bonobos, these activities are quieter and diverse.

Chimpanzees are capable of reproducing during the entire year, but in the wild, the season for reproduction differs for different regions of Africa. The female sexual cycle starts at the age of 7–9 (menarche), and males become sexually mature one or two years later. Under natural conditions, females bear their first infant at 13–14 years of age. The duration of the menstrual cycle in chimpanzees is 37 days, on average. The mean gestation period at the Yerkes Regional Primate Research Center of Emory University (US) was 227 days, but pregnancy may last from 214 to 295 days. As was mentioned above, the first common chimpanzee was born in captivity in 1915; the first pygmy chimpanzee was born in 1962 (the infant died after several weeks from pneumonia). By 1974, 10 sets of twins and two sets of triplets (the infants did not survive) had been born at the Yerkes Center. It is likely that the record longevity in captivity belongs to a female common chimpanzee, Gamma, of the Yerkes Center, where she lived to be 59 years, 5 months (Figure 3.31). Another female at the same center, Jenny, lived to be 2 weeks less than 59 years. (An unverified press release stated that Cheetah, the male chimpanzee known from the films about Tarzan, was alive in 1996 at the age of 64).

Figure 3.31 Gamma (chimpanzee), the oldest ape in the world. Courtesy of Professor H. McClure, Yerkes Regional Primate Center.

As we see, chimpanzees are capable of living and multiplying in captivity, but considerable care must be taken to provide appropriate housing, temperature, and food.

Both species possess a high level of intelligence; it is likely that their intelligence is the greatest in the animal world, other than humans (see above). According to investigators who work with Kanzi, the pygmy chimpanzee, he understands spoken English (Savage-Rumbaugh et al., 1993). Chimpanzees under natural conditions, and to a greater extent when taught in captivity, not only use branches and sticks to fish for ants and termites, but they use leaves and herbs as sponges for removing blood and grime from their bodies, and they use stones to break nuts. Furthermore, as mentioned above, they make elementary tools from various objects, store them away, and they retrieve them when needed.

Taking into account the lesser anatomical specialization and other morphological characteristics of the bonobo, their biochemical (Goodman, 1996), and psychological (Rumbaugh, 1995; de Waal, 1995) affinity to humans, compared to the common chimpanzee (Zihlman, 1979), some scholars consider this primate the most similar to the possible human ancestor (Zihlman, 1989).

Chimpanzees are absolutely unique in experimental studies on animals. The intellect of the chimpanzee, especially that of the bonobo, provides unique conditions to study comparatively, the behavior of humans, its norms, and pathologies. Humans should take good care of their biological sibling and provide them with a long and successful existence. At present, chimpanzee is a declining genus of primate (there are now less than 150,000 world-wide, compared to 2,000,000 at the beginning of the 20th Century), and the peaceful bonobo is a catastrophically declining species; that is, *endangered*. The number of bonobos in the wild (10,000?) is gradually declining, not only due to violent events in Africa, but also as a result of the *peaceful* destruction of its natural habitat. At the same time, investigations indicate that even within a protected population of chimpanzees, given the long period of maturation, mortality is exceeding the birth rate. Hunting chimpanzees for meat and smuggling them to other parts of the world continues, as does the destruction of their forests, cultivation of the land, and other forms of human intrusion into the areas in which other primates have lived.

Experimental use of chimpanzees in medical primatology is allowed only at special primate centers where they are bred for research (chimpanzees are no longer imported for research) and only for those cases in which there is no alternative for solving human problems. In the United States, there is a Committee for the Long-Term Breeding and Use of Chimpanzees in Research.

The *genus Homo* contains the single species of the present day humans, *H. sapiens*. There are no other living subspecies. The species (or subspecies – *H. sapiens sapiens*) is divided into races, which in terms of taxonomic categories, are of course lower than subspecies. Humans are large, tailless, terrestrial, narrow nosed (catarrhines) primates, with plantigrade, vertical locomotion, and flattened nails on their fingers and toes (Thorington and Anderson, 1984). They have an opposable thumb. The length of the body and head is approximately 90 cm and the arms are shorter than the legs. The hand is better adapted for manipulation compared to other primates. The foot is vaulted; it differs greatly from the feet of other hominoids. Judging from the human fetus, the human foot may be derived from that of the chimpanzee or gorilla (Nesturch, 1960). The large toe is not opposable, but it is well-developed as support when walking. The head is well-balanced on the vertebral column and oriented over the physiological horizontal level. The skull is large and rounded, the area of the face is noticeably reduced. The maxillary apparatus is relatively weak, the canines are significantly smaller than those in the other hominoids; the last molars do not always develop. There are no supraorbital ridges, but the chin is prominent. Cranial capacity is 1,085–1,581 ml (Smith and Topkins, 1995), there is a large neocortex, and especially, enlarged associative fields in the parietal, temporal, and frontal lobes of the brain.

There is little hair cover on the body; when humans reach puberty, hair covers the pubis and axillas. In addition, in men, hair covers the chest and parts of the face. Hair grows continuously on the heads of both males and females (baldness is a characteristic of some males). There are eyebrows and eyelashes (the latter also characterize other hominids). Mammary glands are well-developed in women. Sexual dimorphism is apparent, but it is less than in other hominids. The color of the skin varies from black to white.

Humans live everywhere in the world. Reproduction is not seasonal, although it is likely that there are birth peaks in different regions. The age of sexual maturity varies somewhat, depending on the race, which is perhaps explained by geographical location and nutrition. For girls, it is 12–14 years, for boys, one or two years later. The menstrual cycle is 28 days, on average and gestation lasts 9 months. The chromosome diploid number is 46. The mean duration of life in the 'developed countries' exceeds 80 years, but the maximum reaches 100 years and more (the record is about 150).

The sexual life of humans incorporates all the types of sexual relationships found in the other hominoids: monogamy, polygamy, polyandry, and promiscuity (de Waal, 1995). That is why F.B.M. de Waal is correct in writing that none of the present day hominoids, even bonobos, can individually represent the primitive society of humans; only the assembly of behavioural elements of all hominoids (de Waal considers that only three forms are needed: chimpanzee, bonobo, and human) can simulate our own past.

Regardless of common genetic roots, humans have reached a high level of behavioral development and intellect which is above that of any present day hominoid. Thanks to their physical superiority, such as their large and keen brains, specifics of hands and feet, larynx, and associated morphological and physiological complexes, humans have achieved a great break-through in the sphere of behavior. They exhibit completely new properties of social life, civilization, culture, and beauty, which far exceed any possible for other taxa. But even in this great advantage of humans, there are common genetic roots with other primates. Humans have retained some features of nonhuman primate behavior, including leadership, dominance of the strong over the weak, the benefits of sexual dimorphism (to men), territorial conflicts, and even cannibalism. They also preserve certain positive trends, such as protection of the weak, selflessness in war (for the present not eliminated even in human society) and cooperation. As for biology, there is not a very large separation between humans and other primates, especially from the apes and monkeys. Biological similarity with these animals is extraordinarily high, as repeatedly mentioned above. For the present, modern medicine has no choice but to apply this special similarity in experimental research on human problems.

* * *

The present day order Primates (C. Groves in Shoshani *et al.*, 1996) comprises 252 extant species, 61 genera and 13 families (in this text there may be some disagreement with these figures, for reasons which are discussed in the description of individual taxa). In addition, 405 extinct species are known, comprising 218 fossil genera (Ibid.). The primates descended from insectivorous ancestors, probably in Africa, during the late Cretaceous, and spread from there during the Eocene. Formation of the order and its further evolution from the origin to understated taxa, in general, coincides in time with the evolution of other mammals (Shoshani *et al.*, 1996; Goodman *et al.*, 1998).

I mentioned above that the biological history of primates, including humans, in general, supports the ideas and hypotheses of C. Darwin. Modern scientific data also confirmed the important conclusion of T. Huxley regarding the significant affinity of humans and apes compared to the affinity of the latter with monkeys. As we have seen, the distinctions between humans and African apes by fundamental biological characteristics are quantitatively less than 1%. Nevertheless, the lower simians are also close to humans, up to 80% by the same characteristics. The other mammals are much more distant from humans in terms of DNA and immunological biochemistry, to say nothing about other characteristics. During the past 40 years, stronger evolutionary and phenotypic grounds of the greater biological closeness of humans and other primates were obtained than was known in the 1950s. This is a strong foundation for modern medical primatology. On the other hand, the data of experimental pathology and pharmacology of primates are actually a continuation of the theme of this chapter. They strengthen the modern data on evolution and taxonomy of the unique affinity of humans and their congeners in the common order Primates.

Humans must make major efforts to preserve their nearest relatives. Regardless of global plans to save the extant primates, in reality their numbers are steadily declining. Almost half of all species (114 species!) are under the threat of decreasing population, 43 species (each 5th species) are in critical danger of extinction. Approximately 60% of all nonhuman primates are endangered. Such wonderful species in their affinity to humans as gibbons, orangutans, gorillas, and finally, chimpanzees and bonobos, are among the latter (Mittermeier, 1998; Tuttle, 1998). We cannot allow this to happen. We must exert a greater effort to protect these animals and their natural habitats. Primates must be bred in special colonies; this is what Mechnikov and Yerkes called for in the beginning of the 20th century. We should apply various principles and methods of obtaining progeny, including improved artificial breeding. This is needed both to save the valuable fauna of the Earth and to continue experimental research, without which humanity cannot survive into the 21st century.

*   *   *

Part III

Biological foundations and applications of medical primatology

Characterization and possible applications of the affinity between humans and other primates

At the end of the 1960s, information was received from Capetown in South Africa, where Christian Barnard had previously performed the successful transplantation of a human heart, about a second outstanding medical achievement. In this case, a woman who was dying from inadequate liver function was saved by an operation in which her blood was transfused through the liver of a healthy baboon. The Journal of the American Medical Association called this news 'tremendous', and voiced the opinion that, while anthropologists search for the genetic connection between humans and the other primates in as yet unexcavated depths of soil or in impenetrable jungles, for the surgeon this connection is a normal medicinal tube linking the human and simians by means of analogous procedures (Editorial, 1969, p. 143). We can use this concept in a metaphorical sense to evaluate the biological significance of simians in the entire body of research on human illnesses and in medical primatology as a whole. These investigations are yet another illustration of the close consanguinity among primates.

The problems that arose regarding the relationship between humans and the animals which were called *simia* in antiquity, we should recall, were due to their similarities. Ethical and philosophical aspects of this similarity disturbed the most inquisitive minds of the ancients: philosophers and naturalists, medieval free-thinkers, French enlighteners and, no doubt, whole generations of natural scientists. In the context of this monograph, I am interested only in the scientific usefulness of this biological similarity as a basis for conducting medical research on primates. I note that it is not possible, in a short section of this monograph, to present *each* aspect of the affinity that has been demonstrated by science over more than three and a half centuries. I discuss only those characteristics which are either of special interest for biomedical research, or those which were revealed during the last several decades in the intensive development of medical primatology. The characteristics are analyzed under a rubric corresponding to the anatomical and physiological systems of organisms, in accordance with the subdivisions of modern medicine.

It is noteworthy that the theoretical foundations of my objective (the similarities in morphology, physiology and experimental pathology) are not yet fully developed. Despite the theoretical constructs of other sciences, such as homology and analogy, synapomorpha and plesiomorpha, convergent and homology similarity, structural and regulatory genes, and the reassuring concept of neoteny (Mason, 1990), the theoretical basis of affinity within the context of medical primatology is not yet established.

The objectives of this monograph, in general, require us to concentrate on the affinity between human and other primates, although ignoring the differences, which are significant and numerous, would compromise further discussion of what follows.

Humans are the most complex of the social creatures. Whether we like it or not, humans dominate nature; they do not merely adapt to it in the way that the other inhabitants of the planet do, including the nonhuman primates. Humans differ from the other primates in terms of brain development, stable upright posture, articulate speech, powerful intellect, special abilities for abstraction, and various capabilities for work. These distinctions, in turn, determine many others differences. Nevertheless, even the unique human features have a certain comparability among their congeners, the primates. Usually, the comparisons of characteristics are made either with those species for which there are data, for which the affinity with humans is greatest, or for those characteristics which are important for one or another division of medical primatology.

GENERAL INFORMATION REGARDING THE MORPHOLOGICAL CHARACTERISTICS OF AFFINITY BETWEEN HUMANS AND SIMIANS

As was shown above, anatomy was the first discipline to appear in the infrastructure of primatological science. The first scientific publications on primates (Tulp, 1641; Tyson, 1699) and the following ones over one and a half centuries were almost completely anatomical. During the 20th century and continuing to the present time, as I discussed in the previous chapters, fundamental investigations were conducted which allowed us to determine the position of primates, including humans, in evolution, phylogeny, and in the modern taxonomic hierarchy. These investigations revealed the previously mentioned distinctive characteristics of this group of mammals, which are, in general, morphological and partly molecular biological. They also described the extraordinary affinity between humans and the other representatives of the order.

A. Schultz, an anatomist and authority in primatology, was very precise in his assessment of the affinities and distinctions among the higher primates. By the end of his scientific career, Schultz had noted the affinity of humans and simians in the basic properties of fetalization, morphological specialization, ability for adaptation, and intraspecific variability. As a result, he came to the conclusion that humans are not biologically unique. Even the numerical proportion of the sexes and its dynamics are similar in humans and the other higher primates, as is the greater rate of mortality observed with increasing age in male individuals (Schultz, 1971). G. Schaller (1988), who conducted an anatomical examination of a baby gorilla which had died in the forest, was amazed by its similarity with a newborn human child in terms of toothless gums, circularity of the forehead, flat face, and even the capacity of the skull.

With respect to medical primatology, we must consider the unequal longevity of biological life and its stages (growth, maturation, aging) in humans, apes, and monkeys. The extension of each stage of growth and maturation in primates and the increase in the rate of growth during the period of sexual maturity are unique in the animal kingdom. They have evolutionary roots (Schultz, 1969; Watts, 1985) and they provide unique opportunities for biological study of the processes of development and aging, the interrelations between development of the osseous and dental systems, development of the cerebral structures, maturation of the endocrine system, and many other problems.

Despite the differences in timing, the nature of growth in body mass and other properties of organismic development, as a whole are similar in the primates. But one must keep in mind that laboratory research on primates, for example, demonstrated that the osseous system of the hand and wrist becomes completely mature at 6 years of age in rhesus macaques, at 12 years in chimpanzees, and at 18 years in humans (female) (Michejda, 1980). As was previously mentioned, a newborn human baby is the most immature among the primates (Zeveloff and Boyce, 1982). Consequently, maturation of simians occurs more rapidly than it does in humans, while the life cycle and each stage of the cycle are shorter.

One of the principal prerequisites for the use of nonhuman primates in biomedical experimentation is their posture and locomotion which most closely approach the orthostatic (vertical) posture and orthogradity of humans. Comparison of the age at which the dynamics of body development occur in simians under the condition of the earth's gravity provides the best model for clarifying human pathological processes (Belkaniya et al., 1988). Naturally, this prerequisite has an anatomical basis.

It follows from the previous chapters that the anatomical similarity between humans and other primates was the first significant basis for combining them into one group. There is great similarity in the skeleton, especially between humans and apes, in the structure of the vertebral column, the number of vertebra and ribs, and even the flexures, flattening in the anterior-posterior direction, and widening of the thorax with scapulas at the back (Schultz, 1966), and in the anatomy and dynamics of growth in the pelvic bones (Coleman, 1971). The growth and length ratios of the tubular limb bones which are routinely measured on x-ray photographs from two months of age to maturity (radius and humerus bones of arms, tibia and femur bones of legs) has the same tendencies in humans and other present day hominoids. The distinctions are only in a relative shortening of the radius and tibia bones in humans (Buschang, 1982). The hind limbs in bonobos are heavier than the forelimbs, which makes them especially close to humans. The center of gravity in a maturing chimpanzee is located at the same height as it is in human children, although there are differences in the mechanics of the femorotibial joint and similarities in the knee and malleolus joints (Zihlman, 1984; Kimura, 1990). Bipedalism of an immature chimpanzee (as well as of other apes), and the patterns of footstep force and of body inclination are similar to those of human children. The bipedal gait of bonobos is very similar to that of humans and provides the basis 'to study the origin of human habitual bipedalism' (Aerts et al., 1998).

I already mentioned that it is likely that humans and other higher primates evolved from a common ancestor, a brachiator, and brachiation was a prerequisite of bipedalism. Therefore, there is a series of important common features in locomotion, anatomy, and functions of the skeleton, muscles, and biomechanics (Prost, 1979; Yurovskaya, 1991, and others). Interesting, in this regard, was the finding at autopsy of the typical anatomy of a simian foot on an human invalid who had never walked (Davidovsky, 1969). More to the point, as was demonstrated by A. Schultz (1966), the feet and hands of human and fetal macaques can be hardly distinguished (Figure 4.1).

I should point out that the human locomotor apparatus not only resembles that of the apes, but it is also related to that of the monkeys, and even to the quadrupedal primates. It was found (Reynolds, 1985) that 30% to 45% of total body mass is borne by the forelimbs in lemurs, coaitas, guenons, patas, and chimpanzees, along with a high mobility of the forelimbs and a developed hand. This ratio is 55%–65% in the other mammals. A tendency for a decrease in the supporting role of the forelimbs, and consequently, toward bipedalism, is clearly seen in monkeys, to say nothing about the apes (recall that many species of monkey frequently rise up on their hind limbs and

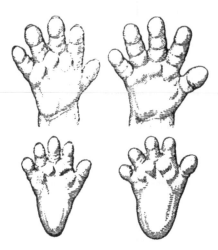

Figure 4.1 Hand and foot of a fetal macaque (left) and fetal human (Schultz, 1966). Reproduced with permission.

they also sleep in a sitting position). This tendency is reflected in the circulatory system of the limbs (for example, in the number and functions of the venous valves) which is similar in humans and even the lowest simians (Thiranagama *et al.*, 1989). The similar type of neural control of locomotion found in primates is unlike that in cats, the traditional laboratory animals for studies of locomotion. Thus, the cat is essentially useless for studies of human locomotion (Vilensky, 1988).

The orthogradic posture in humans and the tendency toward this posture in simians under the condition of the earth's gravity led to significant changes in the vital functions of these primate organisms. The valvular apparatus of the vascular system changed such that blood could be pumped upward; the heart and head were not at the level of the other organs, but much higher. The circulatory system changed, as did breathing, and the cardiovascular system changed as a whole. Energy consumption, neural regulation, and muscular and endocrine regulation of the organism also changed. These common physiological characteristics of primates are, in part, the basis of medical primatology.

Macaques and baboons with bound forelimbs (artificial 'bipeds') can live and travel using only their hind limbs, whereas rabbits, rats, and guinea pigs, unstable in the erect position, quickly fall into orthostatic collapse with complete circulatory distress. Along with other factors, orthogradic posture accounts for the special role of monkeys as research animals, not only in studies on the formation of orthogradity, phylo-ethagenesis, and anthropogenesis, but also in cardiology and space research (Belkaniya, 1982; Belkaniya *et al.*, 1987). As a result of the redistribution of load in erect locomotion, Japanese macaques trained for bipedal locomotion (for 2 km and more) obtained an S-shaped bend of the vertebral column, characteristic of humans. This was similar to the development of lordosis in children at 1–5 years of age (Preuschoft *et al.*, 1989).

The hand played a special role in the biological history of primates, including humans, giving simians and humans a remarkable advantage over other animals (Preuschoft and Chivers, eds., 1993; Westergaard *et al.*, 1998). High mobility of the clavicle, shoulder, radial, and radial-carpal joints, and liberation of the third terminal finger phalanx were crucial events in the alteration of hand function in the higher primates. None of the animals except simians is able to rotate the antebrachium inside (pronation) and outside (supination) as humans do (Roginski, 1969). The thorax or mediastinum of rhesus macaques, moreover, is remarkably similar to that of the human (Silverman and Morgan, 1980). Along with the similarity of bones, muscles, tendons, and blood supply, especially in humans and apes (with differences, of course), similarity is also found in their cerebral processing and finally, in the functioning of the arm and hand. Eighty-seven of the total of 107 characteristics of human hand muscles are found in the other primates (only 15 are found in nonprimates). Sixty-four of them are similar in humans and chimpanzees, and 20 in humans and monkeys (Yurovskaya, 1982). Measurements of the humerus, radius, ulna, metacarpal, and other bones in rhesus macaques indicate a right-sided asymmetry of the forelimbs in simians, similar to humans (Dean *et al.*, 1988). Somatotopic organization of neurons from the hand to the spinal cord and cuneate nucleus is analogous in macaques and humans (Florence *et al.*, 1989). Similarity in configuration of the arterial network in fetal human and macaque hands provides opportunities for clinical research (using a model of neurovascular anastomosis) (Maher, 1990). Furthermore, complex brain-eye-hand coordination played a crucial role in the process of anthropogenesis.

There are many similar properties in humans and other primates, both in the structure, ratio, and the character of growth of the skull bones and in their asymmetry. In this respect, the closest to humans is the bonobo, but the other hominoids and even the monkeys also have affinity with humans (Thoma, 1979; LeMay, 1985). A pig-tailed macaque, for example, is considered an excellent model for the study of growth and development of the skull and face (Sirianni, 1985).

Age changes similar to human ones were found in the bones of monkeys, in terms of their density, mineralization, decrease of the bony tissue mass, in peculiarities of the musculoskeletal anatomy and its dynamics, and in other processes, even though the timing of these processes differs. In particular, such changes occur three times faster in macaques than in humans, providing favorable possibilities for studying the problems of aging of human bone in these primates, and obtaining models of lordosis, kyphosis, osteoporosis, and age osteopenia (Williams and Bowden, 1984; Pope *et al.*, 1989). It is clear that nonhuman primates are also interesting as natural models of pregnancy, lactation, and aging (in chimpanzees) in relation to hormonal, alimentary and other factors (Summer *et al.*, 1989).

The affinity of speech organs in primates is of special interest. Jumping ahead slightly, I note that these organs (lips, mouth, tongue, larynx) occupy the largest surface area in the motor and somatosensory regions of the human cortex. No mammal has such a large cortical area other than the simians, especially the higher ones. In this sense, they resemble humans, although there are significant distinctions even in chimpanzees (Ploog, 1972). The larynx of chimpanzees reveals a certain primitiveness compared to the human one (vocal fold, laryngeal ventricle, laryngeal sacs; muscle fibers do not touch the vocal cords), but the general structure is similar (Barchina, 1973).

Nevertheless, there is a significant difference. The larynx in humans is located lower than the tongue (supralaryngeal tract), even compared to the chimpanzee, which is not advantageous medically, since food can get into the lungs. This was mentioned by C. Darwin. On the other hand, this disadvantage is probably offset by the fact that the larynx is adapted to pronounce sounds (Liberman, 1990). Nerve endings in the primate's lips share common peculiarities with human lips which are not found in any other animal (for example, Meissner's corpuscles). Their evolution in humans and simians does not differ greatly, in comparison to the difference between primates and other mammals (Kadanoff *et al.*, 1980). The nerves of the tongue in primate fetuses have the same characteristics (together with the defining distinctions), even when rhesus macaques and humans are compared (Kozey, 1975). The muscles of the tongue in humans and chimpanzees are practically the same. Both have four layers in the medial part (they are slightly thicker in the posterior part of the fourth layer in chimpanzees) and each has three similar layers in the lateral part (Takemoto, 1996). At the same time, there are quantitative distinctions between humans and chimpanzees in structures of the oral cavity and larynx which indicate that there are mechanical limitations in the chimpanzee's capabilities for distinct speech (Duchin, 1990).

The external structure and number of lobes in the lungs of chimpanzees completely corresponds to those in humans (differences are found in monkeys), though the number of segmental branches varies somewhat. There are 11 such branches in the right lung of humans and 10 in the left one, whereas there are 13 and 12, respectively, in chimpanzees (Bayramian, 1971). The total capacity of the lungs, its correlation with the square area of the lungs and with the characteristics of growth, weight, and span of the limbs coincide to a large degree in baboons and humans (McCullough *et al.*, 1979). The morphology of the tissues in macaques and humans, moreover, are more similar than in any other laboratory animals. Similarity of the trachea in macaques and humans was found in terms of the submucosal membrane gland, in the innervation mechanism of smooth muscles, etc. (St. George *et al.*, 1986). Here I should point to a phenomenon which I shall analyze many times (using other material), and which superbly demonstrates the experimental uniqueness of the primates. The smooth muscle response of the trachea to small amounts of adrenaline in most of the mammals is a clear relaxation. The same catecholamine causes contraction of the similar muscles in simians (*Erythrocebus patas*), as in humans; that is, its action is opposite (Montano *et al.*, 1985).

There is a considerable amount of data on the similarity of the lymphatic apparatus in humans and simians, and on the anatomy of the nodes and vessels and their morphogenesis (Charin, 1979a). Similarity of structure and the correlation of ontogenetic stages were found in the development of the spleen and thymus in baboons (*P. hamadryas*) and humans (Sapin and Charin, 1985). There are further areas of similarity between the other monkeys and humans, such as in the microscopic and histological structure of many internal organs, which I discuss below, for example, the liver, heart, etc. Similarity of the facial muscles in chimpanzees and humans is notable (Goldschmid-Lange, 1976) and the dermatoglyphics in simians and humans are uniquely similar (Gladkova, 1966).

I cannot avoid mentioning the persistent mistake made by scientists who used nonprimates in research, which led to an incorrect conception about the appendix. For many decades, on the basis of research using different animals, the vermiform process of the cecum was considered a rudimentary organ which lost its function in humans. Only the studies carried out on primates (the appendix appears only in certain monkeys of the Old World, and reaches complete homology with humans in the apes) enabled researchers to discover that the appendix is not rudimentary. On the contrary, it is a phylogenetically new structure with active lymphoid, secretory, and incretory functions. It is associated with the bacteriology of the intestines, the activity of the large intestine, and it plays an important role in the immunological processes of the organism. This led to a new understanding of the pathology of the appendix, and appendicitis, in particular (Chromov, 1978; Scott, 1980).

Figure 1.3 Nicolas Tulp(ius). Detail from Rembrant's picture 'Professor Tulp (the anatomy lesson)', 1632. (See Chapter 1, p. 7)

Figure 3.1 Lemur catta (See Chapter 3, p. 73)

Figure 3.2 Eulemur mongoz (See Chapter 3, p. 72)

Figure 3.3 Propithecus verreauxi (See Chapter 3, p. 75)

Figure 3.6 *Callithrix geoffroyi* (See Chapter 3, p. 87)

Figure 3.8 *Leontopithecus chrisomelas* (See Chapter 3, p. 90)

Figure 3.10 Saimiri sciureus (See Chapter 3, p. 93)

Figure 3.13 Pithecia pithecia (See Chapter 3, p. 101)

Figure 3.11 Aotus trivirgatus (See Chapter 3, p. 95)

Figure 3.12 Lagothrix lagothricha (See Chapter 3, p. 98)

Figure 3.15 Cacajao calvus (See Chapter 3, p. 103)

Figure 3.20 Mandrillus sphinx (See Chapter 3, p. 124)

Figure 3.22 Langur monkey (*Presbytis entellus*) wounded by an adult male. Courtesy of S. B. Hrdy. (See Chapter 3, p. 127)

Figure 3.23 Pygathrix nemaeus (See Chapter 3, p. 129)

Figure 3.24 Nasalis larvatus (See Chapter 3, p. 130)

Figure 3.28 Gorilla gorilla beringei (See Chapter 3, p. 164)

BRAIN, NERVOUS SYSTEM, AND BEHAVIOR FROM THE PERSPECTIVE OF MEDICAL PRIMATOLOGY

The brain is a unique substance of living creatures, the essence of its bearer. An anatomist can say that this is fatty tissue economically packed in a bony cavity (the skull). The price of this package, however, is extremely high, not only in a metaphorical sense. The mass of the brain is slightly greater than 2% of the total weight in humans, but it consumes 20% of the total energetic capabilities of the organism (whereas the brain of chimpanzees consumes only 9%, and the brain of marsupials consumes 2%, on average [Hofman, 1983]). It is well-known that the development of the brain in the evolution of humans not only played a crucial role, but the process itself was unusual. In approximately 2 million years, the brain made such a rapid spurt in growth that neurologists consider it the fastest macro-evolutionary process in the history of the animal world (despite the general deceleration in the evolution of hominoid DNA). Over a period of 2 million years, the human brain increased at an average rate of 50 cubic centimeters every 100,000 years, and its total volume grew approximately 3 times (Bielicki, 1987). During this time, the brain capacity of modern *Homo sapiens* increased from that of the African australopithecus (404 ml) to its present volume (1,085 ml–1,581 ml). I note that the brain size of the common chimpanzee is 383 ml on average (all figures are taken from the generalized data of Smith and Topkins, 1995). This means, as mentioned above, that the human brain exceeds the brain of chimpanzee by approximately three times (calculated in relation to body weight).

What was the cost of this increase? No doubt the increase occurred not only as a result of the mechanical addition of new structures, but also because the old animal structures were also subject to change. The basic increase in the brain occurred in the neocortical area, where the human structures are concentrated, providing cognition, speech, and intellect. The area of the temporal lobe (paleocortex) is 93.8 mm^2 in macaques, 324.8 mm^2 in chimpanzees, and 480 mm^2 in humans. It is apparent that the difference between the two latter species is not so great. During the same period, the neocortex increased sharply in its development toward the human state. It is equal to 6.456 mm^2 in macaques (it is incomparably less or absent in other animals), 22.730 mm^2 in chimpanzees, and 80.202 mm^2 in humans (Filimonov, 1949). Thus, it would not be a mistake to assume that in the evolutionary process, the new cerebral structures of higher primates and especially of humans (maximum) separated from those of the rest of the primates, the simians. The ancient structures in humans and simians, which control the somatic and vegetative functions, remained in their original form or one very similar. For example, although the precentral, postcentral, occipital, insular, and limbic areas underwent some transformation, the changes were primarily associated with the specific senses, the regulation of breathing in the medulla oblongata, blood pressure, and the regulation of other vital processes (Preobrajenskaya, 1974). Unlimited opportunities for medical primatology follow from this basis. Neural regulation of the vital functions of the organism, and naturally its pathology, remained especially similar in humans and simians.

I already mentioned the unusual brain similarity of simians and humans, which reflects the initial simian type of structure. Human patients with autosomal imbalance are mentally retarded people. The brain anomalies of these people who are afflicted with trisomy-21 (brachicephalization, narrowing of the supra-temporal gyri and usually hypoplasia of the pons and cerebellum), result in a neural structure which is remarkably similar to that of the simians (Tamraz *et al.*, 1987). During the postmortem dissection of a 13-year-old girl who suffered from microcephaly, one of the more prominent physiologists of this day, academician I.S. Beritaschvili (1974, p. 91), reported that her brain resembled a typical chimpanzee brain. The entire order Primates is sharply distinguished, in the general genealogy of the mammals, by the degree of its neocortical development; it is impossible to differentiate the human evolutionary line from that of the other primates (Kotchetkova, 1973). In structural aspects, the human brain corresponds to the main tendencies in the primate group. The brain grows faster than the weight of the body, the proportion of neocortex increases with the growth of the brain, and the size of the association cortex grows with the growth of the neocortex. Total brain size and the relative size of the association cortex, moreover, is significantly greater in humans (Passingham and Ettlinger, 1974).

According to various investigators, no matter which brain classification of the animal kingdom we consider, humans and other primates, the simians above all, would be assigned to one group. In one classification, humans and simians fall into the highest, ninth 'fronto-temporal' type (Nikitenko, 1967). According to another one, the *primate type of brain* is easily distinguished and its characteristics are well described (Preobrajenskaya, 1974). With respect to the aforementioned increase in the neocortex of the highest primates, neurophysiologists are unanimous in describing the special progressive development of the association cortex in the simians. This is the storage area of the highest cognitive functions, whose size in the other animals is negligible compared to the primates. According to the data of the Institute of Brain of the Russian Academy of Medical Sciences, the association areas account for 10% of the entire neocortex in monkeys, 20% in apes, and 50% in humans (Mering, 1990) (see Figure 4.2, from Penfield, 1956). The leading role in complex integrative activity of insectivores and rodents is played by the limbic cortex, while in primates the main role is played strictly by the association cortex and its frontal cortical areas (Karamyan *et al.*, 1987).

S. N. Olenev, a Russian neuromorphologist, wrote, 'Replying to the question concerning the uniqueness of the human brain compared with the brain of an animal, we can say the difference is not very great structurally, and its character is more likely quantitative than qualitative. Repeated attempts to distinguish nuclear structures or cortical areas which are present only in humans and absent in higher simians produced only modest results. The similarity was distinguished by thousands of characteristics, but the differences by only units' (Olenev, 1987, p. 37; my underline – E. F.). As was mentioned in the previous chapters, there are numerous statements of a similar type. Humans and simians have the same peculiarities of gyrification, location of sulci and convolutions, as well as their correlation with the myelocytoarchitectonics of the brain (Chatchaturyan, 1988; Zielles *et al.*, 1989). S. Blinkov (1955, p. 6) presented the data of E. Beck, who found properties during research on the myelocytoarchitectonics of the supra-temporal cortex convolution (hidden in the depth of the Sylvian fossa), which were missed by Brodmann, the famous German neurologist. Beck distinguished 89 areas in chimpanzees and found the same number in humans with identical structure; naturally, the sizes of the fields were different.

As I mentioned above, macaques are born with a more mature brain than humans and even more mature than chimpanzees (the weight of the brain increases to maturity by 4 times in humans, by 2 times in apes, and only by one-half in rhesus monkeys [Dienske, 1984]). During that time, *there are no changes in their ratios.* Moreover, strange though it may seem, the new structures of cortex (frontal and temporal areas) in human and

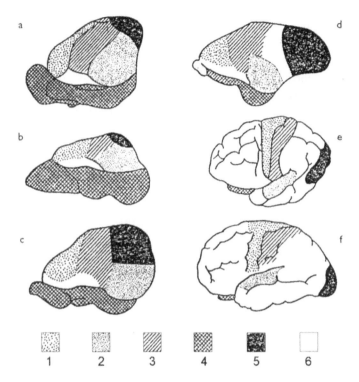

1   2   3   4   5   6

Figure 4.2 Growth of the association cortex in a number of animals (according to W. Penfield, 1956). (a) rat, (b) shrew, (c) treeshrew, (d) tarsier, (e) chimpanzee, (f) human; (1–5): the motor, auditory, somatosensory, olfactory and optic areas respectively; (6) association cortex).

monkey fetuses are formed before the old ones, which is one of the peculiarities of the primate brain. The sequence and relative duration of morphogenesis show some similarity in primates and rodents, but they are practically identical in rhesus macaques and humans (Gribnau and Geijsberts, 1985). During the first half of pregnancy, development of the phylogenetically new portions of the cortex occurs uniformly among primates with correlated dynamics of growth in the same cortical areas. During the second half, it continues to grow in humans and in the postnatal period, it becomes similar in humans and simians (Borisenko and Kesarev, 1986, reviews various research on the topic).

The frontal cortex reaches its most significant stage of development in the primate brain (*aromophosis* – after A. N. Severtsov). This structure is associated with the highest intellectual functions of humans and has essentially no homology in other animals. In the process of evolution, its initial forms appear in predaceous animals. The size of the frontal cortex in cats and dogs, however, does not exceed 3%–8% of the entire area of the cortex, whereas in humans it reaches 24.4%. It is equal to 12.4% in macaques and baboons, and 14.5% in chimpanzees (it is curious that it is equal to 15.2% in a newborn human child) (Kononova, 1962). It is noteworthy that the discovery of homologies of the specific human brain areas 44 and 45 in chimpanzees, the motor area for speech in humans, is not limited to a general or approximate similarity, but rather, has a close morphological likeness. The single characteristic mentioned above of the stratigraphic distribution of pyramidal cell size and even of the absolute values of length and transverse size of the main mass of the pyramidal neurons in analogous layers, significantly surpasses the degree of homology in these unique structures found in other species of the world (Melnik, 1978). Despite the well-known skepticism regarding the claim that simians possess Broca's and Wernicke's centers, associated with speech in humans, there are data supporting the morphological homology of similar structures. There are also data regarding their influence on the vocal and emotional behavior even of monkeys by means of similar neural pathways (Aitken, 1981; Deacon, 1989; Lieberman, 1985, 1990; Gannon et al., 1998). The connections of Broca's center with other structures of the speech complex differ in humans and chimpanzees and the areas themselves are 2–3 times greater in humans than in simians.

The evolved primate brain is not simply a result of progressive development of the neopallium combined with a relative regression of the paleocortex. There is also development of the phylogenetically newest *human* structures of the cortex. In addition to the development of the frontal region, there is further development of the lower parietal regions and the temporal-parietal-occipital subregion associated with thinking, speech, and gnostic functions of the brain. Changes in the ratios of these areas within the cortex itself are also a primate aromorphosis; the projection areas are decreased in favor of the associative regions, especially in apes and humans. Only the primates, including humans, possess direct pathways between the frontal and temporal lobes (Preobrajenskaya, 1974). Large visual fields together with distinct frontal lobes are specific characteristics of the higher primates (Armstrong, 1985). Well-developed sensorimotor and visual cortices, with their specific 'primate' connections constitute a unique natural combination which enables three-dimensional, spatial-optic perception of objects (Dubinin and Schevtchenko, 1976). Analysis and synthesis of various kinds of oral and written speech of humans, grammar, fine arts, mathematics, information and re-information about different modality, and memory take place in the lower parietal and supra-temporal regions. Unlike nonprimates, these specific structures are well-developed in the higher primates. The central sulcus, on the boundary of the motor and sensory areas, is found only at the primate level; the other mammals lack it (Ibid., p. 33). Only in other primates is the structure of the cerebellum, its functions, and its connections homologous to those of humans. Professor G. Khassabov, a well-known neurophysiologist, calculated the ratio of the surface area of the cerebral cortex to the volume of the globus pallidus which participates in the organization of the motor activity. This ratio is quite large in dogs and rabbits (170 and 190) and similarly small in macaques, chimpanzees, and humans (30, 40, and 50, respectively) (see Figure 4.3). The same relations are characteristic of the other components of the striatum, the striapallidal complex (Khassabov, 1978).

(It is interesting that the entire lower parietal region of the cerebral cortex in capuchin monkeys is completely homologous with the phylogenetically new fields 40 and 39. These fields do not reach an area of this size in any of the other lower simians to which capuchin monkeys belong, nor even in gibbons [Dubunin and Schevtchenko, 1976]. These fields are associated with the most complex functions used during manipulation, such as gnosis, stereoisomerism, and singleness of purpose. It is not surprising, therefore, that the behavior of capuchins in these respects is comparable to that of chimpanzees, as was already mentioned [Visalberghi, 1997].)

The cytoarchitectonic characteristics of the brain are similar in humans and simians. Field 17 of the occipital region (area striata), which is directly associated with vision, is distinguished only at the primate level. Layer IV in simians and humans is divided

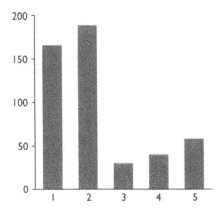

Figure 4.3 The ratio of the surface area of the cerebral cortex to the volume of globus pallidus (Khassabov, 1978). 1, rabbit; 2, dog; 3, rhesus macaque; 4, chimpanzee; 5, human.

into sublayers IV-a, IV-b, and IV-c (Preobrajenskaya, 1974, p. 160). Immunoreactivity to the Ca-binding protein, parvalbumin, and the morphology of neurons sensitive to it are similar in humans and in two species of macaques (rhesus and Java macaques). This similarity suggested that the brain of these monkeys is an adequate model for the study of cortical field 17 in humans (Blumke *et al.*, 1990). Only in monkeys do the basal nuclei of the prosencephalon have a clear structure and cellular population, homologous to Meynert's nucleus in humans. Their degeneration in elderly people leads to senile dementia. Consequently, extrapolation of results of experimental research on this pathological condition is possible only from primates (Dean and Bartus, 1985). Hemispheric asymmetry in the lobes of the frontal cortex with human-like patterns is found in Java macaques (Gannon and Laitman, 1995). Similarity of fine structures at the cellular level and their synaptic organization in primates, including humans, was found in practically every system and structure of the cerebrum (Amunz, 1976; Cowen, 1986; Barbas, 1995; Goldman-Rakic, 1996). Spider-like neurons were not found in the striatum of any mammals other than primates. Unlike other animals, the cells of the striatum are extremely diverse in simians and their axons and dendrites are maximally differentiated in humans. The substantia nigra is considerably enlarged in simians, beginning with the lowest ones, and is most highly developed in humans. The structure of the hypothalamus as well as a series of structures in the mesencephalon are more highly differentiated in primates than in other mammals (Urmancheeva, 1986).

The differences of humans and simians from other animals in terms of the structure of the brain, and the similarity among primates with respect to this structure, are an inexhaustible theme for discussion. Let us focus special attention on the similarity of humans and other primates in brain laterality and the cognitive functions which are associated with this laterality. Numerous papers published in the last 20 years are devoted to this issue (Falk *et al.*, 1988; Bradshaw and Rogers, eds., 1993; Rogers and Kaplan, 1994). Let us also consider the problem of cerebral representation of the senses (I shall deal with this problem in other sections) and the structure of the motor area, which has a unique homology in humans and other primates, including the cellular structure of the motor cortex (Kukuev, 1968). The similarity of connections of the brain with the spinal cord in primates, which differs significantly from those in other animals, was found in the beginning of this century by A. Grinbaum and C. Sherrington (1904), as mentioned above, and was subsequently fully confirmed. Neospinothalamic pathways are found in primates, but not in carnivores (cats). There are also other distinctions in nonprimates. 'Direct cortico-motoneuronal connections are the distinctive feature of the primate cortico-spinal system. This component is found in lower simians and progresses rapidly during the transition to the higher simians and humans' (Schapovalov, 1975, p. 133). Moreover, the motor system is also one of the key areas in the evolution of humans, and it is fundamental to medical primatology.

It is known today that the genes coding brain function and controlling homeotic proteins are similar among the primates. M. Goodman (1996, p. 282) believes that the large human neocortex contains 'not only the anciently conserved features but also some common anthropoid-, hominoid-, and human- specific features' reflected in the genomic DNA of neurogenesis. Goodman, referring to the latest research in molecular phylogenetics, predicted that the ancient and more recent mutations which led during evolution to the present day volume and complexity of the human brain will soon be identified at the DNA level.

(M. Verhaegen [1995], faithful to his concept of the *aquatic ape theory* of human evolution, which is of some interest, concluded that the brain of the modern apes is clearly different from the human brain. For example, there is the smaller size of the cortical areas controlling precise hand movements, breathing, and speech musculature, as well as the association areas, which he believes, strengthens his theory. Without going into the details of Verhaegen's concept, I acknowledge that his statements about the distinctions between apes and humans are accurate, as discussed above. Otherwise, the apes would not be apes but something different, and possibly could themselves reply to the criticism. There are no creatures on earth at present, however, whose brain is so similar to the human brain as is the brain of the apes and other simians close to the apes.)

The similarity of humans and other primates in *brain biochemistry*, neurotransmitter, and receptor systems is particularly interesting for biomedicine and pharmacology. In comparison to other animals, the monoaminergic innervation of the neocortex (the regional distribution of dopamine, noradrenalin, and 5-hydroxytryptamine in the cortex of the cerebral hemispheres) shows more similarities than differences between humans and simians (Lewis *et al.*, 1986). Unlike other laboratory animals, even various characteristics of the enzyme systems in monkeys (cholineacetiltransferase, arylsulfatase, monoamineoxidases) are mainly similar to those of humans (Mathew and Balasubramanian, 1984; Willoughby *et al.*, 1987). An unusual loss of urate oxidase in the process of evolution resulted in the retention of uric acid in the blood of hominoids, which may be a strong stimulant of brain activity (Proctor, 1970).

Cystathionine, an intermediate compound in the synthesis of cysteine, was discovered by Tallan and colleagues in 1958. This compound is found only in the brain and reaches high concentrations only in humans and simians (Promislov, 1974). Histochemical data indicate that the distribution of neuropeptide Y in the human prosencephalon is almost identical to that found in cotton-top tamarins. This peptide is localized with somatostatin, and both are related to normal motor function (Schwartzberg *et al.*, 1990). Enkephalin-like and dynorphin-like immunoreactivity in the globus pallidus and substantia nigra of both monkey and human (which according to the authors cited above have a common morphology), are very similar and their immunoreactivity depends on the total content of peptides in these structures (Haber and Watson, 1983). The concentration of sphingolipids (cerebrosides and sulphocerebrosides) is proportional to the progressive development of the brain in the taxonomic series of placental animals. The brain of the Java macaque is 'similar to the human brain' when studied in terms of the characteristics of these compounds (Levitina, 1982, p. 576). A decrease in norepinephrine and serotonin levels was found in the brains of people who committed suicide during depression, similar to that found in macaques during depression caused by social isolation.

It is very important to the study of biomedical problems that the action of various factors on the process and function of the brain is similar in direction in humans and monkeys. The social environment, which attains special importance in the primates, significantly influences the levels of neurotransmitters in monkeys, analogous to changes in neurotransmitters found during psychopathology in humans, in particular during self-aggression resulting from disruptions in social relationships (Kraemer, 1985). Age changes in the neurotransmitter systems of the rhesus macaque brain coincide with those in humans, which makes simians applicable for the study of similar processes in humans (Wenk et al., 1989).

The energy exchange in brain during malnutrition in young rhesus monkeys leads to the same processes as in humans; the blood supply and oxygen consumption of the brain is reduced and the formation of lactic and pyruvic acids decreases, etc. (Mehta et al., 1980). It was shown that sex hormones in humans and simians influence maturation of the prefrontal cortex and may play a leading role in the differentiation of complex cortical functions in primates (Clark and Goldman-Rakic, 1989). The spread of seizure (convulsive) activity in the cortex and hippocampus system of humans and simians is completely different than that in rabbits, cats, and dogs (Khassabov, 1973). Contraction of the cerebral arteries under the action of norepinephrine and clonidine occurs differently, and even in the opposite direction, in humans and macaques (M. fuscata) than it does in dogs, because it is associated with different types of adrenoreceptors (Toda, 1983). Such differences, as we shall see below, have critical significance for the choice of laboratory animals to test pharmacological remedies and to conduct experimental studies of normal and pathological aspects of the motor system and cardiovascular physiology.

Of course, there is also a *primate type of blood supply to the brain*. The circular route of the cerebral arteries is the basis for this distinction, whereby blood enters from the internal carotid and basilar arteries, a characteristic shared by all primates, including humans (Roskosz, 1984).

During the last 10 years, new data on the affinity of humans and simians also appeared with respect to the morphology of the *aging brain*. These included a decrease in the number and size of neurons, and a similar structure in age plaques of monkeys and humans, as well as other age-related changes. Brain aging in macaques and baboons is associated with the same processes as occur in elderly humans; cortical atrophy, decrease in brain tissue density, distension of the lateral ventricles, and changes in motor function and immunological status (Walker et al., 1988; Fernandez et al., 1994). Formation of senile plaques in the hippocampus, globus pallidus, and neocortical areas of senescent monkeys and other age-related anomalies constitute the neuronal basis for the decrease of memory in primates. I already discussed the similarity of the decrease in levels of catecholamines in cognitive deficits of elderly humans and simians. Investigators who have long-term experience with this problem consider senescent macaques (rhesus, in particular) a good model for studies of the brain, behavioral anomalies, and neuropsychology of elderly humans (Price *et al.*, 1994; Albert and Moss, 1996). I also note that senescent apes (chimpanzee) are the only creatures besides humans in which Biondi bodies are found in the choroid plexus (Oksche *et al.*, 1984).

The similarity in *pathological consequences* of surgical and other operations on the brains of humans and simians is very important and is used in experimental practice. Unlike results obtained in experiments with other animals, the disorders associated with ablation of the cerebral cortex and especially of the frontal cortex are similar in primates. So too are the circulatory disturbances caused by heat stroke, deficient excitatory mechanisms during the development of neurological syndromes, and pathological reactions to self-injury. Even the social changes resulting from injury to certain brain structures are parallel in humans and monkeys, as are the aforementioned manifestations of shock following transection of the spinal cord (Oxbury, 1970; Kling, 1986).

Many *neurological characteristics* are found only in humans and simians, indicating the deep evolutionary affinity of these primates. They share a similar Babinski reflex, the same inability of a newborn to turn from its back to its side without assistance (human and ape), adaptation to flicker fusion, spreading of seizure activity from the archicortex, and the ratio of different brain structures according to data on electrical activity (Mason, 1968; Khassabov, 1973). The form of electrical activity and the components of the electroencephalogram (EEG) pass through the same stages in humans and simians during ontogeny (Urmancheeva, 1986). The same diurnal physiological cycles and circadian rhythms are characteristic of humans and monkeys (excluding *Aotus*) (Scherbakova, 1937; Sulzman, 1983). A chronobiological model with rhesus macaques in transmeridional displacements was recommended as a useful means to study this problem experimentally (Tapp and Natelson, 1989). Apes are subject to hypnosis using the conventional methods of medical hypnosis.

It has long been known that simians possess a monophasic pattern of night sleep, with one long period, similar to the pattern of humans (Wein, 1970). The manifestations of natural sleep on EEGs are very close for the higher primates, but differ in other mammals. Baboons and macaques are used as models for study of the biology of sleep and the pharmacology of sleeplessness due to the electrical properties of their sleep, its periodicity, and phase relations (Balzamo, 1980; Lagarde, 1990).

The neurology of newborn and very young apes and humans is remarkably similar. A comparative study of infant chimpanzees using a standard scale of values during the first month of life showed that the standards of human child development coincide quite well with those of the chimpanzee. The study also demonstrated the role of the environment and indicated that motor activity develops more rapidly in the chimpanzee. The newborns react similarly both to social and nonsocial stimuli and exhibit a striking similarity in manifestation of the orienting response (Hallock *et al.*, 1989; Bard *et al.*, 1991). There was conspicuous similarity to humans in the performance of various reflexes by all newborn apes, although there were differences in the degree to which they performed them as well as some overall differences (Redshaw, 1989). Due to their relatively rapid development, young chimpanzees begin earlier to turn onto their abdomens and backs (at 2 months of age, rather than 4–5 months, as humans), and to sit and stand on their feet while holding onto the fence of a cage, as discussed above (Firsov, 1979). Macaques perform these activities much earlier, consistent with their yet more rapid development during ontogeny (Dienske, 1984).

Piagetian tests of sensorimotor development in orangutans from the age of 1 week to 4 years and in adult individuals (9–11 years) revealed the same 6 stages found in human children. There were differences, in that development occurred more rapidly in the apes during the early stage of ontogeny and more slowly the late stages (Chevalier-Skolnikoff, 1983). Chimpanzees and orangutans ultimately reach stage 6 of object permanence which is comparable to the progress reached by 18-month-old children on the same test (Call and Tomasello, 1996). Other neurological characteristics are presented below, in particular, in the description of the senses in primates.

Memory is one of the main properties of the nervous system which reaches a high level of development in primates, even in the lower simians (*Papio, Macaca*). The memory of primates is superior to that of carnivores, such as dogs and cats, with respect to both short- and long-term memory (Beritaschvili, 1974). The mechanisms and characteristics of memory in simians are very close to the human ones, especially in apes, which are considered the closest model of the critical step to humans (Firsov and Moiseeva, 1989). E. Gower (1990) showed that the Java macaque retains visual information for at least three years. A study conducted at the Sukhumi Center and described by L. N. Norkina, showed that a rhesus macaque maintained a complete set of complicated conditioned reflexes for 8 years. In another case, an anubis baboon recognized the physiologist who had brought it from Africa 6 years earlier (Fridman, 1979b).

Rhesus macaques have the same short-term mechanisms of information processing as people. This was shown, in particular, in a comparative study of image memory in a monkey and a 21-year-old graduate student. Unlike the macaque, the authors noted, the student 'did not get any reward' for performing the tasks equally (Sands and Wright, 1982, p. 1333). Other investigators also report a great similarity in the visual memory of monkeys and humans, and what is more, they have shown that certain mnemonic processes may be more effective in macaques than in humans. It is noteworthy that prominent authorities propose that rhesus macaques are better in learning and memory on certain tests than any other lower simian and in fact, approach the apes in this respect (Washburn *et al.*, 1989). It is likely true that the mechanisms and processes of memory in monkeys are at least similar to the nonverbal part of human memory. We may assume, however, that chimpanzee memory is far superior than that of rhesus macaques due to a more highly developed cortex.

It is interesting that the memory of monkeys and humans is not only comparable following the destruction of certain brain structures, but there is also an absence of differences following destruction of certain elements of the same systems activated in memory processes, for example, in the amygdala-hippocampus (Overman *et al.*, 1990). Injury to other parts of the same hippocampal area led to long-term disturbances of memory in macaques, similar to the data on amnesia in humans with injury to the same brain areas. The action of neuropharmacological agents is consistent among the primates, but may differ in other animals. An example of one difference among many is flumazenil, which improved learning and memory in mice, but did not give such results in *Saimiri* (Rumenik *et al.*, 1989) (I discuss neuropharmacological problems below).

The change in cognitive function with age, and in particular, the senile change for the worse in different types of memory, simian models of human amnesia, the neuro-psychological analysis of these disturbances, as well as studies of the biological and pharmacological aspects of memory, are successful areas of investigation in medical primatology. In this respect, many different species of monkeys are 'unexcelled animal models of human cognition' (Goldman-Rakic, 1996).

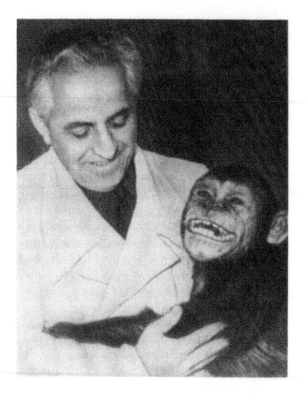

Figure 4.4 Professor Gregory O. Magakyan, Chief of the Clinic (Institute of Experimental Pathology and Therapy) with the laughing male chimpanzee Eman (Sukhumi, 1969).

C. Darwin was the first to report in the scientific literature the similarity of facial expressions in humans and simians (1872). He also related that an ape is able to express almost all human emotions facially, except amazement, surprise, and disgust, a finding since confirmed by J. van Hooff (1981). N. Ladigina-Kohts (1935), moreover, carefully described these emotions in chimpanzees, and L. Firsov, who had more than 30 years experience in research with apes, also confirmed (private communication) that he observed these emotions in chimpanzees. Darwin thought that the similarity of facial expressions in primates was a consequence of their facial musculature. Indeed, anatomists confirm this similarity in general, but there are data that the human muscles used in laughing are not developed in apes (this work was carried out with fetuses) (Goldschmid-Lange, 1976). However that may be, apparently typical laughing, weak and loud, and smiling have been reported by many investigators who study chimpanzees (Goodall, 1971; Firsov, 1982; van Hooff, 1989; see Figure 4.4, Sukhumi Center). At least one original phylogenetic interpretation of the smile and laughter in primates has been given (Preuschoft, 1996). Although smiling and laughing, as well as weeping (without tears in simians), have always been considered specific human characteristics, they have clear equivalents in simians.

It is noteworthy that experimental research on the lower simians also demonstrated elements of similarities in expression with humans, in particular, in the recognition of expressions. In carefully controlled experiments, macaques successfully differentiated various expressions of human and monkey faces in photographs and slides, although the Japanese macaque confused human *disgust* and human *sadness* (Kanasawa, 1994). Apes, and chimpanzees in particular, demonstrate these capabilities in an even more impressive way. It is also well-known that primates recognize familiar people and other simians in slides and photographs. It has been shown, moreover, that their electrocardiograms undergo distinct changes at those times, as does their heart-beat, suggesting that these physiological indices may serve as objective indicators of their cognitive function, a sort of *lie detector*. It is assumed that qualitatively similar processes in humans and even macaques are the basis for similar neural mechanisms of information-processing, such as those used when viewing facial expressions, although there is a difference in the level of hemispheric asymmetry of the species (Boysen and Berntson, 1986; Perrett *et al.*, 1988).

Given the discussion and analysis of the facts above, it is not surprising that nonhuman primates share with us the ability of self-recognition in a mirror. Self-recognition is not merely another common neurological characteristic, it is also very close to reasoning, if not on the same level as it. As I noted above, a normal human child at a very young age (approximately 18–24 months of age) does not recognize itself in a mirror. Intellectually retarded individuals and mentally ill and psychotic people do not recognize themselves in a mirror either. They do not identify themselves and they do not have self-awareness. Gordon Gallup, of the State University of New York, Albany, who was mentioned above, was the first to demonstrate this high intellectual ability in common chimpanzees (Gallup, 1970, 1979, 1997). Regardless of which theoretical lances are broken over this issue, it appears clear that simians, unlike any other animals, share this unusual characteristic with humans.

It is presently known that both species of chimpanzee recognize themselves in mirrors (some individuals from the age of 3.5 years, but usually by 7–8 years of age [Povinelli, 1993]), as do orangutans and gorillas, which may do so with somewhat more difficulty. Gibbons show interest in their images and macaques occasionally attend to a mark on their own bodies (visible only in the mirror). Even among the species with self-recognition, there are individuals that do not demonstrate the ability (Mitchell, 1996). For example, those apes which either could not learn gestural communication in an experiment or were kept in social isolation during their early years were not able to master self-recognition. Macaques, baboons, and capuchins did not identify themselves in a mirror but rather, treated their own image as a congener. On the other hand, macaques are able to distinguish objects and coordinate their activity with mirrors (Itakura, 1987). This subject has generated a large amount of scientific literature and the results are applicable to research in animal behavior, psychology, and psychiatry.

Thus, there is a general evolutionary, phylogenetic and taxonomic affinity between simians and humans. This affinity is based on a specific similarity in general anatomical characteristics, in a unique homology of neural structures and interactions, brain biochemistry, bioelectrical activity, and neural blood supply. These similarities, moreover, undergo parallel changes during early development and aging. This physiological level of functioning creates the conditions for a high level of associative learning and integrative activity in the central nervous system of primates which is unmatched by any other species. Such activities include special sensorimotor perception and reaction, the highest level of intersensory integration of multimodal stimuli (cross-modality), the most complex conditioned reflexive activity in the animal world, and in particular, orienting reflexes and internal conditioned inhibition (Batuev and Kulikov, 1983;

Karamyan and Sollertinskaya, 1990). Simians also resemble humans in development of the central and peripheral mechanisms of the sensory and motor systems, especially of vision and manipulatory skills. This coincidence of neurological characteristics in the higher primates provides surprising human-simian analogues and research possibilities for neurophysiology and general physiology. The development of memory, complex social life, and strong communicative abilities in hominoids requires self-recognition and the mental capacity for preverbal language and abstract thinking. The latter, in turn, provide all simians with unusual capabilities of analysis and generalization. Given this degree of complexity, what level of intellectual *behavior* might we expect?

In several parts of my discussion, I touched on the problem of intellect and behavior in primates, especially the ability to use tools. Let us consider this in more detail. We should recall that the beginning of primatology as a science was associated with the study of anatomy and physiology (see Chapters 1 and 2). Intensive studies of the behavior of primates were initiated in the first decades of the 20th century (Ladigina-Kohts, Keller, Yerkes, Pavlov, and others). Those investigators were struck by the intellect of simians, especially the *anthropomorphic* simians, as apes were called more and more frequently. It became clear that other animals that were thought to have high intelligence (bears, dolphins, wolves, dogs) were, nonetheless, inferior to simians. The information which was obtained on these issues in the second half of this century (from research in both the natural habitat and in laboratories) surpassed all previous knowledge and suppositions.

Object manipulation by monkeys is of considerable interest. This activity progresses into actual tool use in the apes, but even the lower simians surpass all other animals in this respect (Firsov, 1988). A newborn monkey starts to analyze objects in the outside world within several weeks after being born. This process is practically identical for baboons, capuchins, saimiri, guenons, and apes, and even for human children (Ibid., p. 219). Object manipulation by capuchins, which was mentioned above, deserves special attention.

It was well-known for some time that the behavior of capuchins in captivity is unusual. At the Sukhumi Center, they simulated floor-washing with a wet rag and they were able to 'wash clothes', hammer nails with a hammer or stone, and break nuts or other hard objects. Although it appeared that they had been trained to imitate humans, no one taught them these activities; they were learned spontaneously. They facilitated their activities with their prehensile tails, suspending themselves from the cage above the ground and thereby, freeing up their arms and hands for manipulation. Experimental studies confirmed their high level of object manipulation in the cage. Despite their individual skillfulness which can be convergently compared to that of a chimpanzee, capuchins were not successful in protocultural transmission of their experience. Their activity lacked cognitive elements; they did not modify the tools (Visalberghi and Fragaszy, 1990), although they made sponges from paper napkins that they used to soak up water (Visalberghi and Fragaszy, 1987). Capuchins and chimpanzees have similar behavioral capacities, brain laterality, socioecological relationships, and even sexual behavior, but there are significant differences as well (Visalberghi, 1997).

Spontaneous object manipulation by lower simians as well as novel forms of tool use in laboratory experiments is well-known. Other complex forms of behavior by macaques and baboons are also described, such as their ability to extinguish a fire with a cup of water, throw stones at people (but not in group battles among themselves [baboons]), use stones, herbs, and leaves to remove blood stains and dirt (Katz, 1972; Hamilton *et al.*, 1975), and groom their offspring. A male *Papio hamadryas* at the Sukhumi laboratory of N. Goncharov did not wait for treatment of its arm after a blood test was conducted by a laboratory assistant, but instead, took a piece of cotton-wool from the table and wiped off the blood itself. Westergaard (1988) reported that lion-tailed macaques not only used twigs to obtain syrup from containers, but they also modified the twigs for this purpose by changing their size, something which was never done by *Mandrillus sphinx*. A female hamadryas baboon called Keffi had to select the same object it was shown in a choice-reversal situation of a matching-to-sample experiment conducted at the Sukhumi Center by a zoological technician, Vera Tretenko. Keffi quickly mastered this task, for which it received candy as a reward, and did it with virtuosity. When there were no more objects to select for reward, Keffi handed its tail to the technician; when that did not work, it handed the tail of its baby (Fridman, 1979b). One of the first scientists who described tool use in lower simians in detail was professor N. Yu. Voytonis (1949).

Of course apes, and especially chimpanzees, demonstrate a high level of object manipulation. I should add some more detail to the previous description of tool use by chimpanzees, including the ability to use tools precisely and delicately, to construct them in advance, and to know where they are kept, when necessary. It is known that chimpanzees can use many tools to obtain a single objective, i.e., they perform a series of coordinated actions (Derjagina, 1981; Brewer and McGrew, 1990). In one case, a female chimpanzee obtained honey from a bee-hive in an old stump, sequentially using four different wooden tools prepared in advance. Females are more skillful in preparing tools than males, but the latter use stones, sticks, and branches in battles while standing on their hind limbs.

When they use stones to break nuts or other hard fruit, chimpanzees consider the quality and location of the stones, sometimes traveling a distance of 500 m to obtain a particular stone. Each tool (stone, branch, stick, or rod) has an individual purpose, but sometimes they are used in combination. Young chimpanzees learn to obtain their portion of tasty morsels of prey by useful acts of imitation (Boesch and Boesch, 1990). Chimpanzees in different regions of their habitat, moreover, have specific peculiarities. For example, sticks with the ends ground into brushes to gather a maximum quantity of ants were found in Cameroon, but not in other places. They sort through their sets of tools and know very well the purposes for which they are used (Sugiyama, 1985; Sakura and Matsuzawa, 1991). They usually have a plan of action, such as when they use a lever (to move a stone), a pipe or reed (to suck water out of a lake), and of course, in other situations (Firsov, 1988). 'Hygienic' use of tools by chimpanzees is wide spread and includes, picking their teeth with a broken stick, scratching their backs with a stick, and using a stick when it is not feasible to reach an object by hand.

As was already mentioned, pygmy chimpanzees manifest great creativity in manipulating tools such as sticks, rods, and leaves which they construct for themselves. They scoop up water with a cup or plastic box, and they can make a swing by throwing a rope over a crossbar (Jordan, 1982). Less is known about the activity of gorillas and orangutans with tools. They also manipulate objects, both in captivity and in their natural habitats, but it is unlikely that they do it in such diverse ways as chimpanzees (Call and Tomasello, 1994; van Schaik and Knott, 1998). As for chimpanzees, I also mentioned that they have a 'Chimpanzee culture' which is quite comparable to the culture of early humans and even to elements of such culture in modern aborigines of underdeveloped countries (McGrew, 1987, 1992; Wrangham *et al.*, eds, 1994; Joulian, 1996; de Waal, 2001). We may assume that additional forms of tool use by chimpanzees will continue to be found and in fact, the number of examples is increasing (Alp, 1997). This is logical and natural; if hands had not evolved for use as hands, there would only be fangs!

Thus, no animal on the earth reaches the level of object manipulation that simians do. In the apes, tool use becomes so intense that it even inhibits feeding activity in some experiments (I. P. Pavlov). The tool use of chimpanzees is a *critical link* on the path to the performance of labor by people, a primate aromorphosis in evolution. Tool use became possible because of the special structure of the brain and hand in chimpanzees, and due to their memory and ability for preverbal concepts and abstraction (Firsov, 1988). I should add that tool use in apes is one of the best indicators of high intellectual activity.

Robert Yerkes (1943) proposed that chimpanzees are appropriate for the study of thought, symbolic processes, and language. It was a daring proposal, even during his time. It appears that this insightful scientist, who knew primates very well, not only made proposals, but practically drew up the plans for the research of the 1970s–1990s (and possibly for an even longer period). It is possible, however, that even such a prophet would be surprised by the facts that have been uncovered about the cognitive abilities of simians.

Chimpanzees are able to sort objects, not only on the basis of recognition, but by comparing their qualities, identifying characteristics of close similarity, and even by complementation (Tanaka, 1995). When constructing complex objects from blocks, they initially categorize material the way humans do, not only by the quantity of the material, but also by its quality. Black blocks are used to construct a black tower and white blocks are used for a white tower, something which is comparable to the ability of children who are 3.5 years of age or older (Matsutsawa, 1990). Chimpanzees are not only able to add objects, but they also use their symbolic equivalents, such as numbers up to 4 (Boysen and Berntson, 1989). Chimpanzees can learn counting and summation with a computer, which is surprising and the basis for much speculation (Rumbaugh *et al.*, 1989). Chimpanzees can associate half of a jug with half of an apple and a quarter of a disc with a quarter of a fruit. Rhesus macaques and capuchins can also distinguish a characteristic of relative value in an experiment, but not as well as chimpanzees, which generalize, abstract, and perform preverbal activities at a higher level (Malyukova *et al.*, 1990). Nevertheless, macaques can also perform certain numerical tasks and saimiri can distinguish 7 points from 8 and a heptagon from an octagon (Thomas *et al.*, 1990). On the other hand, it is unlikely that Roxellane's monkey (*Rhinopithecus roxellanae*) has the ability to generalize (Lin *et al.*, 1989).

Macaques are able to perform cross-modal transfer in a manner that is comparable to that of human children. They, as well as baboons, can be trained by point gestures, but bonobos do this without training. The development of cognitive abilities in newborn rhesus monkeys, as studied in two complex tests, was similar to that of newborn human children in that performance depended on their sex (and the possible influence of androgens on the level of maturity in certain structures of the brain) (Overman *et al.*, 1996). Rhesus monkeys form a concept 'human' and distinguish categories, not on the basis of remembering specific examples, but by doing it in a human way based on a combination of specific characteristics. Four-year-old rhesus monkeys performed faster (but not as accurate) than children of 4–8 years on a task in an experimental test which examined short-term memory, attention, learning, perception of time, motivation, and color and position discrimination (Gunderson *et al.*, 1990; Paule *et al.*, 1990). The level of cognitive function was higher in rhesus monkeys than in talapoins (*Cercopithecus talapoin*) and of course, higher than in the slender loris (Cai *et al.*, 1990).

Baboons have a very high cognitive potential, one of the highest among monkeys. It is not necessary to refer to ancient Egyptian frescos, where baboons are depicted as assistants of God and humans, to appreciate this. It is not that they merely use stones and other objects (see above). They are also able to conduct cooperative activities in fights, raids on plantations, and in experiments to get food. B. Grzimek, in the review by D. Candland (1987), reported that the absence of two sheep from its group was apparently detected by a baboon-shepherd, a conclusion that was supported by the baboon's frantic searching for its charges.

C. Darwin (1859) described another situation where the leader of a baboon group dashed into a troop of dogs surrounding its youngster, grabbed the youngster, threw it onto its back, as females typically do, and ran away in front of the stunned eyes of its enemies. This is entirely possible. It is known that one baboon leader prevented a boulder from rolling down a mountain until the youngsters in the group had passed the dangerous path. Female baboons at the Sukhumi Center regularly hid sweets given by tourists from the view of the leader. They would hide their hands with the sweet behind their backs or behind the bodies of their youngsters, while maintaining non-committal expressions on their faces.

Baboons, like humans, are definitely able to practice trickery, one of the highest cognitive achievements of humans. The employees of the Sukhumi Center remember very well the history of the escape of all their monkeys, regardless of species, from 'monkey house' #5 (among more than ten outdoor cages). The cages were locked with a T-shaped key. The doors were opened, however, by a hamadryas baboon called Don, who escaped at night through a hole in the fence. The head of the Center, Yury Ivanovich Kondakov, who worked at the vivarium for 42 years, told the author that the baboon opened the cages without a key. In another situation, an anubis baboon that had been wounded by hunters in a Nigerian forest, continuously vocalized in distress as it tried to escape from the local hunters who were chasing it. At some point, the hunters became confused when they began hearing distress vocalizations from a number of different sites. It appeared that other baboons that had not been wounded were imitating the wounded one, thereby, facilitating the escape of their wounded fellow. One of the hunters who lagged behind, apparently saw the escape of the wounded baboon (private communication from Professor G. A. Annenkov, a participant in an expedition to Nigeria).

To continue the theme of cognitive abilities in simians, it is necessary to point out some peculiarities in the natural group behavior of primates. As I mentioned above, the communities of simians reach the highest level of social relations other than humans, something which goes beyond the requirements of 'biological demands' (Panov, 1983; McGrew *et al.*, eds., 1996). The simian community, for the first time in phylogeny, exhibits a complicated social organization with subgroups consisting not only of relatives and sex partners, but also other associations based on interests that are not always clear to human observers. For the first time, grandmothers and grandsons are distinguished here, to say nothing of the intimate mother-infant bond. Groups of youngsters and unimale groups are also distinguished. Social relationships of chimpanzees among themselves may be more important than sex differences; quite often they acquire the social status of their mothers. Consideration of elderly individuals can be seen in the group, as well as widespread attention of grandmothers to the first offspring of their daughters. When there are several daughters, the attention is paid to the offspring of the youngest daughter, for example, as among guenons (Tich, 1970; Hamai, 1994).

Individual patterns of chimpanzee social relationships (and even the associations of hamadryas baboons), including dominance rank relations, are comparable to those of nontechnological societies of people (Bailey and Aunger, 1990) or to not very distant hominid ancestors of humans. Straight bearing, beetle-browed or angry expressions, arrogance, and self-confident behavior (associated with increased levels of testosterone [Mazur, 1985]), are characteristic of both simian and certain human leaders. Many aspects of primate social relations are based on the reproductive needs of the species which, among people, acquire important social significance. Without being too extreme, we can propose that aggression in humans may have a biological foundation with its roots in the primate family tree (Somit, 1990).

Not only are certain social experiences required to survive in such a social environment (which can be fully studied for the benefit of humanity only with the use of primates), which includes the need for communication, but great and varied cognitive problems must also be solved. Both species of chimpanzee and the other apes are well-prepared to master them.

At present the scientific literature on the cognitive abilities of apes is like a shoreless sea. In addition to what was discussed above, I limit myself to an incomplete listing of these abilities from the data of several investigators. Chimpanzees are not only able to evaluate the quantity, perform addition, recognize the middle of an assembly of objects, generalize and analyze, and remember and foresee, they can also make a request (if they are taught a symbolic language, discussed below), ask questions, reply to questions, and make spontaneous statements. They have a sense of humor, they are able to joke, tease, and insult (in particular, they can insult an attendant if he refuses to comply with their requests). Finally, as was mentioned above, they can lie, be cunning, express sympathy, carry out orders, and refer to objects that are absent from view. It is likely that their cross-modal transfer ability is unequaled by any animal other than human beings. They can understand causality and forethought, they distinguish a chisel from a screw-driver, and they use these tools in accordance with their purposes. They can learn to operate a computer, not only under the tutelage of a human instructor, but by merely watching it from outside, as the famous bonobo, Kanzi, did in the studies of Sue Savage-Rumbaugh. The same Kanzi cut a thick rope with a sharp stone of relatively large size and cut a thin rope with a smaller stone, carefully evaluating the degree of sharpness and mass of the stone tool.

Apes share food with their congeners, such as chimpanzees in the forest after a successful hunt. They also do this in a laboratory experiment where they get a reward for successful responses in an experiment. They even share food with people. For example, when in a good mood, the 2-year-old gorilla, Bola, at the Sukhumi Center, would give delicacies to the attendants (the author of this book was not only a witness to these actions, but also the object of the actions). Chimpanzees can force their fellows (or a female can force its child) to deliberate actions, to acquire a habit, or even to learn symbols. They can judge the abilities of their partners to perform an activity (as hamadryas baboons do), and even determine what a person or other creature knows about certain events and the states of knowledge and ignorance in others (Povinelli et al., 1989; Whiten, 1998). It has been known for a long time that chimpanzees learn many household and ethical practices of people when kept in houses.

What was described above regarding the cognitive abilities of higher primates may cause skepticism or suspicion of anthropomorphism in one or another reader, as has already happened many times. My assessments, however, are based on the data of many investigators, among which are very disciplined researchers who documented their results on cinema, photographs, video, and audio records. Some of them devoted their lives to the study of behavior and the psychology of primates. The data that support my conclusions are found in the publications of Goodall, Washburn, Firsov, Teleki, Mason, McGrew, Nishida, Jolly, B. and A., the Gardners, Premack, Rumbaugh, Savage-Rumbaugh, Fouts, Chevalier-Skolnikoff, Reynolds, Menzel, Patterson, Boesch-Archmann, Boesch, Boysen, Byrne, Whiten, DeWaal, Tomasello, Call, and others whose publications are listed at the end of the book.

Nevertheless, sometimes (now significantly less frequently than before) statements appear by some investigators who deny the cognitive abilities of primates, including chimpanzees. They do not find differences in behavior between the apes and monkeys, they consider it completely impossible for simians to have the ability for abstracting, and they declare that all cognitive achievements of primates made in laboratories result merely from training (Goustard, 1987). This archaic scientific disease is overcome by facts, the strongest argument of knowledge. Factual evidence can be interpreted in various ways, but the facts remain facts. Once A. Gardner said with a bit of passion, 'It is not very interesting what was called a language by linguists: what is done by chimpanzees, this is interesting.' (I emphasize that this was said in the heat of a dispute about 'speaking chimpanzees'; at present many linguists consider the language ability of apes to be very high). M. Goustard described an interesting example himself. A chimpanzee with a stick that could not reach a fruit floating in a river, ran downstream with the stick to a narrow area in the river where he waited to get the desired catch. I assume that this example, as well as 'termite-fishing' with a specially prepared stick, and teaching a youngster the hand gestures of hearing-impaired people, reveals more cognitive ability than a chaffinch carrying a stick in its beak.

The same scientist referred to above even denies the prerequisites for drawing ability in chimpanzees, which opposes the opinion of others who find that simians have certain aesthetic perceptions. The latter scientists believe that the drawings of chimpanzees possess proportionality, dynamics, centralization, adjusting the drawing to the sheet format, and finally, that chimpanzees have the desire to draw even without any stimulation by people (Glaser, 1986; Boysen et al., 1987). The problem is not solved as yet, and it awaits solution (Vancatova et al., 1996). Judging from the behavior of young chimpanzees at the Sukhumi Center that were coordinating their behavior with a man's singing (they were moving around and rhythmically slapping their hands on their body), I suggest that they may even have a sense of musical rhythm.

Chimpanzees are not only able to transmit information, they also request transmission of information from others and can transmit information in an experiment using television images. If taught the appropriate symbols, they are able to exchange symbolic information, including symbolic cross-modal transfer, similar to humans (Menzel, 1984; Savage-Rumbaugh et al., 1988). The investigators in this field have nurtured the most exciting cognitive ability of which apes are capable, the ability of apes to use human language with a variety of symbols.

Despite the analogous speech apparatus of humans and other primates in terms of anatomical structure, including brain structure, and despite the similarity in abilities to analyze vocalizations, the anatomical structure of the oral cavity, larynx, and the neural connections of the speech complex are significantly different, as was mentioned above. Hence, the failure of all attempts to teach chimpanzee human speech. With the development of research on primates in the 1960s, new approaches were found to study symbolic language in nonhuman primates. This research caused considerable emotion and discussion in the 1970s–1980s, not only among the investigators involved, but also among the public. The disputes would have been more intense, if not for the fact that new data were available on the behavior, genetics, biochemistry, DNA, and anatomy of apes, which I considered above. Hundreds of publications, many books, and numerous scientific meetings were devoted to the problem of the linguistic competence of apes. The objectives of this monograph limit us to a cursory review of this topic, but one which is necessary to describe the background of this area as it relates to medical primatology.

At the end of the 1960s, Allen and Beatrice Gardner at the University of Nevada shocked the scientific world with news that they had taught a young female chimpanzee, Washoe, signs of the American Sign Language for the deaf (Ameslan). For more than a quarter of a century these scientists have been publishing data, in collaboration with their students, on various aspects of communication between humans and chimpanzees and between two different chimpanzees, using this human language of *signing* (Gardner and Gardner, 1972, 1990, 1994).

Subsequently, in a later experiment, this female chimpanzee actually taught signing to its offspring. It was found that chimpanzees can communicate with each other by signing and even speak to each other by this means as children do (Fouts, 1989). At about this same time, the Premacks began teaching human language to Sarah, a 5-year-old female chimpanzee, using plastic symbols (Premack and Premack, 1972; Premack, 1983). At the beginning of the 1970s, D. Rumbaugh and S. Savage-Rumbaugh started to communicate with a young female chimpanzee, Lana, using a computer and a language of lexigrams called *Yerkish* (Rumbaugh, 1973, 1995; Savage-Rumbaugh, 1986; Savage-Rumbaugh *et al.*, 1993; Savage-Rumbaugh and Levin, 1994). Other investigators also taught symbolic language to chimpanzees (Boysen *et al.*, 1991) and Francine Patterson was successful in teaching signing to the female gorilla, Koko (Patterson *et al.*, 1988). At about this time, others determined that both higher and lower simians had the ability to recognize and respond to human speech, similar to humans (Kuhl, 1988; Sinnott, 1989).

I cannot discuss completely the surprising details of the results these investigators reported, but I must say that these apes displayed much more ability that one might suppose. They showed enormous sensorimotor and cognitive abilities which were qualitatively very close to human ones and which took many scientists by surprise. Using several standardized tests for evaluation (transfer indices, transfer-of-learning skills, etc.), it was found that the apes could be placed at about the intellectual level of children of 2 to 3 years of age. They are able to learn, understand, and use symbols that reflect the functional properties of words and to perform cross-modal transfer using the relative characteristics of the symbols. They demonstrated semantic abilities at the level of children and the syntax and grammar of the symbolic language as a whole at the level of human infants. Moreover, when they were placed in a human language environment at an early age, they eventually began to understand the oral speech of people (English language). They understood not only separate words and simple sentences, but they also understood new sentences that required the syntax of oral speech (at about the level of a 2.5-year-old child). Without external help, the bonobo, Kanzi, understood approximately 70% of a total of 600 new sentences (Rumbaugh, 1995, p. 722). New data of laboratory research of symbolic language in chimpanzees require revision of the communication of this ape in the wild (Fouts and Fouts, 1999).

These investigations once more showed the clear differences in the taxonomic hierarchy of the order Primates. Despite the considerable cognitive abilities of capuchins, baboons, and rhesus macaques, these monkeys are not capable of mastering a symbolic language of people. This has been demonstrated only for the apes, and primarily for the chimpanzee (although the gorilla shows similar abilities), including the understanding of English speech. The pygmy chimpanzee is the most talented in this latter respect and among them – Kanzi (as well as common chimpanzee Washoe). On the other hand, the apes are well behind humans in intellectual ability, in particular, in terms of distinct speech. It has been demonstrated, moreover, that there are exceptionally clever individuals and perhaps, even geniuses among the simians as well as 'block heads' (those of lesser ability). We have also seen that those that recognize themselves very well in a mirror and that successfully solve cognitive problems, master language better than others. This result supports the concept, yet again, that language is a cognitive ability and that it reflects intellect.

The investigations described above provide grounds to propose that the foundations of language are the same for humans and apes, although the latter do not possess the physiological basis for speech. There are no total differences among the humans and apes with respect to language. Though the apes lack speech, they are capable of preverbal conceptual thought. The biological continuum of Darwin really exists, therefore, and is reflected in a psychological (behavioral) continuum (Byrne, 1995). *Speaking* apes have allowed us to look into our evolutionary past. Once again they clearly demonstrated the importance of the language environment and early education for children, as well as the significance of language itself for the development of intellect. D. Rumbaugh (1995) assumes that *understanding* is likely more indicative of intelligence than is speaking. This is a very important assumption which, in my opinion, has direct relevance to the investigation of Alzheimer's disease, a severe scourge of elderly people which is modeled most effectively in simians. The investigations with primates (apes and monkeys) allow us to discover the nature of cognition, speech, hearing, and perception. They provide theoretical approaches for improving the performance of people with retarded speech and mental function and for studying actual brain diseases. They help us to know ourselves (Savage-Rumbaugh et al., 1998).

I note once again that monkeys also have a high level of cognitive ability. In fact, it is sometimes difficult to detect a difference between monkeys and apes (Heyes, 1996). This supports the idea that monkeys can be used successfully in studies on the biology and pathology of the brain as well as in medical primatology as a whole.

Even after this cursory review of the similarities between the brain and behavior of humans and simians, it is not difficult to imagine the possibilities for use of primates in the study of human biology, physiology, and pathology. Let us expand upon this topic by reference to the available statistics. Table 4.1 shows the annual production of publications in the field under consideration during the period for which I quantitatively analyzed the progress of medical primatology (Fridman et al., 1990, p. 105).

It can be seen that the number of publications increased progressively, excluding 1976, when there was a decrease, mentioned above, associated with the prohibition against exporting primates from several countries. The number of publications grew 4 times during a period of less than 20 years. There is no doubt that, at present, the number of publications significantly exceeds 2,000 per year (note that the

Table 4.1 Number of annual publications on research in the areas of neurophysiology, neuroanatomy, neuropathology, and psychology conducted with nonhuman primates during the period, 1967–1985.

1967	1970	1975	1976	1985
480	1077	1405	1130	1836

neuroinfections, neuropharmacology, and physiology of other systems and organs are not included in the figures given in Table 4.1).

It is apparent from the previous discussion and from the history of medical primatology that investigations of normal and pathological neurophysiology, neuroanatomy, and behavior of simians provide more convincing grounds for extending the results to humans than comparable research on any other animals. Significant achievements in this field, moreover, were awarded Nobel Prizes. At present, the function of various brain areas and physiological systems are also being studied with primates. The opportunities for research in these areas are practically inexhaustible. I note that basic research on the brain, which is summarized in various monographs, is becoming more and more *primatological*; the conclusions are more and more frequently based on experimental studies conducted with primates.

In the last several decades, simians have become very important in studies of human *neuropathology*, in large part because it became clear that rodents and other laboratory animals are not suitable models for many conditions or their use is limited biologically. Because of the similarity in cerebral blood circulation of humans and monkeys, it is presently very important to use monkeys effectively to study ischemic diseases in various neural areas and for clarifying the etiology of brain insults and infarcts. Acute local ischemia of the brain in simians is quite similar to massive infarcts in the hemispheres of humans with cranial damage. As a whole, focal experimental ischemia in macaques and baboons is a very convenient and informative model of the corresponding human pathology. The characteristics of sensation, including the EEG data, are very close (Garcia *et al.*, 1983; Vajda *et al.*, 1985). Cerebrovascular lesions during experimental thrombosis and chemical intoxication in monkeys constitute models for the study of similar human disturbances (Sheffield *et al.*, 1981).

In this book, I have repeatedly emphasized the similarity in cerebral and peripheral mechanisms of the motor system in humans and simians. It was noted that the homology of convulsive activity in the brains of primates is unlike that of other animals. I may add the following. The baboon (*P. papio*) is an excellent model for the study of spinal reflexes; the extrapyramidal syndrome in capuchins (*C. apella*) is caused by neuroleptics in the same way that it is in humans (and this effect can be used to screen neuroleptics). Extirpations and other injuries in various brain structures of macaques induce consequences similar to those of injured people (injury to the central frontal gyrus causes paralysis in humans and simians, but not in other animals). Furthermore, we can obtain a model of the human subcortical syndrome only in monkeys. It is not difficult to understand, therefore, that monkeys are extremely useful in experiments designed to study numerous motor disorders, such as hyperkinesis, hypokinesis, dystonia, dyskinesia, and neurodegenerative symptoms and syndromes (Urmancheeva *et al.*, 1966; Eliava *et al.*, 1986; Jenner, 1990).

The baboon (*P. papio*) is probably the best animal model for the study of photo-sensitive epilepsy (Menini and Silva-Barrat, 1990), although there are published data on the successful use of other monkeys to study this convulsive state, for example, various species of macaques (Schroeder *et al.*, 1990) and saimiri (Ronne-Engstrom *et al.*, 1995). Similar investigations were initiated at the Sukhumi Center as early as the 1930s by a prominent scientist, P.V. Botchkarev, because spontaneous attacks occurred in hamadryas baboons. The author of this book saw complete articles written by Botchkarev at the Center, but all the articles of the Sukhumi primatologists were lost during a campaign of political repression in the Soviet Union and thus their research was not published (the original articles are kept in the library of the Sukhumi Primate Center). Further research on epilepsy in baboons was conducted by the French investigators, R. Naquet and his colleagues, and other researchers who obtained valuable data on the etiology, pathogenesis, and therapy for different forms of this ancient human disease. The models were obtained by intermittent photostimulation, injections of aluminum oxide into the sensorimotor cortex, and by other methods (Naquet, 1973; Naquet *et al.*, 1995; Menini and Silva-Barrat, 1998).

I already mentioned that monkeys react to various stimuli similarly to humans, including pharmacological, trauma, intoxication, inborn and aging defects of brain structures, and others. The similarity in pathology of the motor system and the similarity of the behavioral and cognitive spheres are extremely important for biomedicine. Lead intoxication causes disturbances in activity, attention, memory, and the ability to learn in macaques, similar to humans (Rice, 1987). Dyskinesia (and aggression) in rhesus monkeys during treatment with chlorpromazine has common features with people treated with the same drug (McKinney *et al.*, 1980). Injuries or spontaneous degradation of the cortex leads to similar pathology in newborn and elderly macaques, chimpanzees, and humans, up to degenerative dementia (senile mental deficiency) (Pilleri, 1984).

In light of the research described above, it is not surprising that monkeys became the indispensable experimental animals for the study of one of the present day's most distressing pathologies, Parkinson's disease. The functions and dysfunctions of pre-frontal-striatal circuitry, which have unique homologies in humans and monkeys and which are significantly different in other animals, in particular, rodents, are among the cardinal foundations of cognitive dysfunction resulting from neurodegenerative diseases (parkinsonism, Alzheimer's disease). There is an enormous potential in these investigations, therefore, for the use of monkeys, for example, marmosets (Roberts, 1995).

Disturbances in motor mechanisms, memory, and vision, and finally, the cognitive dysfunctions resulting from these illnesses, can be modeled with different species of monkeys. Destruction of the substantia nigra and other areas by narcotics, neuroleptics (tranquilizers) and other substances, death of the corresponding neurons, the 20%–30% deficit of dopamine, which leads to parkinsonism and inevitable death, such as that seen in drug addicts and elderly people. At present, all these lesions are studied successfully in large-scale experiments with monkeys, comparable in dimension, perhaps, only with the study of AIDS. In addition to marmosets, some other species of monkeys are also used for these purposes, namely, *Cebus, Saimiri, Cercopithecus, Macaca,* and *Papio*.

Attempts to model parkinsonism in monkeys and related research aimed at curing motor disturbances are carried out in various ways. The correspondence between injury in the substantia nigra and Parkinson's disease was demonstrated by experimental destruction of this brain area in baboons (Viallet *et al.*, 1981). In 1983, it was found that 1-methyl-4-phenyl-1,2,3,6-tetra-hydropyridine (MPTP), a by-product of the synthesis of a meperidine analog and a neurotoxin, caused a disease similar to parkinsonism (Blume, 1983). Since that time, this compound has been used to obtain experimental models in primates and other animals. Despite some possibility of using rodents and dogs in these experiments, it soon became clear that in this area, monkeys are indispensable. To study this extrapyramidal syndrome, we require not only a similarity in the motor system, but also a species with a sufficiently developed cortex that permits the modeling of cognitive dysfunctions.

It was found that rodents have little sensitivity to MPTP due to specific exchange processes in their brain, such that this toxin and its metabolites vanish from brain tissue in only 24 hours. In macaque brains, however, they are still found after three weeks. It was also found that the threshold value of dopamine and the number of expiring neurons in monkeys following treatment, as well as other changes in the brain, vision (pathology of the retina, saccadic motion of the eyes), behavior, and motor system, are significantly more similar to those in humans than are the comparable results in other animals (Johannessen *et al.*, 1985; Bodis-Wolner *et al.*, 1987; Schultz, 1988; Emborg *et al.*, 1995, and others). There is also the opposite type of example in which a drug that protected mice from MPTP did not produce this protective effect in primates (Perry *et al.*, 1987).

Naturally, after adequate models of parkinsonism were established (which were rather problematic not long ago), and the mechanisms and progress of this disease were revealed, the question of therapy and preventive measures against this threatening pathology arose. At present, this work is in full swing, despite the fact that considerable progress has been made in this field. What has been accomplished thus far could not have occurred without experiments on primates. Experimental transplantation of neural tissue into the brain, including neurons of fetuses and deceased newborn monkeys and humans, became widely discussed as a cure for Parkinson's disease. Despite debates on this complex problem, research in this field, which started in the middle of the 1980s, is continuing, and there are clinical results which offer hope (Wyatt *et al.*, 1986; Bakay *et al.*, 1987; Jiang *et al.*, 1995). It is clear that the search for pharmacological methods of therapy is also being carried out with primates (Andringa *et al.*, 1999). Special research benefits of monkeys are found in the field of gene therapy for Parkinson's disease (Bankiewicz *et al.*, 1998) and in another area of neuropathology which was mentioned above, amyotrophic lateral sclerosis (Aebisher *et al.*, 1996).

Investigation of the role of MPTP in the etiology of parkinsonism showed that natural and exogenous neurotoxins accounted for other neurodegenerative diseases associated with involution of the nervous system in elderly people, such as *Guam dementia*. This disease is associated with the seeds of a local palm which are used as food (monkeys that fed on these seeds also developed a syndrome of brain damage). These seeds can also cause other syndromes which include the characteristics of several diseases (Fowler, 1987). Another dangerous disease of elderly people, Alzheimer's disease, is studied intensively in monkeys. In the opinion of the Director of the National Institutes of Health, Harold Varmus (a Noble Prize winner), this is the most dreadful of diseases ('this is the poliomyelitis of the present time, but much worse'). Although there are forms which affect young people, this disease mostly threatens old people, whose number is growing. As I mentioned above, similar senile plaques were found in the brains of old monkeys. Their amyloid morphology is similar to that of human senile plaques. As in senile dementia of the Alzheimer's type, it is associated with a sharp loss of memory and cognitive and motor functions, and consequently, with pathology of the brain (during this disease, more than 80% of the neurons in the basal Meinert nucleus, which are also well-differentiated in monkeys, are lost). There are clearly good reasons to model Alzheimer's disease in monkeys. The 'aged monkeys are critical to further our understanding of the mechanisms of age-associated neuropathology and the neurobiological basis of cognitive decline with advanced age' (Voytko, 1998, p. 616). New investigations are now taking place into age-related degeneration of primate cortical innervation (Connor et al., 2001).

Cerebral amyloidosis of both types (the senile plaques and associated angiopathies) develops in rhesus monkeys in the same way as in humans. Interestingly, macaques of 16–19 years old (average age) do not have the disease, whereas 100% of those at 33–39 years of age (old monkeys) do have it (Zimbric et al., 1996). Monkeys are the only adequate models available for laboratory research on Alzheimer's disease, its origin, pathogenesis, and therapy, including gene therapy (Terry et al., eds., 1994; Mufson and Kordover, 1998).

In the 1990s, some progress was made in investigations with another valuable model of neuropathology carried out with monkeys (*Papio, Macaca, Cercopithecus, Saimiri*), Huntington's disease. This disease is caused by surgical or chemical insults (Hantraye et al., 1990; Palfi et al., 1998).

Opportunities for large-scale research on human neuropathology with simians are not at all limited to these examples. Various species of macaque are excellent models for reproducing allergic encephalomyelitis, the multiple sclerosis (Alvord et al., 1988). I should also mention the monkey model for sudden infant death syndrome, which has a complex etiology that involves several physiological systems (McKenna, 1986). A valuable model of asphyxia of the newborn with characteristic brain lesions is found in monkeys and a rare model of phenylketonuria, a condition that does not exist in other animals (Waisman and Harlow, 1965; Myers, 1969). Monkeys are also used to develop new technical methods of studying cerebral physiology and pathology. In the history section, I discussed the achievements of medical primatology in the field of neural infections. Lets us add to those the investigation of Creutzfeldt-Jacob disease (Ironside, 1996), Borna disease (Richt et al., 1994), and also the experiments of D.C. Gajdusek's group on human spongiform encephalopathy (Brown et al., 1994). It is very likely that the role of prions in the etiology of neurodegenerative diseases, in particular, will be studied with primates, and they may allow us to initiate new investigations of the diseases mentioned in this section.

Simians have definite advantages over all conventional laboratory animals in the experimental study of *neuroses, psycho-emotional stress, and psychosomatic pathology* of humans. I already discussed the history of monkey models of neuroses and the development of neural pathology in this area, for example, hypertonic disease, ischemia of the heart, myocardial infarction, stomach achylia, pre-diabetic states, amenorrhea, hystero-like paralysis, and hyperkinesis. The leading role played by disorders of the natural vital reflexes and of 24-hour-periodicity disorders, as well as by collision ('sshybka' – after I. P. Pavlov) of the two main neural processes of excitation and inhibition were clarified in these experiments. There was a total number of 59 baboons and macaques engaged in these experiments. Hypertension developed in 16 individuals, coronary insufficiency developed in 19, myocardial infarction was observed in 7, stomach achylia was recorded in 15, amenorrhea developed in 6 females, sexual inhibition was observed in 2 males, and there were three cases of a pathological increase of the tonus of the vagus (Cherkovich and Lapin, 1973). These investigations were continued at a higher pace in the 1970s–1980s because of the emotional peculiarities of primates, especially with respect to the regulation of emotion in monkeys, which are similar to those of humans and not found in any other animals, due to the highly evolved primate brain and behavior, and their neural and biochemical processes (Repin and Startsev, 1975).

The investigation of psycho-emotional stress and the numerous and varied psychosomatic disorders which arise from it have produced important results and conclusions (V. G. Startsev). Emotional stress which is nonpathogenic under nonspecific influences, becomes pathogenic under natural psycho-physiological excitation within a specific functional system and leads to a selective affliction of this system. For example, repeated immobilization (securing the limbs) of a recently fed monkey (with active digestion) for 1–3 hours led to lingering anacidosis, hypoxia, ulcers, tumors, to a precancerous state of the stomach. Repeatedly interrupting coitus in a stressful manner caused psychogenic impotence in male monkeys and neurogenic amenorrhea, disturbances of sexual periodicity, and stillbirths in females.

Hyperactivity in the cardiovascular system (such as occurs when running from danger), induced by repeated intermittent immobilization in 30 hamadryas baboons and rhesus macaques, resulted in arterial hypertension and heart ischemia which persisted for 2–15 years of observation. Activation of carbohydrate metabolism (by sugar load) during immobilization was accompanied by a chronic neurogenic diabetic state with the same vascular and endocrine complications as in humans. These complications include arterial hypertension, diabetic retinopathy, and pathological delivery, as well as such pathognostic indicator of diabetes mellitus as punctate hemorrhage in all serous coats in response to lancing. Psychic interruption of ongoing physiological activity can be destructive for both simians and humans in situations where punishment cannot be avoided, such as when punishing a child who is capricious and offends someone at dinner. Let us conclude this discussion by noting that researchers are successfully developing pharmacological remedies for such conditions with the use of monkey models (Startsev et al., 1987; Startsev, 1991).

In the course of these investigations and combined with clinical data obtained for humans, a major affinity became apparent between the physiological inter-system correlations in monkeys (P. hamadryas) and humans, both under normal conditions and in a state of active neurosis. This discovery is of fundamental value for the experimental study of psychopathology in humans (Jalagoniya et al., 1987).

It is clear that the data presented above do not exhaust either all the research in this field or all the influences that stress exerts on the vital functions of primates. It is known that stress exerts a direct influence on the immunocompetence of macaques (new terms have even been introduced, such as 'psychoimmunology' and 'primate psychoneuroimmunology') (Lubach, 1996; Arnsten and Goldman-Rakic, 1998), on their ovarian function (in a group of subordinate females), and on the mineral density of bones in females with socially depressed ovulation. Stress also leads to other reproductive dysfunctions in macaques and baboons (Shively et al., 1997). The relationship of psychosocial stress to the development of cardiovascular pathology, including atherosclerosis, is the subject of ongoing research with monkeys (Bjorntorp, 1997). Some of the greatest biomedical potential of primate models, in fact, is found in the area of stress research.

The primates are used more and more in one of the most inaccessible fields in experimental pathology of humans, namely, *psychiatry*. There is no doubt that a scientific breakthrough in this field is based on the brilliant research of H. Harlow (see Chapter 2), his students, and other followers. Numerous experiments on depression and conditions close to depression, produced primarily by maternal and social isolation with different species of simians as models, showed significant similarity (although not identity) between these states and their consequences in humans (Harlow and Harlow, 1962; Reite *et al.*, 1989). Simians are naturally highly social animals with distinct and active emotions; they avoid social isolation at all ages. Social isolation represents a shocking trauma to them which leads to psychopathology with neurochemical, immunological, endocrine, and physiological disturbances (Shively *et al.*, 1997). It is well-recognized, for example, that maintaining primates under conditions of chronic boredom and monotony can produce irreversible depression and aberrant behavior (Wemelsfelder and Dolins, 1994).

States of agitation, fear, self-aggression, stereotypy, self-mutilation, dysesthesias, childhood autism, panic disorders, amphetamine psychoses, and phobias have all been modeled in monkeys (Schrier, 1969; Chamove and Anderson, 1981; Levitt, 1985; Castner and Goldman-Rakic, 1996). It is apparent from this list that monkeys have been used for some time in an effort to study experimentally one of the most difficult of human pathologies, namely, schizophrenia. Even as recently as forty years ago, it was difficult to suggest that this disease could be modeled in animals. Nevertheless, these investigations are now carried out routinely (Schlemmer *et al.*, 1996; Breier *et al.*, 1997, and others). Symposia at international congresses of primatologists are also being devoted to the problems of experimental psychopathology (Fuchs and Shelton, 1996).

I anticipate that the most difficult problems of biology, medicine, and pharmacotherapy of brain will be solved using simians. We must remember, however, that the highest simians should not necessarily be used on every research problem, as a rule. On the contrary, they should be reserved for those problems for which they have unique experimental value as *natural models* (under natural conditions and in captivity) and when traumatic elements are excluded. We must also bear in mind that the brain is not only an independent system with its own biology and pathology. It is also a regulatory center for all the vital functions of the organism and, hence, its similarity in humans and other primates is the basic foundation of medical primatology.

OTHER COMMON CHARACTERISTICS OF AFFINITY BETWEEN SIMIANS AND HUMANS: PHYSIOLOGY, BIOCHEMISTRY, GENETICS, AND DERMATOLOGY

This section includes the description of those characteristics of affinity among primates, which either cannot be included in the following specific sections of discussion, are related to several sections, or those for which, in the author's opinion, there is no other appropriate section.

In the middle of the 1930s, a clearly expressed *single phase circadian periodicity* was found in the motor activity and vegetative functions of lower simians. This type of periodicity fully corresponds to human physiology and is unlike that in other animals where it depends on the ecological conditions in which the species exists (Scherbakova, 1937). These features of monkeys were used to study the influence of disturbances in the daily routine of people. Neuroses, as mentioned above, were modeled in baboons and macaques at the Sukhumi Center, for example, by experimentally controlled shifts in 24-hour periodicity. When experimental investigation of problems associated with transcontinental and space flights was required, it became clear that nonhuman primates were the best experimental models for the purpose. Their circadian rhythms are similar to those of humans and the full extent of circadian disruption, which is accompanied by disturbances of phase relations, a condition which is very important for clinical practice, is found only in primates (Sulman, 1983; Rivkees, 1997). Simians, like humans, become sleepy when it rains. The sensory systems of nonhuman primates, moreover, like those of humans and unlike those of certain other species (fish, snakes, cattle, and some mammals), do not react to impending earthquakes (Krusko et al., 1986).

In the 1930s, investigation of *thermoregulation* in primates was initiated in Sukhumi (Slonim, 1952, 1986). It was found that thermoregulation in humans is, in general, very similar to that in monkeys and apes. The day-night periodicity in body temperature is well-defined and unlike other animals, the primates react to warming by increased perspiration. The topography of skin temperature is similar, although the topography of perspiration differs. The 24-hour curves in the body temperature of monkeys and humans are practically identical and gas exchange is similar. According to the data of A. D. Slonim, the mean body temperature of simians is 38°C. Nevertheless, primates, including humans, have no adaptive mechanisms to protect against overheating, they are fearful of it, and they are sensitive to cooling (there are species, however, that are able to withstand rather low temperatures, such as mountain baboons, Japanese macaques, and rhesus monkeys). Humans and rhesus macaques respond in a similar manner to a decrease in temperature, which reflects the similarity in neural regulation of adjustments to temperature variations (Rozsa et al., 1985).

I should especially emphasize the unique affinity in the animal world between humans and simians in their use of perspiration as the main method of heat dissipation. At high external temperatures, 82% of heat loss in baboons is accomplished by sweating through the skin (Hales et al., 1979). Histological and histochemical properties of exocrine glands of rhesus macaques and *E. patas* (as well as the functions of the glands) are qualitatively similar to the data for humans. There are data suggesting that the best laboratory animal to study thermoregulation and its disturbances are patas monkeys, in which the rate of perspiration and the influence of neuroactive agents on perspiration are similar to those of humans (Elisondo, 1988).

Simians have important advantages over other laboratory animals in experimental investigation of the morphology and physiology of *breathing*. The similarity between nonhuman primates and humans in this respect is significant. This similarity relates not only to the apes, in which histology of the lungs (chimpanzee) is the most similar to that of humans (Muggenburg et al., 1982), but is also true with respect to the lower monkeys. The morphology and morphometry of the lungs in baboons (*P. anubis*), which are better adapted for gas exchange than those of the human, provide the grounds for using these primates as models for the human in this area. The mechanics of the respiratory system in various species of macaques are similar to those of the human and unlike those of dogs, cats, and rodents. This similarity makes monkeys the most suitable animal models for the study of corresponding human diseases and the toxicology of breathing (Hakkinen and Witschi, 1985; Maina, 1987). Disturbances in the mechanics of the lungs caused by leukotriene aerosols (C_4 and D_4) were also the same in macaques (*M. fascicularis*) and humans (Krell et al., 1986).

Common features of enzymatic activity and metabolism in the lungs, for example, with respect to acid proteases, are found in humans and monkeys and not in other animals (Moriyama and Takahashi, 1980). The similarity in morphology and enzymatic activity of the bronchoalveolar fluid of humans and Java macaques confirms the value of the latter in the study of experimental silicosis (Hannothiaux et al., 1987). Modeling pulmonary fibrosis (Collins et al., 1982), bronchopulmonary dysplasia, pulmonary hypertension, and other vascular diseases (Weesner and Kaplan, 1987), as well as intoxication with ozone, oxygen, and various toxins (Castelman et al., 1982), has been accomplished in experiments with monkeys. A natural model of nasal polyposis is found in chimpanzees, in which the pathology is completely comparable to that in humans (Jacobs et al., 1984). The similarity of the respiratory system among primates is widely used in surgery and in research on infectious pathology.

I already discussed the unusual similarity in the *biochemistry* of humans and simians, in particular, the similarity in proteins. This theme will be continued in the following sections. I must especially discuss the similarity in enzymes, primarily the enzymes of the liver. These liver enzymes determine the affinity of primates with respect to blood circulation, digestion and other aspects of metabolism, including the metabolism of drugs. We can state categorically that there are no organisms in nature which are so close to humans in this sphere as are the nonhuman primates. This similarity, moreover, has been known for some time. The similarity of esterases and other liver enzymes in monkeys and humans, in contrast to other animals, was found in the beginning of the 1970s (Mendoza and Hatina, 1971). The progress of pharmacological investigations required special attention to the enzymes involved in the metabolism of drugs. It became clear that the activity of azoreductase, nitroreductase, p-hydroxylase, N- and O-dimethylase, and cytochrome-c-reductase, and the content of cytochrome P450 in the liver microsomes of P. anubis are significantly lower than in the majority of other laboratory animals. What is more, the characteristics of the majority of these enzymes are not only comparable to those in humans, but are practically indistinguishable (Autrup et al., 1975). No species differences were found in the structure of hexosaminidases from the livers of humans, chimpanzees, orangutans, baboons, or guenons (Lee et al., 1979). The activity of hepatic glutathion S-transferases, which has been confirmed by various methods, is highest among rats, but approximately equal in rhesus macaques, chimpanzees, and humans (Heinz-Sammer and Greim, 1981; Hoesch, 1988).

The affinity of apes, monkeys, and humans is known with respect to many other enzymes, including glyoxalase, dehydratase, aldehyde oxidase, telomerase, etc. (Chaug et al., 1984; Klots et al., 1998). Alkyltransferase is 8–10 times more active in the livers of humans and macaques than in those of rats (Hall et al., 1985). As a result of the high activity of a series of enzymes, tetrahydrobiopterin products are present only in the livers of monkeys and humans (for example, neopterin, which was measured in 11 species of animals that were studied), whereas isoxanthopterin is present only in the liver of pigs (Hasler and Niederwieser, 1986). Low levels of dihydrofolate reductase in the liver are characteristic of only humans and great apes, in comparison even to monkeys, as well as birds, and all other animals (Whitehead et al., 1987). Alcohol dehydrogenase and its isoenzymes in the livers of nonhuman primates was found for the first time in saimiri (Dafeldecker et al., 1981). As for the enzymes of cytochrome P450 and the chemical properties of cytochrome itself, catalytic and immunological characteristics indicate a similarity in the liver microsomes of humans and simians (even in monkey, Java macaques), which stands in contrast to rats, guinea pigs, and pigs. These characteristics make the simians convenient model systems for the study of cytochrome (Ohmori *et al.*, 1988). No doubt, the examples presented here do not reveal all the characteristics of similarity in the enzymes of humans and other higher primates, and of course, such similarities are not limited to only the liver enzymes. New data are continuously found which indicate that this similarity occurs against a background of obvious differences in a variety of other animals, including more conventional laboratory animals (Habteyesus *et al.*, 1991).

Loss of the two enzymes mentioned above, uratoxidase and gulonolactone oxidase, the corresponding presence of large amounts of uric acid only in humans and higher simians (with one exception, the Dalmatian dog), and the impossibility of synthesizing ascorbic acid *de novo* (also with one exception, the guinea pig), was an extraordinary event in the evolution of the primates (Friedman *et al.*, 1985; Wu *et al.*, 1989). It is possible that these two evolutionary events are somehow interrelated. These data not only confirm the genetic affinity of humans and simians, they also define conditions for the experimental study of an ascorbate metabolism model, in particular, for scurvy (Sato and Underfriend, 1978). The peculiarities of vitamin D metabolism in marmosets, moreover, as mentioned above, allow us to obtain a model of vitamin D-dependent rickets, type II, due to the relationship of certain receptors and functions of steroid hormones (Suda *et al.*, 1986). A determinant specific to the F-antigen in humans, which is important in clinical practice, was found only in rhesus macaques among all the mammals that were studied (Mori *et al.*, 1981). The best model to study a massive and chronic overload of the liver with iron is the baboon (Brissot *et al.*, 1982). The metabolism of lysine is similar in monkeys and humans and is different in rats. The dynamics of calcium metabolism in the young rhesus monkeys and human children is practically indistinguishable and creates unique opportunities for studying this process in simians (Moore *et al.*, 1985).

I should emphasize that the similarity in the biochemical properties of different primate species and humans fully corresponds to the phylogenetic scale described above. This relates both to the hierarchy within the group of monkeys itself and to the distinctions between the latter and the apes. Sulphoxidation of S-carboxymethyl-L-cysteine in macaques is closest to that in humans among the monkeys that were studied (*M. mulatta*, *M. fascicularis*). This affinity is less pronounced in guenons (*C. aethiops*) and lesser still in marmosets (*C. jacchus*) (Mitchel *et al.*, 1986). One of the cyproheptadine glucuronide metabolites, namely glucuronide-I, is excreted in urine in insignificant amounts over 48 hours in various species of monkeys (less than 0.5%), whereas, this amount was 12.4% in humans and 8.6% in chimpanzees. There are other similar examples of the phyletic relationship among the primates (Fischer *et al.*, 1980).

Nevertheless, in many cases even the lower simians demonstrate a significant biochemical and genetic affinity with humans, something that was discussed in part in the chapter on taxonomy. Laboratory research on many enzymes (and, of course, the proteins as a whole, judging from the amino acid sequences and DNA) of widely separated monkeys such as macaques, baboons, marmosets, squirrel monkeys, and capuchins are coded on homologous chromosomes by identical groups of genes (Garver *et al.*, 1977; Creau-Goldberg *et al.*, 1983, 1984). Initially the authors tentatively called this 'extreme conservation' of gene linkage and differential chromosome staining. I consider that this is further scientific evidence of the phenomenal affinity of humans with the other congeners within the order.

It is sometimes said that humans have certain common genes even with bacteria. This is certainly correct, but as we have seen, there are clear differences between primates (including humans), beginning with the prosimians, and other animals. Galagos and lemurs have a unique homology with humans in their *chromosomes*, something that is not found in any other living creatures (Healy, 1995; Apiou *et al.*, 1996). The species of *Cebus* and *Lagothrix*, rather distant from humans, not only demonstrate significant homology with humans in terms of karyotypes, but also exhibit a correspondence in breakable chromosome intervals (Clemente *et al.*, 1987). One particular protein is distinguished in the composition of chromosomes with the help of special monoclonal antibodies. This protein appears only in mitotic cells (but not in the interphase). This 'Flying Dutchman' is found only in primates, but not in other animals (Rao *et al.*, 1985). During the 1990s, numerous homologies have been found between the chromosomes of humans and monkeys (*Colobus, Ateles*, and others) (Bigoni *et al.*, 1997; Seaunez *et al.*, 1997). As to the karyotypes of apes, I already mentioned their unusual relationship with that of humans and additional confirmation of this relationship has been found (McConkey, 1997).

The genotypic affinity of primates is naturally spread at the phenotypic level. This means that the patho-physiological reactions to internal and external pathogens in humans and other primates cannot be only analogous. Hence, there are inexhaustible opportunities to use simians for studies of not only specifically hereditary human diseases, but also for research on other types of pathology whose etiology not only reflects solitary factors, but several others, including environmental influences (Stone *et al.*, 1987). Primates are useful species, moreover, to study the biological properties of proteins and gene mutations. Finally, on the basis of the previous discussion, one can easily understand that simians are exceptional experimental animals for the development of gene therapy, and there are successful examples (Mehtali *et al.*, 1995). J. L. VandeBerg concluded at a symposium where studies on the genetics of primates were presented, 'We stand at the threshold of a new and exciting era in genetic research with nonhuman primates' (1987, p. 7). He meant not only that answers would be obtained regarding the origin of humans, but also that this research would play a critical role in medicine now and in future. There is no doubt that ongoing work today, such as the International Programs 'Genome of Man' and 'Genome of Primates' will give us more new data about the phylogenetic relations and affinity in the order Primates.

In concluding this section, I would like to describe some unpublished papers (unknown to science) of an outstanding scientist at the Sukhumi Primate Center, Peter Viktorovich Botchkarev (1890–1947), Doctor of Medical Science. Working in the 1930s–1940s at the world's oldest primate center, it was practically impossible to publish scientific research because of the strict governmental regulations that were associated with political repression and war. This problem especially affected the most active research scientist at the Center, Professor Botchkarev. Sixty-eight papers, including 6 monographs, remained unpublished from a total of 82 completely finished and appropriately validated studies on the biology of primates (the total number of his scientific studies is no less than 150). One of them, 'Materials for the study of lower simians on the basis of experiments carried out at the Sukhumi Department of VIEM' (this was the name of the Sukhumi Primate Center at that time), was finished in 1938. It consists of 304 pages legibly written by hand in small print. In 1992, the manuscript was kept in the library of the Sukhumi Primate Center. There were about 10 chapters, among which were chapters devoted to taxonomy, comparative anatomy, pathology, reproduction, and behavior of baboons and macaques. The data on physiology include general metabolism, thermoregulation, nitrogen metabolism, carbohydrate metabolism, protein and chemical composition of the blood and urine, mineral metabolism, water balance, and characteristics of cerebrospinal fluid. If we add to this the data on reproduction and endocrinology, in which Botchkarev was a prominent specialist, we may conclude that his unpublished manuscript was an encyclopedia of the lower simian's biology at that time.

Let us briefly discuss the similarity in *the skin* of humans and simians and the possibilities for its application. Primate dermatology became an active research area during the time that medical primatology was developing in the 1960s. The late professor W. Montagna and other specialists made important contributions to this field (Gladkova, 1966; Montagna and Uno, 1968). In principle, the structure of the skin in humans, a primate with vertical locomotion, is unique. The African hominoids, chimpanzee and gorilla, are the closest to humans in this respect (Montagna, 1980). Nevertheless, the other simians, including the lower ones, also have much in common with humans in this area. Montagna proposed that primates, including humans, differ from the other mammals in the following peculiarities of their skin: 1) similar skin patterns, the dermatoglyphics; 2) the presence of Meissner's corpuscles in the volar skin of hands and feet; 3) the presence of nails instead of claws (which gave unusual opportunities for grasping objects); and 4) sweat glands with a complicated vascular system for special thermoregulation in the areas covered with hair. Judging by certain characteristics, the sexual skin of all female catarrhines is comparable to the genital skin of women. As already mentioned, axillary glands which produce odors are found only in humans and the three African hominoids (Montagna, 1985).

The following similarities in characteristics of the skin in humans and simians are known: the system of melanocytes (chimpanzees and gorilla); dendritic cells of the epidermis and dermis; and involucrines of the keratinocitic membranes (Post et al., 1975; Simon and Green, 1989). A series of microelements and their concentration in the hair of the Japanese macaque have the same relationship as in human hair (Folin et al., 1986). Comparative analysis of the hair of simians and humans is of considerable interest. The primates are divided into two groups according to the structural characteristics of the hair medulla. The prosimians and monkeys comprise the first group (vesicular medulla) and the hominoids, including humans, comprise the second group (dense hair medulla); only chimpanzees and gorillas form a transitional group in relation to humans (Clement et al., 1980).

Let us describe certain properties of the skin in humans and simians. The electrical conductivity of the skin is a useful indicator for the assessment of the general emotional state and reactivity of an organism. It is very similar in baboons (*P. hamadryas*) and macaques (*M. mulatta*) to that of humans. Seasonal profiles of minimal and maximal values of this parameter coincide quite well (Neborskiy and Belkaniya, 1986). Unlike the data for rodents, permeability of the skin of humans and rhesus macaques is similar; this is a good model for studying skin absorption and for assessing the impact of chemicals on humans (Wester et al., 1989). The sense of touch in simians, studied at the level of skin receptors to neocortical areas and behavioral effects, are essentially comparable to the data on humans (Darian-Smith, 1982). I already mentioned that monkeys and humans exhibit the same reaction to a change in temperature. In addition, there is a high correlation in the psychophysical assessment of skin pain and hyperalgesia among these primates (La Motte et al., 1983).

It is known that human baldness of the usual type is biologically similar to alopecia in various species of simians, including the highest (Montagna and Uno, 1968). Stumptailed macaques (*M. arctoides*) are of laboratory value in this respect. They are used to study the early aging of hair follicles, the role of androgens, and genetic predisposition, as well as the prevention and curing of human male-pattern baldness (Uno *et al.*, 1997; Imamura *et al.*, 1998). Despite the pessimism with respect to modeling psoriasis in animals, long-term research in this field led to an encouraging application of rhesus macaques and Java macaques (Jayo *et al.*, 1988). Passive passage of severely symptomatic dermatographism (factitious urticaria) to three species of macaques using human serum was described (Murphy *et al.*, 1987). The sequence of restoration and expression of antigens in the basement membrane zone of the skin strongly suggests that macaques (Java), rather than the previously used pigs, may be a suitable model for curing split-thickness wounds (Fine *et al.*, 1987). Finally, using the aforementioned similarity in dermatoglyphics of humans and simians, it is possible to use monkeys to study genetic disturbances and clinical aspects of the corresponding syndromes and constitutional anomalies associated with changes in skin patterns (Newell-Morris, 1981).

BLOOD AND FACTORS OF IMMUNITY

The blood, 'the river of life,' with an unusually high level of polymorphism in primates, represents some of the strongest evidence of phylogenetic links and affinity of humans and their congeners, clearly reflecting their evolutionary past (Ruffie *et al.*, 1982; Erickson and Maeda, 1994). Starting with the previously mentioned research of G. Nuttal (1902), where the blood of 18 species of different groups of simians was compared to human blood using the precipitation reaction, various degrees of affinity with humans were revealed. All subsequent investigations on blood, while strengthening the conclusions of Nuttal, in general, resolved the details of this problem area. This was actually shown in previous sections of this book, given that the arguments presented were based on the similarity in DNA and proteins (the latter of which actually refers to blood proteins).

The chemical and hematological properties of blood in both the apes and monkeys are very close to those of humans in the majority of their characteristics. The relationships for the blood components of *Macaca mulatta* and humans (Bourne *et al.*, 1973) are given in Table 4.2.

The limits of variation for the values are usually nested for various ranges. The authors emphasized that there was a certain discrepancy in the leukocyte values, which were close to the data for a human child. More details on blood for 10 species of monkeys and chimpanzee are given in Table 7 and Tables 10–14, in *Manual of Medical Primatology* (Lapin *et al.*, 1987, pp. 64–74). The hematological parameters are similar in humans and baboons (hemoglobin, hematocrit, thrombocytes, and number of cells), as is the ability of prostacycline to inhibit the aggregation of thrombocytes. The hematological parameters of *Callithrix jacchus* and humans are also similar. The ratio of the glucose levels in erythrocytes and plasma is approximately equal in humans, baboons, and macaques, but it is significantly less in dogs and rabbits; glucose was practically absent in the erythrocytes of pigs (Higgins *et al.*, 1982; McIntosh *et al.*, 1985).

Table 4.2 Comparison of blood chemistry and hematological data of *Macaca mulatta* and humans (Bourne et al., 1973). Reproduced with permission.

	Blood Chemistry	
	Rhesus Monkey	Human
Calcium	9.6–10.2	9.5–11.5 mEq
Phosphorus	5.0–6.2	2.5 mg%
Sodium	141–157	138–146 mEq
Potassium	4.1–6.0	3.8–5.1 mEq
Chloride	97–111	95–106 mEq
CO_2	10–16	20–29 mEq/L
Creatinine	1.1–1.4	0.8–1.3 mg%
Uric acid	1.0–1.4	2.6 mg%
Cholesterol	152–170	150–270 mg%
Total protein	7.8–8.7	6.0–8.0 gm%
Albumin	3.0–4.0	3.5–5.6 gm%
Globulin	4.3–5.2	1.3–3.2 gm%
SGOT	29–44	8–40 units
SGPT	15–20	5–35 units
LDH	393–713	100–350 units
Glucose	71–122	80–120 mg%
BUN	9.0–20.0	8.0–20 mg%
	Hematology	
	Rhesus Monkey	Human
RBCs.	4.46–5.60	4.5–5.5 million
WBCs	5.3–12.3	6–10 thousand
Hemoglobin	11.0–19.0	14.0–16.0 gm
Leukocytes:		
Neutrophils (segmental)	20.0–55.0	65–75%
Lymphocytes	40.0–76.0	20–30%
Monocytes	1.0–2.0	1–2%
Eosinophils	1.0–6.0	2–3%
Basophils	About 1.0	About 0.5%

The close biological relationship between humans and simians is confirmed by the fact that the latter easily tolerate human blood transfusions, unlike all other animals, including dogs and pigs, which only barely tolerate human blood (Hume *et al.*, 1969). The properties of blood are very similar in humans and apes. In addition, the clinical laboratory data on ape's blood, the chart of normative curves for alkali-acidic balance, and the details of blood chemistry in chimpanzees, demonstrate significant similarity, though not identity, with the corresponding human parameters (Clevenger *et al.*, 1971; Takano *et al.*, 1979).

It is well-known that one of the main focuses of the vital powers of the organism, including its immune system, is blood *serum*. I have repeatedly described the affinity of the serum in humans and other primates as a whole, as well as the similarity of its individual components, using various methods of comparison, from the first precipitation reactions to the most modern methods. There is always greater similarity (up to

Table 4.3 Results of the inhibition test performed on human and nonhuman primate plasma samples using anti-human globulin serum (Tsutsumi et al., 1985).

Species	Dilution of plasma (2^{-n})															Control
	0	1	2	3	4	5	6	7	8	9	10	11	12	13	14	
Human	−	−	−	−	−	−	−	−	−	−	−	−	−	+	+	+
Chimpanzee	−	−	−	−	−	−	−	−	−	−	−	−	−	+	+	+
Orangutan	−	−	−	−	−	−	−	−	−	−	−	+	+	+	+	+
White-handed gibbon	−	−	−	−	−	−	−	⊥	+	+	+	+	+	+	+	+
Agile gibbon	−	−	−	−	−	−	−	−	+	+	+	+	+	+	+	+
Gelada	−	−	−	−	−	−	−	⊥	+	+	+	+	+	+	+	+
Hamadryas baboon	−	−	−	−	−	−	−	−	+	+	+	+	+	+	+	+
Patas monkey	−	−	−	−	−	−	−	⊥	+	+	+	+	+	+	+	+
Green monkey	−	−	−	−	−	−	−	−	+	+	+	+	+	+	+	+
Japanese macaque	−	−	−	−	−	−	−	−	⊥	+	+	+	+	+	+	+
Rhesus macaque	−	−	−	−	−	−	−	⊥	+	+	+	+	+	+	+	+
Crab-eating macaque	−	−	−	−	−	−	−	⊥	+	+	+	+	+	+	+	+
Formosan macaque	−	−	−	−	−	−	−	±	+	+	+	+	+	+	+	+
Stumptailed macaque	−	−	−	−	−	−	−	−	+	+	+	+	+	+	+	+
Bonnet macaque	−	−	−	−	−	−	−	⊥	+	+	+	+	+	+	+	+
Assamese macaque	−	−	−	−	−	−	−	⊥	+	+	+	+	+	+	+	+
Pig-tailed macaque	−	−	−	−	−	−	−	−	+	+	+	+	+	+	+	+
Night monkey	−	−	⊥	+	+	+	+	+	+	+	+	+	+	+	+	+
White-throated capuchin monkey	−	−	±	+	+	+	+	+	+	+	+	+	+	+	+	+
Tufted capuchin monkey	−	−	⊥	+	+	+	+	+	+	+	+	+	+	+	+	+
Spider monkey	−	−	±	+	+	+	+	+	+	+	+	+	+	+	+	+
Cotton-top tamarin	−	−	⊥	+	+	+	+	+	+	+	+	+	+	+	+	+
Grand galago	±	±	+	+	+	+	+	+	+	+	+	+	+	+	+	+
Slow loris	±	+	+	+	+	+	+	+	+	+	+	+	+	+	+	+
Ring-tailed lemur	±	±	+	+	+	+	+	+	+	+	+	+	+	+	+	+

100% by specific proteins) in the proteins of humans, chimpanzees, and other hominoids than those of the other catarrhines, less similarity in the platyrrhine monkeys, and less still in the prosimians. This relates to transferrin, albumin, haptoglobin, fibrinopeptides, fibrinogen, fibrinonectin, ceruloplasmin, and to the entire fraction of globulins. The latter represent almost half of the total proteins in serum and determine the most important functions of the organism, such as immune functions, blood coagulation, and transport of iron and copper (Socha *et al.*, 1971; Mahany *et al.*, 1981; Goodman *et al.*, 1974; Tsutsumi *et al.*, 1985, 1989; Erickson and Maeda, 1994). Human monoclonal antibodies to transferrin (Tf-1) reacted only with the transferrin of humans, chimpanzees, gorillas, orangutans, and rhesus macaques, and not with that of the animals lower than the catarrhines; the antibodies to Tf-2 reacted only with the transferrin of humans, chimpanzees, and gorillas (Miller, 1987). We can see the levels of affinity of the globulin serum in human and some other primates in Table 4.3.

The significant similarity in serum enzymes of humans, and even the lower catarrhine simians (green monkeys, macaques, baboons) is impressive, including for example, adenilatkinase, acid phosphatase, A and B peptidases, the inhibitor of C'-1-esterase and many others. They are significantly different, on the other hand, not only from fish, birds, guinea pigs, rabbits, dogs, and horses, but also from the platyrrhine monkeys (Donaldson and Pensky, 1970; McDermid and Ananthakrishnan, 1972). There were quantitative differences from human values in the amino acids in blood plasma of saimiri, rhesus, talapoin, and chimpanzee, but those differences were insignificant (Peters *et al.*, 1971). Serum levels of cholesterol are lower in baboons than in humans and the level of alkaline phosphatase is higher. The majority of serum components, however, have similar values to those considered normal for humans (McGraw and Sim, 1972). Monoclonal antibodies to vitamin-D-binding protein in human serum, assessed by radioimmunoassay or by immuno-precipitation, had a cross-reaction with the similar monkey's (and pig's) protein, but not with that of mice, rats, or chickens (Pierse et al., 1985). The steroid-binding globulins of simians are similar to those of humans and correspond to the scale of primate phylogeny (Robinson et al., 1985). I shall return to discussion of serum components later, but for now I shall discuss the cellular elements of the blood.

As was mentioned above, *erythrocytes* are an excellent example of the close relationship between humans and other primates. They are similar in rhesus macaques and humans in terms of their morphology (observed by scanning electron microscopy) and in the amount and composition of hemoglobin (Lewis, 1977). Let us complement these data with information about the discrete structure of membranes that determine the characteristics of red cells, and especially, by describing the majority of the organism's blood groups and other blood properties. Let us first discuss the varieties of erythrocyte enzymes. The antigenic specificity of mammalian catalase, according to the literature of the 1960s, was divided into 4 groups of organisms: 1) humans, rhesus macaques, and green monkeys; 2) horses, asses, and guinea pigs; 3) sheep, goats, and cattle; and 4) dogs. Aspartate aminotransferase was identical in humans and these monkeys in terms of antigenicity, but it was not similar in any of the other aforementioned animals (Szeinberg et al., 1969a, 1969b). Comparable data were obtained on many other enzymes of erythrocytes (carboxianhydrase, adenilatkinase, acid phosphatase, phosphate dehydrogenase, adenosine deaminase, phosphoglucomutase, etc.) (Goodman et al., 1971; McDermid and Ananthakrishnan, 1972; Biwasaka et al., 1989; and others). Investigations have demonstrated, for example, that the enzymes of erythrocytes are activated by thyroid hormone to an equal degree in humans and monkeys (approximately 1.5 times), whereas, in dogs and rats there was no response to these hormones and in rabbits they were increased 4 times (Davis et al., 1982).

According to the data on electrophoresis with monoclonal antibodies as well as with other methods, various surface proteins of erythrocyte membranes in humans and simians demonstrate a high level of homology. This is known for green monkeys, macaques, baboons, and marmosets and is unlike the data for rats, pigs, and other mammals. Quite often, these proteins in monkeys are equivalent to the corresponding human proteins (Gomperts et al., 1971; Howard and Kao, 1981). The greatest homology with humans in membrane proteins is found in chimpanzees, gorillas, and orangutans (Lucotte and Ruffie, 1982; Nickelis and Atkinson, 1990).

There is a very high degree of homology in the higher primate's membrane sialoglucoprotein, corresponding to the A and B human glycophorines which determine the M-N blood group. This homology may be traced from the level of the genes to actually expressed antigenic determinants. One can clearly see that glycophorine A of chimpanzees always expresses blood group M, glycophorine B expresses either N or O, and the majority of human determinants with glycophorine A is determined in chimpanzees and bonobos with the same glycophorine. In gorillas the amount is about half of this, while in gibbons it is significantly less. This completely corresponds to the phylogeny of the hominoids, as is already known to the reader (Rearden, 1986; Creau-Goldberg et al., 1989). Glycophorine MK of the Japanese macaque, moreover, reveals a 'striking homology' (the authors use this term in the title) in amino acid sequence with human glycophorine A (Murayama et al., 1989).

In the previous sections of this book I described the great similarity in all polypeptide chains of hemoglobin in humans and simians. According to various investigators, cross-immunoreactivity corresponds to the structural homology displayed with humans and ranges from 100% in chimpanzees to 8% in prosimians. There were single differences in the amino acid sequences of the primary structures of α- and β-polypeptide chains among the hominoids, whereas, there were from 3 to 5 substitutions in α-chains and from 5 to 7 substitutions in β-chains among the monkeys (capuchins, marmosets, macaques, baboons). In the γ-chains of fetal hemoglobin, the differences appeared to be even less, approaching identity with humans in *M. mulatta* and *P. cynocephalus* (Garver and Talmage, 1975; Takenaka *et al.*, 1986). Interestingly, the conversion of fetal to adult hemoglobin synthesis during the postnatal period, where a cluster of genes of β-globulin[1] take part, is analogous in baboons and humans (Schroeder *et al.*, 1983), but it occurs more rapidly in the baboon. This can be explained if we recall the chronology of biological life and its stages for these two primates. It is no less interesting that blood identification with a standard commercial preparation of latex sensitized with human hemoglobin gave positive reaction only with the blood of humans and macaques (*M. fuscata*). It was negative with the blood of dogs, cats, rabbits, pigs, rats, and chickens (Okada *et al.*, 1988). The effect of intestinal parasites (trichomonads) on agglutination of erythrocytes is also similar. Erythrocytes of humans and *Saimiri* were the most sensitive to these parasites, unlike rats (low degree agglutination), sheep, rabbits, chickens, and cattle, where this effect was virtually absent (Pindak *et al.*, 1987). A similarity of hemoglobin with the human and a high degree of ability to replenish erythrocyte reserves in *Aotus trivirgatus* makes these primates a valuable model for the study of malaria caused by *Plasmodium falciparum*.

Other properties of erythrocytes are also similar in humans and simians. Saimiri and rhesus monkeys were appropriate animals to study the *in vitro* hemolysis of erythrocytes. Due to analogies with the corresponding regularities in humans, macaques and baboons are convenient models to estimate survival and methods of erythrocyte preservation, including cryopreservation techniques (Rowe, 1995).

In connection with discussion of the similarity in erythrocytes of humans and other primates, it is reasonable to analyze *blood groups*. As mentioned above, simians played an outstanding role as experimental animals in the field. A. Wiener, one of the original researchers in the field, regarded primates as 'ideally suitable subjects' (Wiener and Moor-Jankowski, 1969, p. 39). Antigens of erythrocytes A and B are spread throughout the entire animal world, even in bacteria. The distinct evolutionarily recent system of ABO distinguished in blood, secretions, and tissues, however, is found only at the primate level, or more precisely, no lower than the level of platyrrhine monkeys (although I note that the activity of transferase, which has a direct relationship to ABO, is also high in prosimians [Nakajima *et al.*, 1986]). I should also mention the homologies of apes with humans and the analogies of monkeys (I do not mean convergence, but a lesser level of homology in monkeys) (Socha *et al.*, 1972; Socha and Ruffle, 1983). Thus, despite the peculiarities of antigens in simian species, they are

---

1 In the previous chapter I repeatedly mentioned the similarity of β-globulin genes in human and other primates. There are data of approximately the same level of similarity in the gene cluster of α-globins; their structure is close in all higher primates (monkeys, apes, and humans), but differs from that in lower primates (galagos), where it is closer to the corresponding genes of non-primates (Sevada and Schmid, 1986).

in general closely related to those in humans. This is consistent with the phylogeny of primates, especially the apes, and closest to the bonobo, where the A-antigen, for example, cannot be serologically distinguished from the human. These relationships are also very close to other simians, and primarily to the catarrhines (Socha and Moor-Jankowski, 1979; Blancher and Socha, 1997).

The Old World simians definitely have the elements of the M-N human blood system, but there are no such elements in the blood of prosimians or New World monkeys, at least in terms of the data from investigations with poly- and monoclonal antibodies. A maximum affinity of chimpanzees with humans is also found here (the system V-A-B-D), although not without certain differences (Wiener et al., 1964; Socha et al., 1994). There are other data which suggest that simians have a system related to the Duffy human system, and a very high homology was found in chimpanzees using the Kell human system (Palatnik and Rowe, 1984; Redman et al., 1989). There are also parallels in primates in the Lewis system of blood groups (Ohshima et al., 1988; Blancher and Socha, 1997). Investigation and discovery of the rhesus factor (Rh) of human blood in simians is widely known. At present, it is known that human D-antigens and the corresponding R-C-E-F system in chimpanzees, as well as related Rh systems in other simians, are very complex antigens. Their phenotypes in African apes, however, especially in chimpanzees, can only be regarded as 'counterparts' of the human Rh(D) antigen (Rh-system) (Socha and Moor-Jankowski, 1980; Socha et al., 1987; Blancher and Socha, 1994; Salvignol et al., 1994). The special characteristics of primate erythrocytes have wide application in medical primatology. They are used to study the evolution of antigens, incompatibility in blood transfusions, immunology of transplantations, immunological relations and compatibility of mother and fetus, pathology of the newborn, modeling of sickle cell disease, and gene correction of hereditary blood diseases. They are also used as markers of taxa characteristics and individuals, and to choose pairs of primates for reproduction, as well as for many other purposes.

In general, the discussion about similarities between humans and simians with respect to erythrocytes can be extended to the *leukocytes* of the blood. Again, this similarity, in principle, conforms to the phylogenetic scale of the order Primates and the Class Mammalia. It is highest in the hominoids (sometimes reaching identity), slightly lower in monkeys (with some variations), and lower still in prosimians, as revealed by cross immunological reactions using the method of rosette formation. In the 1970s–1980s, this was demonstrated with monoclonal antibodies and other methods (Brain and Gordon, 1971; Clark et al., 1983). An approximately equal relationship between T- and B-lymphocytes was distinguished in humans, apes, and monkeys (64%–65% in the former and 25%–28% in the latter (Wright et al., 1982). The surface antigens, immunoglobulins, and other proteins of leukocytes are also similar (Haynes et al., 1982; Nooij et al., 1986). Immunoreactive eosinophil granule major basic protein was found in all primates excluding Lemuridae (the other prosimians were not tested) and Callitrichidae. The content of this protein in the plasma of pregnant women and simians exceeded by 4–8 times the content in non-pregnant women, whereas it did not differ at all in pregnant guinea pigs, mice, rats, cats, and dogs. This protein is chemically and immunologically identical in humans and other primates (Wasmoen et al., 1987). The biochemical properties and functions of eosinophils are very similar in humans and rhesus macaques and differ in other animals (in particular, in guinea pigs) (Sun et al., 1989). The aging of T-cells in old macaques occurs in the same way as in elderly people (Eylar et al., 1989). Simians are an appropriate model, therefore, for studying the biology and pathology of leukocytes, the immunology of infections, problems of transplantation, and in particular T-cell depletion of bone marrow transplants, the source of tissue incompatibility, which is vital in transplantation of spinal cord and organs (Phan *et al.*, 1989). Even a common location of fragile sites on lymphocyte chromosomes was found in humans and great apes (Smeets and van de Klundert, 1990).

Since the second half of the 1970s, with the introduction of methods using monoclonal antibodies to science, the phylogenetic positions of primates were confirmed, as was the affinity of monkeys and apes with humans. These methods represented a new and powerful experimental means for studying the leukocytes of simians, especially the catarrhines, where activity of these antibodies against human lymphocytes manifests itself in a quite human way. These antibodies, moreover, do not react with the blood of other mammals. As a result, simians are used for preclinical testing of antibodies and for assessing their properties as suppressors with the objective of preventing rejection of tissue transplants. Simians are also used to study the potential of their immune reaction *in vivo*, including the reaction to pathogenic viruses. Monoclonal antibodies of human leukocytes are used for immunophenotyping lymphocytes and for *in vivo* therapy (Reimann *et al.*, 1994; Klingbeil and Hsu, 1999). As was already mentioned, platyrrhine simians are more distant from humans than other catarrhines in terms of monoclonal antibody probes, with the probable exception of *Aotus* (Friend *et al.*, 1989; Black *et al.*, 1991).

All five classes of *immunoglobulins* in every species of primate were found to be homologous with those in humans (but to a different degree in accordance with phylogeny); IgG, IgM, IgA, IgD, and IgE. The glycoproteins, the major effector molecules of humoral immunity, differ in structure and function. They are rarely found in other mammals and none have all of the forms which are morphologically the same as those in humans. The closest of them to human ones are found in simians. Chimpanzees are closest to humans in terms of this characteristic; each of the 4 subclasses of human IgG are present in chimpanzee serum. Complete cross reactivity with human serum was recorded in baboons (*P. ursinus*) for only IgG and IgM, whereas only partial cross reactivity was observed for IgA. Nevertheless, immunoglobulins similar to those of the human are found in capuchins and marmosets, to say nothing about catarrhines (guenons, macaques, baboons) (Neoh *et al.*, 1973; Mendelow, 1980; Leibl *et al.*, 1986; Schmidt *et al.*, 1996).

The serum *complement system*, a factor of natural immunity against various diseases, is quite similar in various components of humans and some species of monkeys. This makes the latter species appropriate for identifying and tracking these components in primate models of human diseases (McMahan, 1982). There are also data on the phylogenetic affinity of cellular immunity in primates. The antigenic and functional peculiarities of lymphoid cells in chimpanzees, baboons, macaques, and other simians are close to those of humans (De Kretser *et al.*, 1986).

I already mentioned the unusual similarity of the *major histocompatibility complex* (MHC) in humans and chimpanzees, documented both by cross immunological reactions and direct study of the genes of this system. This similarity relates to both classes of MHC. Localization of the complex by chromosome depiction and by DNA analysis was close to that of humans, not only in the great apes, but also in rhesus macaques (Garver *et al.*, 1980; Kasahara *et al.*, 1990). The study of alleles at the classical HLA (humans) and ChLA (chimpanzees) loci – A, B, and C, class I, confirmed their great similarity, but also indicated that they were not identical (Parham, 1996). As for the monkeys, a high degree of affinity to humans was found for several species of macaques, baboons, and marmosets; lemurs are more distant and rats are more distant still. The MHC is an example of conservation in the main structural organization and functions of immunity during evolution of the primates (Balner *et al.*, 1973; Chausse *et al.*, 1984). Besides its theoretical and evolutionary importance, the similarity of MHC in primates is used in practice for investigating this system in biomedical research, experimental transplantation and preclinical assessment of tissue, study of the mechanism of pathogenesis of diseases, and determination of the genetic peculiarities in laboratory colonies of breeding primates.

The discovery, investigation, and application of *interferon* and its recombinant forms is another example of the indispensability of simians in biomedical research, which was mentioned in the first chapter. Interferon, an effective antiviral agent, is very specific to species. The entire set of problems associated with it can only be explored in experiments with primates. Its prophylactic and therapeutic action is similar in humans and other higher primates, including monkeys. Its genetic and structural peculiarities as well as the pharmacokinetics of its two main classes are similar. Both alpha- (leukocytic) and beta- (fibroblastic) interferons circulate in approximately the same way in the blood of humans and macaques. The characteristics of γ-interferon (from splenocytes) of baboons are also similar; the human type 2 adenovirus was more sharply suppressed by monkey interferon than by the human one (Wilson *et al.*, 1983; Dijkmans *et al.*, 1986). The importance of monkeys was again emphasized when it was found that the inducers of interferon (polyI:polyC, tyloron) in rodents and rabbits are not effective in humans and simians due to high serum concentrations of nucleases in the primates (Kaufman, 1971). The experimental development of the inducers was, in general, carried out in experiments with monkeys. As to the creation, confirmation, and toxicity tests of recombinant interferon, this outstanding achievement of biotechnology could not have been carried out without laboratory primates. Overcoming the unwanted functions of these agents, which is still a problem at present, could not have been initiated without investigations with chimpanzees (Schellekens *et al.*, 1984). Nevertheless, experiments with various species of monkeys, especially macaques, advanced the application of interferon against viral infections in human clinical medicine as well as in oncology and AIDS research (Trown *et al.*, 1986; Hobson, 1987). At present, this work is still ongoing (Schellekens *et al.*, 1996).

The primates are indispensable in studies on the other powerful modern stimulators of natural immunity, the *interleukins*, including, of course, their recombinant forms, which provide a broad perspective for curing various human diseases, including AIDS (Geissler *et al.*, 1989; Khan *et al.*, 1996).

Let us analyze the similarity in other properties of the blood in primates, in particular, the *coagulation* of blood. This also conforms to the general tendencies of phylogenetic affinity between humans and simians. The thrombocytes (platelets) of macaques are somewhat smaller in size than human ones, but their fine structure and function are quite similar (Lewis, 1977). Coagulation and fibrinolytic activity, the level and action of blood coagulation factors, thromboplastin, prothrombin and thrombin times, the influence of heparin and its fraction on coagulation, receptors of thromboxane, and thrombus formation are all very close, if not identical in chimpanzees, baboons, macaques, and other simians. This stands in contrast to the blood of dogs, rodents and other mammals. Hence, thromboses, hypercoagulation, and other pathologies of the coagulating system in human blood are best modeled and studied in simians (Shepard *et al.*, 1984; Fareed *et al.*, 1985; Dormehl *et al.*, 1987).

One review (Hanson and Harker, 1987) substantiated the model of acute arterial thrombosis of the brain in baboons, and emphasized that the primary vascular anatomy, the structure, function, and concentration of thrombocytes, and factors of coagulation and fibrinolysis in these monkeys are similar to those in humans. The inhibitors of plasma proteins, moreover, are similar to the human ones by concentration, kinetics, and functions; the action of drugs and their pharmacokinetics are also the same. The authors concluded that baboons are convenient for studies of atherosclerotic changes. A good correlation of thromboelastogram in the blood of humans and monkeys was found during investigations of deep venous thromboses with a baboon model (*P. ursinus*) (Dormehl *et al.*, 1987). The influence of food fats on blood coagulation (and formation of prostaglandins) may be extrapolated to humans, even from experiments with marmosets (McIntosh *et al.*, 1987). The influence of hemorrhagic fever and on the whole, of infectious pathology on the operation of this system, has been studied with primates for some time (Mason and Read, 1971).

The significant similarity in the mechanisms of *hemopoiesis and blood circulation* in simians and humans is well-known (Kooksova, 1972). The genes for erythropoietin in many mammals have a high homology with humans, up to 80%–85%; in monkeys, however, it reaches 91%–94% (Goldwasser *et al.*, 1987; Wen *et al.*, 1993). Other peculiarities of bone marrow make primates, including humans, different from the other animals, for example, in terms of the reproduction of basophils and differentiation of T- and B-lymphocytes (Kirchenbaum *et al.*, 1988; Phan *et al.*, 1989). The micro-architectonics and topography of the vessels in higher primates differs, in particular, from that in dogs. This makes the latter, previously traditional animals for experimental study of endotoxic shock, useless for this purpose (Cavanagh and Rao, 1969; Swan and Reynolds, 1972; MacVittie *et al.*, 1994). Transplantation of bone marrow, the investigation of which, as mentioned, was honored with the Nobel Prize, is very fruitfully used in medicine. It is possible that xenotransplants (in particular, from baboons) may be one of the methods used in the struggle against HIV/AIDS (Relf, 1996).

Thus, the primates, including the platyrrhines, remain the most valuable laboratory animals for research on problems of human blood and immunology and for modeling pathological processes in these fields (Socha *et al.*, 1987; Neubert *et al.*, 1996). An interview with J.-F. Bach, the chairman of the International Congress on Immuno-intervention, in which the discussion touched on the theme of autoimmune diseases, immunology of transplantation, and production of vaccines, was entitled, 'Le singe est l'avenir de l'homme' ('Simians are the future of human health') (Bach, 1991). I have attempted above to demonstrate the clear value of primates in terms of the discrete components of blood and immunity. Blood and immunology, moreover, is a fundamental unity in all the systems of an organism; hence, it is the unity in all fields of medical primatology discussed below.

SENSORY SYSTEMS: SIMILARITY AND MODELS

Vision

The visual system occupies a special position in medical primatology. This is one of the most similar systems in humans and simians. This system in primates is clearly characterized by the presence of a fovea centralis with macula lutea (yellow spot) in the retina and by the presence of maximally clear stereoscopic three-colour vision. The elements of this system are present in other animals, but they take the human form ('a higher level of colour vision: trichomacy' – Bowmaker, 1998, p. 541) only in humans and simians (Young and Farer, 1971; de Valois *et al.*, 1974; Livingstone and Hubal, 1988; Jacobs *et al.*, 1991). Long-term investigations by Professor F. V. Andreev (1990) distinguished three levels of visual development in mammals; 1) monotremes, marsupials, insectivores, chiropterans, rodents, double-toothed rodents, and prosimii; 2) carnivores, pinnipeds, whales, sea cows, elephants, even-toed and odd-toed ungulates, and platyrrhine primates; and 3) catarrhine primates. This gradation among the species reflects quite accurately the evolutionary gamut of the mammalian visual system. Among the primates, in particular, it extends from the galagos, with no color vision (Condo and Casagrande, 1990) and lemurs, with weak color vision (Blakeslee and Jacobs, 1984), to the platyrrhines, with two-color vision (as a rule), and the humans, apes, and other catarrhines which are characterized by stereoscopic vision with four variants of photoreceptors (rods and cones), sensitive to the red, green, and blue colors. As I mentioned above, in primates vision is dominant over all other senses. More than a third of the cortex in primates is occupied with the processing of visual information (Allman, 1987). Ocular development proceeds similarly in macaques and humans, but with a difference in speed of 1:3 (Kiely *et al.*, 1987) or 1:4 (Harwerth *et al.*, 1986). The quality of visual behavior in rhesus macaques is practically identical to that in humans (Bishop, 1983).

I have repeatedly referred to the *primate type of brain structure*; unquestionably, there is a *primate type of visual system*. It is associated with the specific evolutionary destiny of the primates, which started in the Eocene. These characteristics include: frontally directed eyes, a large field of binocular overlap, and sharp stereopsis and focusing adjustments; high visual acuity in the center of the retina, distinctive ipsilateral projection to the lateral geniculate nucleus and ocular operculum (tectum), and increase in projections from the retina to different cortical areas; complexity in the lamination of the lateral geniculate body and its neural connections with the eyes; growth and functional differentiation of the primary visual cortex and extrastriate visual fields; and development of a grasping hand and coordination of the eye and hand; as well as other adaptations (Allman, 1986). I previously discussed the evolutionary increase of the occipital, parietal, and temporal brain lobes, which are related to vision.

The *anatomy* of the visual system in humans has much in common with that in simians, especially in the apes, but also in the monkeys, such that even the latter possess characteristics for developing useful research models. The early development of the lens and the ophthalmic cup in the macaque embryo, long-term organogenesis, and morphological identity to similar stages (11–16) in the human embryo, in particular, create a valid model for correlative molecular investigation (Peterson *et al.*, 1994). Computer tomography of the eye socket in humans and chimpanzees is similar by all parameters and it is similar to that in macaques as well. Immunohistochemical research indicates only insignificant differences between rhesus macaques and humans in the coloring of the corneal tissues, trabecular system, epithelium of the crystalline lens, uveal tract, retina, endothelium of all the eye vessels, and in other respects as well (Tripathi *et al.*, 1987; de Saban *et al.*, 1989). Aging and age pathology in the vision of rhesus macaques represent yet another area in which these monkeys may be used to model analogous processes in humans (de Rousseau *et al.*, 1983).

Neurobiological study of the primate visual system attracted scientists beginning as early as the previous century, but the field has developed especially rapidly since the

1960s. As a result, new data were obtained regarding the unique affinity of humans and simians and the application of this affinity to the welfare of humans. As I already mentioned, the achievements of cerebrologists working with primates in this field (e.g., D. H. Hubel and T. N. Wiesel), were honored with the Nobel Prize. Hubel and Wiesel explained that after working with cats up to 1962, the traditional animals for research on vision, they decided to conduct parallel investigations with macaques; these monkeys became 'the animal for our main studies' (Hubel and Wiesel, 1982, p. 169). It was natural for them to use monkeys, since they are the closest animal to humans with respect to vision. The cats were frequently unsuitable for the study of peculiarities in visual acuity, color vision, stereoscopic vision, ocular nerve path, a 6-layer geniculate nucleus (found only in humans and some species of catarrhines), and architectonics of the cortex with a clear demarcation of the layers and distinct borders between the striatal cortex and the neighboring areas (Hubel and Wiesel, 1968, 1977; Hubel, [1987] 1990).

These investigators made a series of prominent discoveries in the biology of the visual cortex. In particular, they described the columnar structure of the cortex, obtained new information about the mechanisms of visual development, and about cortical plasticity. They showed that the cortex requires ocular stimulation during the period of postnatal development (up to 4 months in cats and up to 4 years in macaques!), and that it is very sensitive to deprivation, at least when a single eye is deprived (which was frequently practiced in clinical ophthalmology without taking account of the age of the child). In these cases, disturbances in cortical neurons and other brain structures and of their interdependence are irreversible, and physiological rehabilitation does not occur; the cortex remains anomalous for life. We may assume that this phenomenon has a more extensive application, as well, such as in the psychological deprivation experiments of H. Harlow. In clinical practice, however, light deprivation distorts ocular dominance and causes amblyopia and strabismus. All this can be avoided, if the procedure of light deprivation takes into account the period of maturation of the cerebral cortex. It is not difficult to understand that the significant results that followed from these outstanding investigations were greatly appreciated by prominent clinical ophthalmologists (Deller, 1979). This line of research continues to this day (Sengpiel et al., 1996). Nevertheless, let us return to the characteristics of similarity in the visual apparatus of humans and simians.

The primate type of vision, mentioned above, relates morphologically to each of the sensory components and to the density and size of the cellular and neural pathways, but it is most clearly represented in the structure of the brain centers (Oshigova, 1981). The external Bellarje's stria and compact radial fibrovascular bundles are a specific system of interrelations that exist only in humans and simians (most clearly in the simians of the Old World). It has been known for some time that central vision, which is characteristic of primates, is projected to cortical field 17, the adjacent field 18, and also to field 19. Only the phylogenetically more recent field 19 is larger in humans and it differs quantitatively in its myeloarchitectonics (Ibid.). Only humans and simians are specific half-optochiasmic.

The homology of many other brain structures of the primate visual system, including subcortical ones (geniculate body, upper corpora bigemina, and pulvinar), is very high. In general, these structures have become models in the study of the central regulation of human vision and its pathology. Examples of the latter include damage to homologous areas of the brain, which causes practically identical pathological consequences in humans and simians (de Courten and Garey, 1981; Oshigova, 1982; Blumcke *et al.*, 1990). The area for color processing (V4), which was found in the prestriatal cortex in rhesus macaques (an Old World monkey) and previously unknown in humans, was subsequently also found in the human cortex (Lueck *et al.*, 1989). On the other hand, ocular dominance columns were not found in the cortex of saimiri (a New World monkey) (Hendrickson and Tigges, 1985).

The primate *retina* is unique in its similarity among humans and simians within the class Mammalia. Some present day investigators described a series of investigations in this field with the compelling title: 'Preface: Why Retina? Why Primate?' (Rapaport and Provis, 1996). These investigators noted the uniqueness of the retina in primates, the only mammals with a fovea centralis, which unlike other animals (rabbits, in particular) possess excellent stereopsis and depth discrimination, together with other advantages. 'As the best, most accessible model of the human, the ape, or monkey retina is the obvious medium for the development of clinical procedures that will eventually find their way into ophthalmology practice' (p. 144). These authors expressed their belief that studies conducted on the simian retina will resolve some current pathological problems of human vision and will eventually clarify problems of which we are not yet aware. What are the specific reasons for this confidence?

The retina of simians is not only anatomically and physiologically similar to that of humans, but it is also similar microscopically at the cellular level. The timing in development of progenitor cells and the onset of their differentiation (in the following cells in the retina: ganglion cells, horizontal cells, cone and rod photoreceptors, amacrine cells, Mueller cells, and bipolar cells) are of fundamental importance to the development of neural structures. Retinal cells do not undergo major differentiation in all mammals and they have a short lifetime. Such human processes can only be studied in primates (even with macaques) in which retinal development 'is a well orchestrated symphony . . . cell genesis, migration, and differentiation is coordinated in a number of ways' (Rapaport *et al.*, 1996, p. 155). Further research with additional animals will enhance our knowledge in this area. The distribution of synapses and dendrites of cells in the layers of the retina and the thickness and density of the yellow macular pigment are also similar among the primates, as are other structures and their functions in the area of perception (Koontz and Hendrickson, 1987). No doubt, there are differences, but these are always less significant that the differences between humans and other animals (Rodieck, 1988). Electroretinograms of cats and macaques differ, but those of macaques and humans are similar (Hess *et al.*, 1986).

The processes of metabolism and blood circulation in the retina of humans and simians are quite similar and they usually differ from those in other animals. On the basis of electrophoresis data, it is apparent that the structure of the interphotoreceptor retinoid-binding protein is practically identical in humans and macaques, but significantly different in nonprimates (Redmond *et al.*, 1986). There is no difference in the metabolism of vitamin A in the eye of the baboon and human. Indolamine-accumulating neurons were found in the retinas of cats, pigs, rabbits, fish, chickens, and even platyrrhine monkeys (*Cebus*). The only species in which these neurons were not found were humans and catarrhine monkeys (*M. fascicularis*). A lack of taurine in the synthetic diet of children leads to the degeneration of retinal photoreceptors. This condition was discovered for the first time in experiments with rhesus macaques and produced the same result (Eninger and Floren, 1979; Sturman *et al.*, 1984; Cammer *et al.*, 1990). There are many such examples. The vascularized retina is similar in humans and simians as well as a few other animals (mice, cats). The retina in rabbits, guinea pigs, and other animals, however, is non-vascularized (the number of astrocytes in the nerve fibers of the retina correlates with the number of blood vessels), but the blood supply is analogous only in human and simians (Young and Farrer, 1970; Zamir and Medeiros, 1982).

In view of these similarities, it is not surprising that pathology of the retina, both spontaneous and experimental, is similar in humans and monkeys. During ophthalmological analysis, human colloid degeneration was found for the first time in animals, 2-year-old baboons (Barnett et al., 1972). Subsequently, it was shown that degeneration of the retina with age in rhesus macaques is similar to hereditary degeneration in people. A natural model of human maculopathy with characteristic pigmentary anomalies in the macula and druse-like damage is also found in rhesus macaques (Monaco and Wormington, 1990). Experimental retinopathies were induced in monkeys by repeated exposure to an argon laser and a model of central serous chorioretinopathy was produced experimentally by repeated injections of adrenaline (Wallow et al., 1974; Yoshioka et al., 1982; Bornes et al., 1990). Transplanted cells of pigmentary epithelium from corpses and fetuses are accepted in the retinas of monkeys (Aotus, Macaca) (Berglin et al., 1997). It was impossible to reproduce a complete model of toxoplasmonic retinitis in rabbits, the traditional animals for studying eye toxoplasmosis, but a valid model was obtained in primates (Culbertson et al., 1982).

The cornea in many species of vertebrates, to say nothing of the mammals, is mainly similar morphologically. Nevertheless, even here, the primates are distinguished for their peculiarities. Unlike the majority of animals, the cornea consists not of 4 layers, but of 5 layers, as in humans, a Bouman's membrane appears in monkeys (Cech, 1964; Fritsch et al., 1990). There are still other peculiarities drawing humans and simians together. The lunula corneae, a milky-white formation on the temporal side of the cornea was found in macaques (M. cyclopis) and newborn children (Japanese), but it disappeared with age (Imai et al., 1975). Hydration of the cornea changes its ability to transmit color (with carboxylic and sulfate groups of mucopolysaccharides) in dogs, pigs, rabbits, and other mammals, but not in primates (Cejkova and Bolkova, 1974). Monkeys were used in numerous experiments to develop the methods of keratoplasties, to study ocular physiology, correct ocular changes, develop techniques for using contact lenses, and to assess the influence of detergents on the cornea, where rabbits, in particular, were useless (Madigan et al., 1987; Boothe, 1994).

The morphology and biochemistry of the lens are very similar in simians and humans and different in other animals, which makes monkeys the most applicable model for studies in this area. This similarity relates to the peculiarities of growth and aging and to the different properties of the lens in macaques and cats. There is also similarity in the structure of each of the three crystalline protein fractions (the previously mentioned alpha-, beta-, and gamma-crystallines), other membrane proteins, cytoplasm, and enzymatic activity. Low levels of the enzymes of glutathione and high levels of gangliosides are characteristic only of humans and simians and are related to the development of cataracts in primates. Thus, it is recommended that only monkeys be used to study cataracts and not mice, rabbits, or cats. Inherited cataracts are also found in monkeys, but not very often (Miranda et al., 1986; Rathbun, 1986; Gellatt, 1994).

Monkeys, apes, and humans differ from all other animals in terms of the apparatus and mechanism of regulation of the outflow of aqueous humor which is directly associated with the modeling of such a well-known human disease as glaucoma. The morphology of Schlemm's canal, trabecular network, the formation of aqueous flow, and ocular hydrodynamics as a whole are surprisingly similar in humans and simians (macaques) and different from those in other animals (Holmberg, 1965; Kaufman and Erickson-Lamy, 1985). The mean value of intraocular pressure in rhesus macaques (14.9 +/− 2.1 mm Hg) corresponds to the pressure in a human eye (Bito et al., 1979). In the 1970s, a large number of investigations were conducted on the peculiarities of normal and pathological aqueous humor in different primates (*Macaca*, *Aotus*, and other genera). After a period of problems and indecision, the simians were used more and more frequently to study glaucoma. The simians provided a completely adequate model for assessing the influence of lasers and in spontaneous diseases, in particular, since they produced the same damage of the optic nerve and other consequences as were found in humans (Gelatt, 1977; Bito, 1997; Gelatt et al., 1998). At present, many experiments in this area are being carried out with primates. This is one of the most intensive fields of research in modern medical pathology.

After this brief review of the affinity between humans and simians in major components, it is reasonable to discuss some common characteristics in particular aspects of their visual systems. As already mentioned, the development of the eye from birth to old age is generally similar in monkeys (macaques, for example) and humans. For this reason, monkeys are excellent animals in which to model visual development and its anomalies in humans, and amblyopia, in particular, if the ratio of their life cycles is estimated as 1:3 (Teller, 1983; Kiely et al., 1987). This relates to *visual acuity*, which is similar in humans and monkeys during infancy (Teller, 1981) and into old age. Visual acuity is 16–31 c/deg in young rhesus macaques and decreases to 15–18 c/deg in old ones, which is characteristic of humans with good ocular health (Ver Hoeve et al., 1996). Numerous experiments conducted on monkeys to clarify the dependence of visual acuity on various factors (illumination, inclination of the object, velocity of the target, etc.) produced the same result; there is great affinity with humans (Barmack, 1970; Bloom et al., 1986).

The psychophysical foundations of vision are practically the same in the higher primates (humans and simians). Information coding by the visual system is analogous, as is the perception of structural images (from global characteristics to local ones), as well as dark adaptation, and increment and spatial luminance contrast sensitivity. The principles of ocular motor function, eye motion, complex eye-head-hand coordination, visual object fixation since infancy, coordination of activity in the eye and upper eyelid, the mechanism of nystagmus and saccadic activity are similar in humans and simians (de Valois et al., 1974; Becker and Fuchs, 1988; Oehler and Shape, 1989). In addition, cholecystokinin causes contraction of the pupillary sphincter in simians (macaques), but not in cats, rabbits, rats, or guinea pigs (Bill et al., 1990).

Monkeys, apes, and humans differ from all other mammals in the peculiarities of *accommodation* because their dioptrical apparatus is characterized by the so-called passive-dynamic type of accommodation (flattened lens, development of three portions of ciliary musculature, cross-cinne connections, and gross and fine ciliary processes). This means that only the higher primates reach 10–12 D (the value is 1–3.5 D in dogs, 2–3 D in cats, and 1–1.5 D in rodents) (Andreev, 1989). Not only is the mechanism of accommodation quite comparable in humans and simians, but so also is its early aging. Accommodation is similar in amplitude and other parameters that were measured, as well as in the pathogenesis of developmental presbyopia. These qualities make macaques (rhesus) the 'first experimental subjects' for studying these problems (Kaufman et al.,

1982, p. 323), as well as other monkeys, which are used in investigations of visual changes associated with aging (Bito *et al.*, 1987).

The specifics of the brain in primates and the special interrelations between binocular and stereokinetic mechanisms also determine a similar high level of *stereoscopic vision* in simians and humans. Some degree of stereopsis in other animals, especially in carnivores (cats), does not preclude important advantages of monkeys. Monkeys have the greatest affinity with humans in this fundamental visual characteristic, one that we take to be representative of the order Primates. Similarity is also found in stereoscopic acuity and depth perception, the range of the eye vergence for vision together, binocular competition, and in the identical consequences of cerebral damage that impacts stereoscopic vision. These characteristics form the basis for using monkeys in experiments which are impossible to perform on humans and for which nonprimate animals are ineffective (Sarmiento, 1975; Harwerth, 1982; Logothesis and Schall, 1990; Poggio, 1995).

As I already mentioned in the beginning of this section, *color vision* is practically the same in humans and catarrhine monkeys. The spectral sensitivity is known for many animals, but not for all of them. Nevertheless, all of them have differences in photoreceptors, compared to humans. Humans and other catarrhine primates have rods and three types of cones (for perceiving red, green, and blue colors). Thus, humans and catarrhine simians are trichromates. The majority of platyrrhines are dichromates (with polymorphism of visual pigments typical for primates of the New World). For example, *Callithrix, Callicebus, Cebus, Saimiri,* and *Ateles* are both dichromates (always in males) and trichromates (females; they can also be dichromates) (Gouras and Zrenner, 1981; Jacobs *et al.*, 1991; Bowmaker, 1998). Peak values of sensitivity in the cones vary in different species of simians (and even in different individuals). Color perception, however, including shade, brightness, and saturation are very similar in humans and other catarrhines, in both photopia and scotopia. Along with color vision, achromatic vision is also similar in humans to macaques, baboons, guenons, and especially to chimpanzees. Even the platyrrhines are very close to human dichromates in their dichromasy, and they are similar in other characteristics of vision, including binocularity. Cats, on the other hand, with a small number of cones in the retina (common mammals have only two spectral classes of cones) and a weak mechanism for color opponetion, are less capable of discriminating color than simians and humans (Orlov, 1972; Matsuzawa, 1985; Merigan, 1989; Latanov *et al.*, 1991, 1997; Bowmaker, 1998).[2]

Taking into consideration these biological prerequisites of the human visual system, it is not surprising that research in ophthalmology during the 1960s–1980s was one of the most dynamic fields in biomedical research in which primates were used. It continues to attract considerable attention to this day. The number of publications in this field from 1967–1985 is listed in Table 4.4 (Fridman *et al.*, 1990).

In describing the biological similarity of vision in monkeys and humans, I mentioned a series of ocular diseases that are appropriate for investigation in primates. No doubt, the list is not exhaustive. The model of amblyopia mentioned above was studied in

---

2 I dare say that a certain confusion in the study of the problem of color vision in primates is related to invalid use of the taxonomic nomenclature of the latter. Some ophthalmologists even today use the term 'anthropoid' only for apes; but humans, apes, and all monkeys are considered to be in the suborder Anthropoidea and all of them must be termed 'anthropoids'. See, for example, Bowmaker, 1998, p. 544 (E.F.).

Table 4.4 Number of annual publications on research of the visual system conducted with nonhuman primates during the period 1967–1985.

1967	1970	1975	1976	1985
80	161	209	194	308

monkeys in its various aspects and with different etiology, especially concerning strabismus. Spontaneous strabismus in macaques is not only identical to the human clinical state, but it also appears in young monkeys at a comparable age to children, and probably with the same frequency as in humans (4%) (Kiorpes et al., 1985). Experimental models of strabismus and amblyopia developed in this manner are quite adequate to the human disease. Macaques of various species (but not cats) are appropriate for these purposes (Joosse et al., 1990; Sengpiel, 1996). Due to the peculiarities of the ocular blood supply (frontal ciliary artery), the study of ischemia in the eye after an operation to correct strabismus can only be performed in primates (Keough et al., 1981).

Experimental reproduction of myopia in primates should be considered the most valuable model of human visual pathology. Spontaneous myopia is widely spread among monkeys and is very similar in young macaques under natural conditions and caged adult macaques in captivity (Young, 1981, and many other publications of this author). Other investigators also described models of myopia in primates (several species of macaques, marmosets, and tupaia) which were produced experimentally by various forms of light deprivation (Raviola and Wiesel, 1985; Troilo and Judge, 1993). In his review of myopia models, however, M. Edwards (1996), while appreciating the use of primates in research in general, proposed that the most suitable animals for these purposes are chickens because of the high cost of primates. Nevertheless, he noted that differences between chickens and humans may be significant compared to those between monkeys and humans. In this book on the biology of animal models, I do not touch on the financial details of research, although I clearly understand their importance. I must point out, however, the contrasts and differences in plasticity of the visual system associated with cortical structures in humans and simians, on the one hand, and all other animals, on the other. This was noted by Hess (1956, cited by Boyko and Manteyfel, 1977) precisely in connection with chickens. Sometimes a model which seems to be similar, in fact may actually be a sham, as I. Russel and S. Pereira (1981) found when they compared the results of studying visual neglect in rats and monkeys.

One more aspect of the problem should be discussed in connection with experiments on myopia. Experiments by E. Raviola and T. Wiesel (1985) showed that atropine and dissection of the optic nerve had a protective effect on the development of myopia induced by light deprivation (by suturing the eyelids) in Macaca arctoides. When the same procedure was used with another representative of the same genus, Macaca mulatta, no effect was found. This result demonstrates the necessity for careful consideration when extrapolating experimental results not only to humans, but even to other species within a single genus of primates.

In such a brief discussion, it is impossible to list all of the unique characteristics of primates that are relevant to ophthalmology when other animals are unsuitable. In addition to those problems that I already discussed, I may also cite a model of optic papilledema (Hedges and Zaren, 1969), Wernicke's disease (Cogan et al., 1985), unilateral blue-blindness (Wright *et al.*, 1987), autoimmune uveitis (Nussenblatt and Gery, 1996), diabetic retinopathy (Laver *et al.*, 1994), saccadic activity (Pare and Munoz, 1996), various surgical procedures, and many others. The condition of the eye in primates is also a valuable indication of diseases in other systems of the organism, such as hypertension, atherosclerosis, diabetes, infections, etc. The primates are indispensable, moreover, in pharmacological and toxicological studies related to vision. Results sometimes obtained with primates not only differ from those found in other animals, but they may even be diametrically opposed (Lynch *et al.*, 1986).

In the middle of the 1990s, there were 150 million people with serious visual problems and 38 million who were totally blind. The assessments of the World Health Organisation show that, if no precautions are taken, the number of blind will reach 75 million by the year 2025. There is no reason to wait until this forecast comes true. We can prevent it with the help of primates.

Other sensory systems

Despite an early interest in studying the *auditory system* of primates (in the 1870s, D. Ferrier determined the localization of the auditory cortex in a monkey), investigation of this sensory system was not carried out intensively until the broader development of medical primatology took place in the second half of the 20th century. W. C. Stebbins (1975) wrote that in 5 years the number of publications on hearing of primates exceeded all that was known in this field up to 1970. The same author stated that the hearing of simians was so close to that of humans that monkeys should be considered an appropriate model to study hearing loss due to the impact of drugs (antibiotics, salicylates, quinine; unfortunately, in the 1990s this list can be increased) or intensive noise (which has not decreased either).

Stebbins (1975) compared absolute and differential sensitivity, their critical ranges, the frequency of oscillations, localization of sound, and the loudness and perception of complex sounds in human speech and in the communication of simians; the acoustic characteristics were comparable in humans and simians (p. 114). In the 1990s, new characteristics of similarity were added. The central (cerebral) regulation of hearing is homologous in humans and simians, particularly, with respect to the auditory cortex of the brain. Inaccuracy of the methods was the reason given for the previous discrepancies between the data from ill people and those from animal experiments. In fact, these discrepancies were due to the use of traditional laboratory animals (for example, rats and cats) in research where simians were required (Heffner and Heffner, 1990).

Because of the unique immobility of the outer ear in humans and apes, among the other mammals, the upper limit of hearing does not exceed 32 kHz (usually 16–20 kHz). In monkeys it reaches 40 kHz and in prosimians 60 kHz (as in the majority of mammals), although frequency discrimination is different in various species (Stebbins, 1975; Sinnott *et al.*, 1987). Nevertheless, despite the different ecological, social, and biological characteristics of the primates (Old and New World), including humans, there is a common pathway for auditory transmission from the cochlea to the auditory cortex (Newman, 1988). The hearing of humans and apes differs from that in other mammals in terms of their narrow bands of high frequency reception.

The similarity of humans and simians in the anatomy of the auditory canal, labyrinth, system of auricular bones and cartilage, air cavities, and their blood supply is well-known (Hohmann, 1969; Axelsson, 1974). The morphology of the middle ear is similar in humans and rhesus macaques. The structure of the tympanic membrane, tympanograms, the complexity of the eustachian tube, biomechanical parameters, and pathology in macaques are very close to those in humans. For this reason, these primates are considered good models for studying normal and pathological hearing of humans. They are also useful for diagnosing clinical problems of the middle ear, functional impassability of the eustachian tube, otitis media, and in particular, for developing techniques of surgical operations (Miller and Donaldson, 1976; Doyle, 1984).

There are other data on the similarity in some aspects of the neural regulation and function of hearing in primates. The time shift in the sensitivity threshold under the influence of auditory stimulation (shot report) in monkeys is equal to that in humans. They have similar induced auditory potentials. Monkeys can identify signals of two complex tones with a common frequency as well as humans, regardless of its presence in one of them. Even acoustic analysis and the central processing of vowel and consonant sounds in monkeys is similar to those in humans (Bary, 1978; Pohl, 1983; Tomiinson *et al.*, 1988). Macaques and baboons may be successfully used in otological trans-plantations, the assessment of ototoxicity of drugs, the study of the phenomenon of oto-acoustical emissions, and in many problems of experimental audiology of humans (Hormann, 1969; Lonsbury-Martin and Martin, 1988).

As already mentioned, chimpanzees are the only species in which no differences were found in *gustatory sensitivity* to sweets, when compared to humans (Hellekant and Danilova, 1996). The similarity of this sensation in humans and the other simians, primarily the Old World monkeys, is significant. Investigations carried out since the 1930s indicated significant affinity with the sense of taste in humans, at least in chimpanzees, but also in other simians (see Glaser, 1970; Kalmus, 1970). Anatomical research demonstrated clear differences from the primates in all the animals that were studied and conspicuous similarities between monkeys and humans in the peculiar-ities of the taste receptors of the tongue (Kadanoff, 1970). Despite previous doubts, detailed investigations on the perception of different substances by primates of various phylogenetic levels generally confirm the correlation of taste sensation in primates with the taxonomic hierarchy already described (with possible deviations on single tests, which is even common among people). The closest animals to humans are chimpanzees and other hominoids; catarrhine monkeys are more distant, and platyrrhines and pro-simians are more distant still.

The affinity of catarrhine monkeys and humans with respect to gustatory sensation is reflected in several areas. The response in the two main taste nerves of the tongue indicates that the sensitive areas of the tongue are similar in rhesus macaque and human. Sweetness and saltiness are tasted on the front and bitterness is tasted on the rear (Danilova *et al.*, 1996). Sweet sensation in man, assessed with seven dipeptide derivatives or analogues, was different in three groups of primates: 1) prosimians; 2) New World monkeys; and 3) Old World simians and humans (Glaser *et al.*, 1996). I note that humans and catarrhine simians are placed in the same group.

There is less information about the similarity of the other sensory systems of humans and simians. As I already mentioned, the *olfactory* system in primates has different importance in various groups, and consequently, the biological structure is different. Olfactory stimuli play a significant role in the lives of prosimians and New World monkeys (feeding, reproduction, and orientation in the environment). On the other hand, this role is greatly decreased in the lives of catarrhine primates, especially the hominoids, including humans. This is reflected in the neural pathways of the olfactory system. There is no vomeronasal system in the neocortex of the Old World primates (with their two olfactory areas), in contrast to rabbits and dogs (which have only one neocortical area). All mammals are divided into two groups according to the neural structure of their olfactory system. Humans and all simians of the Old World are assigned to one group and the New World monkeys and other mammals are assigned to the other (Takagi, 1986). Nevertheless, it is likely that even the platyrrhine, *Cebuella*, uses 'similar olfactory mechanisms for detecting odorants' as human. This does in no way mean, of course, that their sensitivity is similar (Glaser *et al.*, 1994, p. 453).

Limited information about the *sense of touch* was discussed to some extent in the section on dermatology. Data on the *vestibular* apparatus apparently reflect the species-specific form of locomotion and ecology. There is evidence, however, that even in this sensory system, there are areas of similarity among the higher primates, including humans (Matano, 1986, 1987; Sinclair and Burton, 1991).

DENTITION AND MODELS OF DENTAL DISEASES

In the description of evolutionary-systematics, we often touched on the issue of dental characteristics in different groups of primates, this being one of the important criteria of the biological affinity of mammals. The group of catarrhine primates (Old World monkeys, apes, and humans) is assigned a special taxon at the level of the infraorder (Catarrhini) by the number of teeth and the dental formula (and no doubt, by other characteristics). Table 4.5 shows how this group looks in comparison to other laboratory animals.

There are no animals in the world other than catarrhine simians that are so similar to humans in the number of teeth and dental formula. Given the information presented above about the many similarities between humans and other primates, it is not difficult to guess that we also have unique affinity in this area arising from the depths of evolution. Taurodontism, an atavistic peculiarity of the dental pulp ('bovine' molars and 'shovel' incisors), is sometimes seen as a separate characteristic or a component of various syndromes of anomalous human development. It was first described in Neanderthal man, but it is also found in chimpanzees and orangutans (Keeler, 1973; Jorgenson, 1982).

The morphology of the corona dentes differs in monkeys and hominoids (it is more similar in apes and humans [Mahoney and Sciulli, 1983]), although the sequence of calcification of deciduous teeth is the same: central incisor, first molar, lateral incisor, canine, and the second molar. The order of tooth eruption is the same in monkeys (Old World) and chimpanzees, and it differs in humans only by a shift in the sequence of eruption of molars and premolars. In general, the mechanism of tooth formation is practically identical in all catarrhine primates, including humans. Each group is a model for the other groups, taking into account the difference in life-span (Kooksova, 1954; Swindler, 1985). Of course, the eruption of the teeth is associated with maturation of other anatomical characteristics and functions of the organism, which was partly discussed above in relation to the affinity among primates. The speed of dentine formation in humans and rhesus macaques, microstructure of the enamel in humans and great apes, concentration of sodium, potassium, and magnesium in the dentine and enamel in macaques and humans, and even misalignment of the bite in present day humans and chimpanzees, are similar or identical (Melsen *et al.*, 1977; Gordon, 1984).

Table 4.5 Dentition of some animals and humans (adapted from Yankell, 1985, p. 281).

Species	Dental Formula of Permanent Teeth			
	Incisors	Canines	Premolars	Molars
Mouse, rat, hamster	1 — 1	0 — 0	0 — 0	3 — 3
Guinea pig	1 — 1	0 — 0	0 — 0	4 — 4
Rabbit	2 — 1	0 — 0	3 — 2	3 — 3
Ferret	3 — 3	1 — 1	3 — 3	1 — 2
Dog	3 — 3	1 — 1	4 — 4	2 — 3
Pig	3 — 3	1 — 1	4 — 4	3 — 3
Capuchin	2 — 2	1 — 1	3 — 3	3 — 3
Marmoset	2 — 2	1 — 1	3 — 3	2 — 2
Rhesus monkey	2 — 2	1 — 1	2 — 2	3 — 3
Ape	2 — 2	1 — 1	2 — 2	3 — 3
Human	2 — 2	1 — 1	2 — 2	3 — 3

Research showed that monkeys in their natural environment develop the same pathology of the oral cavity as humans; lesions of the dental periosteum, caries, traumatic changes, elimination of teeth, etc. (Gardy et al., 1982). Clinically healthy gums of a monkey do not differ in histology from human gums. Gingivitis, dental plate microbes, and dental calculus in monkeys have the same properties of pathogenesis as in humans, as well as periodontal diseases as a whole. Gingivitis, which can also be induced experimentally, occurs spontaneously in macaques at an average age of 15 or older (Krygier et al., 1973; Offenbacher et al., 1977). Dental microbial flora are practically identical in macaques and humans. Serological reactions to bacteria (for example, *Eusobacterium nucleatum*) in infected teeth are identical in rhesus macaques and humans, but different in dogs. The processes of dental inflammation and recovery of teeth and dental pulp are also similar in simians and humans (Vincent *et al.*, 1983; Warfvinge, 1986).

Such advantages of laboratory primates justify their use in medical experiments, despite their rather high cost. This relates both to catarrhines and the New World monkeys. Since the 1960s, due to dental and oral research on callitrichides and other platyrrhines, a group from England (headed by B. M. Levy) initiated research on dental caries, periodontal disease, oral oncology, and defects in oral-facial development (I note that before this, beginning in 1948, such experiments were carried out on rhesus monkeys at the Institute of Odontological Research in the United States [Hisaoka, 1973]). Levy's group showed why these investigations should not be carried out with the usual laboratory animals, such as mice, rats, hamsters, or dogs, but rather, should be conducted with monkeys (Levy, 1971, 1979). In the 1970s–1980s, numerous experiments were conducted on tooth transplantation with primates.

One of the most serious problems of dental pathology has been dental caries. Experiments carried out in this field with monkeys were especially convenient because the nonhuman primate model was comparable to human disease with respect to many parameters, for example, etiology, clinical condition, radiology, and histology (Yankell, 1985). Immunization against caries was highly successful with monkeys, but it still remains a problem for people (Bowen, 1996). Investigations with primates directed toward other dental and oral diseases, including periodontitis and vaccines against it, are currently underway as well as experiments on osseous transplantation of jaws (Persson *et al.*, 1994; Dodson *et al.*, 1997).

SIMILARITY OF DIGESTION IN PRIMATES AND OPPORTUNITIES FOR EXPERIMENTATION

Interestingly, food is retained in the oral cavity of nonprimate mammals by palatine crests before the stage of chewing, but in simians and humans this is not so. Instead, high speed x-ray photography showed that food is retained by the frontal part of the upper dental arch in these latter species (German *et al.*, 1989). Food consumption is regulated not only by the gustatory apparatus in humans and simians, as it is in rodents (with significant differences in this comparison), but it is also under the control of the visual system (Scott and Giza, 1987). There are also other distinctions from the nonprimate animals. It is common to this day to draw analogies between digestion in humans and dogs (Koto *et al.*, 1994; I emphasize, however, that these investigators certainly appreciated the experimental value of rhesus macaques). This tendency goes back to the second half of the 19th century when classical physiologists (Bernard and later Pavlov) carried out experiments with dogs and extrapolated their results to humans. Experiments with monkeys, however, showed that extrapolation of data from experiments with primates to humans was more justified.

The long-term research of V. G. Startsev (his books were published in the Soviet Union and translated into English in the United States) provided evidence that the physiology and pathology of digestion in humans and lower simians differed from that in dogs. Activity of the gastrointestinal tract in monkeys (*P. hamadryas, M. mulatta, C. aethiops*) and humans is monophasic (unlike dogs), with excitation during the daytime and inhibition at night. Unlike dogs, the inhibitory role of emotional stress in the processes of digestion is greatly increased in simians and humans (during an experiment in which a monkey is exposed to emotional stress, acidity in gastric juice may disappear completely). The saliva of dogs, moreover, essentially lacks salivary amylase activity. In monkeys and humans, this activity is so prominent that it reaches a value of several tens of thousands of units (after Wolgemut). The normal gastric juice of a dog, a carnivorous animal, has a significantly higher value of acidity (0.5% HCl) compared to humans or monkeys (0.1–0.2%). The differences are related to the peptic activity of gastric juice (it is lower in dogs) and continuity in the secretory process (in dogs it starts with the onset of eating). Finally, stomach and duodenal ulcers develop in dogs as a result of experimental neurosis, whereas stomach achylia with precancerous changes of the mucosal coat of the stomach develop in monkeys (*P. hamadryas*) (Startsev, 1972, 1980, p. 244).

There are numerous arguments supporting the basic similarity, and in some respects, even the identity, of the composition, biochemical properties, and physiology of saliva and of the microscopic anatomy of the salivary glands in humans and simians (with certain distinctions in the anthropoids of the Old and New World). This contrasts with the data on other animals (Shestopalova, 1968; Stephens *et al.*, 1986; and others). Similarity in the ultrastructure of various tissues of the digestive tract was found in the epithelium of the stomach, intestines, and lymphoid tissue associated with the intestines. This applies not only to the catarrhine monkeys, but also to marmosets (Spencer *et al.*, 1986; Jablonski *et al.*, 1989). The protein requirement of these monkeys corresponds to the value known for adult people when related to body weight (Zucker and Flurer, 1989). A folic acid deficiency in baboons leads to the same consequences as in humans; loss of body weight, anorexia, diarrhea, leukopenia, and similar biochemical and hematological changes (Siddons, 1974). There is a 96% degree of similarity in nucleotide (and amino acid) sequences of pepsinogen A in rhesus macaques and humans (Evers *et al.*, 1988). Here again, there are conspicuous similarities between simians and humans and differences between them and other laboratory animals.

The concentration of gastrin in plasma and its response to food in rhesus macaques is similar to that observed in other animals (rodents, dogs), but in the duodenum, it is found in high concentrations only in humans and simians (rhesus macaques) (Scallet *et al.*, 1989). The action of soybean products with different amounts of trypsin inhibitor on the pancreas is different in simians, rats and pigs. The utilization of D-methionine in humans and macaques is similar and differs from its absorption and metabolism in rats, rabbits, chickens, and pigs. The secretion of pancreatic juice under the influence of secretin, cholecystokinin, and histamine is different in monkeys and dogs (Stegink *et al.*, 1980; Struthers *et al.*, 1983; Iwatsuki, 1985). Secretin hormone that stimulates gastric secretion causes choleresis in baboons, humans, and dogs. In both primates, choleresis is accompanied by an increased production of cyclic adenosine monophosphate, whereas in dogs (as well as rats), no increase in adenosine monophosphate occurs – there is a different mechanism of bile metabolism (Levine and Hall, 1976).

The content and metabolism of bile in primates, including humans, is also a unique phenomenon of biology. Bile acids and their metabolism are very similar, as is the protein component of bile, conjugates of bilirubin, the ratio of salts (cholesterol) of phospholipids, the ratio of immunoglobulins (IgG/IgA), and the formation of biliary stones. This permits the modeling in simians of cholelithic disease, biliary atresia, kernicterus, biliary cirrhosis, hepatitis, and other human sufferings that cannot be reproduced in other animals. There was confusion in the experimental study of bile and the associated pathology up to the 1960s solely because nonprimate animals were used in the research. Baboons, macaques, saimiri, owl monkeys, and tupaias were used successfully to study these problems (Campbell *et al.*, 1972; Kuvaeva, 1976; Schwaier and Weis, 1982; Cornelius, 1988; Pekow *et al.*, 1995). I already mentioned the indispensability of simians to the study of endotoxic shock. This is associated with a unique affinity of humans and monkeys in terms of the venous blood circulation in the liver. Venous sphincters inherent in dogs (there are no such sphincters in primates) make them inapplicable for the study of this human pathology, as well as pathology of the liver as a whole (Rangel *et al.*, 1970). The experimental transplantation of liver and the modeling of alcoholic cirrhosis of the liver in primates will be discussed below.

The similarity of intestinal microflora in humans and simians is especially important in terms of modeling infection. I shall discuss this in the section dedicated to infectious diseases. Here, I note only that comparison of the resistance to colonization of the digestive tract by bacteria (Enterobacteriaceae) demonstrated statistically reliable differences in the properties of different groups of mammals. Humans do not differ from monkeys, but the two together differ from rodents and dogs (Van der Waaij and van der Waaij, 1990).

It is likely that the kinetics of digestion and the passage of food is most similar in humans and chimpanzees (Milton and Demment, 1988) although, as we saw, the characteristics of digestion are also very similar in humans and monkeys. This allows us to ignore apes in research on these problems. Different species of macaques were effectively used in the study of protein insufficiency, providing the investigator with the main properties and symptoms of the childhood disease of malnutrition, kwashiorkor. Rats (as well as other animals) were previously used in this research, but gave a picture of the disease which was completely different from the one observed in humans (Deo and Ramalingaswami, 1977). The severe consequences of taurine insufficiency in the food of children were demonstrated with young rhesus monkeys (Sturman, 1990). The monkeys provide valid models for studying the influence of long-term parenteral nutrition of humans (Friday and Lipkin, 1990), as well as models of drug-induced nausea and vomiting (Rupniak *et al.*, 1990). Rhesus macaques (but not baboons) were significant experimental subjects for studying the role of the micro-aerophilic bacteria, *Campylobacter pylori*, in the etiology of gastritis and peptic ulcer in humans (Baskerville and Newell, 1988). Monkeys are also useful for modeling human obesity. The mechanism for the increase in fatty tissue and its influence on the reproductive cycle, and social and cognitive factors on growth in primates significantly differ from those in other animals (Kemnitz, 1984; Pond and Mattacks, 1987, 1988).

Primates are indispensable in modern studies on pathology of the large intestine. In the 1960s–1970s common features were found in patho-anatomical characteristics of diseases of the large intestine in humans and simians; changes in the mucosal coat and submucosal layer, the inflammatory and reparative processes, and the disturbances caused by protozoa. The investigators indicated that the frequency of tumors and diverticula in monkeys was comparatively less common than in humans. This result was apparently due to the young age of the animals that were dissected, since monkeys did not survive well in captivity until the 'cancer age' (Lapin and Yakovleva, 1960; Scott, 1979). Now we know that this latter conclusion was correct. In the analysis of 278 autopsies, 48 malignant neoplasms were found in rhesus macaques; 25 of them were cases of colon cancer. Twenty among the latter were found in elderly and old animals (older than 20 years of age) (Kemnitz *et al.*, 1996). If we recall that the frequency of cases of rectal cancer in the United States is second only to lung cancer, it is not difficult to imagine what this statistic means for the congeners closest to humans. In the 1980s, studies of rectal pathology with primates achieved the status of a significant scientific event. The investigation of this problem area continues today.

Nineteen cases of colon adenocarcinoma were found during necropsies carried out on 149 monkeys of the New World, which belong to the tamarin genus (*Saguinus*), *S. oedipus*. There was not a single case of colon cancer in the related species of tamarins, *S. fuscicollis*, which were kept in the same colony. The investigators suggested that there was a genetic basis for these differences (Sayer *et al.*, 1980). It was found that spontaneous carcinomas in *S. oedipus* were preceded by spontaneous ulcerous colitis. Colitis was also observed in *S. fuscicollis* (and even in *Callithrix* [marmosets]), but in fewer numbers, and it did not become malignant (Clapp *et al.*, 1988). As already mentioned, research in this field continues. Investigators have obtained valuable natural models of acute and chronic colitis, the influence of stress on its development, the relationship between inflammation and neoplasm, a model of rectal cancer with metastases, and a model of early carcinogenesis, its diagnosis, immunological, morphological, and other conditions of the transformation of tumors into malignant forms similar to those in humans (Clapp, ed., 1993). From the standpoint of medical primatology, it is apparent that indifference in the choice of the experimental species of monkey, even monkeys of the same genus, is unacceptable. The major peculiarities in pathology of two very close species may be significantly different.

THE URINARY SYSTEM AND EXPERIMENTAL UROLOGY

At the beginning of the 1970s, Dr. James A. Roberts of the Delta Regional Primate Research Center (Tulane University, USA) published an article which was the continuation of his earlier articles, 'Why the Monkey? Use of Nonhuman Primates in Urology Research' (1973). On the basis of the similarity in the urinary system of humans and monkeys with respect to physiology, immunology, and pathology, and taking into account the differences in other animals, in particular dogs, this investigator concluded that the monkey is the most suitable animal for 'most urologic research' (p. 475). After listing the diseases common to humans and simians (hydronephrosis of pregnancy, glomerulonephritis, malignant tumors, immunological disorders), Roberts stated that experimental pyelonephritis could be studied in a primate. For a quarter of a century, this problem has been pursued by this investigator and his collaborators (Roberts *et al.*, 1995).

Specific aspects of the morphological and physiological affinity of the urinary system in humans and simians have been known for some time, but probably not to the same extent as the anatomical data. The primates, like all other mammals (and unlike birds and reptiles), have renal pelvises, but only the simians and humans have monopapillary kidneys (Carneiro de Moura and Pinto de Carvalho, 1958).

Several groups of investigators have noticed a specific structural similarity (multipyramidal point) in humans and spider monkeys (*Ateles geoffroyi*) (Goodman *et al.*, 1977); Szostakiewiez-Sawicka *et al.*, 1980). Blood vessels connecting the kidneys and adrenal glands are similar in humans and simians and these primates differ from dogs in this respect; they are developed to a lesser degree in dogs (Earle and Gilmore,

1982). There are also data regarding a histological affinity in the urethra and penis of humans and baboons, vascular changes in the penis that take place with age (leading to impotence), and a similarity in the other morphological characteristics of the urogenital system (De Kock and Burger, 1985; Bornman *et al.*, 1985). The ultrastructure of the major types of basement membrane in the acellular renal cortex of rhesus macaques is so similar to that in humans that it constitutes a valuable model for studying the pathology of these tissues, especially those involved in diabetes mellitus (Carlson *et al.*, 1986).

Renal physiology in guenons (*C. aethiops*) and baboons (*P. ursinus*) is very close to that of the human with respect to the analysis of blood and urine, cystograms, pyelograms, renograms, and in the impact of diuretics on the renogram. These monkeys are also similar to humans in terms of the influence the menstrual cycle exerts on the function of the kidneys (Goosen *et al.*, 1982). The function of the kidneys in macaques (*M. arctoides*) appears to be similar to that in humans with respect to the clearance of inulin, creatinine, and *p*-aminohippurate, and in the ability of the kidneys to concentrate salt in response to a hypertonic salt solution (Cronin *et al.*, 1972). It is difficult to accommodate these data with those from a 10-year study of creatinine in chimpanzees (Eder, 1996). In that study, serum levels of creatinine, in relation to body weight, were significantly higher in chimpanzees than in humans, and urinary levels of creatinine and the specific density of urine in 24 h samples were lower. If these differences are not accounted for by differences in body weight or the diet in different laboratories (there are other data on creatinine in chimpanzees [Hainsey *et al.*, 1993]), then we must conclude that kidney function in monkeys is more similar to that in humans than it is in chimpanzees, and this is doubtful. I note that other characteristics, such as the influence of mercury diuretics on acid transport in the kidney tubules, are similar in chimpanzees and humans (and differ from those in other animals) (Fanelli *et al.*, 1972).

The other physiological and biochemical parameters of the kidneys in monkeys are analogous with those of humans. Innervation of the kidneys in Java macaques (*M. fascicularis*) is a fair model of the innervation in humans. The kidneys of monkeys and humans contain significantly less renin than the kidneys of any other mammal. Receptors for angiotensin II and the angiotensin-metabolizing enzyme of the adrenal glands have the same function in humans and simians, but they are different in other mammals. The formation of calcium stones in the kidney tubules and the influence that the kidney prostaglandins exert on this pathological process is similar in humans and macaques (Buck *et al.*, 1983; Fiador and Mendelsohn, 1987; Marfurt *et al.*, 1989). The excretion of urates by the kidney is relatively high in pigs and low in simians and humans. The concentration of urates in the blood plasma of mammals decreases from 300 μM–400 μM to 3 μM in the sequence from humans to simians, dogs, rats, rabbits, and pigs (Roch-Ramel and Schali, 1981). Glutamine synthetase activity is high in the kidneys of rabbits, intermediate in the kidneys of rats, and totally absent in humans and monkeys (Horsburg *et al.*, 1978). The peculiarities of uric acid evolution were discussed above.

The electrophoretic picture of the urinary excretion of protein (morning samples) in different simians of the Old World corresponded to that of humans (0.01 mg/ml– 0.02 mg/ml), but conspicuous proteinuria was found (4 mg/ml) in monkeys of the New World (Callitrichidae) (Fuchs *et al.*, 1989). Human urine can be differentiated, on the basis of color (using high precision immunosorbent ELISA with rabbit anti-human uromucoid), from that of various species of animals, but not from chimpanzees and monkeys of the Old and New World (Tsutsumi *et al.*, 1988). The urinary bladder in humans, baboons, and macaques is highly sensitive to atropine, which is unlike other animals; this suggests that there is a similar innervation in the higher primates of the Old World. The sensitivity of the urinary bladder to atropine is relatively low in cats, intermediate in marmosets and capuchins, and most sensitive in the catarrhines (Craggs *et al.*, 1986).

The prostate in nonhuman primates is a potentially useful model for the human gland, but it has rarely been used in experimental studies by urologists. There is a significant homology in the morphology and physiology of this gland in humans, chimpanzees, gorillas, macaques, and baboons, despite certain distinctions (van Camp and van Sande, 1988). The data suggest that the prostate in chimpanzees has the greatest histological similarity to the human prostate (and testicle) among 27 species of wild animals that were assessed. As a whole, the morphology of the prostate gland corresponded to the phylogenetic classification of the species. The antigenic determinants of acid phosphatase from the prostate were identical in humans and simians, but were not found in the glands of dogs and rabbits (Ablin *et al.*, 1970, cited in Shevschenko, 1973). Judging from the antigenic similarity, the proteins in the prostatic fluid of humans and simians are similar and unlike those in other animals (Carter *et al.*, 1985). It is not surprising that the pathological condition of prostatic hyperplasia in humans, including cancer, is adequately modeled in simians, much more adequately than in dogs. This condition not only occurs spontaneously in primates, but it can also be induced experimentally by altering the balance of sex hormones (Lewis, 1984; Ferreira *et al.*, 1995; Waters *et al*, 1998). Spontaneous prostatitis is also known in nonhuman primates (Roberts, 1988).

There are a number of other diseases of the urinary system that occur spontaneously in primates, including inherited anomalies, degenerative changes, nephritis of varying etiology (including viral), pathology of the ureters, infection associated with prolonged diabetes (for example, diffusive glomerulosclerosis), and other diseases (Kaur *et al.*, 1968; Stout *et al.*, 1986). According to the assessment of the World Health Organisation Scientific Group, epidemic nephropathy can only be studied with simians (WHO Report, 1986). Investigation of a neurogenic influence on the urinary apparatus is possible in monkeys. Psychogenic polyuria-polydipsia and neurogenetic disorders of urinary bladder function, in particular irritable bladder syndrome, have been studied in simians (Leomis and Rosenberg, 1980; Ghoniem *et al.*, 1995). I discussed the modeling of pyelonephritis in simians above. There are data indicating that vaccination with a protein conjugate from *Escherichia coli* reduces the progress of this disease in rhesus macaques; in particular, it prevents the formation of scars (Roberts *et al.*, 1993). Primates are a convenient animal for the study of urinary tract infections and for testing vaccines against such infections (Johnson and Russel, 1996), which will be discussed in more detail below. I also note the similarity of microflora in humans and other primates. Ureaplasms detected in monkeys (*Cercopithecus, Macaca*) are similar serologically to those of humans, but differ from those of cattle, goats, sheep, horses, cats, and birds (Ogata *et al.*, 1981).

At the same time, despite the unique advantages offered by simians over other animals, the application of primates to research in experimental urology is rather limited and does not match the opportunities that exist. The number of publications in this area for the years I examined (1967–1985) is fewer that one hundred per year.

HORMONES AND ENDOCRINE STUDIES WITH PRIMATES

I now consider one of the most fruitful fields in medical primatology, the study of the endocrine system in simians. The affinity of the higher primates in this area is very close and forms the basis for studying the biology and pathology of the human endocrine system. I note that two Nobel Prizes were awarded for the study of hormones (using monkeys) (see Chapter 2). The number of papers that have been published reflects quite well the usefulness of primates in this sphere. There were less than 60 papers in 1967, but this increased to approximately 300 in 1985 (Fridman et al., 1990).

The hormone profile as a whole, hormonal biosynthesis, the characteristics of secretion, the number and balance of hormones, their structure and function, and their influence on other physiological systems are very similar in humans and simians, and frequently differ significantly in other animals. This relates to insulin, which was mentioned above, growth hormone, thyroid hormone, and the hormones of the reproductive system, including the sex steroids. There is also considerable similarity in neuroendocrine regulation, especially in the peculiarities of the hypothalamo-hypophysial-adrenal system of primates. The order of hydroxylation and the characteristics of steroid hormone synthesis are strictly species-specific. The main hormone of the adrenal cortex is corticosterone in rabbits and rats and hydrocortisone in dogs and guinea pigs. Both of these compounds are produced in simians and humans, but there is a preponderance of hydrocortisone (Schulster et al., 1976; Udaev et al., 1977). Naturally there are certain differences in the endocrine system of humans and simians, especially the monkeys, but the principles and regulation are close and differ from those in other animals. Growth hormones are divided into two groups on the basis of gene mutations; humans and simians are in one group, pigs, bulls, sheep, horses, and rats are in the other (Chipens et al., 1990).

During the 1950s, outstanding studies were carried out on growth hormone by Li Choh Hao and his colleagues, an investigation that is still pursued by many other investigators. This research showed that growth hormone (somatotropin) in humans and simians, in particular in macaques, shares a similar amino acid composition, close molecular weight, almost identical gene structure (only 4 mutations), analogous stimulation by stress, and comparable metabolic modulation of secretion under the influence of various agents, and it is different in all other creatures – nonprimates. This, in principle, allows one to use somatotropin from monkeys (or its recombinant analogue) to cure human problems, whereas somatotropin from any other animal is not biologically active in humans. It is clear that it is more reliable to test the recombinant hormone (as well as many other recombinant products, which are discussed below) and carry out other experiments using this hormone, specifically in primates (Li Choh Hao and Parkoff, 1956; Lazarev, 1973; Wheeler and Styne, 1988; Leone-Bay et al., 1996).

Simians have been used in various endocrine investigations, in particular, for modeling colloid goiter (Andrus et al., 1964) and other disturbances, endocrine reactions during emotional stress, and also as noted, reproductive endocrinology.

One of the most critical problems of modern medicine is diabetes (it is the fourth highest cause of death in the world. In the middle of the 1990s, there were 135 million people suffering from this disease and the number predicted by the WHO for 2025 is twice that amount). At present, numerous investigations on diabetes are carried out with simians. To gain some perspective on the validity of this choice, we need only recall the affinity of humans and other primates, discussed above. Not only is there a general affinity, but there are also specific similarities in neurology, hematology, immunology, vision, insulin secretion, biochemistry of paratrophy, diabetic retinopathy, and diabetic damage of the kidneys. In the 1980s, data confirmed the genetic, and consequently phenotypic, similarity of insulin (and proinsulin) in humans and monkeys (Naithani et al., 1984; and others). It was shown that the influence of food on the periodicity of oscillations in the basal level of insulin, glucose, and glucagon were practically the same in rhesus macaques and humans (Hansen et al., 1982).

Spontaneous diabetes develops in various mammals (Chinese hamsters, rats, guinea pigs, and mice of certain strains), including many species of simians of the Old and New World. The finding of a special susceptibility to diabetes mellitus in Celebes black macaques (M. nigra) was likely a stimulus for studying this disease in a variety of monkeys (Howard, 1971). Many aspects of human diabetes were studied with the natural model of diabetes mellitus in black macaques, although the model was not identical to the human disease. The areas include hormone metabolism during the early stages of prediabetes, glucose tolerance, some anomalies of glycemia, the role of amyloid in damage to the islets of Langerhans, pathogenesis of the islets themselves, the magnitude of autoantibodies to the cells of the islets, obesity, the influence of age on the onset and course of the disease, microangiopathies and their relationship to atherosclerosis, development of other complications of diabetes, and opportunities for therapy (Howard and Fang, 1988; Howard Jr., 1995).

Diabetes in rhesus macaques is 'an excellent model' for the study of Type II noninsulin-dependent diabetes mellitus in humans (Bodkin et al., 1996). Two years after the appearance of hyperglycemia, the early characteristics of diabetic neuro-pathy were observed in monkeys (Cornblath et al., 1989). Primate models of diabetes were analyzed in detail and found quite adequate in a series of publications in the 1990s (Bodkin, 1996; Hansen, 1996; and others). As I already mentioned, models of various forms of diabetes, including insulin-dependent diabetes, have been studied in monkeys. In addition to the model of spontaneous diabetes, a model of diabetes mellitus is produced in monkeys by the administration of streptozocin or by ablation of the pancreas (Jones et al., 1984). Monkeys were effectively used to study diabetes of pregnancy and its influence on the development of fetal defects (Kemnitz et al., 1985). Spontaneous and experimental diabetes were also studied in chimpanzees, various species of macaque, baboons, guenons, marmosets, tamarins, owl monkeys, and ring-tailed lemurs.

In the 1990s, primates were used in many such investigations. Opportunities for preventing the early indications of diabetes have been studied in association with the fine mechanisms of pathogenesis of the disease. These studies cannot be conducted with sick people or any other laboratory animal. Antidiabetic drugs are being tested and other methods of curing diabetes are being sought, including xenotransplantation of the islets of Langerhans (Thomas et al., 1999). Pathological disorders associated with diabetes have been studied in the nervous system, visual system, kidneys, and the cardiovascular system (Zhou et al., 1994; Watson et al., 1997).

I mentioned above that primates are used in various endocrine investigations. Simians play a special role in endocrine studies on the biology, pathology, and toxicology of the human reproductive system. The physiological action of hormones is so interlaced with the physiology of reproduction that I discuss reproductive endocrinology in a separate section, immediately below.

SIMILARITY OF THE REPRODUCTIVE SYSTEM IN PRIMATES AND ITS EXPERIMENTAL APPLICATION IN MEDICAL PRIMATOLOGY

There are certain similarities in the reproductive systems of all mammals, but they reach their greatest similarity in humans and other primates, especially among the simians. We should recall that, according to Darwin, the reproductive system plays a fundamental role in the origin of species, including humans. This was reflected in the title of his book and in its main thesis regarding the role of sexual selection (Darwin, 1871). Darwin devoted a special paper to sexual selection in monkeys (Darwin, 1876). There is a long list of reproductive characteristics that demonstrate the affinity of humans and other higher primates. They include late and prolonged development of sexual maturity (against a background of a relatively long biological life-span), the presence of an ovarian menstrual cycle in females, practically identical hormonal and neural regulation of fertility and sexual behavior, relative independence of mating from the time of ovulation and phase of the menstrual cycle, pervasive sexual responsiveness of females (throughout the reproductive phases, including pregnancy), an absence of seasonal reproduction in many species (not only the apes, but also some monkeys), the influence of social status on reproductive success, a comparatively long gestation period, which is predominantly monocystic in the majority of species (single births), and undoubtedly, great anatomical and physiological homology (but without identity) of many structures and all phases of the reproductive cycle. These, together with the common biological similarities between humans and simians, make the latter indispensable subjects for experimental investigation of the corresponding human problems, and their embryo-toxicology and pharmacology. Chimpanzees, the species of ape that was studied most in this area, were found to be most similar to humans (Graham and Hodgen, 1979).

It is likely that no field of medical primatology attracts as much attention on the part of national and international organizations, including WHO, as primate reproduction and associated fields (with the exception of poliomyelitis and AIDS). This may be due in part to the tragedy of thalidomide which shocked humanity. Many scientific meetings, including those under the aegis of WHO, have been held in this field. One of the first symposia of this type was organized by WHO in Sukhumi (1971). At that time, no data were available on the reproductive systems of many species of simians, but the symposium focused on those species of monkey that had been studied (primarily macaques and baboons). Nevertheless, the symposium ascertained that nonhuman primates are the *only species* in which it is possible to elucidate the causes of ectopic pregnancy, to study endocrinology of the menstrual cycle, secretion of placental hormones, metabolism of steroid hormones during pregnancy, anatomy and physiology of the placenta, and to investigate gametogenesis, monozygotic (uniovular) twins, and the use of intrauterine contraceptives (Concluding remarks, 1972). In a short time, this list was supplemented with the problems of gamete transport, implantation, fetal development, pathology of the climacteric, development of vaccines against pregnancy, and other important issues. Beginning in the 1960s, considerable research with primates has been pursued in the fields of pharmacology and embryo-toxicology of pregnancy (Delahunt and Lessen, 1964; Hendrickx, 1972b; Goldzieher et al., 1974; Wilson, 1978). The number of publications on the reproductive system of primates grew from 136 in 1968 to 420 in 1975 and remained at approximately this level up to 1985, the last year of my own quantitative analysis (Fridman *et al.*, 1990). Let us briefly discuss the foundations of research in this area.

Primates are of great importance in the study of *sexual maturation and puberty* and its correlation with other aspects of maturation (the general growth of the organism, ossification of the skeleton, development of teeth, maturation of the brain and endocrine system) because of the significant affinity of this process in humans. Various factors that are studied in association with maturation include the environment, ecology, food, and sexual dimorphism. Similarities between humans and nonhuman primates were distinguished for both male and female individuals, including the same increase in pubertal growth, adolescent sterility, the same correlation of hypothalamo-hypophysial hormone secretion with development of the testicles, and the increase in testosterone levels in males, as well as many other comparable relationships (Harrison *et al.*, 1979; Watts, 1985).

Simians are no less valuable in the study of *senescence of reproductive function*. Specific human features of age-related changes in the reproductive system are characteristic only of these primates. The females of many primate species, and apes in particular, are subject to menopause, with many features of the preclimacteric and climacteric period in women, including the loss of ovulation long before the termination of menstruation (several years in chimpanzee), while preserving sexual activity even after the complete termination of menstruation which occurs before old age (20–30 years in advance in chimpanzees). The main features characterizing these processes in women are observed in simians, including monkeys, such as the change in level and character-istics of hormone metabolism, morphology of the ovaries, decrease in folliculogenesis, intermittent bleeding, and postmenopausal osteoporosis (Jelinek *et al.*, 1984; Bowlez *et al.*, 1985; Galloway, 1997).

The senescence in male sexual function is also similar to the human; the decrease in the level of sexual behavior is especially similar in apes and humans. Human sexual behavior also has common features with macaques and baboons in terms of hormonal regulation, the significant role of social factors, the relative independence of sexual behavior (of females) from the phase of the sexual cycle, and in the joint stimulation of male and female individuals (Nadler, 1986; Nadler and Phoenix, 1991). The age-related changes of steroidogenesis in human and simian males are identical. Excretion of steroid hormones from a senescent male baboon (*P. hamadryas*) essentially coincides with similar processes in a senescent man, regardless of some quantitative distinctions. The adrenal androgen, dehydroepiandrosterone, and its sulfate play a significant role in the prevention of atherosclerosis, diabetes mellitus, and tumor development. These hormones are secreted in significant amounts only in primates and decrease during the process of aging. It is not difficult to understand, therefore, the significance of studying these problems in monkeys (Lapin, 1996; Goncharova, 1997).

All the processes of senescence cited above, including menopause, are simulated in laboratory models using simians. Rhesus macaques were the main subjects of these studies for many years. Later, scientists more frequently shifted from seasonally repro-ducing rhesus macaques (which can also bear young at an elderly age, greater than 25 years) to Java macaques which do not exhibit seasonality in their reproductive cycles. This was also encouraged by the state of the market for importing primates. Nevertheless, by the middle of the 1990s, the genus of *Papio* baboons was newly 'discovered' for these purposes. As I already mentioned in relation to the baboons, this genus is the most appropriate animal to model the problems of reproduction in humans

(Carey and Rice, 1996; Johnson and Kapsalis, 1996; Hendrickx and Peterson, 1997). Also as noted above, investigations in the field of reproduction were carried out with baboons at the Sukhumi Primate Center throughout its entire history, beginning in the 1930s (Botchkarev, 1933; Goncharov *et al.*, 1978; Goncharova, 1997).

Similarities in *morphology* of the reproductive organs were found in the primates long ago, and they represent the main foundations that distinguish this group as a separate taxonomic order. Histological and cytological investigations carried out in the last 50 years fully confirm and widen the foundations of this similarity in humans and simians. This relates to the structure of the vagina and cervix of the uterus (the transitional zone of epithelium in female baboons and women is identical, which is important for the study and diagnosis of cervical tumors), interstitial cells of the ovary, and the morphology of the endometrium during the menstrual cycle and at all phases of the sexual cycle (baboons) (Kraemer *et al.*, 1977; Micha and Quimby, 1984).

The system of tubules in the nucleus of cells in the endometrial epithelium, which is associated with preparation of the endometrium for implantation, was found only in humans and simians; there is no such structure in the epitheliocytes of rats, guinea pigs, cats, or dogs (Volkova, 1983, p. 204). Protein synthesis in the endometrium and deciduous layer of the uterus in humans and baboons is the same, as is the metabolism of a series of enzymes and the pattern of receptors for the main sex hormones. This relates not only to the endometrium, but also to the biochemistry of the corpus luteum and other components of the reproductive system (Fazleabas *et al.*, 1989; Rojas *et al.*, 1989). It was thought for a long time that all primates except humans lack a hymen, but some doubt has recently been cast on this position, at least for chimpanzees and gorillas (Evans *et al.*, 1987). Simians are valid species for modeling the central neural regulation of ovarian function and female sexual pathology, including a valuable model of endometriosis with baboons, which is quite different in primates from that in rodents (Pohl and Knobil, 1982; Su *et al.*, 1988; D'Hooghe, 1997).

There is much in common between men and the males of various simian species in terms of the microscopic structure, endocrine function, and biochemistry of the testicles, the ultrastructure and physiology of the epididymis, the process of spermatogenesis, and the morphology and biochemistry of spermatozoids (Hinton and Setchell, 1980; Katsiya *et al.*, 1986; Dadoune and Alfonsi, 1989). This similarity extends to the apes, baboons, macaques, and other primates, even including the tree shrews (*Tupaia belangeri*) (Collins *et al.*, 1987), but it is not found in any nonprimate animals. Prolonged immobilization of a male macaque leads to a delay in spermatogenesis as it does in sick people confined to bed. Monoclonal antibodies against sialoglycoprotein of the sperm head plasma membrane in humans recognize the antigens in the sperm of the Java macaque, but not in the spermatozoids of 10 other species of animals. Local x-ray radiation of the testicles even in low doses leads to significant depletion of the seminiferous epithelium in rhesus macaques, as it does in humans. This does not occur in rodents, where rapid regeneration of the tissues is observed (van Alphen and de Rooij, 1986; Villarroya *et al.*, 1987).

The fundamental characteristic of female reproduction in humans and many primates is the *ovarian menstrual cycle*. This peculiarity of the primates is so striking that it even confused the specialists, who for a long time called the sexual cycle of simians an 'estrous cycle,' based on behavioral data. Regardless, it was the monkeys (rhesus macaques) in which the mechanism of women's menstruation was found, as I mentioned in the second chapter of this book. In 1971, Professor Ya Kirschenblat wrote, 'The erroneous concept is widely spread that menstruation in primates corresponds to the estrus in females of other mammals. On the contrary, these are two different phenomena ...' (p. 319). No doubt, the specialists understood the difference in these concepts, but it is likely that they did not have the courage to admit the equivalence of this phenomenon in humans and other creatures.

In 1987, J. Loy considered this problem in terminology with respect to sexual behavior and appealed to his colleagues to reduce 'the terminology of estrus with respect to all catarrhines' (p. 176). It is known that menstruation also occurs in female capuchins (*Cebus*) (Mahoney, 1985) and is probably found in an implicit form in other platyrrhines when a distinct ovarian cycle is observed. As was shown in the previous chapter, an ovarian menstrual cycle is also recorded in female tarsiers. There are data that species which are distant from primates are also subject to menstrual cycles, such as elephant shrews (*Elephantus myurus*) and bats (*Glossophaga sorcina*), something which is difficult to explain (Finn, 1987).

Different opinions on this problem have been stated recently with respect to sexual behavior (Nadler and Dahl, 1989; Goy, 1990). The majority of investigators agree, however, that the reproductive endocrinology and physiology of the menstrual cycles in women and female simians (not only apes) are principally similar; the distinctions are only quantitative. There are clearly considerable differences between primates and other mammals in terms of the periodic bleeding of females at menstruation (approximately once a month). The blood loss in relation to body weight is very close in female monkeys, apes, and healthy women (Shaw et al., 1972). There are also clear distinctions from other animals with respect to the emancipation of copulatory activity from fertility, as a consequence of the moderation in the hormonal control of sexuality. As such, there is the possibility of mating during nonfertile phases of the sexual cycle, especially under certain circumstances, and a significant influence of social factors on sexual motivation. These factors could have had an important influence on the evolution of primates as well as on the origin of humans themselves.

Homology of the *hormonal picture of the sexual cycle* is unusual in humans and simians. This relates primarily to the species of hominoids, that is, to the apes. Practically all the sexual hormones, protein and steroid, have been studied at all phases of the female cycle, including luteinizing hormone (LH), follicle-stimulating hormone (FSH), 17β-estradiol and other estrogens, progesterone, testosterone, aldosterone, pregnanediol, and others. Their concentration in blood, urine, and bile was studied at different phases of the cycle. Their metabolism and metabolic products were investigated, as were their interactions, transformations, and variability under the influence of pharmacological agents. One species of ape may be closer to humans in one hormonal characteristic while another species may be closer in other characteristics. All of the four great ape species, however, are closer to humans than any other primate (Nadler et al., 1979, 1985; Czekala et al., 1987). It is clear that these species represent the best models for the study of corresponding human problems (as long as the research does not adversely affect the health of these rare and valuable animals). This does not mean that monkeys are not suitable to study the problems of human reproduction. Monkeys are widely and effectively used in these investigations and this application is well-justified.

Two graphs from the paper of W. C. Hobson et al. (1976) clearly demonstrate the similar dynamics of the two gonadotrophins in the serum of humans, chimpanzees, and rhesus macaques during the menstrual cycle (despite quantitative differences) (Figure 4.5).

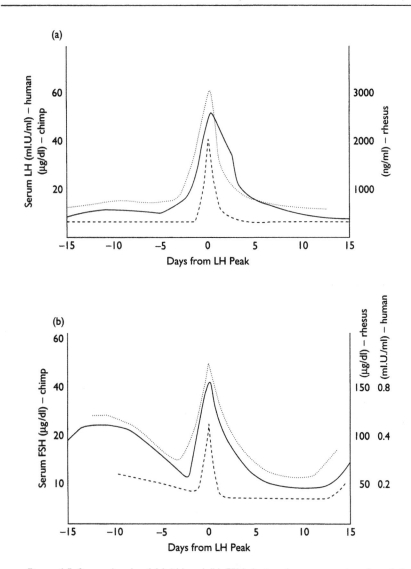

Figure 4.5 Serum levels of (a) LH and (b) FSH during the menstrual cycles of rhesus macaques, chimpanzees, and humans (Hobson et al., 1976).

    Endocrine disorders associated with maturation of the corpus luteum in ovulatory and anovulatory menstrual cycles of rhesus macaques provide unique opportunities to study and correct these conditions in women, not only during maturity, but also during puberty and in senility (Wilks *et al.*, 1979). This also applies to the other species of macaques, baboons, and the other monkeys. Hypothalamic control of the simian menstrual cycle differs from control of the estrus cycle in rats (Knobil, 1981). The levels of gonadotrophins during the menstrual cycle (and gestation period), and their variations in various catarrhine monkeys, are similar to those in humans. The influence of other hormones on LH and FSH are also analogous and unlike the reactions in rats (Yoshida, 1983; Finfscheidt *et al.*, 1990). Numerous investigations on estrogens, progesterone, progestins, and other hormones, the regulation of their receptors in the uterus, ovary, and oviducts, and their conversion provide further grounds of the close relationship between humans and monkeys. They also support use of the monkeys in models of human medical problems (Stevens et al., 1970; Knobil, 1995). The liver of adult simians does not inactivate estrogens, which is similar to the data for humans, but differs considerably from investigations with rats, guinea pigs, rabbits, and dogs (Donovan and van der Werften Bot, 1974).

Variation in the balance of sex hormones is also similar in humans and simians with respect to pathology of other systems and organs. For example, hormonal function of the adrenal and sexual glands in baboons (*P. hamadryas*) during hemoblastosis coincides with the types of disorders that occur in sick humans (Goncharova and Goncharov, 1985).

The systems that mediate sexual behavior under the influence of pharmacological preparations differ significantly in primates and rodents; stimuli that clearly change the behavior of rats may have no effect in rhesus macaques (Chambers and Phoenix, 1989). Induction of ovulation with drugs and synchronization in the development of follicles and steroidogenesis have been studied successfully in monkeys. Follicular stimulation in female rhesus monkeys has been accomplished with recombinant human gonadotrophin (Kenigsberg et al., 1986; Zelinski-Wooten et al., 1994).

Nonhuman primates are used as models in investigations on the role of different brain structures in the manifestation of proceptive and receptive components of sexual behavior. The influence of menstrual cycle phase (follicular vs luteal) on eight categories of behavior in stumptailed macaques (*M. arctoides*) (aggressive, submissive, affiliative, sexual, self-directed, etc.) is comparable to the data on women (Kendrick and Dixson, 1986; Ochoa-Zarzosa et al., 1994). Disturbances of the menstrual cycle during stress, mentioned above, also attracted the attention of medical primatologists. The influence of chronic self-injection of narcotics on the menstrual cycle was studied in macaques (Mello et al., 1997). It is likely that baboons, as mentioned above, are the best experimental subjects among the monkeys for studying sexual cycles, but macaques, guenons, and other catarrhines as well as marmosets, tamarins, and saimiri are also suitable (ovulation and ovarian function).

It is natural that the hormone profile for the reproductive system in male simians is no less similar to that in humans than was shown for females. It is also natural that the greatest affinity is found in great apes. Changes in the levels of testosterone throughout life are similar in men and chimpanzees, gorillas, and orangutans (Kingsley, 1988). This also relates to the other sex hormones. Nevertheless, monkeys are also effectively used in the study of male reproductive endocrinology. A male rhesus macaque is an adequate model of androgen and estrogen metabolism in men. The influence of testosterone on mRNA of the sex steroid-binding protein (which is highly homologous in humans and monkeys) is quite similar in the higher primates, including macaques (Kottler et al., 1990).

There is no better model than monkeys (possibly excluding the apes) for studying the influence of androgens on aggressive behavior, the role of circadian-ultradian cycles on the level of sex hormone secretion, and the importance of orchidectomy and other experimental treatments on the physiology of the hypothalamo-hypophysial-adrenocortical axis and the hypothalamo-hypophysial-gonadal axis. Other animals, and rodents in particular, give mainly different results in experiments on these problems (Dixson, 1980; Smith and Worman, 1987; and others). Agents that increase sexual activity in male rats are not effective in male macaques. The personal choice of a sexual partner is very important in these simians, to say nothing about the apes (Phoenix and Chambers, 1986). There are also data suggesting that orgasm is characteristic not only of apes, which was already mentioned, but also macaques and even guenons. Some assume that orgasm in females is a response to orgasm in males (Dobroruka, 1985; Linnankovski and Lelnonen, 1985).

The organizer of a symposium on the neuroendocrinology of reproduction at the 16th International Primatological Society wrote, 'Research in monkeys is extremely important for understanding diseases and treatment of human patients. For example, the finding of the hypothalamic GnRH pulse generator by Knobil and his colleagues has been widely used in treatment of infertility and precocious puberty' (Terasawa, 1996, #254).

The model of *pregnancy* and associated problems in simians is a valuable source of scientific knowledge. Starting in the middle of the 19th century and continuing through the 20th century, embryologists, gynecologists, obstetricians, biochemists, endocrinologists, pharmacologists, toxicologists, and other specialists in this field focused their attention on the affinity of humans, apes, and monkeys. There is still a unique potential for experimental research in biology, pathology, and pharmacology of human reproduction. I briefly mentioned authoritative opinions about these areas while describing the various species at the beginning of this section. Now let us discuss them in detail.

In addition to what was already said about the affinity of primates, new arguments were found which further confirm the importance of research on the simians. Not only was homology found between humans and other hominoids, but also between humans and monkeys. Fetal-placental-uterine anatomy, embryology, and physiology are very close in humans and baboons, although there are also distinctions. One of these differences is in the ratio of total and intracellular fluid and in the amount of body fat in females (baboons). The similarity of pregnancy in monkeys (macaques, baboons) and humans is characterized by similar steroid production throughout the course of the pregnancy, comparable functional anatomy of the uterus, and an absence of recognizable endocrine signals that precede delivery. Disorders associated with pregnancy and which are similar to human pathological disorders are also found in simians (Elygulashvili, 1955; Brans et al., 1985; Germain, 1990). Timing of the birth of an infant (predominantly night and early morning) also coincides in humans and simians (Jolly, 1972).

Biochemical and physiological parameters of the cardiopulmonary system and changes in the blood and urine of pregnant baboons (*P. cynocephalus*), in general, differ from those in nonpregnant females, similar to the finding in women. Pregnancy-associated plasma protein-A is found in pregnant women, chimpanzees, gorillas, baboons, and macaques. There is no such protein in nonpregnant women, men, or males of these primate species, and it is also lacking in marmosets, rats, sheep, mares, and cows, *regardless of pregnancy*. Monospecific antiserum against steroid-binding proteins in human plasma at the late stage of pregnancy show cross immunological reaction with the similar proteins of nonhuman primates. The reaction ranges from complete in chimpanzees and gorillas to a weak one in prosimians, but there is no reaction with the plasma of other mammals (Mercier-Bodard et al., 1981; Sinosich, 1986; Bischof et al., 1989). I already discussed the similarity of thyroxin-binding globulin during pregnancy of catarrhines in the section on blood. Isoforms of myosin and actin have similar spectra in female monkeys and women at different stages of pregnancy, but this differs in pregnant rats. The same should be said about the concentration of pregnancy-specific β_1-lipoprotein and the activity of a number of enzymes for hormone biosynthesis and metabolism which change during pregnancy (Braunstein and Asch, 1986; Cavaile et al., 1986; Rojas et al., 1989).

Chorionic gonadotrophin (CG) is secreted by the placenta only in humans and some other primates. Rodents only produce a substance that is similar to CG. The other gonadotrophins, their releasing hormones from the hypophysis, and their receptors are species-specific and are very similar in humans and simians. As already mentioned, apes are the closest relatives to humans in this respect, especially chimpanzees. According to certain parameters, however, gorillas and orangutans may be closer. Nevertheless, monkeys, primarily baboons and macaques, are also similar to humans in their gonadotrophins, and thus they are effectively used in this research area as well. The reactivity of the hypophysis to gonadotrophins during pregnancy and the puerperium, their morphology, the properties of CG, its excretion in urine, and its disappearance from the blood after pregnancy is interrupted, are all quite similar in humans, baboons, macaques, and in certain respects, in platyrrhines. Very important, pregnancy in rodents and rabbits is supported by hormones produced by the ovaries, whereas in humans and other primates, this function is performed by the placenta (Hobson et al., 1976; Asch et al., 1984; Korte et al., 1987; Garcia et al., 1989).

As I mentioned previously, similar data regarding an affinity with steroid hormones in humans were obtained in pregnant apes, baboons, macaques, other catarrhines, and also in marmosets, despite differences from humans in the absolute values (Solomon and Leung, 1972; Goncharov et al., 1980; Goland et al., 1990). There are also data on the similarity of lactogens in the placenta and fetuses of catarrhines, including humans (Gusdon et al., 1970; and others).

Comparative morphology and physiology of the primate *placenta* have been studied for some time, as we saw in the previous chapter. Similarity in this area is used as one of the taxonomic foundations for determining the order and its high taxa. Despite the specific peculiarities in each of the subgroups of primates, comparison with other placental mammals indicates that the former are the closest to humans to an extent that corresponds to the phylogenetic scale (although monodiscoidal, hemochorial, and villous placentae, as was shown above, are observed not only in lower simians, but also in tarsiers). The fine structure of the placenta and its layers in fetal apes, baboons, macaques, guenons, and even in marmosets, are close to those in humans (Soma, 1983; Owiti et al., 1989). Blood circulation in the placenta of baboons has more tortuous arterial vessels, but in general, is very similar to the human system (Lee and Yeh, 1983).

The immunological affinity of plasma proteins in humans and simians is always significantly greater than in other animals. Immunoreactive eosinophil granule major basic protein (mentioned above) in the plasma and placenta is chemically and immunologically identical in humans and other primates. Its content in the plasma of pregnant women and simians increases 4–8 times compared to nonpregnant individuals, whereas its analog is unchanged in guinea pigs, mice, rats, cats, and dogs (Gusdon et al., 1970; Wasmoen et al., 1987). The same is observed in the comparison of gonadotrophin concentrations in the placenta and the metabolism of testosterone by its microsomes (baboon). As I stated in the description of macaques, the placental gonadotrophin-releasing hormone is identical to that in humans. This permits the investigation of different aspects of human reproductive endocrinology, including the development of synthetic gonadotrophins (Duello and Boyle, 1996).

Due to significant affinity of humans and simians at all stages from gametogenesis to the postnatal period, the *embryology of primates* provides rich opportunities in the areas of biology and experimental medicine. Starting with the first comparisons of fetuses of different species of ape with human fetuses, carried out in the 1880s by J. Deniker (1886) (awarded the Brocá Prize), investigators discovered many aspects of the affinity of humans and other primates in this area.

Humans have their own peculiarities of embryology, but many features are uniquely similar in humans, other living hominoids, and monkeys (taking into account the ratio of the life-span and the gestation period in the latter). There are similarities in implantation, formation of the structures connecting the embryo and the mother, development of the embryo and differentiation of the fetus, prenatal and postnatal changes, the birth process, and many other anatomical and physiological characteristics and events of the gestation period. There is no such similarity in any other animal (Schmidt, 1969; de Lemos and Kuehl, 1987; and many others).

I already mentioned the similarity in the initial formation of the cerebral structures in humans and simians. The concept that gestation periods are related to the phylogenetic scale in humans and nonhuman primates is, no doubt, reasonable in general. There are specific differences, however, which is especially important for toxicologists. The literature indicates that the period of organogenesis is 2–3 times shorter in rodents than humans, but many parameters in macaques show that they practically or completely coincide. The neural plate appears on the 8th day of pregnancy in the mouse embryo, the 18th to the 20th day in humans, and the 19th to the 20th day in rhesus macaques. The first somite appears on the 8th day in mice, the 20th to the 21st day in humans, and the 20th to the 21st day in rhesus macaques. The primordial forelimb appears on the 9th day in mice, on the 28th day in humans, and on the 27th to the 29th day in rhesus macaques. There are yet other coincidences and similar figures (Hendrickx *et al.*, 1983), but the picture is clear. Embryonic stages in the common marmoset (*Callithrix jacchus*) differ from those in humans in terms of the *speed* of development (although there are also coincidences), but embryonic development in these two species, in general, is quite comparable. Given the small size of marmosets, they are especially suitable for teratological investigations (Merker *et al.*, 1988a; Stahlmann and Neubert, 1995).

The fine structure and cytological peculiarities of the endoderm and mesothelium of the rhesus yolk sac are similar to those of the human yolk sac. These monkeys can be used, therefore, to model functional maturation of the human yolk sac. I already discussed the similarity of teeth in primates. This similarity also extends to the development of the primordial teeth in humans, apes, and monkeys, as well as the primordial craniofacial skeleton (*M. arctoides*) and the development of the face in general. Monkeys can provide a valid model of the development of organs in the fetus and newborn, sexual differentiation and its anomalies, including intrauterine and postnatal infections, disorders of metabolism, and respiratory insufficiency (King and Wilson, 1983).

I have repeatedly emphasized the validity of primates for experimental research on biomedical issues. In light of my purposes, I continue to do this in the context of pregnancy, the fetus, and newborns. V. Sopelak and G. Hodgen (1984) wrote that 'primates are invaluable models' for investigation of these problems and named some that are already known to the readers of this book. Even their list, while very competent, does not include all the opportunities for experimental study of pregnancy and associated problems in primates. To those already cited, I add the early loss of the fetus due to environmental pollution, which accounts for 20%–25% of human abortions. Biological markers of this pathology can be successfully studied in macaques (Hendrickx et al., 1996). Retardation of intrauterine growth accounts for 25% of perinatal mortality in humans. Models in common laboratory animals are inadequate, but in monkeys (saimiri) they are quite adequate (Brady et al., 1996).

The problems of *lactation* can also be studied in simians because there is much in common with respect to these processes between humans and other primates. The role of lactation in the suppression of fertility is a special problem of considerable interest in medical primatology, if we consider the reproductive variety of these species that are so closely related to humans (seasonality and its absence, variation in sexual cycles, different forms of social and sexual behavior). The data demonstrate significant similarity (despite differences) among humans, chimpanzees, and macaques in terms of the macro-anatomy, functional mechanisms and innervation of the lactiferous glands, and the morphology of blood supply (Macpherson and Montagna, 1974). The chemical composition of milk and its components (lysozyme, casein, lactose, lipids, triglycerides, etc.) are similar, but not identical (Nishikawa et al., 1976; Davidson and Lonnerdal, 1986). The response of prolactin in blood serum to various influences (stress, ketamine, thyrotropic hormone, etc.) in simians does not differ from comparable reactions in humans (Aidara et al., 1981). The mechanism of lactational anovulation in monkeys was successfully studied by Ordog et al. (1998).

Methods for *artificial reproduction* were developed in research carried out on primates (Wolf et al., eds., 1993). This problem has two main objectives; experiments designed for application in human clinical practice, and those designed to enhance the breeding of primates for scientific purposes. As we have seen, the latter objective became especially important when limitations were placed on the importation of primates. Research on artificial reproduction has been carried out since the beginning of the 1970s and has used various species of simians, including apes. In general, it consists of cultivating embryos *in vitro*, with subsequent transfer into the uterus of an adult sexually mature simian female (which sometimes has one or another type of fertility problem or has been specially castrated). There are now very interesting investigations into the cloning of primates.

Beginning in the 1980s, live births were obtained from female recipients in macaques and baboons. At about the same time, methods of artificial impregnation were developed in chimpanzees and other apes. It also became possible to overcome the species barrier (rhesus – Java macaque). Methods for cryopreservation of gametes and embryos, ovarian stimulation with pharmacological preparations, and other procedures were used, which are also important in human clinical practice. As mentioned in the previous chapter, there was a successful birth of a gorilla after conception *in vitro* and further transplantation of the embryo (Gould, 1983; Hodgen, 1983; Pope et al., 1996; Younis et al., 1998).

No doubt, simians are one of the best laboratory animals in which to develop and test *contraceptives*. Chimpanzees may be the best animals for this purpose because they are closest to humans, but for most investigators they are difficult to obtain in sufficient numbers (Nadler, 1994b). Nevertheless, from the 1960s to the present time, monkeys have been effectively used in this field. Due to the advantages discussed above, different species of baboons and macaques appear to be much more appropriate for this research than rats, guinea pigs, rabbits, or dogs. Various primate species (not only those named above) have proved to be adequate laboratory subjects in the development of various contraceptive methods, including endocrine, chemical, immunological, and surgical (female and male). Many of the methods tested in the laboratory have been used with humans (Prasad and Diczfalusy, 1983; Jayaprakash *et al.*, 1994). Subsidiary functions of contraceptives were also studied in monkeys, in particular, the influence of steroids on atherosclerosis of coronary vessels. In addition, the impact of contraceptives on the rhythm and secretion of sex hormones was analyzed. The investigation of antifertility agents in monkeys is continuing (Wallen *et al.*, 1984; Koritnik *et al.*, 1986; Luo *et al.*, 1995).

Wide scale research on a vaccine against pregnancy has been conducted with monkeys because they are indispensable in this area. The principle action of antifertility vaccines is active immunization with the objective of neutralizing the biological substances required for fertilization and the initiation of pregnancy. The investigations with simians involve different versions and modifications of vaccines whose actions are directed against gonadotrophins, sperm proteins, or other antigens. This work has been ongoing for many years. It received a new stimulus at the beginning of the 1980s as a result of the revolution in vaccine production. This revolution was associated with the discovery and application of methods of hybridomas and recombinant products. Monoclonal antibodies targeted to impact DNA and other reliable instruments of modern science required animal models that were most similar to human biology. Simians were again important here, since they were specifically required in some directives of WHO. Simians (chimpanzees, baboons, macaques, marmosets, saimiri) are indispensable for this research because of the similarity of the antigens in their spermatozoids and ova as well as the similarity of their topography, the mutual recognition of gametes and their obstruction to suppress fertility, and the similarity of subsidiary functions of vaccines (Talwar, 1980; Rao *et al.*, 1988; Bellinger *et al.*, 1998).

The characteristics of affinity discussed above also enable *the modeling* of biological and pathological processes of human reproduction in experiments with nonhuman primates. Such models were described, in general, in the description of the separate components of the reproductive system. The opportunities for biomedical research in this area are inexhaustible and the list of these models is increasing. Numerous problems of obstetrics and gynecology can be pursued in simians, including biology and pathology of the endometrium, prolapse of the uterus, bleeding during late pregnancy, prolonged gestation, placental anomalies, abnormality of the fetus, and retention of a dead fetus. Additional areas include the study of inherited diseases, intrauterine gene correction, prenatal stress of mothers, the influence of maternal stress on offspring, the role of the mother's protein nutrition on the fetus, maternal diabetes and the health of the fetus, risk factors and therapy for the newborn, environmental pollution and health of the offspring, hypoxia and asphyxia of the fetus and newborn, and many other problems. Models of such inherited abnormalities as harelip and cleft palate have also been studied in monkeys (Hinshaw *et al.*, 1981; Lue *et al.*, 1983; Harrison *et al.*, 1995).

One of the main branches of medical primatology is *embryonic and fetal toxicology*, which received wide social attention in connection with the tragedy of thalidomide. Toxicity during intrauterine development had been studied in pregnant animals as early as the 1920s–1930s. The traditional laboratory animals were rodents, but a catastrophe at the end of the 1950s and the beginning of the 1960s revealed the danger of using this inadequate animal model. The tranquilizer, thalidomide (which had many synonyms), produced by a well-known West German company, Chemi-GruneThal, was thoroughly tested in rodents (and other nonprimate mammals) without any detected problems. When it was used by pregnant women, however, it had devastating effects on the limbs, brain, eyes, ears, cardiovascular system, pulmonary system, urogenital system, other components of the skeletal and muscular system, and the alimentary canal of their newborn children. Eventually, when thalidomide was tested in monkeys, it produced practically the same consequences in rhesus macaques as it had in humans. It was then found that thalidomide caused abnormalities in almost all of the 10 species of monkeys that were studied; macaques, baboons, guenons, and marmosets. A prosimian, the galago (*G. crassicaudatus*) (more distant biologically from humans than the other monkeys), was not sensitive to this teratogenic influence.

Monkeys are not only better in these experiments than rodents or rabbits, they are also better than dogs and any other animals that were tested in the past. They are better because of their similarity with humans in terms of the metabolism of drugs, placental transport, sensitivity to different teratogens, and other parameters discovered in studies on toxicity in the embryo and fetus. In certain cases catarrhine monkeys were more suitable than platyrrhines (saimiri) (Wilson, 1972; Hendrickx et al., 1983). The conclusion to be drawn from this research is clear; if tests of thalidomide had been carried out in primates, the tragedy could have been averted. I note that investigation of thalidomide has continued into the 1990s. It is useful, in fact, in the treatment of a number of dangerous human diseases, but its embryotoxicity was confirmed yet again in marmosets (Neubert et al., 1996).

The data on thalidomide forced scientists, leaders in public health, and regulatory agencies to pay more attention and show more responsibility when testing drugs for humans, especially those used during pregnancy. WHO and other competent organizations (FDA, 1966; WHO, 1975; EPA, 1986) issued guidelines which recommended that preclinical tests not be limited to the usual laboratory animals, but that primates should also be used. Other embryotoxic preparations were subsequently tested whose action in simians and humans was different from that in other laboratory animals (triamcinolone acetonide, a number of antibiotics, fungicides, hormonal compounds, etc.). Teratogenic activity of the other preparations on fetuses, including those used in human clinical medicine, was revealed in monkeys (antinausea bendictine, anticonvulsant valproic acid, varieties of retinoids, aspirin, some viruses, x-rays, etc.). Thanks to these investigations in primates, some dangerous preparations are not used at all for people, while warnings are included with some others against their use during the course of pregnancy (Hendrickx and Binkerd, 1990; Neubert et al., 1990; Hendrickx and Peterson, 1997). It is likely that baboons, macaques, and marmosets are the best species for experiments in reproductive toxicology, including the assessment of biotechnological products which at present are used more and more widely.

Nevertheless, a review of the guidelines for preclinical tests in reproductive toxicology carried out in the middle of the 1980s indicated that rats, mice, and rabbits were most frequently used in the tests (Korte et al., 1987) (in general, this was done due to the higher cost of primates). The latter investigators requested that tests with a 'third species', one of the primates, preferably the Java macaque (*M. fascicularis*), be mandatory. I should acknowledge the well-founded opinion of A. Hendrickx, a prominent specialist in the field of reproductive toxicology. He believes that these valuable models should only be used in exceptional situations where other animals and *in vitro* tests do not work, and in which no other alternative exists for conserving the life and health of people (Hendrickx and Binkerd, 1990). I believe that this should be the basic principle for all of medical primatology.

I shall conclude this section with data regarding the *influence of alcohol* on the reproductive system of primates. Again, we have here another valuable model for studying human pathology. It was shown that the systematic administration or self-administration of ethanol in macaques (*M. mulatta, M. nemestrina*) has the same influence on their reproductive apparatus as in humans. It caused amenorrhea in females (although it did not decrease the levels of LH and estradiol as in women) and it significantly depressed the production of testosterone in males. It produced the fetal alcohol syndrome during pregnancy, anomalies of the face, and retarded development overall. It also increased the number of spontaneous abortions and the rate of stillbirths (Mello *et al.*, 1984; Clarren *et al.*, 1987).

Phylogenetic affinity and experimental pathology of the cardiovascular system

Having discussed the characteristics of affinity in the main biological systems in humans and other primates, I now address the system which not only attracts the special interest of present day humanity (at least because it is the leading cause of death in humans), but also because of its many interactions with, and influences on, the other systems considered above.

It is appropriate to mention here one of the peculiarities associated with the use of primates in laboratory research, the extraordinary reactivity of their physiological systems to the environment and experimental conditions, although this relates to other sections as well. This peculiarity was uncovered in the 1930s at the Sukhumi Primate Center by the immediate followers of I. P. Pavlov while they were studying conditioned reflexes in simians, and also in experiments of some other scientists (P. V. Botchkarev, L. N. Norkina, D. I. Miminoshvili, I. A. Utkin, Yu. P. Butnev, M. I. Kuksova, V. G. Startsev, G. M. Cherkovich, B. A. Lapin, and others).

For a long time, it was thought that heart rate, vagal tonus, the number of leukocytes in the blood, and the pH of gastric juice differed significantly in humans and monkeys. Later it became clear that these parameters were distorted by the stress that the monkeys experienced while the measurements were being made, by their flight response to novelty, and their highly reactive physiological processes. If one structures the laboratory conditions so as to exclude factors that psychologically traumatize the monkeys (using the same room, the same experimenter or assistant, and a quiet atmosphere) or if one uses telemetry while the animal remains 'free', then the anomalies disappear and the values acquire stability and many of them are similar to human ones. Heart rate is recorded at approximately 70–80 beats per minute and rarely exceeds 100 (not 160–200 as was considered 'normal' before). Vagal tonus is well-defined, judging from the sinusoidal arrhythmia on the electrocardiogram (I note that this ECG is essentially similar in humans and simians), the number of leukocytes is 6,000–7,000 per mm^3 (not 12,000–43,000 per mm^3), and the gastric juice is quite acid, not alkaline or neutral, as it was when the animals were under stress. The approach of people or any other danger immediately caused high tachycardia, the disappearance of sinusoidal arrhythmia, leukocytosis, and increased pH of the gastric juice (Chercovich, 1994).

Anatomy and biometrics of the baboon heart and its microscopic characteristics (*P. hamadryas, P. cynocephalus*) are similar to those of the human heart. The energetic metabolism of the myocardium differs from that of the human cardiac muscle (in terms of O_2, glucose, pyruvate, lactate, and free fatty acid utilization) only by a greater dependence on carbohydrate. As was shown above, the normal blood parameters of macaques and humans agree quite well. The coronary blood supply to the heart in different species of monkeys more closely resembles the human condition than it does in any other animal, especially, dogs. Collateral blood circulation is well-developed in dogs, whereas terminal arteries predominate in monkeys, similar to humans. Simians are the only mammals (9 species were studied) in which dimorphism in the endothelial cells of vessels was found with a similar topography to that of the human (Brink *et al.*, 1970; Roussel, 1985; Teofilovski-Parapid *et al.*, 1988).

Let us cite another example of the unique affinity of humans and simians at the finest level of heart morphology, the level of myocardial cells, or cardiomyocytes. The regeneration of these cells is critically important for healing infarctions of the cardiac muscle. During postnatal cardiogenesis, when ploidness (the number of chromosomal haploids) in the muscle cell nucleus in all nonprimate mammals stops after birth, in humans and other primates it paradoxically continues to increase. Polyploidy in the nucleus of cardiomyocytes in primates is associated with peculiarities of DNA synthesis, which is blocked under similar conditions in other animals. This is an intriguing field of activity for cardiologists in their attempts to stimulate regeneration of the myocardium. This finding is additional evidence for a primatologist of the profound genetic relationship between humans and simians (Rumjantsev, 1982).

The content of catecholamines (adrenaline, norepinephrine, dopamine) in the myocardium of the heart ventricles in monkeys and humans is so similar that experimental data on the simians readily extrapolate to humans. The nature and the degree of histopathology in the arteries and veins of primates are practically the same. Specific heart receptors in primates react to the influence, for example, of leukotriene D_4 and blocking agents, with similar changes in the hemodynamics and activity of the heart. Cardiac glycosides (digitalis) and alkaloids (caffeine) have a cytotoxic influence on various types of cell cultures, but the toxic action of these agents on the cell cultures of humans and monkeys is 100 times greater than it is in mice or Syrian or Chinese hamsters. The species differences in sensitivity of the cells and the influence of other compounds that regulate heart function (in particular, the cardiotonic steroids) are similar as well (Pierpont *et al.*, 1985; Gupta *et al.*, 1986; Mitchell *et al.*, 1986; Hahn and MacDonald, 1987). These differences in response to the same influences, their similarity in humans and simians, and their different (and sometimes opposite) response in other laboratory animals, represent one of the fundamental advantages of medical primatology.

Let us amplify this discussion. According to pharmacological data (Van Stee and Back, 1969, with a reference to Meek, 1941), cyclopropane caused an *acceleration* in the heart rate of dogs, but a *deceleration* in rhesus macaques and humans. These investigators showed that the response of the central nervous system to another agent, bromotrifluoromethane, was opposite in dogs and monkeys (but similar in macaques and baboons). After inhalation of this substance, monkeys clearly developed cortical depression, whereas dogs showed excitation. Such differences are not isolated instances. Intravenous injection of leucine enkephalin significantly decreased heart rate in rabbits and dogs (somewhat less so in dogs), but increased it in simians. The increase in arterial blood pressure and heart rate in monkeys during bilateral occlusion was expressed to a significantly greater degree than in rabbits or dogs (Koyama *et al.*, 1985).

An agonist of α-adrenoreceptors (xylasine) did not induce vasoconstriction in the isolated femoral arteries of dogs, but it did in monkeys (Kawai and Chiba, 1986). Norepinephrine (in the Russian literature it is frequently called 'noradrenaline') induced a dose-dependent constriction of the arteries in the skeletal muscles of both dogs and simians. Selective blocking agents of α-adrenoreceptors (bunazosin, ketanserin), however, suppressed the norepinephrine-induced vasoconstriction in monkeys, but only slightly influenced vasoconstriction in dogs. The investigators concluded that there are species-specific differences in the adrenergic innervation of dogs and primates (Sinanović and Chiba, 1988). Arteries in the brains of humans and monkeys (macaques) show the same response to vasoactive agents, suggesting that they have the same functional mechanisms. Reactivity of the arteries in dogs, however, differs significantly from that in primates (Toda, 1985).

Such paradoxes are rather frequent in the physiology of the coronary arteries, and they also occur with respect to the β-adrenoreceptors (Toda and Okamura, 1990). Intracoronal injection of acetylcholine induces coronary vasodilatation in dogs. The same treatment reduces coronary blood flow in baboons. A decrease of oxygen utilization by the myocardium in baboons leads to metabolic vasoconstriction (which can be corrected). Cocaine, moreover, induces constriction of the coronary vessels in baboons to a greater degree than in dogs (Shannon and Shen, 1996). Vasoconstriction mediated by muscarine receptors is a peculiarity of the coronary vessels in primates. The adrenoreceptors in these vessels are much more important in simians and humans than in dogs (Satoh et al., 1982; van Winkle et al., 1988).

This difference in response applies to various autonomic functions, but not only to them. Thanks to the experiments on simians (Papio), it was possible to avoid clinical tests of the initial version of an isosorbide dinitrate preparation for treatment of myocardium infarctions. The primate research demonstrated that although coronary vasodilatation occurred in dogs, vasoconstriction occurred in simians (and humans). Baboons, among all the monkeys, exhibit the clearest affinity with humans in the pharmacokinetics of this preparation (Seljeskog et al., 1963; Doyle and Chasseaud, 1981) (I note that occlusion of the coronary arteries also leads to manifestly different disorders of the myocardium in pigs, dogs, and baboons [Shen and Vatner, 1996]). Let us remember that the purpose for which MPTP was intended was to decrease arterial blood pressure in humans. It did not reach the clinic because of the experimental foundation established in the research with monkeys. The primate research demonstrated (as suggested by the findings in human drug addicts), that this compound was implicated in the development of Parkinson's disease (King et al., 1988).

Besides the agents mentioned above, including thalidomide, additional ones such as norbomide, isoniaside, aminoxytrophene, methylfluorocetate (Coulston, 1966), the antidepressant Abbott-25794 (Winfield, 1971), metamphetamine (Ellinwood, 1971), aminazine (Byrd, 1974), atropine, scopolamine (Haude and Ray, 1968), chlordiazepoxide, and melipramine (Weiss and Laties, 1967) caused similar effects in humans and simians, but different and sometimes opposite effects in other animals. The previously popular antihypertension preparation, reserpine, caused depression and sleepiness in humans and monkeys, but produced excitation in all other animals if used in comparable dosages (Reite et al., 1969; Sulser et al., 1978). Many modern medicines were approved for human practice because they proved acceptable when tested in laboratory primates. Many others that were first tested in rodents and other traditional laboratory animals were blocked from use in humans thanks to experiments in simians which suggested that the drugs were dangerous to humans.

Renin and the entire 'renin-angiotensin' system for blood pressure regulation are entirely alike in humans and simians and different in other laboratory animals which

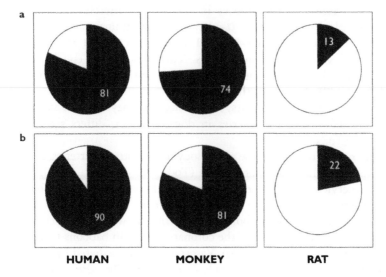

Figure 4.6 Comparison of the hemodynamic response (percent, black part of the circle) to (a) obsidian and (b) nitroglycerin in humans, monkeys, and rats (from Belkaniya *et al.*, 1987).

were studied (dogs, rats, pigs, etc.). One can effectively immunize marmosets, for example, with purified human renin (this is, however, associated with the development of autoimmune kidney disease in monkeys). The hypotensive effect of monoclonal antibodies against human renin is found in monkeys (*Callithrix jacchus*), but the same antibodies do not block renin in rats, dogs, or pigs. Correspondingly, other inhibitors of renin which influence primates (for example, imidazole alcohol inhibitors), do not show inhibitory activity in rats, pigs, or dogs (Poulsen and Burton, 1976; Cazaubon *et al.*, 1986; Michel *et al.*, 1987; de Forest *et al.*, 1989). Theoretical prerequisites for these phenomena are found in the singular evolutionary history of primates, in the similarity of their neural regulation (Miminoshvili *et al.*, 1956), the peculiarities of blood physiology and biochemistry, the orthostatic adjustment of blood circulation to a semi-vertical posture, as well as the adaptation of blood circulation to gravity, discussed above. The relationship of primates is well-illustrated by their hemodynamic response to obsidian and nitroglycerin (Figure 4.6).

These prerequisites to the study of cardiovascular physiology and pathology of primates closely correspond to the research developments in this field during the years 1967–1985 (Table 4.6) (Fridman *et al.*, 1990).

Above, I discussed the modeling of neurogenic hypertension, cardiovascular insufficiency, and myocardium infarcts in primates. These models are produced by various methods. It is likely that normal blood pressure does not differ significantly in macaques,

Table 4.6 Number of annual publications on research of the cardiovascular system conducted with nonhuman primates during the period 1967–1985.

1967	1970	1975	1976	1985
76	125	180	185	244

baboons, chimpanzees, and humans. It increases with age in the simians, similarly to the human condition. G. O. Magakyan (1953) reported blood pressure in monkeys as 115–135 (systolic) and 65–85 (diastolic), depending on the species, sex, and age. This was confirmed, in general, by other investigators. A female baboon (*P. hamadryas*), Vesta, at the Sukhumi Center, had a stable blood pressure of 190/110 while caged. Within a year of it being placed in an aviary (relatively free-ranging), the blood pressure had dropped to 135/85. Magakyan stated that no animals can replace simians in the experimental study of hypertensive disease and coronary insufficiency. He concluded that primates (macaques and baboons) fully reflect the pathology of the corresponding human disease, 'reflecting different stages and degrees of disturbance' (Magakyan, 1977, p. 24). Both experimental and spontaneous hypertension are well-modeled in primates (Gentili and Tasker, 1978; Giddens *et al.*, 1987). As already mentioned, models of hypertensive disease in simians are used to study different disorders of heart and brain function, the influence of salt on blood pressure, inhibitors of renin, and other pharmacological effects (Chercovich *et al.*, 1977; Dyer *et al.*, 1995; Kemper *et al.*, 1996). Modeling of thrombotic diseases was reported above, investigations that are still being conducted. In particular, new anticoagulant inhibitors of thrombin are being sought (Downing *et al.*, 1997). A model of human cardiac arrhythmia in macaques is also available (Agus Lebana *et al.*, 1997).

Starting in the 1970s, no less than half of all research on the cardiovascular system of primates was conducted on atherosclerosis, which at present is still a dominant problem in this area. There are many similar features between humans and simians that serve as prerequisites for modeling atherosclerosis and many investigators in this area of research favor monkeys and apes. G. H. Bourne and M. Sandler (1973, p. 262), for example, noted that the study of atherosclerosis in chimpanzees '. . . is as important as the application of humans themselves.' It is apparent, despite these favorable indications, that primates were not recognized in the past as better models than other laboratory animals. A review by T. B. Clarkson and colleagues, some of the most experienced specialists in the study of atherosclerosis in primates (1970), is very instructive. After completing a comprehensive analysis of spontaneous and experimental atherosclerosis in different animals, including simians (*Pan, Papio, Macaca, Saimiri, Cebus*, and others), it seemed as if the authors did not consider that primates had special value in this field among the other animals (pigs, cats, pigeons, and turkeys). In the course of time, however, investigators increasingly realized the effectiveness of experiments with simians. A group of investigators, including T. B. Clarkson, stated in the abstracts of a Congress of the International Primatological Society, that macaques (*M. fascicularis, M. nemestrina*) 'represent excellent models for study in pathogenesis of human atherosclerosis' and associated disorders (Adams *et al.*, 1994, p. 290).

Much research has been devoted to the study of low density lipoproteins (LDL) in simians, the main form of cholesterol transport in the organism, and high density lipoproteins. Their protein components (apolipoproteins), various forms of the latter and their genetic products and metabolites were analyzed in a clear and comprehensive way in order to compare the data for humans and laboratory animals. A maximum affinity with humans was found in every case between apes and monkeys in comparison to other organisms. The relationship corresponded to the phylogenetic level of the specific taxa, as was already noted in the description of taxonomy. The study of LDL with immunological methods in different classes and species showed that the similarity to humans ranges from 1%–10% in reptiles and fish, 10% in birds, 26%–37% in guinea pigs, 35%–58% in pigs, 80%–88% in simians of the Old World, and 90%–97% in chimpanzees (Goldstein *et al.*, 1978). The primary structure of the DNA sequence of apolipoprotein A-I (apo-A-I) was characterized in the Java macaque by 97% identity with the human apo-A-I (Polites *et al.*, 1986).

Numerous studies were conducted with primates in this area, including research on the composition and properties of lipids, cholesterol, triglycerides, phospholipids, fatty acids, their relationships involved in atherogenesis and their interaction with hormones, and finally, their atherogenic function in primates. The results in these areas usually showed a remarkable similarity between humans and chimpanzees, baboons, macaques, and other simians. There was also great similarity (82%–87%) between human apoliprotein (A-I and other fractions) and that in marmosets (*C. jacchus*). Given the phyletic status of marmosets, this was unexpected, but it means these monkeys are also valid models for the study of human atherosclerosis (Blaton *et al.*, 1972; Williams *et al.*, 1991; Clarkson and Tansey, 1995; and others).

Feeding simians an atherogenic diet with a high level of cholesterol and saturated fats (the 'diet of western countries', and especially the diet of 'middle America' in the 1970s), caused atherosclerosis in a comparatively short time (1–24 months), with pathology that was very similar to the human condition. The simians developed a variety of pathological conditions, including hypercholesterolemia and atheromatous plaques. There were various adverse effects on the coronary arteries, the thoracic and abdominal portions of the aorta, the iliac artery, vessels of the limbs, and catastrophic damage to the tissues of the myocardium and liver. When the amount of cholesterol and fats in the diet was reduced, there was regression in the experimentally-induced atherosclerosis. The influence of various factors on the development of this pathology (age, heritability, various diets, and pharmacological preparations) was studied in primates, especially those factors that cannot be studied adequately in other animals (psychosocial stress, simultaneous hypertension, hormonal therapy, and the postmenopausal period). The similarity of risk factors for coronary artery disease associated with atherosclerosis is unique in humans and simians (Prathap, 1975; Howard *et al.*, 1979; Wissler *et al.*, 1983; Gallo, ed., 1987; Williams *et al.*, 1991, 1994; Fincham *et al.*, 1998).

In the 1990s, the problem of atherosclerosis in its various experimental aspects became one of the most actively pursued fields of medical primatology. The specific topics that were studied in primates include the risk factors for development of atherosclerosis and heart ischemia, the relationship of aging to atherosclerosis, cerebral ischemias, models of restenosis against a background of atherosclerosis, prophylactics and consequences of atherosclerosis, cholesterol absorption inhibitors, estrogen therapy, drug therapy in general, the effect of monoclonal antibodies, and other factors that influence the regression of coronary atherosclerosis and many other questions in this area (Clarkson *et al.*, 1996; Wissler, 1996; Cefalu and Wagner, 1997; Williams *et al.*, 1997; Clarkson, 1998; Wagner *et al.*, 1998).

I can say without overstatement that experimental studies in primates played a significant role in many of the modern discoveries pertinent to the struggle against atherosclerosis. They were instrumental in encouraging people in the developed countries to reduce the content of cholesterol in their diet, which in turn decreased the mortality attributable to atherosclerosis. These investigations represent another strong confirmation of the great value of medical primatology as an integral branch of medical science. An adequate understanding of cardiovascular pathology can be achieved only with experimental models in primates. Only with these animals is it possible to study

the separate and interactive influences on atherogenesis of female social status, contraceptive hormones, and the premenopausal period in a single experiment (Kaplan
et al., 1995). Nonhuman primates are clearly indispensable when research involves the
primate type of neural regulation, human biochemical peculiarities in blood chemistry,
thrombogenesis and blood circulation, tissue similarities at the cellular level, and other
conditions inherent only in humans and simians. Nevertheless, this does not preclude
specific supplementary experiments in other laboratory animals.

There are other pathological conditions of the cardiovascular system studied in
primates which I only mention in passing, including myocarditis, embolism, injuries,
complications related to infectious diseases, parasitic cardiomyopathies, surgical innova-
tions, and harmful environmental effects on the heart, including those caused by
chemical compounds, industrial noise, etc.

*   *   *

I have concluded my description of the affinity between the individual physiological
systems of humans and simians and the models of pathology (natural and experimental)
that have a direct bearing on these systems. There are other models and research
opportunities that I have not covered. These are discussed in a concise form in the
next chapter.

Chapter 5

Other models and opportunities for research with primates

INFECTIOUS PATHOLOGY

Having now demonstrated that simians have great experimental value as research subjects, as discussed in the previous chapters, we cannot but notice that there is no field in the medical and biological sciences in which the laboratory use of primates was so popular and intense as in the study of human infections. Historically, research in this area eradicated many dangerous human diseases for which laboratory primates were indispensable. Academician B.A. Lapin wrote, 'no animal other than a simian can reproduce such a variety of infectious human diseases in the appropriate form' (1996, p. 246). According to a report by WHO (1996), in the middle of the 1990s, 17 million people of the 52 million that died annually of disease, including 9 million children, died from infectious diseases. The same study indicated, moreover, that despite undoubted progress in the struggle against infection, humanity is on the verge of a crisis in outbreaks of contagious diseases (including those which return from time to time). It is clear, therefore, that monkeys and apes have not only played an important role in experimental medicine in the past, but they will continue to do so in the future.

According to our data, there are more than 40 infectious human diseases (bacterial, viral, parasitic) which cannot be studied experimentally without the use of simians. Production and testing of vaccines (poliomyelitis, measles, rubella, hepatitis, malaria, etc.), moreover, as well as the preparation of tissue cultures, which is essentially mandatory for growing viruses in modern laboratories, cannot bypass experiments with primates. Although there are many other infectious diseases which can be reproduced in the more common laboratory animals, they can be more effectively, and even sometimes more justifiably, investigated in monkeys (for example, pneumonia, tuberculosis, yellow fever, and others). In fact, primate studies are sometimes required. Finally, there are diseases (or outbreaks that occur from time to time) of unknown, but clearly infectious nature. In these cases, there is no alternative to the use of simians in the research (kuru and other neuropathies, and AIDS). Sometimes after modeling a new infection in primates, one can use less valuable laboratory animals (as in the case of epidemic typhus), but even here the initial impulse for research on the etiology of the disease should be conducted on simians.

The theoretical basis for the advantages of simians in the study of human infections is the common evolutionary past of primates, the joint evolution of their parasitic fauna, and especially, the relationship between their tissues and immune factors, as discussed earlier. This relationship was noted initially by Darwin and later used by Mechnikov. There are also similarities among interrelations of primate micro- and macro-organisms

Table 5.1 Number of annual publications on research of infectious diseases conducted with nonhuman primates during the period 1967–1985.

1967	1970	1975	1976	1985
408	718	623	521	627

and the fact that 'pathological processes that develop in simians under the influence of the same type of etiologic agent, are similar, in principle, to analogous processes in humans in their pathomorphological characteristics' (Krylova, 1986, p. 4).

Modeling infectious pathology in simians generally facilitates study of the etiology and transmission of the pathogen, clinical development of the disease, pathoanatomical insight, immunological changes, opportunities for vaccine prophylaxis and treatment, and the epidemiology of the disease. All these areas can be analyzed in relation to the corresponding processes in humans and are examined for the most appropriate extrapolations. These advantages cannot be found in any of the traditional laboratory animals, even when one or another model can be reproduced in them. Quite often when working with traditional animals, the experimenter must be satisfied with only a part or a single aspect of the research opportunities listed above. Of course, this does not mean that the other animals should be rejected altogether. As for the production and testing of vaccines and the treatment of infections, it has long been accepted that monkeys are 'ideal' subjects for this type of research. They allow us to study the safety of preparations and duration of the protective effect after immunization, and the specific and nonspecific factors of immunity in a model that is closest to the human.

We conducted, as mentioned above, a statistical analysis of all the publications in medical primatology throughout the world during the period from 1967 to 1985. The results with respect to the area of *infections* (bacterial, viral, parasitic) are shown in Table 5.1 (Fridman *et al.*, 1990).

The depression during the 1970s (Table 5.1) is explained by the previously mentioned restrictions on importation of macaques (especially *M. mulatta*, the species with which most infectious disease research has been conducted). I should point out that when the absolute numbers are all relatively high, there is a well-known quantitative saturation effect. Within the period under study, articles on infections were the second or third most frequent publications in medical primatology. We may assume that in the 1990s, the number of publications in this area will increase due to intensive research on AIDS, viral hepatitis, and others.

Together with my two dear colleagues, one of whom is a pathologist (B.A. Lapin), and the other a specialist in communicable diseases (E.K. Dzhikidze), we constructed a table of infectious human diseases that required only simians as experimental subjects, based on research in medical primatology and the practice at the Sukhumi Primate Center (Lapin *et al.*, 1987, pp. 136–137). The present author slightly supplemented this table with data obtained during the 10 years that followed publication of the earlier joint book. The data are presented in Table 5.2.

I should mention that this table is in all probability far from complete. Moreover, as was already mentioned, simians have been used in many investigations of infectious human diseases which are also reproduced in other laboratory animals. Among them are tuberculosis, cholera, epidemic typhus, plague, staphylococcal, and streptococcal

Table 5.2 Infectious diseases of humans that are reproduced only in nonhuman primates.

Type of pathogen	Disease	Susceptible species (genera)	Remarks
Bacterial infections	Abdominal typhoid	*Pan troglodytes, Hylobates spp.*	Simians are susceptible to almost every type of salmonella pathogen (*S. tiphimurius, S. enteritidey, S. stanly*) that causes enterocolitis and the other types of salmonellosis
	Paratyphoid-B	*Macaca mulatta*	
	Schigellosis	*Macaca (mulatta, nemestrina, fascicularis), Cercopithecus aethiops, Papio hamadryas*	*M. mulatta* and *M. nemestrina* are most sensitive to dysentery
	Escherichiosis	*Pan troglodytes, Macaca (mulatta, fascicularis), Papio spp.*	
	Syphilis	*Pan, Pongo, Hylobates, Papio spp., Macaca spp., Mandrillus sphinx*	Chimpanzee is the most susceptible
	Gonorrhea	*Pan, Papio spp.*	Chimpanzee is the most susceptible
	Leprosy (Hansen's disease)	*Pan, Hylobates spp., Cercocebus atys, Macaca mulatta, Cercopithecus aethiops, Tupaia belangeri*	*Pan, C. atys* are the most susceptible; 9-belt armadillos are susceptible to the pathogen
	Trachoma and the other chlamydiosis (in particular, chlamydiosis of the sexual organs)	*Pan, Papio spp., Macaca spp. Aotus trivirgatus*	*Pan, M. nemestrina, A. trivirgatus* are the most susceptible
	Helicobacter pylori infections	*Macaca (mulatta, fascicularis, fuscata), Papio spp., Saimiri, Nasalis larvatus*	'New' pathogens of the gastrointestinal tract
Mycoplasma infections	Atypical mycoplasm pneumonia	*Macaca mulatta*	
	Infections of the urinary system and sexual tract	*Macaca (mulatta, arctoides), Papio hamadryas, Cercopithecus sabaeus, Saguinus spp., Pan*	Acholeplasms, mycoplasms, ureaplasms are the pathogens
	Septic arthritis	*Pan troglodytes*	*Mycoplasm hominis* is the pathogen
Viral infections	Poliomyelitis	*Pan, Macaca spp., Cercopithecus aethiops, Papio spp.*	*Pan* is susceptible by peroral introduction
	Measles	*Pan, Macaca spp., Papio spp., Cercopithecus sabaeus*	

Table 5.2 (cont'd)

Type of pathogen	Disease	Susceptible species (genera)	Remarks
	Chicken pox (or varicella)	Macaca spp., Cercopithecus aethiops, Erythrocebus patas	
	Smallpox	Pan, Macaca mulatta	Pan is the most susceptible
	German measles (or rubella)	Macaca mulatta, Cercopithecus sabaeus	
	Epidemic parotiditis	Macaca (mulatta, fascicularis, nemestrina, nigra), Papio hamadryas	
	Hepatitis A (pathogen – hepatitis virus A or HAV)	Pan, Macaca spp., Papio spp., Cercopithecus aethiops, Saguinus spp., Aotus spp., Galago senegalensis	Is the pathogen the same in humans and simians? (Lapin, 1996)
	Hepatitis B	Pan, Hylobates spp., Macaca (mulatta (?), assamensis (?)), Tupaia belangeri (?)	For information regarding the two latter species see Ge et al., 1990; Yan et al., 1996. Hepatitis B and C are associated with cancer of the liver. 90% of the AIDS-HIV-positive persons are also seropositive for HBV, and 80% for HCV. In USA, HCV accounts for 150,000 cases per year with 10,000 deaths annually (St. Clair, 1997)
	Hepatitis C	Pan, Macaca spp., Saguinus spp.	
	Hepatitis D (associates with the HBV co-infection)	Pan	
	Hepatitis E	Pan, Macaca fascicularis, Aotus spp.	
	Hepatitis G	Pan, Macaca mulatta, different species of New World monkeys	
	Hepatitis GB	Pan, Saguinus spp.	The pathogen has 3 forms which can cause hepatitis in Saguinus (GB-b and GB-c) or exist without symptoms in different species of platyrrhines (GB-a).
	Infectious mononucleosis	Pan, Macaca mulatta, Cercopithecus aethiops	

Table 5.2 (cont'd)

Type of pathogen	Disease	Susceptible species (genera)	Remarks
	Dengue fever	Pan, Macaca spp., Cercopithecus aethiops., Saimiri sciureus, Cebus spp, Ateles spp.	Infection is not apparent in all species
	Chikungunya fever	Macaca spp., Papio spp., Cercopithecus sabaeus	Human hemorrhagic fever is investigated in primates and is also reproduced in other laboratory animals.
	Ebola fever	Macaca (fascicularis, mulatta), Papio hamadryas, Cercopithecus sabaeus	
	Venereal lymphogranuloma	Macaca mulatta	
	Viral encephalomyocarditis	Erythrocebus patas, Cercopithecus aethiops, Saimiri	
	Acquired immunodeficiency syndrome (AIDS)	Pan, Hylobates spp., Macaca spp., Papio spp., Cercocebus atys, and other simians	Taking into account the undesirability of using apes in dangerous experiments, two subspecies of macaques (Macaca nemestrina and Macaca leonina) are the most appropriate to investigate AIDS; susceptible to HIV-1 and HIV-2 (Lapin, 1996).
Viruses of slow infections (or prions?)	Kuru	Pan troglodytes, Aotus trivirgatus, Saguinus spp., Macaca spp.	
	Alzheimer's disease	Pan, Papio cynocephalus, Macaca (mulatta, fascicularis), Cercopithecus aethiops, Saimiri	There is an assumption that prions can shed light on the etiology of the neurological diseases listed below which are reproduced only in simians.
	Creutzfeldt-Jacob disease	Pan, Saimiri, Cebus spp.	
	Amyotrophic lateral sclerosis	Macaca (mulatta, fascicularis)	
Parasitic diseases	Tropical malaria	Aotus (trivirgatus, nancymae (?)), Saimiri sciureus, Pan	Pathogen: Plasmodium falciparum
	Tertian malaria	Aotus trivirgatus, Saimiri sciureus, Pan	Pathogen: Pl. vivax, Pl. ovale

Table 5.2 (cont'd)

Type of pathogen	Disease	Susceptible species (genera)	Remarks
	Quartan malaria	Aotus trivirgatus, Macaca mulatta	Pathogen: Pl. malariae
	Trichomoniasis	Saimiri sciureus	Trichomoniasis in women (Tripanosoma vaginalis)
	Chagas' disease	Cebus apella, Saguinus spp., Macaca mulatta	Trypanosoma cruzi; other species of trypanasomes are studied in primates, as well.
	Schistosomiasis	Cebus spp., Macaca arctoides, Papio spp., Patas, Pan, Cercopithecus aethiops, Saimiri sciureus	Schistosoma haematobium, S. mansoni and S. japonicum were studied
	Filariasis:	Presbytis (cristata, entellus)	Single models (Misra et al., 1997)
	– wuchereriasis (elephantiasis) – loiasis	Mandrillus (sphinx, leucophaeus), Papio spp., Erythrocebus patas, Macaca mulatta	The best model is Mandrillus (Wahl and Georges, 1995)
	Leishmaniasis (visceral leishmaniasis, cutaneous l., muco-cutaneous l.)	Aotus trivirgatus, Saimiri sciureus, Papio cynocephalus, Macaca mulatta, Cercopithecus aethiops, Presbytis entellus	All species of human Leishmania are best modeled in monkeys; L. donovani, L. major, L. braziliensis. Other species of Leishmania are also studied in monkeys.

infections, legionnaires' disease, tetanus, diphtheria, rabies, and many others. The pressing problem in medical primatology, however, is the use of monkeys to create and apply vaccines and treatments when other animal models, on the whole useful for study of one or another disease, appear inadequate for testing vaccines. This happened when yellow fever, pertussis, plague, tuberculosis, influenza, and other vaccines were investigated (Schmidt, 1961; Galindo, 1973; Preston and Stanbridge, 1976; Beisel and Stokes, 1983; Rimmelzwaan et al., 1996). It goes without saying that all the prophylactic and therapeutic agents for the diseases listed in Table 5.2 were developed in primates, including long-term efforts at prophylaxis for schigellosis, hepatitis, and malaria (Shroyt, 1961; Dzhikidze, 1973, 1996; Heppner and Ballou, 1998; Pride et al., 1998). Simians become ill spontaneously with many human diseases, they can be infected with these diseases from sick people (measles, poliomyelitis, hepatitis, and other diseases), and sometimes they become a dangerous source for infecting other people (Marbourg disease, viral hepatitis, monkey pox, and others). The only species which is susceptible to each of the 7 hepatitis viruses is P. troglodytes.

I already mentioned that monkeys are indispensable for the study of a number of bacterial toxins. In recent decades, science and clinical medicine obtained a number of models in simians that simulate the diseases caused by bacteria. We may cite leprosy as a disease with a four thousand year history. For 100 years, attempts had been made to obtain an animal model of leprosy (starting with S. Neisser). We may also cite recent investigations which were revolutionary in terms of the understanding of gastrointestinal pathology (gastritis, gastric ulcer, and duodenal ulcer), and related to species of *Helicobacter* (Turanov *et al.*, 1973; Euler *et al.*, 1990; Meyers *et al.*, 1991; Gormus *et al.*, 1995; Fox, 1998). Many methods and all generations of vaccines against intestinal infections (schigellosis, salmonellosis) were studied in simians. As already mentioned, in the Soviet Union antibiotics were tested in simians under the supervision of Z. V. Ermolyeva (Dzhikidze, 1996). In various countries, antibiotics were also being invest-igated in primates in the 1990s.

It became clear in the beginning of the 1970s that all the main types of human viruses could be studied in simians (Kalter, 1973). This list has been increasing continuously in the decades since that time. Families of viruses were studied in primates (and their taxonomy was corrected and improved): Adenoviridae, Arenaviridae, Coronaviridae, Filoviridae, Flaviviridae, Hepadnaviridae, Herpesviridae, Lentiviridae, Orthmyxoviridae, Paramyxoviridae, Picornaviridae (including of course, the *Enterovirus* genus with the poliomyelitis virus), Poxviridae, Reoviridae, Retroviridae, Rotaviridae, Togaviridae, and Toroviridae (Soave, 1981; Blenden and Adinger, 1986). It is possible that other families of viruses studied in primates may have been omitted from this list. We must remember that in addition to the unique viral infections presented in Table 5.2 (in the context of exclusively primate models), the list of families comprises almost all other modern human viral diseases, including the most dangerous ones.

Generally, half of the publications in the entire area of experimental infections in simians concern virological investigations. Many viruses have such biological similarity in humans and simians that they can be distinguished solely with special methods, while others are absolutely identical or are *twin-viruses*. This means that humans and simians can *interchange* viruses (for example, yellow fever, chikungunya, and dengue viruses). It has been hypothesized that the human AIDS virus has a simian origin (Gao *et al.*, 1999). There are other monkey viruses which are mortally dangerous to humans (for example, the herpes B virus and others already mentioned). This relationship enables the study of human viral diseases in monkeys because '. . . not only the viral flora, but the entire spectrum of viral infections of simians and humans are very close, and what is especially important is that these diseases develop in monkeys in the same way as in humans both clinically and morphologically' (Shevtsova and Lapin, 1987, p. 71; Kalter and Heberling, 1990; Kalter *et al.*, 1997). The investigation of infectious human diseases, the development, testing, and ensuring the quality and safety of vaccines is a great service of medical primatology to humanity. Experience indicates that infections are an ever present threat to human health and we must continuously perfect the means of eradicating them. It is likely that monkeys are indispensable at the present time for testing poliomyelitis vaccines, but that transgenic mice could be substituted for certain types of neural investigations (Levenbook *et al.*, 1997).

We already discussed the complete eradication of small pox from the planet. The poliomyelitis virus is awaiting the same fate. According to WHO, it should have been completely eliminated by the year 2000. The measles virus will be next. All this research is conducted with laboratory primates. At present, numerous virological invest-igations in simians, as mentioned, are carried out to study AIDS and viral hepatitis. An avalanche of primate investigations in these fields gives justified hope, regardless of the special difficulties (references to AIDS comprised 80% of all publications in virology, according to the July 1998 issue of *Current Primate References*. Moreover, there were additional AIDS references in the areas of Pharmacology and Therapeutics, as well as in Immunology, Blood, and others). Present day AIDS specialists have determined that there are 'striking similarities between simian AIDS induced by SIV and human AIDS induced by HIV.' They reported 'In conclusion, it is obvious that the study of the various SIV/primate systems available have contributed significantly to our understanding of how the immunodeficiency virus interacts with the host ... There are, however, still large gaps in our understanding of precisely how and why HIV and SIV infections result ultimately in AIDS. Hopefully, with the new technologies becoming available, the combination of knowledge gleaned from the studies of HIV in humans and SIV in primates will allow these gaps to be filled in the near future' (Norley and Kurth, 1997, p. 401). Our ability to eradicate AIDS, and in particular, our ability to develop an effective vaccine against this disease, is directly related to the use of non-human primate models (Almond and Stott, 1999; Geretti, 1999). More than ten annual symposia have been held on *Nonhuman Primate Models for AIDS* and, no doubt, these meetings will continue.

On the topic of parasitic invasions, I must point out that the table is perhaps incomplete. It is quite possible that amoebiasis (dysenteric and visceral), besiosis, bartonellosis, blastomycosis, histoplasmosis, coccidiomycosis, and other diseases investigated in monkeys should be included in the table.

Primate parasitology is another area which provides evidence of human evolution, an area which was also noted by Darwin. Humans and apes are hosts of ectoparasites that are not found in any other animals on Earth, including any other primates (head louse, body louse, crab louse) (Fiennes, 1967). On the other hand, there are related species of lice in humans and all simians, including monkeys. This is explained by the joint evolution of parasites in primates and their inheritance by humans from simian-like ancestors (as is the case with the previously mentioned vuhererioses and malaria) (Zhdanov and Lvov, 1984). Twenty of the almost 30 protozoan parasites found in humans are the same in simians and they are not found in any other animals. The helminths are relatively rare related in kinship, but 30 of these species exist and they are common to humans and other primates.

Primates were not used very often as subjects in experimental parasitology until the 1970s. T.C. Orihel, a prominent specialist in parasitology, wrote that the usual laboratory animals 'are not an adequate model of humans.' They do not react to invasions in the human way, and only the primate 'is an ideal model for the study of parasitic infections' (Orihel, 1971, p. 772). R.E. Kuntz was of the same opinion (1973, p. 194) and thought that researchers '. . . only start to understand the potential value of primates.'

The history of the struggle against malaria, one of the most dangerous human diseases, is quite dramatic. At present, this disease causes annual losses of more than 2 million human lives. We saw that research on this disease is impossible without monkeys. We should recall the great affinity of erythrocytes in humans and simians, which is relevant to the similar pathogenesis of this disease in primates. The blood of *A. trivirgatus* contains all the classes of malarial precipitins (*Pl. falciparum*) found in humans (Young, 1973). It is clear that the common evolutionary origin of plasmodia in humans and simians plays the most important role here (Escalante *et al.*, 1995). The species of malaria pathogens in humans and simians are so close that the results of experiments in monkeys can be extrapolated almost directly to humans. Humans are susceptible to a number of simian plasmodia, and conversely, simians may be susceptible to plasmodia of humans.

The fight against a lethal form of tropical malaria caused by *Pl. falciparum*, which was rampant in the United States Army in Vietnam, occupies a special place in the history of medical primatology. There was no nonprimate animal model of this disease, but chimpanzees and gibbons and, since the end of the 1960s, owl monkeys (and later saimiri) were used most widely in experimental research on this disease. This research with nonhuman primates permitted the testing of hundreds of thousands of treatment compounds. The compound, chloroquine, was discovered and research on a vaccine was initiated. This was very difficult due to the instability of *P. falciparum* strains (Modell, 1968; Siddiqui, 1977; Gysin, 1990). In the 1990s, extensive research has been carried out with simians in an effort to develop an anti-malarial vaccine. Live vaccines, DNA vaccines, recombinant vaccines, and virus-vector vaccines were tested. Genetic therapeutics and prophylaxis with interleukins are currently being studied (Gardner *et al.*, 1996; Tine *et al.*, 1996; Heppner and Ballou, 1998; Perera *et al.*, 1998). Monkeys are also used in the search for a vaccine against schistosomiasis, filariasis, leishmaniasis, and other infections which were not included in Table 5.2 (see also Chapter 2) (Yole *et al.*, 1996; Leroy *et al.*, 1997; Dube *et al.*, 1998).

The study of infections from the perspective of medical primatology revealed several important facts. The modeling of various diseases in simians is very species specific. Sometimes rare and exotic species of primates should be enlisted in these investigations (e.g., *Mandrillus, Presbytis, Aotus*). Not only the species, but also the subspecies may be important in the research. For example, the Guyana phenotype of saimiri differed from the Bolivian and Peruvian phenotypes of the same species in its reaction to the Indochina strain of *Pl. falciparum* (Campbell *et al.*, 1986).

OTHER MODELS OF HUMAN PATHOLOGY

Allergy

At present, allergy is one of the most pressing problems of human health. According to press coverage, more than 20% of the population in the United States alone suffers from allergies, not including the 10 million Americans that suffer from allergic bronchial asthma. Experiments with primates contribute to the scientific solution of this problem. To get an impression of the advantages of simians in research on allergies, we should recall the previously mentioned homology of immunoglobulin E (IgE), mast cells, the similarity of immunology as a whole, and of neural regulation in human and nonhuman primates.

Since the end of the 1960s, United States scientists (L. Leyton, R. Patterson, K. Harris, K. & T. Ishizakas, and others) have actively carried out investigations on allergy in primates (and in other laboratory animals as well). Although antibodies with properties resembling those of human reagins were found in different animals, a special similarity to the human was shown only for simians. Human IgE, for example, sensitizes the skin of humans and macaques, but does not sensitize the skin of guinea pigs (Ishizaka and Ishizaka, 1971). It was shown that tissues of macaques have receptors to human reagins and to inhibitors that suppress the release of histamine in humans (Malley *et al.*, 1971). It was also found that monkeys are 'ideal experimental subjects' to study the function of mast cells and their similar heterogeneity to those of the human, which is directly related to allergic reactions (release of histamine under the influence of IgE) (Barret and Metcalfe, 1986). It was found that anaphylaxis in primates differs significantly from that in other mammals in terms of the type of mediators released, the mechanism of their release, and the sensitivity of the tissues to different mediators (Lichtenstein, 1983).

Nonhuman primates are susceptible to the same allergens (vegetable, animal, bacterial, and pharmacological origin) as humans. Atopic human allergy was easily transferred by intracutaneous introduction of serum from a sick human to Old World and New World simians as well as to prosimians, with the same pathological consequences as in humans. L. L. Layton (1965) proposed that passive sensitization of skin by reagins may be a specific peculiarity of the order Primates because it was not possible to induce such sensitization in representatives of five other orders or in *Tupaia glis*. Simians (with fair skin) were recommended for tests with a skin probe to human allergens because of the ability of their skin to fix human reagins (Perlman, 1969).

Clinical allergic reactions to penicillin found in *M. arctoides* were the same as in humans; edema of the face and eyes, and vomiting (Mladinich *et al.*, 1987). Various allergic diseases have been studied in monkeys, predominantly in various species of macaques, including bronchopulmonary aspergillosis, seborrheic dermatitis (a type of psoriasis), hypersensitivity pneumonitis, etc. (Slavin *et al.*, 1978; Newcomer *et al.*, 1984; Perez Arellano *et al.*, 1992). A significant number of investigations were carried out in macaques (*M. mulatta, M. fascicularis*) on experimental allergic encephalomyelitis that closely resembled human multiple sclerosis (Ravkina, 1972; Rose *et al.*, 1994), as well as studies of experimental pollinosis (Chiba *et al.*, 1990). Such investigations, moreover, frequently include testing and assessment of antiallergenic substances.

I must especially discuss the modeling of bronchial asthma in monkeys. This ancient human disease has been studied intensively for more than 100 years, but it and a number of other allergic diseases are far from completely understood, although some success has occurred (Patterson and Harris, 1997). In the past, there were significant difficulties in the study of asthma associated with the lack of an adequate experimental model. Different laboratory animals were used, and for a time, dogs seemed the most applicable. A new perspective on the problem of asthma was achieved with the advent of medical primatology. It appeared that as a result of significant similarity in sensitivity of the respiratory system in macaques (*M. mulatta*) and humans, the former, as well as human asthmatics, and different from dogs, were highly responsive to pharmacological treatments, in particular, to aerosols of carbachol, histamine, and prostaglandin F_2. Since hypersensitivity of the respiratory tracts is the defining characteristic of clinical asthma, its model in macaques may be assumed to be very valuable (Krell, 1979). Various investigators used some approaches to model asthma in simians. It is likely that the most successful approach was the method of inhalation of aerosols containing antigens of ascarides (*Ascaris suum*) to sensitize rhesus monkeys.

Thus, primates play the dominant role as laboratory subjects in the study of asthma, although other animals, and dogs, in particular, may be used in certain cases. The scientists who have been successfully investigating this problem for about 40 years wrote, 'Perhaps the best animal model for studying certain features of asthma – both in theory and for evaluating potential pharmacologic agents – is the rhesus monkey model of asthma' (Patterson and Harris, 1997, p. 46). These authors cited among the advantages of this model the naturally occurring IgE antibody against the *Ascaris* antigen in 20% of adult rhesus monkeys, and an asthma-like response in about 1%, 'an incidence rate similar to the human population.' The interrelations of IgE antibodies and antigens in these monkeys are similar to those in humans. Their bronchial lumen mast cells respond with a release of histamine which depends on the antigen challenge, their airway responses to each of the allergens are well-studied, and the receptors to the antagonists of some allergens are also well-studied and quite homologous to the human ones.

Roy Patterson in the Appendix to the article, 'Pearls for Practitioners', wrote about the similarity of naturally-developing IgE-mediation in monkeys and humans, about an allergy which persists for years in the majority of monkeys as well as humans, and about the opportunity in human practice for a course of treatment with IgE-mediated allergy in monkeys. A disadvantage of the model is that its benefit mainly occurs in experimental practice; individual rhesus monkeys, like people, are very variable in the severity of their asthmatic reactions. The authors conclude that at present, neural modulation of the immune system during allergy has become an important field of investigation, and that 'the implications of extending this observation from allergic subhuman primates to the human allergic state illustrates the great potential for research in atopic diseases' (Ibid., p. 47). We should point out that therapeutic approaches to this complicated disease continue to be pursued with simian models (different species of macaques and chimpanzees).

I conclude this section on allergy by recalling the great similarity between auto-immune mechanisms of humans and simians and the opportunities that arise because of this similarity for reproducing systemic lupus erythematosus in monkeys (macaques) by feeding them alfalfa seeds and sprouts. It is apparent that this rare but extremely dangerous human disease can be reproduced more adequately in monkeys than in mice or dogs (A new model . . . , 1982). There is also a model of autoimmune malignant myasthenia gravis in rhesus monkeys (Toro-Goyco et al., 1986).

Oncology

In this book, it has been repeatedly mentioned that simians have been used in experimental oncology for some time. According to data from pathologists at the Sukhumi Primate Center (in the middle of the 1980s more than one-third of all spontaneous tumors known in monkeys were described here), the onconosological structure in humans and simians are quite different. Hemoblastosis and non-malignant tumors of the digestive system are frequently observed in monkeys. A significant number include malignant tumors of the intestines, kidneys, and upper parts of the alimentary tract. Unlike the data for humans, however, cancer of the lungs, stomach, and uterine cervix are only rarely observed (Krylova, 1986).

As already mentioned, the concept that malignant tumors are rare in nonhuman primates was rejected because the statistical data were derived from primates at vaccine centers where they do not usually live into old age. Since the end of the 1940s, following the first successful experiments of N. N. Petrov with his colleagues, other osseous sarcomas, as well as carcinomas of soft tissue (stomach, lungs, tongue), were induced (Melnikov and Barabadze, 1968; Ohgaki et al., 1986; Beniashvili, 1994). These studies demonstrated that malignant tumors could be reproduced in the closest relatives of humans using chemical and radioactive carcinogens (methylcholantren, nitrous compounds, isotopes of silver, cobalt, plutonium dioxide), and products of tobacco smoke. The morphological similarity in humans and simians was significant. The similarity of carcinoma of the large intestine in primates was discussed in the section on digestion.

Fruitful results in studying the importance of the oncogenic viruses were obtained in primates. At present, the viral etiology of certain forms of human malignant structures is accepted (carcinoma of the liver resulting from viral hepatitis B and C, carcinomas caused by the Epstein-Barr virus [EBV], carcinoma of the uterine cervix induced by the papilloma virus, and probably some malignant tumors of blood). These results, to a significant extent, reflect the research advantages of monkeys that were the subjects of decisive experiments in oncovirology. Since the end of the 1960s and during the entire 1970s, there was not a single issue of the information bulletin on medical primatology that lacked publications on viral oncogenesis, which was still unresolved in humans. (I note that, according to our data, there were 57 articles published in 1967 alone on cancer in primates, 176 in 1976, and 114 in 1985.)

It is obvious that the stimulus for this research was the demonstration that the chicken virus of the Rauss sarcoma can overcome the species barrier and induce similar neoplasms in rodents. This was shown in the 1950s by academician L. A. Zilber, the author of the viral-genetic theory of cancer, and his colleagues. It was simultaneously shown by the group of G. Ya. Svet-Moldavsky. In the beginning of the 1960s, I. Monroe and W. Windle at the University of Puerto Rico Primate Center and the group of L. Zilber, B. Lapin, F. Ajigitov, and I. Obuch, showed that the same Rauss chicken virus induces fibro- and rhabdosarcomas in macaques (rhesus). Thus, a bird virus appeared capable of inducing sarcomas not only in mammals in general, but also in the order to which humans themselves belong! (Lapin, 1996).

Beginning in 1965, experiments on modeling blood tumors in monkeys were initiated at the Sukhumi Primate Center with the blood of humans who were suffering from leukemia. As we already mentioned, the experiments led to an outstanding result; B- and T-cell lymphomas associated with two types of viruses developed in baboons (*P. hamadryas*) (After a long-term study, the investigators found two new subfamilies of primate oncogenic viruses, EBV-like herpes viruses of primates and a subfamily of Human-Simian T-lymphotropic viruses, or HTLV-1/STLV-1). The other sensation was the horizontal transfer, in a group of baboons, of not only these viruses, but the disease itself, a malignant lymphoma. The etiologic role of STLV-1 and EBV-like simian viruses in the development of the disease was demonstrated and confirmed using the most modern methods of immunology and molecular biology. Retrovirus STLV-1 was found in baboons in three forms containing different degrees of homology with the human virus HTLV-1; monkeys with lymphoma had only one of these forms.

Before these experiments began, no more than 20 cases of lymphoma had been reported in simians. According to the data of B. A. Lapin (1996), about 400 baboons died at the Sukhumi Center as a result of a lymphoma outbreak. Virus-associated hemopoietic malignancies were also modeled in macaques (Lapin and Yakovleva, 1970; Lapin *et al.*, 1975, 1996; Yakovleva *et al.*, 1997). At present, research on these viruses and related forms is carried out in many laboratories of the world and in many species of simians. The phylogenetic link between HTLV-1 and STLV-1 was demonstrated, their biology is studied, and precautionary measures are taken against diseases associated with them (Yamashita *et al.*, 1997).

As described in Chapter 2 of this book, a vast amount of research has been carried out on monkeys to study other oncogenic viruses, especially in the United States, England,

Japan, and other countries. A similarity between humans and simians was shown not only for the etiology and pathogenesis of hemoblastosis, but also for other proliferative processes. Monoclonal antibodies distinguished a universal antigen in the proliferative tissues of humans and other primates, but the cells of other mammals did not express this antigen (Dubey et al., 1987). A cytogenetic similarity between humans and simians was found in several oncogenic viruses at the level of chromosomes and individual genes, as well as at the DNA level. These results provided the grounds to assume a common origin of these viruses in humans and simians (Komuro et al., 1984; and others). Since the discovery of the oncogenes, more and more data have accumulated on the similarity of these intimate structures of oncogenesis in primates. The homology of the c-mos proto-oncogene in humans and an Old World monkey (E. patas), for example, was 97% (Paules et al., 1988). Moreover, monkey virus DNA has been found in tissues of malignant tumors of humans (Pennissi, 1997).

Due to the fact that simians are extremely susceptible to certain forms of oncogenic viruses, there are grounds to study, in simians, human neoplasms whose viral etiology is likely to be high (Kaposi's sarcoma and malignant tumors of the uterus and rectum) (Lapin, 1996). Of course, monkeys are also used in other oncological investigations. In concluding this section, I note that the achievements of primate research in this field are not as obvious practically as they are for example in the area of infectious diseases. The problem lies not with the primates, but in the fact that oncological problems appear to be multifaceted in many respects. Nevertheless, the theoretical aspects of these problems cannot be resolved without laboratory primates.

Radiobiology

One of the pressing medical problems of the current time is associated with the technical progress of recent decades; I refer to radiobiology. The first 'experimental' subject in this field was a human being. As already mentioned, certain investigations on the medical aspects of radiation were even classified as secret. The only information available in the open press suggested that simians were also the most valuable experimental subjects in this field. It is likely that the initial problem investigated by medical radiology in primates (at least in the Soviet Union) was radiation disease in its acute and chronic forms, prophylaxis, treatment, and the long-term consequences of radiation caused by ionizing (gamma-radiation) and roentgen (x-ray) radiation.

It was found that the pathogenesis of radiation disease is far from identical in different animals. This is especially evident in the so-called 'infectious phase' of radiation disease, more precisely, with respect to the gastrointestinal syndrome. It was concluded that '. . . radiation disease in humans is less similar to the type of sickness that occurs in guinea pigs and dogs, and most similar to the pathologic processes in monkeys' (Semenov and Yakovleva, 1965, p. 57). Large-scale radiobiological investigations were conducted on simians in laboratories throughout the world (Figure 5.1). The problems under study included the influence of radiation on the function of the vital systems of primates, such as immunity against infection, hemopoiesis, hormone production and state of the adrenal glands, activity of the digestive tract, germinal apparatus, and the central nervous system, and cytogenetic alterations. It was shown that even a long time after radiation disease developed in monkeys, the number of tumors continued to increase, from 1%–2% initially to as high as 50% (Yakovleva, 1968); even 3–5 years after exposure to radiation in the majority of clinically healthy monkeys, the

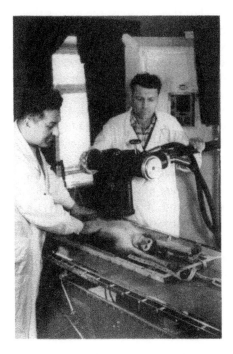

Figure 5.1 X-ray research with a monkey by Dr. Vartan G. Simavonyan (left) and his assistant Evgeniy P. Demenkov (Sukhumi Primate Center).

chromosomal defects in bone marrow cells were 2–3 times greater than those in un-radiated monkeys (personal communication, O. T. Movchan, 1965). The nervous system is also severely affected by radiation (Kimeldorf and Hunt, 1969).

The germ cells of primates are especially vulnerable to the impact of radiation. According to the data of N. P. Dubinin, primate chromosomes appear to be two times more sensitive to radiation than those in mice. It was found that there is no actual lower threshold of radioreceptivity in simian chromosomes because even minimum doses of radiation compromised reproduction (and would likely do so in humans). This was mentioned previously in Chapter 2.

Methods of protection against radiation with chemical agents were developed and tested in monkeys (after initial tests in rodents). They included sulfur-containing agents (mercamine, cystamine), amino-compounds (mexamin, serotonin, thriptamine associated with acetylcoline), combinations of neurotropic amino-compounds (adrenaline with acetylcholine), as well as antibiotics and many other pharmacological preparations. The combined action of several of these so-called radioprotectors significantly exceeded the arithmetic sum of their separate actions. The protective action of anoxia was also investigated (Semenov, 1957, 1967). In the 1950s, experiments were initiated with primates on autotransplantation (from shielded parts of the body) and homotransplantation of bone marrow (Barkaya *et al.*, 1964).

Further investigation of radiation effects in primates fully confirmed the pioneering research in this field. Humans and baboons (*P. papio*) are equally susceptible to radiation injury from plutonium (^{239}Pu), whereas dogs are only one-quarter as sensitive (Metivier *et al.*, 1974). Absorption of uranium radionuclides from the alimentary tract of baboons was 6–7 times higher than in mice. For these reasons, the figures that were recommended for the absorption of uranium and plutonium in humans were extrapolated directly from the experiments on baboons (Bhattacharyya et al., 1989). The time required for excretion and clearance of inhaled cobalt oxide particles from the lungs of baboons also corresponds to the figures obtained in humans, unlike the data for other animals (Andre et al., 1989).

Total x-ray radiation of young rhesus macaques at 35–40 months age with doses of 750–900 rad led to a decrease in bone growth over three years (by radiography) after experimental treatment. The decrease was 11% greater than when macaques were treated at 41–51 months age. On the basis of these data, it was concluded that x-ray radiation of children with doses greater than 750 rad would lead to a risk of decreased growth (Sonneveld and van Bekkum, 1979). Cytogenetic injury of somatic cells (lymphocytes) under the influence of x-ray and ©-radiation (200 and 300 rad, respectively) was not clearly different in humans and chimpanzees. Neither was it different in several species of guenons (Cercopithecus) exposed to acute x-ray radiation (200 rad); these species were recommended to model chromosomal injury in somatic cells of humans (Decat et al., 1982; Caballin et al., 1983). Such investigations continue in other species of catarrhines as well (Borrell et al., 1998).

Chromosomal aberrations in the germ cells of males and females under the influence of ionizing radiation appeared to be no less critically dependent on the level of biological organization of mammals than other relationships we have described. A comparison based on the maximum number of chromosomal injuries in mice, guinea pigs, hamsters, rabbits, monkeys, and humans indicated that a radiation level of greater than 600 rad was required to produce injury in mice, as little as 100 rad in monkeys (Macaca, Cercopithecus), and only 78 rad in humans (Brewen, 1979). The same general relationship was found for x-ray radiation. In this case, the response of testicular cells in rhesus monkeys significantly differed from that in mice, and fully corresponded to the condition in humans (Van Buul, 1983) (as was already mentioned in the section on reproduction).

Simians were also used in tests conducted on weapons of mass destruction. At present simians remain the most valuable subjects for investigating the problems of protection from x-ray and in the search for methods of radiation treatment for metastases of malignant tumors (Goncharenko and Kudryashov, 1996).

Diseases of metabolism

Simians are also used in experimental investigations of human diseases for which models are comparatively rare. These diseases include diseases of metabolism. Above, we mentioned the spontaneous development of rickets in South American monkeys whose diet lacked one form of vitamin D. The response to this deficit in saimiri, for example, is very similar to the one in humans (Lehner et al., 1968). Osteomalacia, which is characteristic of human type II rickets and resistant to vitamin D, was observed in marmosets (Flucker et al., 1990). The possibilities for modeling protein-calorie deficiency in monkeys was already discussed. Aminoacidopathies ('the Falling disease' in children) was reproduced in monkeys (Gerritsen and Siegel, 1972). Calcium pyrophosphate deposition diseases, similar to human metabolic diseases of the articular cartilage that leads to degenerative arthropathies, was observed in a colony of macaques (M. mulatta) (Roberts et al., 1984).

Arthritis is induced in simians (usually in macaques; *M. mulatta, M. fascicularis*) using the method of immunization with collagen. Not long ago, this model could hardly be imagined, despite the information on spontaneous polyarthritis in primates (including apes). At present, these models permit the experimental study of many aspects of various forms of arthritis in humans and the search for new methods of treatment (Bakker *et al.*, 1990; 't Hart and Bontrop, 1998).

Surgery

Scientists have long been attracted to primates for the objectives of experimental surgery. Originally, this interest was based on the anatomical similarity of humans and simians and on the *a priori* concept of their general biological affinity. Since the 1960s, however, this research has obtained strong scientific support from comparative investigations in hematology and blood circulation, immunology, genetics, transplantation, and other fields. I previously discussed (Chapter 2) the successful experimental development of the lobotomy operation in chimpanzees (Yerkes Primate Center), which earned a Nobel Prize for the scientist. In 1940, A. A. Vishnevsky, a prominent Soviet surgeon, at the Sukhumi Primate Center, developed the operation of double-sided ablation of the lobes of the lungs in baboons, which was subsequently applied in human clinical practice.

In the 1950s, intensive experiments were started in primates on resuscitation of the individual from the condition of clinical death. This work was carried out by a group of Moscow scientists at the Sukhumi Center under the supervision of Professor V. A. Negovsky. These investigations yielded an important practical result. Within a short period of time, studies of resuscitation and intensive therapy were introduced into clinical medicine and widely applied. The participants in this research were awarded the State Award of the Soviet Union. On September 27 1960, Negovsky and his colleagues carried out the famous experiment on resuscitation of a female baboon (*P. hamadryas*) 30 minutes after clinical death was induced under the condition of deep hypothermia. The female, called Sarjenta, lived for 14 years after resuscitation and gave birth to 8 young (Fridman, 1979b). This investigation (as well as cephalic transplantation in simians in the United States [White *et al.*, 1971]), was very important theoretically, and stimulated new approaches in surgery and other fields of knowledge.

Many other experiments and clinical studies on surgical procedures were carried out on simians (not all of them were published in the scientific literature). We previously mentioned cross-hemoperfusion of the liver in an operation using a baboon ('the best treatment method for fulminating hepatic insufficiency' [Lie *et al.*, 1983, p. 224])[1] and the first transplantations of bone marrow in medical radiology. Monkeys were also the best experimental animals in development of methods of neural angiographic surgery, and in research on the methods of protection against thermal shock and the effects of scalding. Due to the peculiarities of diffusion and blood supply in the flexor profundus tendons of primates in comparison to dogs, rabbits, and other laboratory animals, it was recommended that only simians should be used in experimental operations on finger tendons (it is noteworthy that Professor V. S. Krylov and his

---

1 I emphasize that a similar operation through the liver of a pig which, unlike a baboon, hardly tolerates human blood transfusion, caused an increase of antibodies against the proteins in serum (both in pigs and humans) and led to diathesis, an anaphylactic reaction (Abouna *et al.*, 1974).

colleagues were honored with the State Award of the Soviet Union in 1985 for the development and introduction into practice of the first toe to hand transplant operation, which was performed in monkeys). Experimental scoliosis as well as the physiology and mechanisms of pain were also studied in monkeys (Thomas and Dave, 1985; Halpern and Dong, 1986; and others).

Primates were successfully used and still remain valuable experimental subjects for developments in the field of heart surgery. In the 1990s, baboons were used as the most appropriate models in operations on cardiopulmonary bypass (Hiramatsu *et al.*, 1997) and in other similar areas as well.

As mentioned above, at the end of the 1960s and the beginning of the 1970s, due to species differences in the hemodynamics of internal organs, it was found that simians are indispensable to the study of experimental shock. The applicability of theories of pathophysiology of shock to humans that were worked out in dogs was suspect mainly because endoxin had an opposite action on the vessels of the internal organs in primates and dogs (Brobmann *et al.*, 1970; Swan and Reynolds, 1972). Later it was shown that the so-called pancreatic shockogeneous factor (supernatant of pancreatic tissue) decreased arterial pressure, increased cardiac output, and induced portal venous hypertension one hour after introduction. When aprotinin was introduced simultaneously, it completely blocked these changes in the hemodynamics of pigs, partly blocked it in dogs (decreased hypertension), but had absolutely no influence on the hemodynamics of monkeys (*M. mulatta*). It was concluded that transitory hemodynamic shock in response to auto-transplantation of a pancreatic suspension (which was adopted in patients who required pancreatectomy) is not prevented by aprotinin (or heparin) (Traverso and Gomez, 1982). Models of septic (endotoxic – *E. coli*), anaphylactic, hemorrhagic, and other forms of shock were successfully studied in both macaques and baboons (Bar-Joseph *et al.*, 1989; Premaratne *et al.*, 1995).

The most numerous investigations carried out with simians were those in one of the most popular and successful fields of modern surgery, *organ transplantation*. In this area, primates were used not only as experimental subjects but also as direct donors of organs for humans. It is apparent that modern surgeons in this field became more careful than their predecessors, who started clinical transplantations in the 1960s, in their forecasts with respect to the possibilities of xenotransplantation within the order Primates. Nevertheless, at present, we are pinning our hopes in this area specifically on research with simians. These species stand as 'concordant systems' with respect to the human organism, and transplantation of organs from the former to the latter is called 'concordant xenografting' (Hammer, 1996; Taniguchi and Cooper, 1997). This fully reflects the state of affairs predicted from the details of biological affinity of humans and other primates.

Shigeki Taniguchi and David Cooper (1997) presented summary tables of xenotrans-plantations in a historical perspective (p. 14). According to these tables, 10 of the 14 total cases of xenotransplantations of kidneys to humans were made with simian kidneys (1905–1966). The maximum duration of life in a patient with a complete kidney transplant from a nonprimate (sheep) was 9 days (1923, Neuhof). The record for the duration of function with a chimpanzee donor was about 9 months (1964, K. Reemtsma) and with a baboon donor about 2 months (1964, T. E. Starzl). Five of the 9 total cases of xenotransplantation of a heart to a human (1964–1992) used chimpanzee and baboon donors. The maximum duration of life in a patient with a nonprimate organ is no more than one day; the record for a man with a baboon heart is 20 days (1984, L. Bailey). Eleven of the 12 total cases of liver transplantations to humans (1966–1993) were carried out with chimpanzee and baboon donors. The maximum duration of life in a patient with a nonprimate liver (pig donor) is no greater than 1 day. The record for a man with a baboon liver is 70 days (1992, T. E. Starzl). At present, these data can be supplemented.

It is not surprising that we are pinning our hopes in this field mainly on primates. A rat with a heart from a mouse lived about 3 days (Hammer, 1996, p. 1349), while a man with a heart from a baboon lived 20 days! No doubt, the reliability of these comparisons was exaggerated, although L. Bailey (1987) and earlier J. J. W. van Zyl and his colleagues (1968) had specifically selected baboons among the other experimental subjects for organ transplantation. This selection was based on the following assumptions: baboons can easily withstand surgical operations; they clearly have blood groups 'of the human type' (the problem of primate blood groups and serological criteria for transplantations was especially considered by Socha et al. [1987] and Socha and Moor-Jankowski [1989]); the anatomy of their heart and blood vessels is similar to the human anatomy; their responses to the transplant are similar to those of humans, including rejection, and they have similar effects of immunosuppression. Bailey assumed that specially selected baboons would be the sources of xenogenetic 'bridging' transplants before allotransplantation of hearts to newborn and older child recipients, transplants that would probably become permanent heart substitutes. He was basing his assumptions on the comparative availability of primates, low risk of thromboembolism, similar anticoagulation, and a decrease in the likelihood of infection. He emphasized, however, that there was an inherited vulnerability in the transplanted heart to rejection by the host and to other adverse effects. Other specialists in the field, including D. Cooper, had essentially the same opinion.

It is likely that the most appropriate biological subjects for xenotransplantation to humans would be chimpanzees (*Pan*) (although rare attempts at heart transplantations from chimpanzees were not successful [Hardy et al., 1964; Barnard et al., 1977]). It is my opinion, however, that the relatively small number of these animals in captivity and the ethical problems associated with their use are barriers to such practices.

Primates (primarily monkeys) have played an enormous role in the experimental development of allotransplantation. As a result, significant achievements have been made in organ transplantation in human clinical practice. Monkeys have been and remain the most appropriate experimental subjects for transplantation of kidneys, lungs, heart, the heart-lung complex, liver, pancreas and pancreatic islets, gall-bladder and bile ducts, fallopian tubes, and blood vessels (Hitchcock, 1969; Cooper et al., 1994; Kawauchi et al., 1996). Since the 1960s, when clinical transplantation of liver was not available, researchers were actively developing operations for the allotransplantation of liver in monkeys. One should keep in mind that practically all means of immunosuppression were tested in the final stages in monkeys (cyclosporin A, azathioprine, methylprednisolone, deoxyspergualine, total lymphoid irradiation, antithymocyte globulin) (Hammer, 1996). In due time, the Scientific Group of the World Health Organisation recommended that 'each of these preparations (immunosuppressants – E. F.) should be tested in simians' regardless of the costs (WHO Report #496, 1973, p. 49). New suppressive agents are also being tested in monkeys at present.

Nevertheless, in the 1990s, xenotransplantations (xenografts) of organs became the main branch of research in experimental transplantation with primates. This was encouraged by the development of powerful means of preventing transplant rejection, including transgenetic procedures, and the paucity of human organs required for transplantation. By the middle of this decade, numerous experiments on xenotransplantation had been conducted.

It is likely that pigs and baboons are the main organ donors for humans. In this monograph on primates, there is no need (nor competence by the author) to give recommendations on the value of xenotransplants from different species as substitutes for nonfunctional human organs nor to comment on the prospects of the work as a whole. This is the domain of the surgeons who are engaged in the research and who have advanced technical and medical methods of modern surgery. But many of the biological comparisons between humans and animals that are presented in this book, as well as others that are not presented here, suggest that we should be wary of using the organs of pigs for transplantation to humans. We have written about the differences between pigs and primates in terms of DNA, genetic 'programming' of the length of biological life, genetics as a whole, immunology, biochemistry, endocrinology, physiology and pathology of vital organs, blood and hemodynamics, and finally, the responses to pharmacological and other influences.

Nevertheless, numerous experiments on the use of pig organs in transplantation are being carried out because of the significant interest in their potential use in humans. Of course, the main experimental subjects in these operations are simians, providing the last threshold on the way to xenotransplantation to humans. By 1998, the majority of vital organs had been transplanted from pigs to monkeys, including the heart, lungs, kidneys, liver, and pancreas (Kaplan et al., 1995; Matsumiya et al., 1997; Kawauchi et al., 1998; Lin et al., 1998; and many others). Attempts have also continued on the direct transplantation of organs from other primates and nonprimates to humans. Pigs and baboons are still the donors that are used most frequently (Luo et al., 1995; Borie et al., 1996). When irreversibly adverse effects occur in human organs and the only solution is transplantation, the organs of animals, and especially the organs of another primate, may become the last hope.

Aging

When describing the opportunities for medical primatology in the last two chapters, I repeatedly mentioned the value of laboratory primates for research into the biology and pathology of aging in different physiological systems of the human organism. A present day scientist gave the following title to his fundamental work on this problem: 'Primate gerontology: An emerging discipline' (De Rousseau, 1994). By the use of this title, the author emphasized the profound value of this field of science. Earlier, C. J. De Rousseau and other scientists carried out many experiments on gerontology with simians. We should recall that I. I. Mechnikov (1915) had early on proposed to establish colonies of laboratory primates precisely for the purpose of conducting such human-related research.

I previously discussed one of the peculiarities in the evolution of primates that is manifested by an increase in duration of the prepubertal period and the process of sexual maturation itself, and which is associated with an increase in maximum longevity. The compatibility of the dynamics of growth, development, and aging in humans and simians was also considered. In fact, several colonies of old monkeys and apes are maintained in the Regional Primate Research Centers of the United States for research on aging (Monkey colonies . . . , 1982; Understanding aging, 1990; Erwin, 1998).

Figure 5.2 Murray, an old (38?) male baboon, *Papio hamadryas*, 5 days before death (Sukhumi Primate Center).

It was shown, in particular, that macaques are an 'excellent model' for studying the processes of human aging (De Rousseau *et al.*, 1983). On the other hand, growth and development of the usual laboratory animals, rodents, for example, differ significantly from the similar processes in macaques and baboons, even taking into account the differences in the duration of biological life and its stages (Bourliere, 1982) (we should recall that if we exclude the apes, the most appropriate subjects for the experimental study of aging in the reproductive function of females are baboons).

As mentioned previously, aging of the brain and cognitive processes, visual recognition, various forms of senile dementia, special disturbances of the brain and memory, as well as somatic and biochemical changes, all have been studied successfully in simians (Maloney *et al.*, 1986; Bartus, 1993; Ervin, 1998). A symposium was held at a congress of the International Primatological Society which was directly concerned with human gerontology: 'Slowing aging by calorie restriction: Studies in rhesus monkeys.' This issue had been studied predominantly in rodents, but required experiments with subjects that were more closely related to humans. Research on primates was required to confirm the positive results of calorie restriction on glucoregulation, body weight, prevention of obesity and related disturbances up to diabetes and changes in behavior, and mineral metabolism in bone and energetic balance (Lane, 1996). No doubt the problems of gerontology, which are becoming more and more apparent with the steady aging of humanity, will find their solution in experiments with primates.

Space medicine and biology

After World War II, medical experiments in the United States were initiated with simians in the field of high-speed aviation and weight-less flights. Different species of mammals including monkeys were launched into the stratosphere to study the influence of high-speed acceleration and weightlessness on living organisms and to create methods for controlling vital systems at high altitudes and over great distances. Shortly afterwards (in the 1950s), specialists at the National Aeronautics and Space Administration (NASA) became interested in these investigations. A new branch of biomedicine was born, which we now call space medicine and biology. In the 1950s, monkeys as well as chimpanzees were involved in these investigations in the United States.

Taking into account information on the affinity of humans and simians already known to the reader, I must say that using primates in this new synthetic field of science was entirely natural and even inevitable. Actually, even before the first manned flights by humans, simians provided answers to the initial questions regarding the physical and psychological condition of cosmonauts and astronauts in unexplored space. The flights of apes into space in the 'Mercury program' became sensations. The first chimpanzee was called Ham (January 31, 1961), the second, Enos (later in November of the same year). These experiments immediately preceded the suborbital flight of the astronaut, Allan Shepard (May, 1961), and the orbital flight of John Glenn (February, 1962), both of which followed the first orbital flight of Yury Gagarin (April 12, 1961). After these sensational achievements, research with monkeys in special programs on space research began in the United States, the Soviet Union, France, and other countries that joined these programs. Unprecedented cooperation was initiated between the Soviet Union and the United States in this field (considering the state of the 'cold war'). This cooperation has continued to the present day and increased in scope (Hines and Skidmore, 1994; Dai et al., 1996). When bans were imposed in the United States on launching monkeys into space (land experiments were allowed as before), these flights were continued within the framework of the joint programs of the Soviet and Russian biosatellites of the 'Cosmos' type.

It is the opinion of the Russian co-chairman of the program, Professor Innessa Koslovskaya (and no doubt, the opinion of other specialists), that monkeys are indispensable in research on biomedical problems of humans in space. The investigators are interested not only in the medical characteristics and mechanisms of adjustment in the extreme conditions of space flight, with its inevitable hypokinesis and other negative factors, but also in the possibility for physical labor and intellectual activity in the state of weightlessness. Live humans, the main subjects of the research, cannot answer many of these questions; however, our closest relatives can be successfully tested both on land and in orbital space.

Before the first space flight of a woman, monkeys at the Sukhumi Primate Center were tested in centrifuges to assess the influence of acceleration and spatial orientation of the body, in particular, on physiology of the ovarian menstrual cycle. Experiments preceding and following the flight were systematically carried out on primates (blood and hemopoiesis, immunology, cardiovascular system, respiration, digestion, reproduction, vestibular physiology, etc.). In addition to the results of the research on antiorthostatics and hypokinesia mentioned above, simians provided valuable data on the influence of space radiation on the organism, biodynamics under conditions of vibration, vomiting, and other reactions of motion sickness. These data on primates are most applicable to humans and frequently differ from the data obtained on other mammals, fish, birds, or insects (Belkaniya, 1982; Wood *et al.*, 1986). At present, according to our data, such investigations (land experiments) in the United States are carried out only on simians.

In the second half of the 1970s, when a long-term space station were launched into orbit, new problems appeared associated with prolonged weightlessness in humans. Naturally, the main experimental information here was also obtained from laboratory primates. Prominent specialists in various fields of science are involved in the research, in particular, D. Washburn, B. Lapin, D. Rumbaugh, and others. Young rhesus macaques are usually used more often than other animals (the 'Bion' program). The most quick-witted, resolute, and enduring males are chosen (to avoid the complications of menstrual cycle phase on the experiments). Usually there is one pair of candidates (and two pairs of reserve substitutes), that undergo about 10 psychomotor tests and instrumental training, including performance on a computer. Following their return from space, the monkeys undergo a postflight check-up using the same program as that used for the cosmonauts. In the 1990s, important data were obtained in rhesus monkeys on selected problems for assessing behavior during a space flight and its telemetric control (Rumbaugh *et al.*, 1996).

During the space flights with simians the following were recorded: the state of skeleton muscle; phases of sleep (similar to humans in space); immunological effects of weightlessness; and other data (Bodine-Fowler, 1994). Important physiological indicators of primate adaptation were found, in particular, characteristics of the circulatory system, adjusted for vertical blood supply to the head under the condition of the earth's gravity and normal orthostatics, but positively modulated during the initial period of adaptation to weightlessness (which excludes brain edema). Changes in ocular control and spatial orientation of the vestibular apparatus were found and methods for their correction were sought, as well as other related problems. The simian flights were used to modify construction of the space and antigravity suits of the astronauts (Dai *et al.*, 1994, 1996). Comprehensive investigations with nonhuman primates have already given, and no doubt will continue to give in the future, valuable scientific information to facilitate the persistent efforts of humans to conquer space.

In this conviction, we conclude the description of animal models in medical primatology. In the final section, I will illustrate the opportunities that nonhuman primates provide in several areas for the benefit of human health.

PRIMATES IN PHARMACOLOGICAL RESEARCH

Having demonstrated the fundamental characteristics of affinity between humans and other primates, and having described the opportunities provided by biomedical investigations on simians, few would doubt the usefulness and need for the preclinical use of monkeys in pharmacology and toxicology. This paradigm, however, was not immediately accepted. In the beginning, scientists sometimes doubted the special value of primates for the study of medical preparations. In the 1970s–1990s, however, these doubts were due mostly to the difficulties in maintaining monkeys in laboratories and their high cost, rather than a scientific foundation for research on the problem. The idea of testing pharmacological agents on monkeys was strongly reinforced by industry, by testing vaccines for safety, and as was previously mentioned, by the history of thalidomide.

Table 5.3 Number of annual publications on pharmacological investigations carried out with nonhuman primates.

	1967	1970	1975	1976	1985
Pharmacology, toxicology, and anesthesiology	116	335	560	585	683

Since the end of the 1950s, there has been an explosive increase in the testing of new preparations on primates. Some pharmaceutical companies advertised that their products were, 'Tested on simians!' Psychopharmacology encompassed a special agenda (there were reasons for this). A scientific review on psychopharmacology in the United States was prepared by the Moscow Institute of Medical Information, under the editorship of two pharmacology professors. In the context of proposing the use of psychotropic agents on laboratory animals, the review concluded with the maxim, 'The time has come, however, to carry out experimental studies on simians' (Avrutsky and Stepanyan-Tarakanova, 1970, p. 24).

The rapid development of pharmacological investigation in primates is demonstrated by a quantitative analysis of world publications in this field from the 1960s–1980s (Table 5.3). I note that not all research performed by private companies is published (Fridman et al., 1990).

In the last year of my analysis (1985), the number of publications in medical primatology on pharmacology moved from sixth place (in 1967) to second place, overtaking, in terms of number, such diverse areas as infection, reproduction, vision, and endocrinology. In 1976, when the number of publications in almost all areas decreased due to the reduction in exportation of primates, there was no decrease in the publications on pharmacology. Testing pharmacological preparations on monkeys was supported and was sometimes even required by international and national organizations. No single animal model, including a primate model, however, is considered fully sufficient for predicting all possible effects in humans (WHO, 1991). Special colonies of monkeys were established by the pharmacological industry, which in the United States required 40% of all laboratory primates (70% in Canada) (Walcroft, 1983).

Despite the fact that only 6% of all studies with laboratory animals was devoted to psychopharmacological investigations on simians in the 1960s, by the end of that decade almost all the basic psychotropic preparations had been tested in primates, including amphetamines, aminasine, atropine, caffeine, meprobamate, morphine, pentobarbital, perphenasine, reserpine, scopolamine, tetrabenasine, tryptamine, phenazocine, phencyclidine, and others (Reynolds, 1969). As mentioned above, some of these agents caused completely opposite effects in the commonly used laboratory animals to those observed in humans and simians.

In various sections of this book I have mentioned the indispensability of primates to a number of pharmacological investigations, in particular the toxicology of reproduction and cardiovascular pathology. Sometimes one might doubt the advantages of primates over the other animals in extending data to humans, for example, in terms of pharmacokinetics (Doyle, 1981) (A paper was published at the same time, however, in which a prominent pharmacologist stated that it was precisely their special affinity to humans that made monkeys especially appropriate in pharmacokinetic research [Campbell, 1981]). A majority of specialists, moreover, agree that the metabolism of medicinal plants, which is slower in humans than in nonprimate animals, nevertheless is closest in humans and simians. In this connection, one may compensate for differences in metabolic rate of pharmacological preparations by varying the dosage (with calculation of the pharmacokinetic characteristics), but it is impossible to compensate for species differences in metabolism *per se* (Smith and Caldwell, 1976).

The similarity between humans and simians in the liver enzymes that metabolize pharmacological preparations was mentioned above. The activity of microsomal and lysosomal liver enzymes in baboons is quite similar to that in humans, but is essentially different from that in other laboratory animals (Autrup *et al.*, 1975; Holm, 1978). The catalytic and immunological similarity of microsomes in humans and in some species of monkeys is well-known, at least with respect to metabolites of the cytochrome P450-IID subfamily (Jacqz-Aigrain *et al.*, 1991). It is the opinion of some specialists that the disparity between the metabolism of pharmacological preparations in the livers of primates and rodents makes the extrapolation of data from rodents to humans of doubtful value (Maloney *et al.*, 1986) (I note that the modern rule for preclinical testing of new medicine requires no less than 3 different species of mammal, one of which must be a primate. In countries of the European Community, only two species are required, one of which should be a rodent and the other either a dog or a simian [Weber, 1997]).

R. L. Smith and J. Caldwell (1976) considered the metabolism of pharmacological preparations in detail, comparing humans, several species of simians (most frequently rhesus macaques), and various other species of animals. They compared the mechanisms of conjugation (glucuronic acid, sulfate, amino acids, mercapturic acid), methylation, and acetylation using their own data and data in the literature. Monkeys (different species to a different degree) displayed similarity or close correspondence to humans in every metabolic reaction, whereas the other laboratory animals almost always demonstrated differences from humans. The authors summarized data on the adequacy of rhesus monkeys and other mammals as models of human metabolism without respect to kinetics and excretion of the compounds. Table 5.4 shows that metabolism in monkeys was similar to human metabolism in 71% of the compounds (there was not a single case in which it was invalid), whereas in dogs there was similarity in only 19% and in rats 14% (Ibid., p. 351). I note that while the authors possessed data for only 10 species of simians (from a total exceeding 150), a lesser number is presented in Table 5.4.

It is clear that this table could be significantly supplemented, but the general trend would not be changed. We can see the basis for the conclusion of the pharmacologists when they advocated the advantages of simians for testing human medicines. Similar conclusions were expressed in the reviews of other specialists (Weber and Madoerin, 1977; Wilson, 1978; Mazue and Richez, 1982; Weber, 1997).

The half-life for the excretion from plasma of the antiphlogistic nonsteroidal preparation, fenclorac, was 1.5 hours in rats, 6 hours in dogs, and 3 hours in humans and monkeys. The main route for excretion from dogs, moreover, was in bile, whereas, the route for excretion from humans and rhesus monkeys was in urine (de Long *et al.*, 1977). The main path for metabolism of 5-nitroimidazole in rats and dogs was the alimentary canal and excretion was observed mainly in faeces (>60%). More than 60% of the marker in rhesus monkeys, however, was excreted in urine (Barrow, 1987). The same was observed in a study on the metabolism of femoxetine (Larson, 1981). Metabolism of 1,2,3,4-tetrachlorobenzene, sulphadimethoxine,

Table 5.4 Nonhuman primates as metabolic models for humans (Smith and Caldwell, 1976).

Compound	Species		
	Rat	Other nonprimate*	Rhesus monkey†
Amphetamine	invalid	fair (D)	good
Phenmetrazine	invalid	poor (G)	good (M)
Chlorphentermine	invalid	good (G)	good
Norephedrine	invalid	invalid (R)	good (M)
Phenylacetic acid	invalid	invalid (D)	good
Indolylacetic acid	invalid	invalid (D)	good
1-Naphthylacetic acid	poor	fair (R)	good
Hydrotropic acid	good	good (R)	good
Diphenylacetic acid	good	good (R)	good
4-Hydroxy-3.5-diiodobenzoic acid	invalid	invalid (R)	good
Sulphadimethoxine	poor	poor (D)	good
Sulphadimethoxypyridine	poor	invalid (R)	good
Sulphamethomidine	invalid	invalid (R)	good
Sulphasomidine	fair	good (D)	good
Isoniazid	–	poor (D)	good
Indomethacin	poor	poor (D)	fair
Halofenate	poor	poor (D)	good
Oxisuran	good	fair (D)	fair
2-Acetamidofluorene	good	poor (D)	fair
Nalidixic acid	–	fair (D)	fair
Methotrexate	poor	good (D)	good
Phencyclidine	poor	fair (D)	poor
Morphine	fair	fair (D)	fair

* D = Dog, G = Guinea pig, R = Rabbit, † M = Marmoset

4,4-dichlorobiphenyl, carbazeran, and many other compounds in humans and simians was different from that in rodents, rabbits, and dogs (Sipes *et al.*, 1980; Dulik *et al.*, 1987; Schwartz *et al.*, 1987). The disposition of metabolism was generally similar in humans and monkeys.

During research on the 40 compounds mentioned above (in various aspects of this discussion), the phenomenon of unidirectional metabolism of psychotropic preparations was observed in humans and simians. This list could be enlarged. This does not mean that the metabolism was exactly the same in every respect, but it was closer, at least, than the analogous process of any other animals that were studied. Moreover, the most valuable neuro- and psycho-pharmacological preparations introduced into modern medicine were included among the compounds that gave similar results in humans and simians. Clearly, it would have been impossible to conduct completely adequate experiments on these preparations without primates.

Sometimes, after the research with primates, it was necessary to review the results obtained earlier with other animals before extrapolating the data to humans (Garrick and Murphy, 1980). Monkeys are indispensable to the study of marijuana, marijuana-like isomers of the cannabinoles, and other narcotics, all of which represent a serious social problem for humanity. It was found that rhesus monkeys exhibit psychological dependence on the same compounds as humans, including morphine, codeine, cocaine, caffeine, amphetamines, barbiturates, and ethanol. They do not develop any dependence on nalorphine, chlorpromazine, or the other compounds for which humans develop no dependence. There is great similarity among primates (humans and simians) in development of the abstinence syndrome (Seevers, 1968; Deneau et al., 1969). When evaluated on the same scale, humans and rhesus monkeys obtained the same results for 29 of the 33 total morphine-like analgesics (Griffiths and Balster, 1979). At present, primates remain the most valuable models for experimental study of addiction (Wilner, 1997).

I previously discussed the opposite effects that are obtained between primates, including humans, on the one hand, and the usual laboratory animals. Let us supplement these examples with some on pharmacological aspects. The reaction of the cerebral arteries in humans and monkeys, for example, differs significantly from that in dogs. Norepinephrine (noradrenaline) caused constriction of the arteries in primates, while there was practically no constriction in dogs. On the contrary, dopamine produced vasodilation of the cerebral arteries in primates and vasoconstriction of the arteries in dogs (Toda, 1985).

The unique similarity of brain function in humans and simians, including the regulation of behavior, as described above, provides the opportunity to study the effects of neurotropic agents on group and individual behavior, areas that cannot be studied adequately in other laboratory animals. Serotonin agonists and antagonists, for example, induce effects of different intensity in rats and the more sensitive marmosets; the latter have a richer behavioral repertoire. The operant responses of rhesus monkeys to anticholinergic drugs closely approximate those of humans. The antidepressant, clomipramine, and the neuroleptic haloperidol influenced the abnormal behavior of rhesus monkeys in essentially the same way that it influences human mental disturbances. Groups of 4–5 stumptailed macaques (*M. arctoides*) can be a useful model for studying the action of some hallucinogens on human social behavior. The sedative action of diazepam on social behavior was also successfully studied with this same species of monkey. Baboons (*P. papio*) are quite adequate subjects for studying the influence of pharmacological preparations on the central neurotransmitter system of humans. Macaques are frequently used for research in neurotoxicology. They are susceptible to low doses of the preparations and provide valid data, with respect to humans, on sensory and cognitive functions which correlate quite well with the morphological and biochemical changes (Schlemmer and Davis, 1986; Evans, 1990; Valin et al., 1991).

The influence of chronic treatment with antipsychotics on dopamine neurons in the basal ganglia and mesocortex, and consequently, on the motor system of humans and other primates, was proved experimentally in monkeys (*Cercopithecus aethiops*) (Roth et al., 1980). Currently, monkeys occupy a special position among laboratory animals when neuropharmacologists seek new antipsychotics with minimal or no concomitant complications such as extrapyramidal syndrome (EPS). Unlike the model in rodents, which is widely used in preclinical investigations, the 'nonhuman primate models of dystonia, dyskinesia, and parkinsonism qualify as homologous animal models of EPS. Symptoms in patients and monkeys are very alike, and they occur over similar time courses' (Arnt et al., 1997, p. S14). The investigation of antipsychotic and other neurotropic preparations (neuroleptics, tranquilizers, antidepressants, psychostimulants), analgesics, narcotics, soporifics, antispasmodics, and other neuropharmacological preparations is one of the most active and effective fields of modern pharmacology and medical primatology (Casey, 1996; Lidow et al., 1997).

The use of simians in research is clearly not limited to neuropharmacology. Besides investigations on vaccines and antibiotics, monkeys provided unique results, with respect to the application to humans, in research on medical preparations used in cardiology (vasodilators, hypotensives, inhibitors of thrombin, antiarythmics, nitroaromatics, agents for atherosclerosis reduction, etc.), oncology, gynecology, urology, and other sections of medicine (Zacchey and Weidner, 1974; Reindel et al., 1994; Refino et al., 1998). The choice of laboratory animals for preclinical testing was discussed at a symposium on the safety of biotechnology products intended for human use. The majority of the participants agreed that 'the most relevant experimental animal is the monkey' (Symposium discussion, 1987, p. 198).

Monkeys remain indispensable subjects for preclinical tests of new preparations. The pharmacokinetics of the new quinolone antibacterial pro-drug, prulifloxacin, was investigated in three animal species. Only the Java macaque (M. fascicularis) displayed one of the five forms of the preparation (oxo form) in urine. This form was not detected in rats and dogs (Okuyana and Morino, 1997). The metabolism of a new antagonist of glycoprotein IIb/IIIa-TAK-02, an antithrombotic preparation, is somewhat different and more pronounced in macaques than in guinea pigs (due to a higher concentration in monkey plasma). Although this result was obtained in three species, including dogs, the finding in monkeys strengthens the confidence of investigators that the preparation is appropriate for humans (Kawamura et al., 1997).

Data presented in different sections of this book clearly indicate that primates are indispensable for solving the problems of toxicology, including both the effects and the safety of medical preparations. It is especially dangerous when the effects of the same preparation in primate (humans and simians) are different from those in all other laboratory animals. Rats, rabbits, and dogs did not display any effects that might be problematic to humans when they were tested with such dangerous substances as aflatoxin B_1, products of DDT (mutagenic and carcinogenic activity), the toxic compound soman, or a zidovudine preparation and others. These compounds were dangerously toxic, however, in humans and monkeys (Markaryan and Adjigitov, 1980; Smith and Wolthuis, 1983; Ayers, 1989). Conversely, coumarin, nabilone, and technical chlordane adversely affected the organs of rats and dogs, but were harmless to humans and monkeys (Cohen, 1979; Sullivan et al., 1987; Khasawinah et al., 1989). Methanol produced toxic effects on the brain and eyes of rhesus macaques and humans (extravasation into the putamen and edema of the optic nerve), but it did not cause these effects in rats and dogs (Koizumi, 1989).

Similar examples are not at all limited to the data presented. Due to these fundamental species differences, primates have been used to evaluate the toxicity of lead and oxygen, to clarify the action of methylmercury, an overload of iron in the liver, poisoning with organic phosphates, the absorption of tobacco smoke, the influence of pesticides and pollution in the environment, the toxicity of drugs on audition, ancillary effects of anticarcinogenic preparations, aspirin, vitamins, indomethacine, butadiene, and many other substances (Siddal, 1978; Stinson et al., 1989; Schumann et al., 1996). Nevertheless, we should remember that care must also be taken when extrapolating experimental data from simians to humans.

It is likely that the greatest number of experimental studies on the effects of alcohol have been carried out with rats. Nevertheless, the achievements obtained with simians in this field could not have been obtained with any other laboratory animal. Rhesus macaques and chimpanzees (as well as baboons and other monkeys) respond to ethanol similarly to humans, including the response to withdrawal. Other animals respond differently (the speed of alcohol oxidation in chimpanzees corresponds to that in humans, i.e., 100 mg/hr/kg body weight) (Lester, 1968; Pieper *et al.*, 1972). An outstanding result in the determination of alcohol toxicity was obtained in baboons (*P. cynocephalus*). Regardless of the data obtained with rats, it was found that chronic drinking of alcohol leads to alcoholic hepatitis even though nutrition is adequate (Rubin and Lieber, 1973). Histological changes in the liver of humans and baboons were identical. Alcoholic cirrhosis and fibrosis were induced despite an adequate diet. The data on simian models were very close to those in people and they were much clearer in simians than in rats (Lieber and Decarli, 1989; Mezey, 1989).

As mentioned previously, experiments with monkeys permitted the experimental verification of the teratogenic action of alcohol on the fetus during pregnancy. It was also shown that blocking alcohol intoxication with derivatives of benzodiazepine, which inhibits the behavioral effects of alcohol, is harmless in rats. The same preparation, however, induces harmful side-effects in saimiri in terms of strong tremor and distinct convulsive activity with tonic and clonic phases and full convulsions. Thus, preclinical tests should not be limited to experiments in rodents (Miczek and Weerts, 1987). Experiments in simians demonstrate the strong negative influence of alcohol on memory and learning, metabolism and hormone balance, perinatal stress, and the condition of newborns. Chronic use of alcohol by baboons leads to biochemical disturbances, which for a number of reasons cannot be demonstrated in other animals, but is very dangerous to humans (Lieber, 1997).

Thus, primates are indispensable in studies of pharmaceutical, biological, and re-combinant (biotechnological) preparations. As we have seen, various species of simians are applicable for these purposes. It is apparent that chimpanzees would be the most valuable research animal biologically, but because of an unwillingness to conduct

Figure 5.3 Baboons in Sochi-Adler Primate Research Center.

experiments on the closest relative of the human species and because of the high efficacy of monkeys in preclinical pharmacology, we can restrict ourselves to the use of monkeys for most purposes, i.e., *Papio, Macaca, Cercopithecus, Saimiri, Callithrix, Cebus*, and the other genera. Of course, this does not mean that we should exclude other orders of animals from these experiments. On the contrary, one should always seek reasonable opportunities for experimenting with other laboratory animals, not only for reducing the costs of the research, but also for the conservation of primates, the laboratory twins of humans, the unique fauna of the Earth.

Conclusion

The historical path of primatology is tortuous and complex due to many reasons, not the least of which are the ideological ones, because research in primates inevitably touched on the origin and biology of humans themselves. The slow progress of primatology left an imprint on the development of biomedical investigations with simians, i.e., on the development of medical primatology, whose origin and outstanding achievements only began in the 20th century (see Chapter 2). Beginning in the 1960s, the rapid rise of medical primatology was associated with the strengthening of its biological foundations; sometimes the course of experimental investigations themselves demonstrated the complementary contributions of medicine and biology.

One may skeptically question use of the term, 'medical primatology.' We do not call biomedical research with dogs 'medical caninology', or biomedical research with mice and rats 'medical rodentology.' In the case of primates, however, we make an exception. The reason we make this exception is the special experimental subjects we study, representatives of an order in which humans themselves belong; we are conducting research with creatures from the 'Order of Man' (Oxnard, 1984). This sets the science of medical primatology apart from the others because it embodies a number of characteristics which are inherent only in this field.

With the firmly established concept of the affinity between humans and simians (which is predominantly anatomical, the others being *a priori* and generally unrelated), we try to exploit this affinity with the concrete data of biomedical research relevant to humans, while also trying to avoid the pitfalls of anthropomorphism. The conceptual basis for this affinity was established by the modern data of general biology (evolution, phylogeny, taxonomy, molecular biology, genetics, and others), by advances in particular biomedical sciences, and by data obtained in the course of medical experimentation itself. This affinity, based on the common evolutionary history of primates and the closest roots of origin, is unusually great. It does not correspond to the relationship of humans with any other representatives of the living world. It is even greater, moreover, than we imagined before the 1960s.

Of course, there are different degrees of affinity when we compare one species (human) with the 250 other species within the order. Thus, research can concentrate on both the similarities ('the glass is half full'), and the differences ('the glass is half empty'), which indeed sometimes takes place. Nevertheless, the affinity of humans with other primates in important biological characteristics (which can be quantified) reaches 90%–99% in the hominoids (most closely in *Pan*) and 50%–80% in other simians (*Papio*, *Macaca*, *Cebus*, and others). It is always less in other animals. The glass is more than half full. The primate that is most distant from *Homo sapiens* is biologically closer to humans than any other species of animal. One may be assured that these relationships will only be further supported, and that the analytical methods developed for this purpose will be improved. New comprehensive data on this topic may still require some modification in the taxonomy of hominoids.

The other important characteristic of medical primatology naturally follows from the previous one, the uniqueness of the 'laboratory material'. Simians allow us to create models of human diseases and other experimental conditions on the basis of an actual homology, something that is impossible with other laboratory animals (the number of simian models for infectious diseases alone exceeds 40; Table 5.2). Owing to this property, noteworthy achievements were obtained in medical science precisely because of experiments with primates. The use of primates in biomedical research now forms the basis of modern public health policy (in research on syphilis, classical typhus, tuberculosis, yellow fever, poliomyelitis, measles, malaria, and kuru, and in the areas of surgery, ophthalmology, pharmacology, etc.). As a result, primate research has contributed significantly to fundamental discoveries in the biomedical sciences (physiology, neurology, endocrinology, immunology, and others).

As has been demonstrated throughout this book, one of the main advantages of medical primatology is the similar direction of experimental results on simians with data on humans, in contrast to that in other laboratory animals (in which an 'opposite effect' may occur). Consequently, experimental data from the latter animals may be inapplicable for use in clinical practice with humans.

The special quality of experimentation with primates, in addition, consists of the essentially complete interrelation of the physiological systems and conditions, as in reality, where it is necessary not only to have nosological and systemic similarity, but also an accounting of the dynamics of many attendant biological parameters (neuro-regulational, endocrinological, immunological, biochemical, genetic, etc.). In other words, primate research on atherosclerosis of humans not only assumes an approximate similarity in the cardiovascular system, which may also exist in certain rodents, dogs, and pigs (including convergent similarity or similarity of 'mosaic evolution'), but also benefits from a similarity in the structure and circulation of hormones (estrogens, testosterone, and others), similar age-related changes in blood parameters, including the mechanisms of coagulation, a similar immune system, neural regulation, tendency toward orthogradic stature, and even similar indications of binocular vision. One cannot expect to find this *multitude of closely related characteristics* in research with nonprimates. Thus, research on the 'effects of estrogen on cardiovascular responses of premenopausal . . .' in females (Williams *et al.*, 1994) would be more effective and valuable in monkeys than in rodents. The same may be said about study of the 'correlation of behavioral, physiological, virological, and neuropathological variables associated with SIV infection . . .' (Marcario *et al.*, 1997).

It is not surprising that many biomedical scientists consider laboratory primates to be 'excellent', 'wonderful', or 'ideal' models for studying human diseases. Although these terms were included in some parts of this book, they were more often deliberately omitted in order to avoid total conformity and possible accusations of bias. At the same time, we must remember that in some cases the processes modeled in simians are not identical to those in humans. The model does not have to be an exact copy. Thus, it is necessary to be cautious even when extrapolating experimental data from nonhuman primates to humans.

An investigator who intends to conduct research with simians should possess certain preliminary training with primates, because there are specific peculiarities in this field, not only in terms of zoological technology (maintenance of primates, experimental preparations, and epidemiological safety, etc.), but also in scientific considerations as well. Experiments with primates require, among other things, calculation of the extraordinary reactivity of physiological parameters and consideration of the well-established emotional component of behavior in the closest human relatives. Failure to acknowledge these special characteristics may lead to a significant distortion of experimental results.

Among the other idiosyncracies of medical primatology, one should be aware that taxonomic and specifically species variations of the laboratory primates are of great importance. Hundreds of examples demonstrate that the results of experiments depend on the choice of an adequate species or even the choice of a particular subspecies of monkey. It is difficult to explain why two closed related species in the same genus have such different responses to the same influence, why their nosology and pathologic process, or the metabolism of a pharmacological preparation is so different. Clarification of these peculiarities would require significant scientific effort. As we demonstrated above, species of the genera of *Macaca*, *Papio*, *Cercopithecus*, *Callithrix*, *Saguinus*, *Saimiri*, and *Aotus* are the most widely used genera as subjects of investigation. It is important to focus more attention on application of the species of *Papio*, whose breeding in captivity is still limited, but which according to experience has great potential value and has probably been underestimated so far.

The technical aspects of acquiring, maintaining, and breeding primates, as well as commercial problems, for the most part, have not been discussed in the book. We must note, however, that despite the relatively high cost of research with primates, their rearing and maintenance (but no longer any 'acquisition' costs!), monkeys are not more expensive than dogs if the issue is analyzed in detail. This was demonstrated by a prominent medical primatologist, Professor Jan Moor-Jankowsky (1978), the founder and long-time director of a primate center near New York City. The most effective approach for the use of primates in biomedical research, in our opinion, is a research center which is designed specifically for research with primates.

Examples of primate research on human diseases, which are impossible to reproduce in other animals, were repeatedly given in this book. In 1968, the head of the Medical School of the University of California, Dr. R. Prichard, published a list of human diseases for which there were no adequate animal models at the time. Pox, leprosy, gonorrhea, myocardial infarction, baldness, podagra, asthma, gallstone disease, leukemia, glaucoma, appendicitis, multiple sclerosis, and other diseases of unknown etiology were included in this list. It will be apparent to anyone who reads this book that each of these diseases, with rare exceptions, is modeled in simians.

Modern fundamental monographs on biology and medicine are more and more frequently based on primate data and, in that sense, are becoming more and more 'primatological' (as the primates themselves, with increasing investigation become, according to the view of one scientist, more and more 'human-like'). This is an objective process, especially stimulated by medical primatology; it is very useful for such a multidisciplinary science as primatology proper, in which there is great need for interaction with various other disciplines.

Does all this mean that we must exclude all lower animals from laboratory practice? Not at all. I especially emphasize in this book that traditional laboratory animals should be used in experiments as much as possible. Such animals are not only less expensive to acquire, but their use helps to preserve the relatively small number of Earth's primates, which are scientifically priceless because of their special affinity to humans. We must agree with the leading specialists in medical primatology who call for the use of primates only in those cases where a medical or behavioral problem that is important to the life and health of people cannot be solved with other laboratory animals. Even in these cases, one should use primates that have been bred and reared in captivity and are not among the species endangered with extinction. Moreover, the most humane methods of experimentation should be observed in such research (Lapin, 1996).

The use of primates in research is an especially sensitive issue with respect to the apes, and more precisely, with respect to common chimpanzees (*Pan troglodytes*), one of the nearest relatives of humans. Unlike the other hominoids, chimpanzees are not so rare in nature, do not as yet come under official prohibitions, and are still used, though not very frequently, in medical experiments, mostly in the United States. If a model does not appear 'naturally' in captivity or during observations in the wild, it is preferable to exclude these primates as laboratory subjects. Unfortunately, there are several pressing problems in modern medicine that directly threaten the life and health of many people, and we cannot avoid using chimpanzees in the solution of these problems. We must admit that we wish it were different.

As to the demonstrations by those who press for 'the rights of animals', one can understand, in one sense, the motivation behind them. On the other hand, we must consider, for example, the 500,000 people in the United States that become infected with viral hepatitis annually, and the feelings of relatives of the 16,000 people who die from this disease *annually* (St. Claire, 1997). This ignores, moreover, the millions of people world-wide who escaped poliomyelitis and small pox, and all those who were saved thanks to investigations with primates (according to data of the World Health Organisation, more than one million people died of hepatitis B in 1996 alone). In 1952, there were 58,000 cases of poliomyelitis in the United States; hundreds were lost and thousands were crippled. After 30 years, thanks to the introduction of a vaccine in 1984 which was developed in research with primates, there were only 4 cases of this disease. There was also a plan for the complete eradication of the poliomyelitis virus. In the 1990s, 50,000–55,000 nonhuman primates were used annually in the United States for scientific purposes; this is less than 0.3% of the total number of laboratory animals involved in research (Bowden and Johnsen-Delaney, 1996). For the time being, we must be resolved to this important use. The contribution of medical primatology to science and public health is more significant than this ratio would suggest.

The outstanding achievements of medicine and biology in research with primates are described in this book, some of which were awarded the Nobel Prize. There are also several investigations which deserve this high award, but were not honored in this way for various reasons. The following discoveries also deserve the highest gratitude of humanity: the rhesus factor in blood; clarifying the regulation of the female menstrual cycle; the victory over poliomyelitis; the scientific foundation for prohibition of nuclear weapons; experimental studies on the development of neuroses and depression; research in atherosclerosis and other human suffering; and finally, radical changes in the concept of primate phylogeny and our own origins, obtained in recent decades from the comparisons of DNA, proteins, and the karyotypes of simians.

At the present time of instability caused by the recurrent pressure of mysticism, when even some intellectuals speak about the theory of evolution as an 'already' refuted 'whimsy' of Darwin's followers (despite the fact that it was precisely this theory which provided the foundation for biological and medical centers throughout the world), the latest data of medical primatology represent powerful evidence (by the principle of positive feedback) of the validity of the evolutionary approach. The beneficial practice of conducting research on simians with the objective of preserving human health and longevity is much stronger than any argument.

Afterword by the author

I would like to relate some information about the events that preceded my writing of this book. I believe that the 32 years that I worked at the famous scientific center, the Sukhumi Primate Center, the Institute of Experimental Pathology and Therapy (IEPT) of the Academy of Medical Sciences of the Soviet Union, was a great gift that was given to me. When I arrived there in 1960, a strong team of investigators, as we would say now, was being formed under the leadership of a future great academician, Boris Lapin; the reputation of this oldest medical primate research center in the world was becoming more and more widely acknowledged.

My teachers and colleagues, B. Lapin, B. Petrov, M. Nesturch, P. Skatkin (the latter three are Moscow's scientists), I. Dzheliev, G. Magakyan, E. Dzhikidze, G. Cherkovich, G. Khassabov, and G. Annenkov were always open to my numerous questions. I began with pleasure to specialize in the history of primate research, systematics, and the nomenclature of the order, using biomedical data and analysis of various types of research on simians. From the initial days of my research, the problem of the affinity between humans and the other primates stimulated my greatest interest. I began accumulating a database of the various characteristics of this affinity in my archive. The data were taken from biology, especially primatology, and what was new at that time, experimental biomedical investigations.

Beginning in the autumn of 1966, I encountered rich opportunities to pursue this work, and my personal scientific disposition became my official professional responsibilities. Due to a significant intensification of research with simians in the developed countries, the Director of the IEPT entrusted me with the task of organizing the Primate Information Center (PIC) with a laboratory of scientific analysis in medical primatology for investigators of the Soviet Union and the countries of Eastern Europe. From the beginning of its existence, our Center operated in close cooperation with a similar center in the United States (Primate Information Center of the Regional Primate Research Center, University of Washington, Seattle, USA), with the cordial assistance of the Director of that research center, a prominent primatologist, Professor Theodore Ruch. Scientific publications that were associated with any type of research conducted with primates began arriving at the Sukhumi IEPT from many countries around the world. Here they were translated or reviewed and then introduced into a database using a special information storage and retrieval system, initially entered on punched cards and later using computers. For two and a half decades, no fewer than 130,000 publications (including earlier studies) were collected at our Center.

The collection of this amount of scientific information in different languages required a significant effort by numerous people. The number of employees at the Sukhumi PIC exceeded 20 individuals, including scientists, interpreters, bibliographers, and technicians. I recall the names of my best assistants with excitement: Victoria Popova, Lalya Fitozova, Gulya Murtazina, Emma Chalyan, Emma Gablaya, Tanya Shokareva, Luisa Ryabova, and Sveta Meleta. Despite their miserable salaries, these dedicated people gave all their heart and enthusiasm to create the richest scientific archive on primates in Europe. This archive and personal investigations by the author and his colleagues established the basis for a systemic analysis on the development of medical primatology throughout the world and took the next step in building the biological foundations of this active field of science.

My career at the IEPT ended with the beginning of the war in Abkhazia (August, 1992). The State of Israel organized the evacuation of refugees from Sukhumi. I was allowed to take only 75 kg of luggage on the airplane. By that time, my personal scientific archive weighed more than 120 kg, as eventually became clear. A rare collection of scientific material on primates, which I had been collecting for more than three decades could have been lost. The representative of the Ministry of Foreign Affairs of Israel, Mr. Izar Hardan, saved the situation. He allowed me to rescue my archive and personally controlled the registration of my luggage at the airport.

It was my good fortune to meet Dr. Benjamin Sklarz, a multilingual chemist in Israel, who acquainted me with a scientific publishing house and helped me to fulfill all the necessary formalities. My colleagues from Adler also helped me. Academician Lapin organized here a new scientific research center with some of the former employees of IEPT, the Institute of Medical Primatology, within the reorganized Russian Academy of Medical Sciences. My American friends, Drs. J. Pritchard, J. Moor-Jankowsky, D. Bowden, M. Goodman, J. Schrier, L. Jacobsen, C. Johnson-Delaney, my colleague from the Czech Republic, Dr. Marina Vancatova, and of course, my editor, Professor Ronald D. Nadler, have all helped me. Professor Nadler's editorial work on this book was supported in part by NIH grant RR-00165 from the National Center Research Resources to the Yerkes Regional Primate Research Center of Emory University. The Center of Absorption in Science (Jerusalem) provided me with a computer and other technical means, and Dr. Michael Rabinovich taught me to work with a new program. I bow deeply in my gratitude to all these people, living and passed away, and to agencies that have facilitated my research.

I am grateful to all whom we cite in this book. The reader will find their names in the bibliographic list of references at the end of the monograph. I am grateful to thousands of other scientists, whose works I cannot cite; their research on primates created the scientific climate that permitted this book to be published. I am grateful, as well, to all the investigators who have worked and who continue to work in one of the most exciting and useful spheres of modern scientific knowledge.

Eman P. Fridman, D. Biol.
Hadera, Israel.

References

Abbreviations

IPS: International Primatological Society
APS: American Society of Primatologists
WHO: World Health Organization
NY: New York
M: Moscow

Abouna GM, Amemiya H, Andres G, Porter KA, Hamilton D, 1974. Immunological studies in patients receiving repeated allogenic and xenogenic liver hemoperfusions. *Transplantation* 18 (5): 395–408.

Adachi J, Hasegawa M, 1995. Improved dating of the human/chimpanzee separation in the mitochondrial DNA tree: Heterogeneity among amino acid sites. *J Mol Evol* 40: 622–628.

Adams MR, Kaplan JR, Williams JK, Clarkson TB, 1994. Primate models of coronary artery atherosclerosis. *Congr IPS* 15: 290.

Aebisher P, Pochon NA-M, Heyd B, Deglon N, Joseph J-M, Zurn AD, Baetge EE, Hammang JP, Goddard M, Lysaght M, Kaplan F, Kato AC, Schluep M, Hirt L, Regli F, Porchet F, De Tribolet N, 1996. Gene therapy for amyotrophic lateral sclerosis (ALS) using a polymer encapsulated xenogenic cell line engineered to secrete hCNTF. *Human Gene Ther* 7 (7): 851–860.

Agus Lelana RP, Hayes ES, Hasibuan IR, Walker MJA, Budiarsa IN, Ungerer T, Sajuthi D, 1997. [Preliminary study: The longtailed macaque as an animal model for cardiac arrhythmias]. *J Primatol Indonesia* 1 (1): 9–21 (Indonesian w/English sum).

Aidara D, Tahiri-Zagret G, Robyn G, 1981. Serum prolactin concentrations in mangabey (*Cercocebus atys lunulatus*) and patas (*Erythrocebus patas*) monkeys in response to stress, ketamine, TRH, sulpiride and levodopa. *J Reprod Fertil* 62 (1): 165–172.

Aitken P G, 1981. Cortical control of conditioned and spontaneous vocal behavior in rhesus monkeys. *Brain Lang* 13 (1): 171–184.

Albert MS, Moss MB, 1996. Neuropsychology of aging: Findings in humans and monkeys. In: *Handbook of the Biology of Aging.* 4 ed. E. L. Schneyder, J. W. Rowe, eds. San Diego, Acad Press: 217–233.

Alexander R M, 1992. Human locomotion. In: *The Cambridge Encyclopedia of Human Evolution.* S. Jones, R. Martin, D. Pilbeam, eds. Cambridge, Cambr Univ Press: 80–85.

Allman J, 1986. Evolution of the visual system in the Eocene primates. *Amer J Primatol* 10 (4): 386.

Allman J, 1987. Evolution of the visual system in primates. *Amer J Primatol* 12 (3): 326.

Almond N, Stott J, 1999. Live attenuated SIV: A model of a vaccine for AIDS. *Immunol Lett* 66 (1–3): 167–170.

Alp R, 1997. 'Stepping-sticks' and 'seat-sticks': New types of tools used chimpanzees (*Pan troglodytes*) in Sierra Leone. *Amer J Primatol* 41 (1): 45–52.

Alvord EC, 1977. Demyelination in experimental allergic encephalomyelitis and multiple sclerosis. In: *Slow Virus Infect Centr Nervous Syst.* New York e.a.: 166–185.

Alvord EC, Rose LM, Hruby S, Richards TL, Petersen R, Shaw CM, Clark EA, Ericsson LH, Stewart WA, Paty DW, Kies MW, 1988. Experimental allergic encephalomyelitis in nonhuman primates: An excellent model of multiple sclerosis. In: *Biomedical Research in Primates*. Rijswijk, TNO Primate Center: 31–49.

Amlinsky IE, 1964. [Scientific treatise on Man's nature and on means of its change. In: *Mechnikov I. I: Etudes of Optimism*]. M: 294–333 (Russian).

Amunz VV, 1976. [Development of reticular formation of the truncus cerebri in monkey ontogene with comparison to man]. *Arch anat gistol i embriol* 71 (7): 25–29 (Russian).

Anderson JF, Goldberger J, 1911. The period of infectiolity of the blood in measles. *JAMA* 37: 113–114.

Andre S, Metivier H, Masse R, 1989. An interspecies comparison of the lung clearance of inhaled cobalt oxide particles. Part III. Lung clearance in baboons. *J Aerosol Sci* 20 (2): 205–217.

Andreev FV, 1989. [Some characteristics of dioptric apparatus of mammals eyes]. *Dokladi AN SSSR* 306 (2): 508–511 (Russian w/English sum).

Andreev FV, 1990. [*On Phylogenetic Changes of Mammals Vision Organs*] vol 1. M (Russian w/English sum).

Andrews J, 1998. Infanticide by a female black lemur, *Eulemur macaco*, in disturbed habitat on Nosy Be, north-western Madagascar. *Folia Primatol* 69 (Suppl 1): 14–17.

Andrews P, 1987. Aspects of hominoid phylogeny. In: *Morphology in Evolution: Conflict or Compromise?* C. Patterson, ed. Cambridge, Cambr Univ Press: 23–53.

Andrews P, Martin L, Whybrow P, 1987. Earliest known member of the great ape and human clade. *Amer J Phys Anthropol* 72 (2): 174–175.

Andringa G, Lubbers L, Drukarch B, Stoof JC, Coools AR, 1999. The predictive validity of the drug-naive bilaterally MPTP-treated monkey as a model of Parkinson's disease: Effects of L-DOPA and the D1 agonist SKF 82958. *Behavioural Pharmacol* 10 (2): 175–182.

Andrus SB, Roach AM, Fillios LC, 1964. Experimental production of colloid goiter in the Cebus monkey. *Proc Soc Exp Biol Med* 116 (4): 963–967.

A new model of systemic lupus erythematosus (SLE), a disease that affects mainly young women, has been developed by feeding alfalfa seeds and sprouts. *Res Resour Rep*, 1982, 6 (9): 12–13.

Annenkov GA, 1974. [*Proteins of Blood Serum in Primates*] M, Medicine. (Russian w/English sum).

Apiou F, Rumpler Y, Warter S, Vesuli A, Dutrillaux B, 1996. Demonstration of gomeologies between human and lemur chromosomes by chromosome painting. *Cytogenet Cell Genet* 72 (1): 50–52.

Ardito G, 1980. Interesse filogenetico dello studio die parassiti nei primati (nota preliminare). *Antropol contempor* 3: 325–328.

Argaut Ch, Rigolet M, Eladari M-E, Galibert F, 1991. Cloning and nucleotide sequence of the chimpanzee c-myc gene. *Gene* 97 (2): 231–237.

Arita J, 1979. Virological evidence for the small-pox eradication programme. *Nature* 279 (5711): 293–298.

Armstrong E, 1985. Allometric considerations of the adult mammalian brain. In: *Size and Scaling Primate Biol*. New York–London: 115–146.

Armstrong C, Lillie RD, 1934. Experimental lymphocytic choriomeningitis of monkeys and mice produced by a virus encountered in studies of the 1933 St. Louis encephalitis epidemic. *Publ Health Rep* 49: 1019–1027.

Arnsten AFT, Goldman-Rakic PS, 1985. Catecholamines and cognitive decline in aged nonhuman primates. *Ann New York Acad Sci* 444: 218–234.

Arnsten AFT, Goldman-Racic PS, 1998. Noise stress impairs prefrontal cortical cognitive function in monkeys – Evidence for hyperdopaminergic mechanism. *Arch Gener Psychiat* 55 (4): 360–362.

Arnt J, Skarsfeld T, Hyttel J, 1997. Differentiation of classical and novel antipsychotics using animal models. *Intern Clin Psychopharmacol* 12 (Suppl. 1): S9–S17.

Asch RH, Cotoulas IB, Smith G, Eddy CA, Balmaceda JP, 1984. Pituitary responsiveness LH-RH during gestation and puerperium in the Rhesus monkey. *Acta Europ Fertil* 15 (1): 15–23.

Asratyan EA, 1970. [About one deep idea of I. P. Pavlov]. *J vish. nerv. deyateln.* 20 (2): 269–279. (Russian).

Atassi MZ, Tarlowski DP, Paull JH, 1970. Immunochemistry of sperm whale myoglobin. VII. Correlation of immunochemical cross-reaction of eight myoglobins with structural similarity and its dependence on conformation. *Biochim et biophys* 221 (3): 623–635.

Autrup H, Thurlow BJ, Wakhisi J, Warwick GP, 1975. Microsomal drug-metabolizing enzymes in the olive baboon (*Papio anubis*). *Compar Biochem Physiol* B 50 (3): 385–390.

Avrutsky GY, Stepanyan-Tarakanova AM, eds, 1970. [Some aspects of information activities on psychopharmacology in USA. In: *New Psychopharmacological Preparates*]. M, VNIIMI: 3–29 (Russian).

Axelsson A, 1974. The vascular anatomy of the rhesus monkey cochlea. *Acta Oto-Laringol* 77 (6): 381–392.

Ayala EJ, 1980. Genetic and evolutionary relationships of apes and humans. *Life Sci Res Rep* 18: 147–162.

Ayers KM, 1988. Preclinical toxicology of zidovudine. An overview. *Amer J Med* 85 (2A): 186–188.

Azimov A, 1967. [*Short history of biology*]. M, Mir (Russian).

Bach J-F, 1991. Le singe est l'avenir de l'homme. *J intern. med.* # 200: 61, 65.

Bailey LL, 1987. Biologic versus bionic heart substituties. Will xenotransplantation play a role? *ASAIO* 10 (2): 51–53.

Bailey RC, Aunger R, 1990. Humans as primates: The social relationships of Efe pygmy men in comparative perspective. *Intern J Primatol* 11 (2): 127–146.

Bailey WJ, Fitch DH, Tagle DA, Czelusniak J, Slightom J, Goodman M, 1991. Molecular evolution of the Ψ_η-globin gene locus: Gibbon phylogeny and the hominoid slowdown. *Mol Biol Evol* 8 (2): 155–184.

Bailey WJ, Hayasaka K, Skinner CG, Kehoe S, Sieu LC, Slightom JL, Goodman M, 1992. Reexamination of the African hominoid trichotomy with additional sequences from the primate B-globin gene cluster. *Mol Phylogenet Evol* 1: 97–135.

Bakay RAE, Barrow DL, Fiandaca MS, Iuvone PM, Schiff A, Collins DS, 1987. Biochemical and behavioral correction of MPTP Parkinson-like sindrome by fetal cell transplantation. *Ann N. Y. Acad Sci* 495: 623–640.

Baker TG, 1972. Gametogenesis. *Acta endocrinol* 71 (Suppl 166): 18–41.

Bakker NPM, van Erck MG, Zurcher C, Faaber P, Lemmens A, Hazenberg M, Bontrop RE, Jonker M, 1990. Experimental immune mediated arthritis in rhesus monkeys. A model for human rheumatoid arthritis? *Rheumatol Intern* 10: 21–29.

Baldini A, Miller DA, Miller OJ, Ryder OA, Mitchel AR, 1991. A chimpanzee-derived chromosome-specific alpha satellite DNA sequence conserved between chimpanzee and human. *Chromosoma* 100 (3): 156–161.

Balner H, Gabb BW, Toth EK, Dersjant H, Vreeswijk W, 1973. The histocompatibility complex of rhesus monkeys. I. Serology and genetics of the Rhl-A system. *Tissue Antigens* 3 (4): 257–272.

Balzamo E, 1980. *Papio anubis*: Un primate parmi les primates. Etats ole vizilance et achvites pontogeniculo-corticales (PGC). *Electroenceph Clin Neurophysiol* 48: 694–705.

Balzamo E, 1981. Criteres (neurophysilogiques) utiles de differenciation des especes chez les primates: activites EEG et organisation des etats de vigilance. *Sci Tech Anim Lab* 6: 53.

Bankiewicz KS, Bringas JR, McLaughlin W, Pivirotto P, Hundal R, Yang B, Emborg ME, Nagy D, 1998. Application of gene therapy for Parkinson's disease: Nonhuman primate experience. *Advances in Pharmacol* 42: 801–806.

Barbas H, 1995. Anatomic basis of cognitive-emotional interactions in primate prefrontal cortex. *Neurosci Behavior Rev* 19 (3): 499–510.

Barchina TG, 1973. [Polish morphological journal 'Folia Morphologica' (the organ of Polish Anatomical Society) in 1971]. *Arch anat. gistol. i embriol.* 64 (7): 115–122 (Russian).

Bard KA, Platzman KA, Lester BM, Suomi SJ, 1991. Orientation to social and nonsocial stimuli in neonatal chimpanzees and humans. In: *Infant Behavior and Development* (USA), MS, 27 pp.

Bar-Joseph G, Safar P, Stezoski WS, Alexander H, Levine G, 1989. New monkey model of severe-volume controlled hemorrhagic shock. *Resuscitation* 17 (1): 11–32.

Barkaya VS, Lapin BA, Semenov LF, Strelin GS, Schmidt NK, 1964. [Therapy of the radiation disease in monkeys with autotransplantation of bone marrow from the protective shield site. In: *Pathogenesis,*

Clinics, and Therapy of the Acute Radiation Disease in Experiments on Monkeys]. Sukhumi (USSR): 39–46 (Russian).

Barmack NH, 1970. Dynamic visual acuity as an index of eye movement control. *Vis Res* 10 (12): 1377–1391.

Barnabas J, Goodman M, Moore JW, 1971. Evolution of hemoglobin in primates and other threian mammals. *Compar Biochem Physiol* **B** 39 (3): 455–482.

Barnabas J, Goodman M, Moore GW, 1972. Descent of mammalian α-globin chain sequences investigated by the maximum parsimony method. *J Mol Biol* 69: 249–278.

Barnabas S, Usha R, Guru Raw TN, Barnabas J, 1987. General relationships of mammalian order and evolutionary development of primates inferred best-fit α-globin phylogenies. *J Biosci* 12 (3): 165–174.

Barnard CN, Wolpowitz A, Losman JG, 1977. Heterotopic cardiac trans-plantation with a xenograft for assistance of the left heart in cardiogenic shock after cardiopulmonary bypass. *South Afr Med J* 52: 1035.

Barnett KC, Heywood R, Haguo PH, 1972. Colloid degeneration of the retina in a baboon. *J Compar Pathol* 82 (2): 117–118.

Barnicot NA, 1969. Comparative molecular biology of primates. A review. *Ann N. Y. Acad Sci* 162 (1): 25–36.

Barnicot NA, Wade PT, 1970. Protein structure and the systematics of Old World monkeys. In: *Old World Monkeys*. New York–London, Acad Press: 227–260.

Barret KE, Metcalfe DD, 1986. Mast cell geterogeneity: Studies in non-human primates. In: *Mast Cell Differentiation and Heterogeneity*. NY, Raven Press: 231–238.

Barriel V, Darlu P, Tassy P, 1993. A propos des conflits entre phylogenies morphoques et moleculaires: Deux exemples empruntes aux mammiferes. *Ann Sci Nat Zool Paris 13 series*, 14 (4): 157–171.

Barrow A, Burford SR, Forrest TJ, Hawkins AJ, Rose DA, Stevens PM, Vose CW, Walls CM, 1987. Studies on the disposition of a 5-nitroimidazole in laboratory animals. *Europ J Drug Metabol Pharmacokin* 12 (2): 85–90.

Barton N, Jones JS, 1983. Mitochondrial DNA: New clues about evolution. *Nature* 306 (5944): 317–318.

Bartus RT, 1993. General overview: Past contributions and future opportunities using aged nonhuman primates. *Neurobiol of Aging* 14 (6): 711–714.

Baru AW, 1978. [*Acoustic Centers and Identification of Sound Signals*]. Leningrad, Nauka (Russian).

Baskerville A, Newell DG, 1988. Naturally occurring chronic gastritis and C. pylori infection in the rhesus monkey: A potential model for gastritis in man. *Gut* 29 (4): 465–472.

Bassin FV, 1971. [An open letter to E. A. Asratyan]. *Voprosi phylosph* 4: 159–163 (Russian w/English sum).

Batuev AS, Kulikov GA, 1983. [*Introduction to Physiology of Sensory Systems*] M, Vishaya shkola (Russian).

Bauer K, 1974. Cross-reactions between human and animal plasma proteins. VI. An assay method for ape and monkey plasma proteins using antihuman antisera. *Humangenetik* 21 (3): 273–278.

Bayramyan EA, 1971. [To the anatomy of simian lungs and bronchi]. *Trudi Erevan med instituta* 15 (1): 75–79 (Russian).

Beard JM, Goodman M, 1976. The gemoglobins of *Tarsius bancanus*. In: *Mol Anthropol*. New York–London: 239–255.

Bechterew VM, 1906. [*The Bases of Studies of Brain Functions*]. St. Petersburg, Issues I–VII (Russian).

Beck BB, 1977. Köhler's chimpanzees – how did they raelly perform? *Zool Gart* 47 (5): 352–360.

Becker W, Fuchs AF, 1988. Lid-eye coordination during vertical gaze changes in man and monkey. *J Neurophysiol* 60 (4): 1227–1252.

Beevor CE, Horsley V, 1890. A record of the results obtained by electrical excitation of the so called motor cortex and internal capsule in an orang outang (*Simia satyrus*). *Proc Roy Soc* 48 (159).

Begun DR, Ward CV, Rose MD, eds, 1997. *Function, Phylogeny, and Fossils: Miocene Hominoid Evolution and Adaptation*. NY, Plenum Press.

Beisel WR, Stokes WS, 1983. Experimental infections in primates and human volunteers. In: *Experimental Bacterial and Parasitic Infections*. NY, Elsevier Sci Publish: 11–15.

Belkaniya GS, 1982. [*Functional System of Antigravitation*]. M, Nauka (Russian).

Belkaniya GS, Dartsmeliya VA, Galustyan MV, Demin AN, Kurochkin YuN, Sheremet IP, 1987. [Anthropophysiological basis of a species-specific stereotype of cardiovascular system reactivity in primates] *Vestn Akad Med Nauk SSSR* 10: 52–60 (Russian).

Belkaniya GS, Kurochkin YuN, Demin AN, Djemilev ZA, Gvindjiliya I, 1988. [Simians in medico-biological research of phylo-etagenesis] In: *5 All-Union Congr Gerontol and Geriat*. **Part 1**, Kiev: 64 (Russian).

Belkin RI, 1958. [I. I. Mechnikov's creative work in Darwinism. In: *Mechnikov. Selected Works on Darwinism*]. M: 307–374 (Russian).

Bellinger DA, Williams JK, Adams MR, Honore EK, Bender DE, 1998. Oral contraceptives and hormone replacement therapy do not increase the incidence of arterial thrombosis in a nonhuman primate model. *Arterioscl Trombos Vascul Biol* 18 (1): 92–99.

Beniashvili DS, 1994. *Experimental Tumors in Monkeys*. Boca Raton, CRC Press.

Benjamin LS, 1968. Harlow's facts on affects. *Voices. The Art and Sci of Psychother* 4 (1): 49–59.

Berglin L, Gouras P, Sheng Y-h, Lavid J, Lin P-K, Cao H-y, Kjeldbye H, 1997. Tolerance of human fetal retinal pigment epithelium xenografts in monkey retina. *Graefe's Arch Clin Exp Ophtalmol* 235 (2): 103–110.

Beritashvili IS, 1968. [*The Memory of the Vertebrates, Its Characteristics and Descent*]. 2 edition, M (Russian).

Bernard C, 1866. [*Introduction to Study of Experimental Medicine*]. St. Peterburg – Moscow (Russian).

Bhattacharyya MH, Larsen RP, Cohen N, Ralston LG, Moretti ES, Oldham RD, Ayres L, 1989. Gastrointestinal absorbtion of plutonium and uranium in fed and fasted adult baboons and mice: Application to humans. *Radiat Protect Dosim* 26 (1–4): 159–165.

Biedermann V, 1982. Serumproteine bei makaken – interspezifischer vergleich. *Versuchstierk* 24 (1–2): 88–89.

Bielicki T, 1987. [Evolution of hominid's brain]. *Kosmos* 36 (3): 545–562 (Polish).

Bigoni F, Stanyon R, Koehler U, Morescalchi AM, Wienberg J, 1997. Mapping homology between black and white colobine monkey chromomes fluorescent in situ hybridization. *Amer J Primatol* 42 (4): 289–298.

Bill A, Andersson SE, Almegard B, 1990. Cholecystokinin causes contraction of the pupillary sphincter in monkeys but not in cats, rabbits, rats and guinea-pigs: Antagonism by lorglumide. *Acta Physiol Scandin* 138 (4): 479–485.

Bischof P, Germain G, Cedard L, 1989. Pregnancy-associated plasma protein-A in pregnant cynomolgus monkeys (*Macaca fascicularis*): Radioimmunoassay, normal levels, effect of Ru-486, and preliminary characterization. *Biol Reprod* 4: 853–859.

Bishop PO, 1983. Vision with two eyes. *J Physiol Soc Jap* 45 (1): 1–18.

Bito LZ, 1997. Prostaglandins: A new approach to glaucoma management with a new, intriguing side effect. *Survey Ophtalmol* 41 (Suppl 2): S 1–S 14.

Bito LZ, Kaufman PL, DeRousseau CJ, Koretz J, 1987. Presbyopia: An animal model and experimental approaches for the study of the mechanism of accomodation and ocular ageing. *Eye* 1 (2): 222–230.

Bito LZ, Merritt SQ, DeRousseau CJ, 1979. Intraocular pressure of rhesus monkeys (*Macaca mulatta*). *Invest Ophtalmol Vis Sci* (St Louis) 18 (8): 785–793.

Biwasaka H, Saito S, Sasaki Y, Kumagai R, 1989. [Phosphoglucomutase types of animal red cells by isoelectric focusing]. *Res Pract Forens Med* 32: 23–29 (Japanese w/English sum).

Bjorntorp P, 1997. Stress and cardiovascular disease. *Acta Physiol Scand* (Suppl 640): 144–147.

Black CM, Mcdougal JS, Evatt BL, Reimer CB, 1991. Human markers for IgG2 and IgG4 appear to by on the same molecule in the chimpanzee. *Immunology* 72 (1): 94–98.

Blakeslee B, Jacobs G, 1984. Color vision in the ring-tailed lemur (*Lemur catta*). *Proc Austral Physiol Pharmacol Soc* 15 (1): 74.

Blancher A, Socha WW, 1994. Chimpanzee and gorilla counterparts of the human Rh blood group system. In: *Curr Primatol 1: Ecol and Evol*. B. Thierry, J. R. Anderson, J. J. Roeder, N. Herrenschmidt, eds. Strasbourg, Univ Louis Pasteur: 349–358.

Blancher A, Socha WW, 1997. The ABO, Hh and Lewis blood group in humans and nonhuman primates. In: *Mol Biology and Evol of Blood Group and MHC Antigens in Primates*. Blancher/Klein/Socha, eds. Berlin-Heidelberg, Springer-Verlag: 30–92.

Blanckaert C, 1989. L'anthropologie en France le mot et l'histoire (XVI–XIX) siecle). *Bull et mem Soc anthropol Paris* 1 (3–4): 13–44.

Blaton V, Vandamme D, Peeters H, 1972. Chimpanzee and baboon as biochemical models for human atherosclerosis. In: *Med Primatol 1972. Proc 3 Conf Exp Med Surg Primates 1972*, **Part III**. Basel, S. Karger: 306–312.

Blenden DC, Adldinger HK, 1985. Transmission and control of viral zoonoses in the laboratory. In: *Lab Safety: Princ and Pract*. Washington, DC: 72–89.

Blinkov SM, 1955. [*Peculiarityes of the Cerebrum Structure of Human. The Temporal Lobe of Man and Simians*]. M, Medgiz (Russian).

Bljacher LYa, 1971. [Georg Zeydlizc and his course of Darwinism in Derpt Univer. In: *From the History of Biology*]. **Iss3**. M: 5–58 (Russian).

Bloom KR, Zwick H, Houghton PW, 1986. The rhesus as a model for human spectral dynamic visual acuity. *Prim Rep* 14 (July): 20.

Blumcke I, Hof PR, Morrison JH, Gelio MR, 1990. Distribution of pervalbumin immunoreactivity in the visual cortex of Old World monkeys and humans. *Compar Neurol* 301 (3): 417–432.

Blume E, 1983. Street drugs yield primate Parkinson's model. *JAMA* 250 (1): 13–14.

Bodine-Fowler ST, 1994. Adaptation of skeletal muscle to spaceflight: Cosmos rhesus project, Cosmos 2044 and Cosmos 2229. *Report NO. NASA-CR-197041*. Univ of California San Diego Sch of Med, 19 pp.

Bodis-Wolner I, Ghilardi MF, van Woert M, Chong E, Glover A, Marx M, Onofrj M, Mylin LH, 1987. Visual processing in the MPTP treated monkey. *Neurosci* 22 (Suppl): S 299.

Bodkin NL, 1996. The rhesus monkey: Providing insight into obesity and diabetes. *Lab Anim* 25 (2): 33–36.

Bodkin NL, Ortmeyer HK, Hansen BC, 1996. Aging and glucoregulation in monkeys: Evidence for positive effects of calorie restriction. *IPS/ASP Congr Abstr*: # 754.

Boesch C, Boesch H, 1990. Tool use and tool making in wild chimpanzee. *Folia Primatol* 54 (1–2): 86–99.

Boesch-Achrmann H, Boesch C, 1994. Homization in the rainforest: The chimpanzees's piece of the puzzle. *Evol Anthropol* 3: 9–16.

Bonch-Osmolovsky GA, 1940. [New data about descent of man]. *Priroda*, 3: 63 (Russian).

Boothe RG, Kiorpes L, Gammon JA, Smith EL, Harwerth RS, Crawford ML, Eggers HM, 1986. Symposium: Primate models of abnormal visual development. *Prim Rep* 14 (July): 53.

Borges J, Zong Y, Tso MO, 1990. Effects of repeated photic exposures on the monkey macula. *Arch Ophtalmol* 108 (5): 727–733.

Borie DC, Poynard T, Hannoun L, 1996. [Xenotransplantation in man. Part 2: Control of rejection and current aspects of porcine liver transplantation in man]. *Gastroenterol cliniq et biologiq* 20 (11): 982–990 (French).

Borisenko OV, Kesarev VS, 1986. [Development of structure bases of the brain integrative activities. In: *Adaptive and Compensating Processes in Brain*]. M, Institute of Brain, Acad of Med Sci USSR: 7–8 (Russian).

Bornman MS, du Plessis DJ, Ligthelm AJ, van Tonder HJ, 1985. Histological changes in the penis of the chacma baboon – A model to study aging penile vascular impotence. *J Med Primatol* 14 (1): 13–18.

Borrell A, Ponsa M, Egozque J, Rubio A, Garcia M, 1998. Cromosome abnormalities in peripheral blood lymphocytes from *Macaca fascicularis* and *Erythrocebus patas* (Cercopithecidae, Catarrhini) after X-ray irradiation. *Mutation Res – Fundamental and Mol Mechanisms of Mutagenesis* 403 (1–2): 185–198.

Botchkarev PV, 1932. [*Simians in Sukhumi*]. Sukhumi, Abgiz (Russian).

Botchkarev PV, 1933. [Materials to study physiology of the female sexual secretion in simians]. *Arch biolog nauk* 33 (1–2): 263–269 (Russian).

Bourne GH, 1965. The move to Atlanta. *Yerkes Newslet* 2 (1): 3–9.

Bourne GH, 1971. *The Ape People*. NY, G. P. Putnam's Sons.

Bourne GH, 1974. The Yerkes Primate Research Center, Emory University. *Yerkes Newslet* 13 (2): 8–19.

Bourne GH, Bourne NG, Keeling ME, 1973. Breeding monkeys for biomedical research. *AGARD Conf Proc* 110: C6-1–C6-6.

Bourne GH, Sandler M, 1973. Atherosclerosis in chimpanzee. In: *The Chimpanzee* Vol 7. Basel, S. Karger; Baltimore, Univ Park Press: 248–264.

Bourliere F, 1982. Existe-t-il un modele animal du vieillissement humain? *Sci. et. tech. lab* 7 (2): 83–85.

Bowden DM, Johnson-Delney C, 1996. U.S. primate research is alive and well in the 1990s. *Contempor Topics* 35 (6): 55–57.

Bowen WN, 1996. Vaccine against dental caries: A personal view. *J Dental Res* 75 (8): 1530–1533.

Bowlez EA, Weaver DS, Telewski FW, Wakerfield AH, Jaffe MJ, Miller LC, 1985. Bone measurement by enhanced contrast image analysis: Ovariectomized and intact *Macaca fascicularis* as a model for human post-menopausal osteoporosis. *Amer J Phys Anthropol* 67: 99–103.

Bowmaker JK, 1998. Evolution of colour vision in vertebrates. *Eye* 12 (3b): 541–547.

Boyko VP, Manteyfel YuB, 1977. [The influence of the eyes turning on the visual perception of the tortoise *Emys orbicularis*]. *J evol biochim i physiol* 13 (1): 97–98 (Russian).

Boysen ST, Berntson GG, 1986. Cardiac correlates of individual recognition in the chimpanzee (*Pan troglodytes*). *J Compar Psychol* 100 (3): 321–324.

Boysen ST, Berntson GG, 1989. Numerical competence in a chimpanzee (*Pan troglodytes*). *J Compar Psychol* 103 (1): 23–31.

Boysen ST, Berntson GG, Prentice J, 1987. Simian scribbles: A reappraisal of drawing in the chimpanzee (*Pan troglodytes*). *J Compar Psychol* 101 (1): 82–89.

Boysen ST, Raskin LS, Berntson GG, 1991. Comprehension of spoken English by common chimpanzees (*Pan troglodytes*). *Amer J Primatol* 24 (2): 91.

Bradshaw JL, Rogers LJ, eds, 1993. *The Evolution of Lateral Asymmetries, Language Tool Use, and Intellect.* San Diego, Acad Press.

Brady AG, Gibson SV, Williams LE, Manatekar PA, 1996. Is the squirrel monkey (*Saimiri boliviensis boliviensis*) a model for intrauterine growth retardation? *IPS/ASP Congr Abstr*: # 648.

Brain P, Gordon J, 1971. Rosette formation by peripheral lymphocytes. II. Inhibition of the phenomenon. *Clin Exp Immunol* 8 (3): 441–449.

Brans YW, Kuehl TJ, Shannon DL, Reyes P, Menchaca EM, Puleo BA, 1985. Body water content and distribution in nonpregnant adult female baboons. *J Med Primatol* 14 (5): 263–270.

Bräuer G, Rimback K, 1990. Late archaic and modern *Homo sapiens* from Europa, Africa, and Southwest Asia: Craniometric comparisons and phylogenetic implications. *J Human Evol* 10 (8): 789–807.

Braunstein GD, Asch RH, 1986. Pregnancy-specific β_1-glycoprotein concentrations throughout pregnancy in the rhesus monkey. *J Clin Endocrinol Metabol* 62 (6): 1264–1270.

Brede HD, Murphy GP, 1972. Bacteriological and virological considerations in primate transplants. In: *Primates in Medicine* 7. E. Goldsmith, J. Moor-Jankowski, eds. Basel, S. Karger: 18–28.

Breier A, Su T-P, Saunders R, Carson RE, Kolachana BS, de Bartolomeis A, Weinberger DR, Weisenfeld N, Malhotra AK, Eckelman WC, Pickar D, 1997. Schizophrenia is associated with elevated amphetamine-induced synaptic dopamine concentrations: Evidence from a novel positron emission tomography method. *Proc Natl Acad Sci USA* 94 (6): 2569–2574.

Brewen JG, 1979. Cytogenetic studies on the effects of ionizing radiation on mammalian germ cells. *Proc 6 Intern Congr Radiat Res Tokyo*, 1979: 510–518.

Brewer SM, McGrew WC, 1990. Chimpanzee use of a tool-set to get honey. *Folia Primatol* 54 (1): 100–104.

Brink AJ, Lewis CM, Bosman AR, Lochner A, 1970. The baboon (*Papio ursinus*) heart (coronary blood supply, muscle function and metabolism). *Folia Primatol* 13 (1): 11–22.

Britten RJ, 1986. Rates of DNA sequence evolution differ between taxonomic groups. *Science* 231 (4744): 1393–1398.

Britten RJ, Kohne DE, 1968. Repeated sequences in DNA. *Science* 161: 529–540.

Brobmann GF, Ulano HB, Hinshaw LB, Jacobson ED, 1970. Mesenteric vascular responses to endotoxin in the monkey and dog. *Amer J Physiol* 219 (5): 1464–1467.

Brown P, Gibbs CJ, Jr, Rodgers-Johnson P, Asher DM, Sulima MP, Bacote A, Goldfarb LC, Gajdusek DC, 1994. Human spongiform encephalopathy: the National Institutes of Health series of 300 cases of experimentally transmitted disease. *Ann Neurol* 35 (5): 513–529.

Brown S, 1888. Experiments on special sense localizations in the cortex cerebri of the monkey. *Med Res (NY)* 34: 113–115.

Brown WM, Cann RL, Ferris SD, George M, Jr, Wilson AC, 1980. The assessment of genetic variation and evolutionary relationships using mitochondrial DNA analyses: Studies with higher primates. In: *ICSEB – II: 2 Intern Congr Syst and Evol Biol, Vancouver, 1980.* S1: 34.

Bruce EJ, Ayala FJ, 1979. Phylogenetic relationshps between man and the apes: Electrophoretic evidence. *Evolution* 33 (4): 1040–1056.

Buck AC, Lote CJ, SampsonWF, 1983. The influence of renal prostaglandins on urinary calcium excretion in idiopathic urolithiasis. *J Urol* 129: 421–426.

Buffon G, 1749–1788. *Histoir Naturelle* 1–36, Paris.

Buffon G, 1766. Nomenclature des singes. In: G. Buffon: *Histoir Naturelle* 14. Paris.

Buschang PH, 1982. The relative growth of the limb bones for *Homo sapiens* – as compared to anthropoid apes. *Primates* 23 (3): 465–468.

Bynum WF, 1973. The anatomical method, natural theology, and the functions of the brain. *ISIS* 64 (224): 445–468.

Byrd KE, 1981. Sequences of dental ontogeny and callitrichid taxonomy. *Primates* 22 (1): 103–118.

Byrd LD, 1974. Modification of the effects of chlorpromazine on the chimpanzee. *J Pharm Exp Ther* 189 (1): 24–32.

Byrne R, 1995. *The Thinking Ape: Evolutionary Origins of Intelligence.* Oxford, Oxford Univ Press.

Byrne R, Whiten A, eds, 1988. *Machiavellian Intelligence: Social Expertise and the Evolution of Intellect in Monkeys.* Clarendon.

Byrne RW, 1998. Chimpanzee and human cultures: Comment. *Curr Anthropol* 39 (5): 604–605 (up to 703).

Caballin MR, Miro R, Egozque J, 1983. Rearrangements in lymphocytes from man and *Pan troglodytes* induced by X and γ-radiation. *Clin Genet* 23 (3): 226.

Caccone A, Powell JR, 1989. DNA divergence among hominoids. *Evolution* 43 (5): 925–942.

Cai J, Tian Y, Ma Y, 1990. A comparative study on cognitive function of the brain in the slow loris (*Nycticebus coucang*) and rhesus monkey (*Macaca mulatta*). *Prim Rep* 26: 73.

Call J, Tomasello M, 1994. The social learning of tool use by orangutans (*Pongo pygmaeus*). *Human Evol* 9 (4): 297–313.

Call J, Tomasello M, 1996. Object permanence in chimpanzee, orangutans and 18-month-old human children. *IPS/ASP Congr Abstr:* # 274.

Calmette A, 1924. Conference sur l'utilisation des singes en médicine expérimentale. La laboratoire Pasteur de Kindia (Guinee Française). *Bull Soc pathol exot* 17: 10–19.

Cammer TJ, Malsbury DW, Tsin ATC, 1990. Vitamin A metabolism in the baboon eye. *Brain Res Bull* 24 (6): 755–757.

Campbell CB, Burgess P, Roberts SA, Dowling HR, 1972. Rhesus monkeys to study biliary secretion with and intact enterohepatic circulation. *Austral New Zel J Med* 2 (1): 49–56.

Campbell CC, Collins WE, Milhous WK, Roberts JM, Armstead A, 1986. Adaptation of the Indochina I/CDC strain of Plasmodium falciparum to the squirrel monkey (*Saimiri sciureus*). *Amer J Trop Med Hyg* 35 (3): 472–475.

Campbell DB, 1981. The use and abuse of human pharmacokinetics. *Medicographia* 3 (1): 27–30.

Campos M, Pal S, O'Brien TP, Tailor HR, Prendergast RA, Whittum-Hudson JA, 1995. A chlamydial major outer membrane protein extract as a trachoma vaccine candidate. *Investigative Ophtalmol Visual Sci* 36 (8): 1477–1491.

Candland DK, 1987. Tool use. In: *Comparative Primate Biology 2B: Behavior, Cognition and Motivation.* NY, Alan R. Liss: 85–100.

Carey KD, Rice K, 1996. The aged female baboon as a model of menopause. *IPS/ASP Congr Abstr:* # 079.

Carlson EC, Surerus KK, Hinds D, 1986. Ultrastructural analysis of major basement membranetypes in rhesus monkey *Macaca mulatta* acellular renal cortex. *Acta Anat* 125: 14–22.

Carmichael L, 1969. The past, present and future of primatology. In: *Proc 2 Intern Congr Primatol, Atlanta 1968,* 1. Basel/New York. S. Karger: XI–XIV.

Carneiro de Moura A, Pinto de Carvalho A, 1958. Pielografia comparada nos mamiferos. *Gaz med Portug* 11 (2): 211–223.

Carpenter CR, 1940. Rhesus monkeys (*Macaca mulatta*) for American laboratories. *Science* 92: 34.

Carpenter CR, 1972. Breeding colonies of macaques and gibbons on Santiago Island, Puerto Rico. In: *Breeding Primates.* Basel/New York, S. Karger: 76–87.

Carter DB, Timmins JG, Adams LD, Lewis RW, Karr JP, Resnick MI, Buhl AE, 1985. The antigenic relatedness of proteins from human and simian prostate fluid. *Prostate* 6 (4): 395–402.

Casey DE, 1996. Behavioral effects of sertindole, risperidone, clozapine and haloperdol in *Cebus* monkeys. *Psychopharmacol* 124: 134–140.

Castleman W, Dunqworth D, Tyler W, 1976. Lesions hyperplasiques et inflammatoires des voies aeriennes pulmonaires de primates, suivant une exposition subaigue a de faibles taux d'ozone (0, 2 ppm). *Bull Union intern contr tuberc* 51 (1/2): 599–601.

Castner SA, Goldman-Rakic PS, 1996. Amphetamine psychosis in male and female rhesus monkeys. *Biol Psychiat* 39 (7): 590–591.

Castracane VD, 1996. Late luteal rescue in the baboon: A surrogate for human reproductive endocrinology. *IPS/ASP Congr Abstr*: # 075.

Cavaille F, Janmot C, Ropert S, d'Albis A, 1986. Isoforms of myosin and actin in human, monkey and rat myometrium. Comparison of pregnant and non-pregnant uterus proteins. *Europ J Biochem* 160 (3): 507–513.

Cavanagh D, Rao PS, 1969. Endotoxin shock in the subhuman primate. I. Hemodynamic and biochemical changes. *Arch Surg* 99 (1): 107–112.

Cazaubon C, Carlet C, Richaut JP, Nisato D, Corvol P, Gagnol JP, 1986. Effet hypotenseur d'un anticorps monoclonal anti-réñine humaine (4G1D8) chez le marmouset vigile désodé. *Arch malad coeur et valss* 79 (6): 840–846.

Cěch S, 1964. Srovnávaci mikroskopická anatomie rohovky. *Scripta medica* 37 (6–7): 265–281.

Cefalu WT, Wagner JD, 1997. Aging and atherosclerosis in human and nonhuman primates. *Age* 20 (1): 15–28.

Cejková J, Bolkova A, 1974. Differences in hydration characteristics of corneas in various animal species. Histochemical study of acid mucopolysaccharides. *Albrecht v. Graeve's Arch klin exp Ophtalmol* 190 (4): 353–360.

Chambers KC, Phoenix CH, 1989. Apomorphine, deprenyl, and yohimbine fail to increase sexual behavior in rhesus males. *Bechav Neurosci* 103 (4): 816–823.

Chamove AS, Anderson JR, 1981. Self-aggression, stereotype, and self-injurious behaviour in man and monkeys. *Curr Psychol Rev* 1: 245–256.

Chang CS, Sassa S, Doyle D, 1984. An immunological study of S-aminolevulinic acid dehydratase specifity consistent with the phylogeny of species. *Biochim Biophys Acta*, 797: 297–301.

Chanock RM, 1996. Reminiscenes of Albert Sabin and his successful strategy for the development of the live oral poliovirus vaccine. *Proc Ass Amer Physic* 108 (2): 117–126.

Charin GM, 1979a. [Morphologic reconstruction of monkey lymphatic nodes in ontogeny. In: *Macro-Microstructure of Tissue in Norm, Pathol, and Experim* 6]. Cheboksari: 30–34 (Russian).

Charin GM, 1979b. [Age morphology of baboon hamadryas timus] *Arch anat histol i embryol* 6: 34–39 (Russian).

Chatchaturyan AA, 1988. [*Comparative Anatomy of the Cerebrum Cortex in Man and Simians*] M, Nauka (Russian).

Chausse A-M, Muller JY, Ruffie J, Lucotte G, 1984. Comparison of EcoRI restriction MHC genes in man and several primate species. *Biochem Syst Ecol* 12 (2): 245–250.

Chercovich GM, 1994. Some biological normals in monkeys (macaques and baboons). *Congr IPS* 15: 221.

Chercovich GM, Elinek I, Capek K, Efremova ZK, 1977. [Effect of chronic excessive salt intake on arterial blood pressure in monkeys]. *Vestn Acad Med Nayk SSSR*, 8: 25–31 (Russian).

Chercovich GM, Lapin BA, 1973. Modeling of neurogenic diseases in monkeys. In: *Nonhuman Primates and Medical Research.* New York–London, Acad Press: 307–327.

Chevalier-Skolnikoff S, 1983. Sensorimotor development in orang-utans and other primates. *J Human Evol* 12 (6): 545–561.

Chiarelli B, 1962. Comparative morphometric analysis of primate chromosomes. 1. The chromosomes of anthropoid apes and of man. *Caryologia* 15 (1): 99–121.

Chiarelli B, 1967. Caryological and hybridological data available for a taxonomic revision of tha Old World primates at a supergenetic level. In: *Progress in Primatol.* Stuttgart: 160–163.

Chiarelli B, 1973. I cromosomi del primati e l'origine del kariotipo umano. *Sapere* 74 (765): 13–15.

Chiarelli B, 1974. The study of chromosomes. In: *Perspect Primate Biol.* New York–London, Acad Press: 151–176.

Chiarelli B, 1985. Chromosomes and the origin of man. In: *Hominid Evol: Past, Present and Future.* NY, Alan R. Liss: 397–400.

Chiba N, Tamura T, Koizumi K, Tanigawa M, Shiba M, 1990. Experimental cedar pollinosis in rhesus monkeys. *Intern Arch Allergy and Appl Immunol* 93 (1): 83–88.

Chipens GI, Gnilomedova LE, Ievinya NG, Kudryavzev OE, Rudzish RV, Sclyarova SN, 1990. [Profiles of homologic proteins in philogenic near animal species reflect of the genetic code simmetrical structure]. *J evol biochim i physiol* 26 (2): 252–258 (Russian).

Chism J, Rowell TE, Richards SM, 1978. Daytime births in captive pates monkeys. *Primates* 19 (4): 765–767.

Chlenov, 1902. [Current status of the syphilis microbiology study]. *Russ arch patol clin med bacteriol* 14: 709–727 (Russian).

Christiaens GCM, 1982. J. E. Markee: Menstruation in intraocular endometrial transplant in the rhesus. *Europ J Obstetr Gynecol Reprod Biol* 14: 63–65.

Chromov BM, 1978. [Physiological role of appendix vermiformis]. *Clin chirurg* # 4: 65–69 (Russian).

Ciochon RL, 1983. Hominoid cladistics and the ancestry of modern apes and humans. In: *New Interpretation of Ape and Human Ancestry.* R. L. Ciochon, ed. NY, Plenum Press: 783–843.

Ciochon RL, Corruccini RS, 1979. Morphometric analysis of platyrrhine femora with taxonomic implications and notes on two fossil forms. *J Human Evol* 4 (3): 197–217.

Ciochon RL, Corruccini RS, 1982. Miocene hominoids and new interpretations of ape and ancestry. In: *Advanced Views in Primate Biology.* Berlin–Heidelberg–New York, Springer Verlag: 149–159.

Clapp N, ed, 1993. *A Primate Model for the Study of Colitis and Colonic Carcinoma: The Cotton-Top Tamarin (Saguinus oedipus).* Boca Raten, CRC Press.

Clapp NK, Henke ML, Lushbaugh CC, Humason GL, Gangaware BL, 1988. Effect of various biological factors on spontaneous marmoset and tamarin colitis. A retrospective histopathologic study. *Digest Dis Sci* 33 (8): 1013–1019.

Clark AS, Goldman-Rakic PS, 1989. Gonadal hormones influence the emergence of cortical function in nonhuman primates. *Behav Neurosci* 103 (6): 1287–1295.

Clark EA, Martin PJ, Hansen JA, Ledbetter JA, 1983. Evolution of epitopes on human and nonhuman primate lymphocyte cell surface antigens. *Immunogenetics* 18 (6): 599–615.

Clark Le Gros WE, 1958. *Early Forerunners of Man: A Morphological Study of the Evolutionary Origin of the Primates.* Chicago.

Clarkson TB, 1998. Nonhuman primate models of atherosclerosis. *Lab Anim Sci* 48 (6): 569–572.

Clarkson TB, Antohny MS, Klein KP, 1996. Hormone replacement therapy and coronary artery atherosclerosis: The monkey model. *Brit J Obstet Gynecol* 103 (Suppl 13): 53–58.

Clarkson TB, Anthony MS, Jerome CP, 1998. Lack of effect of raloxifene on coronary artery atherosclerosis of postmenopausal monkeys. *J Clin Endocrinol Metabol* 83 (3): 721–726.

Clarkson TB, Prichard RW, Bullock BC, Lehner NDM, Lofland HB, St Clair RW, 1970. Animal models of atherosclerosis. In: *Animal Models for Biomedical Research. III. Proc of a Sympos.* Washington, Natl Acad Sci: 22–41.

Clarkson TB, Taunsey G, 1995. Cynomolgus monkey for the study of hormonal influences on atherosclerosis. In: *Atherosclerosis* X, E. P. Woodford, J. Davignon, A. Sniderman, eds. NY, Elsevier Sci Publ: 132–139.

Clarren SK, Bowden DM, Astley SJ, 1987. Pregnancy outcomes after weekly oral administration of ethanol during gestation in the pig-tailed macaque (*Macaca nemestrina*). *Teratology* 35 (3): 345–354.

Clement J-L, Sottan M, Gastaldi G, le Pareux A, Hagege R, Connet J, 1980. Etude comparative des poils des singes anthropoides et de l'homme. *L'Anthropologie* (Paris) 84 (2): 243–253.

Clemente IC, Garcia M, Ponsa M, Egozcue J, 1987. High-resolution chromosome banding studies in *Cebus apella, Cebus albifrons* and *Lagothrix lagothricha*: Comparison with the human karyotype. *Amer J Primatol* 13 (1): 23–26.

Clevenger AB, Marsh WL, Peery TM, 1971. Clinical laboratory studies of the gorilla, chimpanzee, and orangutan. *Amer J Clin Pathol* 55 (4): 479–488.

Cogan DG, Witt ED, Goldman-Rakic PS, 1985. Ocular signs in thiamine–deficient monkeys and in Wernicke's disease in humans. *Arch Ophtalmol* 103, August: 1212–1220.

Cohen AJ, 1979. Critical review of the toxicology of coumarin with special reference to interspecies differences in metabolism and response and their significance to man. *Food Cosmet Toxicol* 17 (3): 277–289.

Coimbra-Filho AF, Mittermeier RA, 1981. *Ecology and Behavior of Neotropical Primates* 1. Brazil Acad Sci, Rio de Janeiro.

Coleman H, 1971. Comparison of the pelvic growth patterns of chimpanzee and man. In: *Proc 3 Intern Primatol, Zurich 1970.* 1. Basel, S. Karger: 176–182.

Collins JF, de los Santos R, Johanson WG, 1986. Acute effects of oleic acid-induced lung injury in baboons. *Lung* 164 (5): 259–268.

Collins PM, Pudney J, Tsanq WN, 1987. Postnatal differentiation of the gametogenic and endocrine functions of the testis in the tree-shrew (*Tupaia belangeri*). *Cell Tiss Res* 250 (3): 681–686.

Colyn M, Gautier-Hion A, Thys van den Audenaerde D, 1991. *Cercopithecus dryas* Schwartz, 1932, and *C. salongo* Thys van den Audenaerde, 1977, are the same species with an age-related coat pattern. *Folia Primatol* 56 (3): 167–170.

Concluding remarks, 1972. In: *The Use of Non-Human Primates in Research of Human Reproduction.* Stockholm, WHO-Centre: 513–519.

Condo GJ, Casagrande VA, 1990. Organization of cytochrome oxidase staining in the visual cortex of nocturnal primates (*Galago crassicaudatus* and *G. senegalensis*). 1. Adult patterns. *J Compar Neurol* 293 (4): 632–645.

Congress IPS 17, 1998. *Abstracts.* Antananarivo.

Connor JM, Darracq MA, Roberts J and Tuszynski MH, 2001. Nontropic actions of neurotrophins: subcortical nerve growth factor gene delivery reverses age-related degeneration of primate cortical cholinergic innervation. *Proc Natl Acad Sci USA* 98: 1941–1946.

Conroy GC, 1990. *Primate Evolution.* NY, W. W. Morton.

Coolidge HJ, 1984. Historical remarks learning on the discovery of *Pan paniscus*. In: *The Pygmy Chimpanzee: Evolutionary Biology and Behavior.* New York – London, Plenum Press: IX–XII.

Cooper DKC, Ye Y, Niekrasz M, 1994. Heart transplantation in primates. In: *Handbook of Animal Models in Transplantation Research.* D. V. Cramer, L. G. Podesta, L. Makowka, eds. Boca Raton, FL, CRC Press Lewis Publ: 173–200.

Coppenhaver DH, Dixon JD, Duffy LK, 1983. Prosimian hemoglobins. I. The primary structure of the β-globin chain of *Lemur catta. Haemoglobin* 7 (1): 1–14.

Corbey R, Theunissen B, eds, 1995. *Ape, Man, Apeman: Changing Views since 1600.* Leiden, Leid Univ.

Cornblath DR, Hillman MA, Striffler JS, Herman CN, Hansen BC, 1989. Peripheral neuropathy in diabetic monkeys. *Diabetes* 38 (11): 1365–1370.

Cornelius CE, 1988. Animal models. In: *The Liver: Biology and Pathobiology.* NY, Raven Press: 1315–1336.

Corruccini RS, 1982. Dentino-enamel junction and primate relationships. *Intern J Primatol* 3 (3): 271.

Coulston F, 1966. Qualitative and quantitative relationships between toxicity of drugs in man, lower mammals, and nonhuman primates. In: *Proc Conf on Nonhuman Primate Toxicol.* Warrenton, Virginia: 3–23.

Cowen D, 1986. The melanoneurons of the human cerebellum (*Nucleus pigmentosus cerebellaris*) and homologues in the monkey. *J Neuropathol Exp Neurol* 45 (3): 205–221.

Craggs MD, Rushton DN, Stephenson JD, 1986. A putative non-cholinergic mechanism in urinary bladders of New but not Old World primates. *J Urology* 136 (6): 1348–1350.

Créau-Goldberg N, London J, Cochet C, Rahuel C, Cartron JP, Turleau C, de Grouchy J, 1989. Evidence of genic gomology between human and primate glycophorins and localization on homologous chromosomes in the capuchin monkey. *Cytogenet Cell Genet* 51 (1–4): 981.

Créau-Goldberg N, Turleau C, Cochet C, de Grouchy J, 1983. New gene assignments in the baboon and new chromosome homoeologies with man. *Ann génét* 26 (2): 75–78.

Créau-Goldberg N, Turleau C, Cochet C, Huerre C, Junien C, de Grouchy J, 1984. Conservation of the human COLIAI-TK-GAA synteny and homoeologous assignment in the African green monkey and the baboon (Cercopithecidae). *Human Genet* 68 (4): 333–336.

Cronin JE, 1989. Molecular clocks and the fossil record: Predictions for Old World monkeys. *Amer J Phys Anthropol* 78 (2): 208.

Cronin JE, Meikle WE, 1979. The phyletic position of Theropithecus: Congruence among molecular, morphological, and paleontological evidence. *Syst Zool* 28 (3): 259–269.

Cronin JE, Meikle WE, 1982. Hominid and gelada baboon evolution: Agreement between molecular and fossil time scale. *Intern J Primatol* 3: 469–482.

Cronin JE, Sarich VM, 1975. Molecular systematics of the New World monkeys. *J Human Evol* 4 (15): 357–375.

Cronin RJ, Solomon S, Klingler EL, Jr, 1970. Renal function in *Macaca speciosa*. *Compar Biochem Physiol* 37 (4): 511–516.

Culbertson WW, Tabbara Kf, O'Connor GR, 1982. Experimental ocular toxoplasmosis in primates. *Arch Ophtalmol* 100 (2): 321–323.

Cushing H, 1969. Letter from Harvey Cushing. *JAMA* 207 (10): 1862.

Czekala NM, Shideler SE, Lasley BL, 1988. Comparisons of female reproductive hormone patterns in the hominoids. In: *Orangutan Biology*. NY, Oxford Univ Press: 177–122.

Czelusniak J, Goodman M, 1995. Hominoid phylogeny estimated by model selection using goodness of fit significance tests. *Mol Phylogenet Evol* 4: 283–290.

Dadoune J-P, Alfonsi M-F, 1989. Nuclear changes during spermiogenesis in man and the monkey. In: *Developments in Ultrastructure of Reproduction (Progress Clin Biol Res 296)*. NY, Alan R. Liss: 165–170.

Dafeldecker WP, Pares X, Vallee BL, Bosron WF, Li T-K, 1981. Simian liver alcohol dehydrogenase: Isolation and characterization of isoenzymes from *Saimiri sciureus*. *Biochemistry* 20 (4): 856–861.

Dagosto M, 1984. Rev: Passingham R. E. The Human Primate. Oxford, Freeman, 1982. *Folia Primatol* 42 (1): 84.

Dai M-j, Mc Garvie L, Kozlovskaya I, Raphan T, Cohen B, 1994. Effects of spaceflight on ocular counter-rolling and the spatial orientation of the vestibular system. *Exp Brain Res* 102 (1): 45–56.

Dai M-j, Raphan T, Kozlovskaya I, Cohen B, 1996. Modulation of vergence by off-vertical yaw axis rotation in the monkey: Normal characteristics and effects of space flight. *Exp Brain Res* 111 (1): 21–29.

Daiger SP, Goode ME, Trombridge BD, 1987. Evolution of nuclear families in primates. Copy-number variation in the argininosuccinate synthetase (ASS) pseudogene family and the anonymous DNA sequence, D1S1. *Genetica* 73 (1–2): 91–98.

Damian RT, Lichter EA, 1972. Immunological studies in chimpanzee plasma proteins. *Primates in Med* 6. Basel, S. Karger: 1–66.

Dandelot P, 1959. Note sur la classification des cercopithèques du groupe aethiops. *Mammalia* 23 (3): 357–368.

Dandieu S, Lucotte G, 1984. Cartographie de restriction comparee de la sequence d'ADN hautement répétés coupee par BamHl chez *Papio papio* et *P. cynocephalus*. *Biochem Syst Ecol* 12: 441–449.

Dandieu S, Rahuel C, Ruffie J, Lucotte G, 1984. Comparaisons des séquences d'ADN hautement répétés chez l'homme et diverses especes de primates. *Biochem Syst Ecol* 12 (2): 237–244.

Danilova V, Hellekant G, Ninomiya Y, Roberts T, 1996. The sense of taste in *Macaca mulatta*: Responses of single taste fibers of chorda tympani and giossopharyngeal nerves to an extended array of stimuli, including ethanol. *IPS/ASP Congr Abstr*: # 708.

Darian-Smith I, 1982. Touch in primates. *Ann Rev Psychol* 30 (Palo Alto, Ca): 155–194.

Darwin C, 1859. *On the Origin of Species by Means of Natural Selection, or the Preservation of Favoured Races in the Struggle for Life*. London: J. Murray. (The same in Russian, St. Petersburg, 1864).

Darwin C, 1871. *The Descent of Man and Selection to Sex*. London. (The same in Russian, St. Petersburg, 1871).

Darwin C, 1876. Sexual selection in relations in monkeys. *Nature* 15: 18–19.

David F, Ruffié J, Lucotte G, 1987. Comparaison des séquences de fragments de restriction répétés spe-cifiques du chromosome Y humain chez le chimpanze et le gorille. *Biochem Syst Ecol* 15 (4): 511–514.

Davidson LA, Lonnerdal B, 1986. Isolation and characterization of rhesus monkey milk lactoferrin. *Pediatr Res* 20 (2): 197–201.

Davis FB, Kite JH, Davis PJ, Blas SD, 1982. Thyroid hormon stimulation in vitro of red blood cell Ca^{2+}-ATPase activity: Interspecies variation. *Endocrinology* 110 (1): 297–298.

Davidovskiy IV, 1969. [The problem of modelling in pathology. In: *Modelling in Biology and Medicine*]. Leningrad, Medicine: 28–40 (Russian).

De Bonis L, Jaeger J-J, Coiftait B, Coiftait P-E, 1988. Découverte du plus ancien primate Catarrhinien connu dans l'Eocène superier d'Afrique du Nord. *C. r. Acad sci Paris* 306 (ser II): 929–934.

De Campos MF, Rodrigues F, 1987. Rats and marmosets respond differently to serotonin agonists and antagonists. *Psychopharmacol* 92: 478–483.

De Courten C, Garey LJ, 1981. Similarities in neuronal development in the lateral geniculate nucleus of monkey and man. *Proc Intern Congr, Zürich 1981 (Neurogenet and Neuro-Ophtalmol)*. Amsterdam e.a.: 125–128.

De Forrest JM, Waldron TL, Oehl RS, Scalese RJ, Free CA, Weller HN, Ryono DE, 1989. Pharmacology of novel imidazole alcohol inhibitors of primate renin. *J Hypertension* 7 (2, Suppl): 15–19.

De Grouchy J, 1979. L'evolution des etres organises et la naissance des especes. *Biomedicine* 30 (3): 129–134.

De Grouchy J, 1987. Chromosome phylogenies in man, great apes, and Old World monkeys. *Genetica* 73 (1–2): 37–52.

De Grouchy J, Turleau C, Roubin M, Klein M, 1972. Evolutions caryotypiques de l'homme et du chimpanzé. Etude comparative des topographies de bandes apres denaturation menageé. *Ann génét* 15 (2): 79–84.

De Jong WW, Goodman M, 1988. Anthropoid affinities of *Tarsius* supported by lens αA-crystallin sequences. *J Human Evol* 17 (6): 575–582.

De Kock MLS, Burger EG, 1985. A histological study of the urethra of the male baboon: Is it similar to man's? *J Urology* 134 (3): 617–619.

De Kretser TA, Lee GFH, Thorne HJ, Jose DG, 1986. Monoclonal antibody CI-panHu defines a pan-human cell-surface antigen unique to higher primates. *Immunology* 57: 579–585.

De Lemos RA, Kuehl TJ, 1987. Animal models for evalution of drugs for use in the mature and immature newborn. *Pediatr* 79 (2): 275–280.

De Lond AF, Smyth RD, Polk A, Nayak RK, Martin G, Douglas GH, Reavey-Cantwell NH, 1977. Comparative metabolism of fenclorac in rat, dog, monkey, and man. *Drug Metabol and Disposit: Biol Fate Chem* 5 (2): 122–131.

De Queiroz K, Gauthier J, 1992. Phylogenetic taxonomy. *Ann Rev Ecol Syst* 23: 449–480.

De Saban R, Cabanis EA, Iba-Zizen M-T, Rinjard J, Villiers PA, Leclerc-Cassan M, Lopez A, 1989. Biometrie in vivo de la cavite orbitaire des catarrhiniens en tomodensimetrie. *Can. anthropol. et biom. hum.* 7 (1–2): 81–108.

De Valois RL, Morgan HC, Polson MC, Mead WR, Hult EM, 1974. Psychophysical studies of monkey vision. I. Macaque luminosity and color vision test. *Vis Res* 14 (1): 53–67.

De Valois RL, Morgan HC, Snodderly MD, 1974. Psychophysical studies of monkey vision. III. Spatial luminance contrast sensitivity tests of macaque and human observers. *Vis Res* 14 (1): 75–81.

De Waal FBM, 1995. Bonobo sex and society. *Scientific American*, March, 272: 50–64.

De Waal FBM, 2001. *The Ape and Sushi Master*. NY, Basic Books: 154, 194.

Deacon TW, 1989. The homologues to human language circuits in monkey brains. *Amer J Phys Anthropol* 78 (2): 210–211.

Dean RL, Bartus RT, 1985. Animal models of geriatric cognitive disfunction: Evidence for an important cholinergic involvement. In: *Senile Dementia of the Alzheimer Type*. J. Traber, W. H. Gispen, eds. Berlin-Heidelberg, Springer-Verlag: 269–287.

Decat G, Leonard A, de Meurichy W, 1982. Ginétique cellulaire et radiosensibilité chromosomique des lymphocytes de quatre espéces de cercopithèques. *C. r. Soc biol* 176 (3): 373–377.

Delahunt CS, Lessen LJ, 1964. Thalidomide syndrome in monkeys. *Science* 146: 1300.

Deller M, 1979. Are amblyopic man and ape related? *Trends in Neurosci* 2 (Special Vis Issue, August): 216–218.

Deloison Y, 1985. Comparative study of calcanei of primate and *Pan-Australopithecus-Homo* relationship. In: *Hominid Evolution: Past, Present and Future*. NY, Alan R. Liss: 143–147.

Delson E, 1977. Catarrhine phylogeny and classification: Principles, methods and comments. *J Human Evol* 6 (5): 443–459.

Delson E, 1982. (On the nomenclature of *Papio, Mandrillus*, Colobinae . . .). *Lab Primate Newslet* 21 (4): 8–9.

Delson E, Dean D, 1991. The systematics of *Theropithecus*: Major lineages and relationships within the Papionini. *Amer J Phys Anthropol* (Suppl 12): 67.

Delson E, Rosenberger AL, 1984. Are there any anthropoid primate living fossils? In: *Living Fossils*. NY e.a.: 50–61.

Deneau G, Yanagita T, Seevers MH, 1969. Self-administration of psychoactive substances by the monkey. A measure of psychological dependence. *Psychoparmacol* 16: 30–48.

Deniker J, 1886. Recherches anatomiques et embryologiques sur les Singes Anthropoïdes. *Arch de Zool expér.*, 3 Sér. T. III, bis supplem. Thése de doctorat ès Sciences, in 8°, 265 pp, 9 pl. et fig.

Deo MG, Ramalingaswami V, 1977. Non-human primate kwashiorkor. In: *Use of Non-Human Primates in Biomedical Research, New Delhi, India 1975*. New Delhi, Indian Natl Sci Acad: 270–287.

Deriagina MA, 1981. [Etological analysis of ape manipulate activitys in relation to the problem of origin the instrument activity]. *Voprosi anthropol*, # 7: 100–112 (Russian).

Deriagina MA, Efremova NS, 1998. Ethological approach in analysis of taxonomy of *Tupaja* and Strepsirhini. *Congr IPS 17 Abstr*. # 375.

DeRousseau CJ, 1994. Primate gerontology: An emerging discipline. In: *Biological Anthropol and Aging: Perspect on Human Variat over the Life Span*. D. E. Crews, R. M. Garruto, eds. NY, Oxford Univ Press: 127–153.

DeRousseau CJ, Bito LZ, Kaufman PL, 1983. Age-dependet impairment of the rhesus visual and musculoskeletal systems and possible behavioral consequences. *Amer J Primatol* 4 (4): 329–330.

Devor EJ, Dill-Devor RV, Magee HJ, Waziri R, 1998. Serine hydroxyme-thyltransferase pseudogene, SHMT-ps1: A unique marker of the order Primates. *J Exper Zoology* 282 (1–2): 150–156.

D'Hooghe TM, 1997. Clinical relevance of the baboon as a model for the study of endometriosis. *Fertil and Steril* 68 (4): 613–625.

Diamond JM, 1988. DNA-based phylogenies of the three chimpanzees. *Nature* 332 (6166): 685–686.

Dienske H, 1984. Early development of motor abilities, daytime sleep and social interactions in the rhesus monkey, chimpanzee and man. *Clin Develop Med* 94: 126–143.

Djian F, Green H, 1989. Vectorial expansion of the involucrin gene and the relatedness of the hominoids. *Proc Natl Acad Sci USA* 86 (21): 8447–8451.

Dijkmans R, van Damme J, de Ley M, Billiau A, 1986. Characterization of baboon interferons (α, β, γ) and isolation of an interferon-γ cDNA clone. *Arch intern physiol et biochim* 94 (1): B 18.

Dixson AF, 1980. Androgens and aggressive behavior in primates. *Aggress Behav* 6 (1): 37–67.

Dobroruka LJ, 1985. Abnormales Verhalten, Orgasmus und Trachtigkeitsdauer beim Vervet-Weibchen. *Zool Garten* (Jena) 55 (5/6): 341–342.

Dodson TB, Bays RA, Pfeffle RC, Barrow DL, 1997. Cranial bone graft to reconstruct the mandibular condyle in *Macaca mulatta. J Oral Maxillofac Surg* 55 (3): 260–276.

Domen RE, 1986. Discovery of the Rh red blood cell antigen. *Arch Pathol Lab Med* 110 (2): 162–164.

Donaldson VH, Pensky J, 1970. Some observations on the phylogeny of serum inhibitor of C'1 esterase. *J Immunology* 104 (6): 1388–1395.

Donovan BT, van der Werften Bot DD, 1974. [*Physiology of Sex Development*]. M, Pedagogika (Russian).

Doolittle RF, Mross GA, 1970. Identity of chimpanzees with human fibrinopeptides. *Nature* 225 (5253): 643–644.

Dormehllrene C, Jacobs DJ, Pretorius JP, Maree M, Franz RC, 1987. Baboon (*Papio ursinus*) model to study deep vein thrombosis using 111-indium-labeled autologous platelets. *J Med Primatol* 16 (1): 27–38.

Dorozynski A, 1982. Une machine a remonter le cours de l'evolution. *Sci et vie* 131 (774): 36–41.

Downing LJ, Wakefield TW, Strieter RM, Prince MR, Londy FJ, Fowlkes JB, Hulin MS, Kadell AM, Wilke CA, Brown SL, Wrobleski SK, Burdick MD, Anderson DC, Greenfield LJ, 1997. Anti-P-selectin antibody decreases inflammation and thrombus formation in venous thrombosis. *J Vascul Surg* 25 (5): 816–828.

Doyle E, 1981. Comparative pharmacokinetics in nonhuman primates. In: *1 Congr Europ biopharm et pharmacokinet*, Clermont-Ferrand, 1981. Paris: 272.

Doyle E, Chasseaud LF, 1981. Pharmacokinetics of isosorbide dinitrate in rhesus monkey, cynomolgus monkey, and baboon. *J Pharm Sci* 70 (11): 1270–1272.

Doyle WJ, 1984. Functional Eustachian tube obstruction and obitis media in a primate model. A review. *Acta Otolaryngol* (Stockholm), Suppl 414: 52–57.

Dube A, Sharma P, Shrivastava JK, Misra A, Naik S, Katiyar JC, 1998. Vaccination of langur monkeys (*Presbytis entellus*) against *Leishmania donovani* with autoclaved *L. major* plus BCG. *Parasitology* 116 (Pt 3): 219–221.

Dubey DP, Staunton DE, Parelch AC, Schwaming GA, Antoniou D, Lazarus H, Yunis EJ, 1987. Unique proliferation-associated marker expressed on activated and transformed human cells defined by monoclonal antibody. *J Natl Cancer Inst* 78 (2): 203–212.

Dubinin NP, 1974. [Man's social and genetic program in the light of scientific-technical revolution problems. In: *Philosophy and Natural Sciences*]. M, Nauka: 191–203 (Russian).

Dubinin NP, Gubarev V, 1966. [*The Thread of Life (Essays of Genetics)*]. M, Atomizdat (Russian).

Dubinin NP, Shevchenko YuG, 1976. [*Some Questions about Biosocial Nature of Man*]. M, Nauka (Russian).

Duchin LE, 1990. The evolution of articulate speech: Comparative anatomy of the oral cavity in *Pan* and *Homo*. *J Human Evol* 19 (6–7): 687–697.

Duello TM, Boyle TA, 1996. Placental pro-gonadotropin releasing hormone in rhesus monkeys: Sequence analysis. *IPS/ASP Congr Abstr*: # 223.

Dugoujon JM, Arnaud J, Loirat F, Hazout S, Constans J, 1989. Blood markers and genetic evolution in Cercopithecinae. *Primates* 30 (3): 403–422.

Dugoujon JM, Hazout S, 1987. Evolution of immunoglobulin allotypes and phylogeny of apes. *Folia Primatol* 49 (3–4): 187–199.

Dukelow WR, 1972. Reproductive physiologists and nonhuman primate research. In: *7 Intern Kongr fur Tier Fortpflansung, Munchen 1972*. Band III: 2311–2312.

Dulik DM, Fenselau C, 1987. Species-dependent glucuronydation of drugs by immunobilized rabbit, rhesus monkey, and human UDP-glucuronyltransferases. *Drug Metabol Disposit: Biol Fate Chem* 15 (4): 473–477.

Dunlap SS, Thorington RW, Jr, Aziz MA, 1985. Forelimb anatomy of New World monkeys: Myology and the interpretation of primate anthropoid models. *Amer J Phys Anthropol* 68 (4): 499–517.

Dunn FL, 1966. Patterns of parasitism in primates: Phylogenetic and ecological interpretations, with particular reference to the Hominoidea. *Folia Primatol* 4 (5): 329–345.

Dutrillaux B, 1975. *Sur la nature et l'origine des chromosomes humains*. Monographie des Annales de Génétique. L'Expansion Scientifique Française, edn. Paris.

Dutrillaux B, 1985. Etude chromosomique des ancetres l'Homme. *L'Anthroplogie* (Paris) 89 (1): 125–133.

Dutrillaux B, Couturier J, 1986. Principes de l'analyse chromosomique appliquee a la phylogenie: l'exemple des Pongidae et des Hominidae. *Mammalia* 50 (num. spec.): 22–37.

Dutrillaux B, Fosse A-M, Chauvier G, 1979. Etude cytogénétique de six espéces ou sous-espéces de mangabeys (Papinae, Cercopithecidae). *Ann génét* 22 (2): 88–92.

Dutrillaux B, Lombard M, Carroll JB, Martin RD, 1989. Chromosomal affinities of *Callimico goeldii* (Platyrrhina) and characterization of a Y-autosome translocation in the male. *Folia Primatol* 50 (3–4): 230–236.

Dutrillaux B, Rumpler Y, 1989. Absence of chromosomal similarities between tarsiers (*Tarsius syrichta*) and other primates. *Folia Primatol* 50 (1–2): 130–133.

Dutrillaux B, Viegas-Pequignot E, Dubos C, Masse R, 1978. Complete or almost complete chromosome banding between the baboon (*Papio papio*) and man. *Humangenetik* 43: 37–46.

Dyer AR, Stamler R, Elliot P, Stamler J, 1995. Dietary salt and blood pressure. *Nature Med* 1 (10): 994–996.

Dzhikidze EK, 1973. [The role of monkeys in study infections of humans. In: *Biology and Acclimatization of Simians*]. M, Nauka: 49–51 (Russian).

Dzhikidze EK, 1996. [History of medical primatology in native land]. *Vestn Rossiyskoy Akad Med Nauk* (10): 40–47 (Russian).

Earlie AM, Gilmore JP, 1982. Species differences in vascular connections between the adrenal and the kidney in the monkey, human and dog. *Acta Anat* 114 (4): 298–302.

Eck GG, Jablonski NG, 1984. A reassessment of the taxonomic status phyletic relationships of *Papio baringensis* and *Papio quadratirostris* (Primates: Cercopithecidae). *Amer J Phys Anthropol* 65 (2): 109–134.

Eder G, 1996. A longitudional study of the kidney function of the chimpanzee (*Pan troglodytes*) in comparison with humans. *Europ J Clin Chem Clin Biochem* 34: 889–896.

Editorials, 1969. Bleeding into a baboon. *JAMA* 207 (1): 143–144.

Edwards MH, 1996. Animal models of myopia. A review. *Acta Ophtalmol Scand* 74: 213–219.

Ehinger B, Floren I, 1979. Absence of indolcamine-accumulating neurons in the retina of humans and cynomolgus monkeys. *Albrecht v. Graefe's Arch Ophtalmol* 209 (3): 145–153.

Eliava VM, Dzjokua AA, Kolpakova NF, 1986. [Locomotory hypokinesia ifluence on various motor reactions in simians]. In: *Actual probl kosm biol i med*. M: 183–185 (Russian).

Elizondo RS, 1988. Primate models to study eccrine sweating. *Amer J Primatol* 14 (3): 265–276.

Ellinwood EH, 1971. Comparative methamphetamine intoxication in experimental animals. *Pharmakopsychiat-Neuro-Psychopharmacol* 4 (6): 351–361.

Elliot-Smith G, 1924. *Evolution of Man*. London, Oxford Univ Press.

Elygulashvili IS, 1955. [*Pregnancy and Delivery in Simians. An anatomy and Physiology Research*]. Moscow – Leningrad.

Emborg ME, Tetrud JW, McLaughlin WW, Bankiewicz KS, 1995. Rest tremor in rhesus monkeys with MPTP-induced parkinsonism. *Soc Neurosci Abstr* 21 (Pt 2): 1257.

Enders JF, Weller TH, Robbins FC, 1949. Cultivation of the Lansing strain of poliomyelitis in cultures of various human embryonic tissues. *Science* 10: 85–87.

EPA (Environmental Protection Agency), 1986. Final guidelines for the health assessment of suspect developmental toxicants. *Fed Registr* 51: 34028–34040.

Epstein MA, 1974. The use of South American primates in viral oncology. In: *Intern Sympos Use Monkeys in Exp Med*, Novemb 1974, Ms, 6 pp.

Erickson LM, Maeda N, 1994. Parallel evolutionary events in the haptoglobin gene clusters of rhesus monkey and human. *Genomics* 22 (3): 579–589.

Erwin J, 1983. (About the term 'primatologist'). *Lab Primate Newslet* 22 (1): 1–2.

Erwin J, 1998. The great ape aging project: Behavior, cognition, and neurobiology. *Congr IPS* 17 *Abstr*: # 268.

Escalante AA, Barrio E, Ayala FJ, 1995. Evolutionary origin of human and primate malarias: Evidence from the circumsporozoite protein gene. *Mol Biol Evol* 12 (4): 616–626.

Ettlinger G, 1984. Human, apes, and monkeys: The changing neuropsychological viewpoint. *Neuropsychologia* 22 (6): 685–696.

Euler AR, Zurenko GE, Moe JB, Ulrich RG, Yagi Y, 1990. Evaluation of two monkey species (*Macaca mulatta* and *Macaca fascicularis*) as possible models for human *Helicobacter pylori* disease. *J Clin Microbiol* 28 (10): 2285–2290.

Evans HL, 1990. Nonhuman primates in behavioral toxicology: Issues of validity, ethics and public health. *Neurotoxicol Teratol* 12 (5): 531–536.

Evans S, Moysan F, Short R, 1987. Evidence for the existence of a hymen in chimpanzees and gorillas. *Intern J Primatol* 8 (5): 553.

Evarts E, 1982. [Mechanisms of the brain which rule the motions. In: *Brain*]. M: 199–217 (Russian).

Evers MPJ, Zelle B, Bebelman JP, Pronk JC, Mager WH, Planta RJ, Eriksson AW, Frants RR, 1988. Cloning and sequencing of rhesus monkey pepsinogen. A cDNA. *Gene* 65: 179–185.

Eylar EH, Molina F, Quinones C, Zapata M, Kessler N, 1988/1989. Aging: Deficient mitogenic response et old rhesus monkey T cells. *Aging: Immunol Infect Dis* 1 (4): 219–226.

Falk D, 1982. Primate neuroanatomy: An evolutionary perspective. In: *A History of American Physical Anthropology, 1930–1980*. NY, Acad Press: 75–103.

Falk D, Pyne L, Helmkamp CR, DeRousseau CJ, 1988. Directional asymmetry in the forelimb of *Macaca mulatta*. *Amer J Phys Anthropol* 7 (1): 1–6.

Fanelli GM, Jr, Bohn DL, Reilly SS, 1972. Effects of mercurial diuretics on the renal tubular transport of p-aminohippurate and diodrast in the chimpanzee. *J Pharmacol Exp Ther* 180 (3): 759–766.

Fareed J, Kumar A, Rock A, Walenga JM, Davis P, 1985. A primate model (*Macaca mulatta*) to study the pharmacokinetics of heparin and its fractions. *Seminars Thrombos Hemostas* 11 (2): 138–154.

Fariello RG, 1998. Pharmacodynamic and pharmacokinetic features of cabergoline. Rationale for use in Parkinson's disease. *Drugs* 55 (Suppl 1): 10–16.

Fazleabas AT, Verhage HG, Bell SC, 1989. Secretory endometrial and decidual products of the baboon (*Papio anubis*) and human. *J Cell Biochem* 13 (Suppl 13, Pt B): 193.

FDA (Food and Drug Administration), 1966. *Guidelines for Reproduction Studies for Safety Evaluation of Drugs for Human Use*. Washington, DC: Drug Review Branch, Division of Toxicological Evaluation, Bureau of Science, FDA.

Fedigan LM, Strum SC, 1999. A brief history of primate studies: National traditions, disciplinary origins, and stages in North American field research. In: *The Nonhuman Primates*. P. Dolhinow and A. Fuentes, eds. Mayfield Publish Co. Mountain view, California: 258–269.

Felsenstein J, 1987. Estimation of hominoid phylogeny from a DNA hybridization data set. *J Mol Evol* 26: 123–131.

Fernandes CI, Alvarez L, Robinson MA, Macias R, Gonzalez O, Molina H, 1994. Brain aging in nonhuman primates: A similar process to humans. *Congr IPS 15*: 102.

Ferreira BR, Bechara GH, Pissinatti A, Cruz JB, 1995. Benign prostatic hyperplasia in the nonhuman primate *Leontopithecus*. *Folia Primatol* 65 (1): 48–53.

Ferrier D, 1874. The localisation of function in the brain. *Proc Roy Soc* 22: 229–232.

Ferrier D, 1875a. Experiments on the brain of monkeys. *Proc Roy Soc* 23 (1): 409–432.

Ferrier D, 1875b. Experiments on the brain of monkeys (second series). *Phylosoph Transact* 165: 43–488.

Ferris SD, Wilson AC, Brown WM, 1981. Evolutionary tree for apes and humans based on cleavage maps of mitochondrial DNA. *Proc Natl Acad Sci USA, Biol Sci* 78 (4): 24–32.

Fiador R, Mendelsohn FA, 1987. Angiotensin II and converting enzyme receptors in mammalian adrenals. *Proc Austral Physiol Pharmacol Soc* 18 (1): 39.

Fiennes R, 1967. *Zoonoses of Primates*. London.

Filimonov NI, 1949. [*Comparative Anatomy of the Mammals Cerebrum Cortex*]. M, Izd. Acad Med Nauk SSSR (Russian).

Finaz C, Cochet C, de Grouchy J, 1978. Identite des caryotypes de *Papio papio* et *Macaca mulatta* en bandes R, G, C et Ag-NOR. *Ann génét* 21 (3): 149–151.

Fincham JE, Benade AJS, Kruger M, Smuts CM, Gobregts E, Chalton DO, Kritchevsky D, 1998. Atherosclerosis: Aortic lipid changes indused by diets suggest diffuse disease with focal severity in primates that model human atheromas. *Nutrition* 14 (1): 17–22.

Findllay GM, McCallum FO, 1937. An interference phenomenon in relation to yellow fever and other viruses. *J Pathol Bacteriol* 44: 405–424.

Fine J-D, Redmar DA, Goodman AL, 1987. Sequence of reconstitution of seven basement-membrane components following split-thickness wound induction in primate skin. *Arch Dermatol* 123 (Sept): 1174–1178.

Fingscheidt U, Schult P, Robertson DM, Nieschlag E, 1990. Regulation of LH and FSH by inhibin and testosterone in rats and non-human primates. *Acta Endocrinol* 122 (Suppl 1): 43.

Finn CA, 1987. Why do women and some other primates menstruate? *Persp Biol Med* 30 (4): 566–572.

Firsov LA, 1972. [*The Memory of Apes. Physiological Analysis*]. Leningrad, Nauka (Russian).

Firsov LA, 1977. [*The Behavior of Great Apes in Natural Environment*]. Leningrad, Nauka (Russian).

Firsov LA, 1979. [In isolation from the one's people. *Sci and Life*] (M), # 12: 70–75 (Russian).

Firsov LA, 1982. [*I. P. Pavlov and Experimental Primatology*]. Leningrad, Nauka (Russian).

Firsov LA, 1988. [Darvinism: History and the present. In: *Materials of a Conf*]. Leningrad, Nauka: 216–222 (Russian).

Firsov LA, Moiseeva LA, 1989. [Memory as a factor of the anthroposociogenesis. In: *Biol Preconditions of Anthroposociogenesis* 2.] M: 3–24 (Russian).

Fischer M, Erhardt W, Schmahl W, 1979. Pavianhaltung zur xenogenen leberperfusion beim akuten leberzerfall des menschen. *Primaten-Inform* # 6: 5–7.

Fitch WM, 1971. Toward defining the course of evolution: Minimum change for a specific tree topology. *Syst Zool* 20: 406–416.

Fitch WM, Margoliash E, 1967. The construction of phylogenetic trees – A generally applicable method utilizing estimates of the mutation distance obtained from cytochrome C sequences. *Science* 155: 279.

Fladerer FA, 1991. Der este Fund von *Macaca* (Cercopithecidae, Primates) im Jungpleistozan von Mitteleuropa. *J Saugetierkunde* 56 (5): 272–283.

Fleagle JG, 1986. The fossil record of early catarrhine evolution. In: *Major Topics in Primate and Human Evolution*. Cambridge, Cambr Univ Press: 130–149.

Fleagle JG, Kay RF, eds, 1994. *Anthropoid Origins*. NY, Plenum Press.

Florence SL, Wall JT, Kaas JH, 1989. Somatotopic organization of inputs from the hand to the spinal gray and cuneate nucleus of humans. *J Compar Neurol* 286 (1): 40–70.

Fluker CI, Wetzel A, Rambeck WA, Zucker H, 1990. Osteomalazie bei Marmosetaffen. Ein Modell fur die Pseudomangel Rachitis Typ II. *J Animal Physiol*, Suppl 20: 74–77.

Folin M, Parnigotto PP, Montesi F, Grandi C, Darin S, 1986. Determinazione di elementi in traccia indicatori della dieta in pell di *Macaca fuscata*. *Anthropol contempor* 9 (3): 211–215.

Fooden J, 1969. *Taxonomy and Evolution of the Monkeys of Celebes (Primates: Cercopithecidae)*. Basel-New York, S. Karger.

Fooden J, 1980. Classification and distribution of living macaques (*Macaca* Lacepede, 1799). In: *The Macaques: Studies in Ecology, Behavior, and Evolution*. NY etc, Van Nostrand Reinhold Co: 1–6.

Fooden J, 1986. *Taxonomy and Evolution of the Sinica Group of Macaques: Overview of Natural History*. Fieldiana Zool, # 29.

Fooden J, 1991. *Systematic Review of Philippine Macaques (Primates, Cercopithecidae: Macaca fascicularis subspp)*. Fieldiana Zool, # 64.

Foran DR, Hixson JE, Brown WM, 1988. Comparisons of the ape and human sequences that regulate mitochondrial DNA transcription and D-loop DNA synthesis. *Nucl Acid Res* 16 (13): 5841–5861.

Ford EW, 1986. Obstetrical problems of nonhuman primates. *Amer Ass Zoo Veter Ann Proc*: 155–172.

Fouts DH, 1989. Signing interactions between mother and infant chimpanzees. In: *Understanding Chimpanzees*. Cambridge, Harvard Univ Press: 242–251.

Fouts R, Mills ST, 1997. *Next of Kin: What Chimpanzees have Taught Me about Who We are*. NY, Morrow.

Fouts RS, Fouts DH, 1999. Chimpanzee sign language research. In: *The Nonhuman Primates*. P. Dolhinow and A. Fuentes, eds. Mayfield Publish Co, Mountain view, California: 252–256.

Fowler S, 1987. Plant toxin linked to 'Guam dementia'. *New Sci* 115 (1573): 31.

Fox JG, 1998. Review article: *Helicobacter* species and in vivo models of gastrointestinal cancer. *Aliment Pharmacol and Therapeut* 12 (Suppl 1): 37–60.

Fox JG, Lee A, 1997. The role of *Helicobacter* species in newly recognized gastrointestinal tract diseases of animals. *Lab Anim Sci* 47 (3): 222–225.

Franklin RA, Heatherington K, Morrison B, Ward T, 1981. Metabolism of 4-benzamido-1-[4-(indol-3-yl)-4-oxobutyl]-piperidine in rats and monkeys. *Xenobiotica* 11 (3): 159–165.

Friday KE, Lipkin EW, 1990. Long-term parenteral nutrition in unrestrained nonhuman primates: An experimental model. *Amer J Clin Nutr* 51 (3): 470–476.

Fridman (Linin) EP, 1963. [The mountain from which a lot can be seen. *Sci and Techn*] (Riga), # 10: 11–13 (Russian and Latvian).

Fridman EP, 1967a. [Ilia Illich Mechnikov: The founder of modern experimental primatology]. *Voprosi anthropol* # 25: 35–44 (Russian).

Fridman EP, 1967b. [Development of knowledge about primates in Russia before 1917]. *Voprosi anthropol* # 27: 167–178 (Russian).

Fridman EP, 1972. [*Laboratory Double of Man*]. M, Nauka (Russian).

Fridman EP, 1974. [History, modern scale and biological basis of medical investigations on primates. In: *Using Monkeys in Experimental Medicine. Intern Sympos, Sukhumi 1974*]: 59–61 (Russian).

Fridman EP, 1977a. [Biological prerequisites and quantitative characteristics of medical investigations on monkeys]. *Vestn Acad Med Nauk SSSR*, # 8: 72–80 (Russian).

Fridman EP, 1977b. [History of development of primatology. In: *All-Union Conf 'Modelling in Monkeys the Most Important Diseases of Humans', Sukhumi 1977*]. Sukhumi, Alashara: 52–54 (Russian).

Fridman EP, 1978. [History of primatology: Ideological aspects. In: *Historical-Biological Investigations 6*]. M, Nauka: 97–113 (Russian).

Fridman EP, 1979a. [*The Wisest Simians*]. 3rd Supplem Edition. Sukhumi, Alashara (Russian).

Fridman EP, 1979b. [*Primates: Living Prosimians, Monkeys, Apes and Humans*]. M, Nauka (Russian).

Fridman EP, 1989. [Evolutionary, comparative-biological and historical basis of horizontal taxonomy of Hominoidea. In: *Biological Prequisites of Anthropo-Sociogenesis*] 1, Part I. M, Inst Ethnograpy, Acad Sci USSR: 23–80 (Russian).

Fridman EP, 1991. [*Etudes of the Nature of Simians*]. M, Znanie (Russian).

Fridman EP, Khassabov GA, 1972. [Primates in investigations of brain and behavior. In: *Sympos on Frontal Lobes Physiol, Sukhumi 1972*]: 93–96 (Russian).

Fridman EP, Lapin BA, Shatberashvili OB, Cherkovich GM, Chobanyan LA, Nemet GK, Alekseeva ZP, Popova VN, 1990. [Database of abstracts on medical primatology. In: *Problems of Information Theory and Practice # 59*]. M, VINITI: 102–114 (Russian).

Fridman EP, Popova VN, 1983. Species of the genus *Macaca* (Primates, Cercopithecidae) as research subjects in modern biology and medicine. *J Med Primatol* 12 (6): 287–303.

Fridman EP, Popova VN, 1987. [Biological and quantitative characteristics of medical research for species of three genera of lower Catarrhini (*Macaca, Papio, Cercopithecus*)]. *Vestn Acad Med Nauk SSSR*, # 8: 32–36 (Russian).

Fridman EP, Popova VN, 1988. Species of the genus *Papio* (Cercopithecidae) as subjects of biomedical research: I. Biological basis of experiments on baboons. *J Med Primatol* 17 (6): 291–307.

Friedenthal H, 1902. Neue versuche zur frage nach der stellung des menschen im zoologischen system. In: *Sitzungsber der Koniglich Preussisch Akad der Wissensch*: 830–835.

Friedman TB, Polanco GE, Appold JC, Mayle JE, 1985. On the loss of uricolytic activity during primate evolution. I. Silencing of urate oxidase in a hominoid ancestor. *Compar Biochem Physiol* B 81 (3): 653–659.

Friend PJ, Waldmann H, Hale G, Calne RY, 1989. Evaluation of therapy with monoclonal antibodies in kidney and liver transplantation. In: *Transplant and Clin Immunol*. Amsterdam: 51–56.

Fritsch P, Lepage M, Courant D, Bernard F, Moulin G, Mayolini P, 1990. Wound healing after laser photoablation of the rabbit and primate corneas. *Biol Cell* 69 (2): 24a.

Fuchs E, Hensel J, Boer M, Weber MH, 1989. Urinproteine bei primaten. *Klin Wochenschr* 67 (Suppl 17): 19–22.

Fuchs E, Shelton SE, 1996. Experimental strategies in primates to understand human psychopathology and neurology. *IPS/ASP Congr Abstr*: # 183.

Fukui K, Takada N, Takatsu A, 1994. Species identification, especially between human and chimpanzee, by DNA probe. *Jikeikai Med J* 41: 339–344.

Fuller GB, Burnett B, Hobson WC, 1986. A primate model for assessing estrogenicity. The castrate female rhesus monkey. In: *Current Perspectives in Primate Biology*. NY, van Nostrand Reinhold Co: 24–31.

Fulton JF, 1937–1939. *The chimpanzee in experimental medicine*. Trans Kans City Acad Med.

Fulton JF, 1941. Introduction. In: Th. C. Ruch: *Bibliographia Primatologica, Part I*. Springfield, Ch. C Thomas: XI–XIII.

Fulton JF (about him), 1962. John Farquhar Fulton memorial number. *J History Med* 17: 1–51.

Gadsby EL, Jenkins PD, 1996. Africa's endangered drill-conservation and in situ captive breeding. *IPS/ASP Congr Abstr*: # 78.

Gajdusek DC, 1977. Unconventional viruses and the origin and disappearance of kuru. *Science* 197 (4307): 943–960.

Galat G, Galat-Luong A, 1976. La colonisatio de la mangrove par *Cercopithecus aethiops sabaeus* au Senegal. *Ext de la Teet la Vie* 30: 3.

Galindo P, 1973. Monkeys and yellow fever. In: *Nonhuman Primates in Medical Research*. G. H. Bourne, ed. New York – London, Acad Press: 1–15.

Gallo LL, ed., 1987. *Cardiovascular Diseases. Molecular and Cellular Mechanism, Prevention and Treatment*. NY, Plenum Press.

Galloway A, 1997. The cost of reproduction and the evolution of postmenopausal osteoporosis. In: *The Evolving Female: A Live-History Perspective*. M. E. Morbeck, A. Gallovey, A. L. Zihlman, eds. Princeton, Princ Univ Press: 132–146.

Gallup GG, Jr, 1970. Chimpanzees: Self-recognition. *Science* 167: 1277–1286.

Gallup GG, Jr, 1979. Self-recognition in chimpanzees and man: A developmental and comparative perspective. In: *Child and Family*. New York – London: 107–126.

Gallup GG, Jr, 1987. Self-awareness. In: *Comparative Primate Biology. 2B. Behavior, Cognition, and Motivation*. NY, Alan R. Liss: 3–16.

Gallup GG, Jr, 1994. Self-recognition: Research strategies and experimental design. In: *Self-Awareness in Animals and Humans: Developmental Perspectives*. S. T. Parker, R. W. Mitchell, M. L. Boccia, eds. Cambridge, Cambr Univ Press: 35–50.

Gallup GG, Jr, 1997. On the rise and fall of self-conception in primates. In: *The Self Across Psychology: Self-Recognition, Self-Awareness, and the Self Concept*. J. G. Snodgrass and R. L. Thompson, eds. *Ann N. Y. Acad Sci* 818: 73–82.

Gannon PJ, Holloway RL, Broadfield DC, Braun AR, 1998. Asymmetry of chimpanzee planum temporale: Humanlike pattern of Wernicke's brain language area homolog. *Science* 279 (5348): 220–222.

Gannon PJ, Laitman JT, 1995. Human-like patterns of hemispheric asymmetry in the frontal lobe of macaques. *Amer J Phys Anthropol*, Suppl 20: 95.

Gantt DG, 1986. Enamel thickness and ultrastructure in hominoids: With reference to form, function, and phylogeny. In: *Comparative Primate Biology 1. Systematics, Evolution, and Anatomy*. NY, Alan R. Liss: 453–475.

Gao F, Bailes E, Robertson DL, Chen YL, Rodenburg CM, Michael SF, Cummins LB, Arthur LO, Peeters M, Shaw GM, Sharp PM, Hahn BH, 1999. Origin of HIV-1 in the chimpanzee *Pan troglodytes troglodytes*. *Nature* 397 (6718): 436–441.

Garcia JH, Mitchem HL, Briggs L, Morawetz R, Hudetz AG, Hazelric JB, Halsey JH, Conger KA, 1983. Neuronal injury as a function of local cerebral blood flow. *J Neuropathol Exp Neurol* 42 (1): 44–60.

Garcia M, Wood R, Balmaceda JP, Rojas FJ, 1989. Studies on chorionic gonadotrophin hormone in rhesus – disappearance rates. *Human Reprod* 4 (1): 39–49.

Gardashyan AM, Chondkarian OA, Bunina TL, Popova LM, 1970. [Experimental data to study of amyothrophic lateral sclerosis etiology]. *Vestn Acad Med Nauk SSSR* # 10: 80–83 (Russian).

Gardner MJ, Doolan DL, Hedstrom RC, Wang R-b, Sedegah M, Gramzinski RA, Aguiar JC, Wang H, Margalith M, Hobart P, Hoffman SL, 1996. DNA vaccines against malaria: Immunogenicity and protection in a rodent model. *J Pharmaceut Sci* 85 (12): 1294–1300.

Gardner RA, Gardner BT, 1972. Communication with a young chimpanzee: Washoe's vocabulary. *Coloq Intern CNRS* # 198: 241–260.

Gardner RA, Gardner BT, 1990. A vocabularary test for cross-fostered chimpanzees. *Seminar Speech Lang* 11 (2): 119–131.

Gardner RA, Gardner BT, 1994. Ethological roots of language. *NATO ASI (Adv Sci Inst)* Series D 80: 199–222 (*The Ethol Roots of Culture*). R. A. Gardner *et al.*, eds. Dordrecht, Netherlands, Kluwer Acad Publ).

Gardy AG, Hatchuel DA, Sher J, 1982. The oral pathological conditions of the hard tissues of the vervet monkeys in their natural environment. *Diastema* 10: 43–49.

Garrick NA, Murphy DL, 1980. Species differences in the deamination of dopamine and other substrates for monoamine oxidase in brain. *Psychopharmacol* 72: 27–33.

Garver FA, Talmage DW, 1975. Comparative immunochemical studies of primate hemoglobins. *Biochem Genet* 13 (11–12): 743–757.

Garver JJ, Estop AM, Meera KP, Balner H, Pearson PL, 1980. Evidence of similar organization of the chromosomes carrying the major histocompatibiluty complex in man and other. *Cytogenet* 27 (4): 238–245.

Garver JJ, Estop AM, Pearson PL, Dijksman TM, Wijnen LM, Meera KP, 1977. Comparative gene mapping in the Pongidae and Cercopithecidae. In: *Chromosomes Today 6. Proc 6 Intern Conf, Helsinki 1977*. Amsterdam e.a.: 191–199.

Ge X, Chen J, Huang G, Huang D, Liu W, Jiang Y, Li R, Wang S, 1990. [Experimental infection of *Macaca assamensis* with human virus hepatitis B. *Chin J Virol*] 6 (1): 19–26 (China w/English sum).

Gebo DL, Simons EL, 1987. Morphology and locomotor adaptations of the foot in early Oligocene anthropoids. *Amer J Phys Anthropol* 74 (1): 83–101.

Geissler K, Valent P, Mayer P, Leihl E, Hinterberger W, Lechner K, Bettelheim P, 1989. RhiL-3 expands the pool of circulating hemopoietic stem cells in primates – synergism with RhGM-CSF. *Blut* 59 (2): 190.

Gelatt KN, 1977. Animal model for glaucoma. *Invest Ophtalmol Vis Sci* (St Louis) 16 (7): 592–596.

Gelatt KN, 1994. Inherited congenital cataracts in animals. In: *Congenital Cataracts*. E. Cotlier, S. Lambert, D. Taylor, eds. Austin, R. G. Landes Co: 17–32.

Gelatt KN, Brooks DE, Samuelson DA, 1998. Comparative glaucomatology-II: The experimental glaucomas. *J Glaucoma* (Philadelphia) 7 (4): 282–294.

Gentili F, Tasker RR, 1978. A quantitative study of decorticate hypertonia in the squirrel monkey and decerebrate rigidity in the cat. *Surg Forum* 29: 505–508.

Geoffroy-Saint Hilaire E, 1812. Tableau des quadrumanes, ou des animaux composant le premier orde de la classe des mammiferes. *Ann Mus Hist Nat Paris* 19: 85–122, 156–170.

Geretti AM, 1999. Simian immunodeficiency virus as a model of human HIV disease. *Rew Med Virology* 9 (1): 57–67.

Germain G, 1990. Les primates dans l'étude des mécanismes de la gestation. *Pathol-Biol* 38 (3): 159–165.

German RZ, Saxe SA, Crompton AW, Hiiemae KM, 1989. *Amer J Phys Anthropol* 80 (3): 369–377.

Gerritsen T, Siegel FL, 1972. Use of animal models for the study of aminoacidopaties. In: *Proc 3 Intern Congr Neuro-Genetics and Neuro-Ophtalmol, Brussels 1970*. Basel, S. Karger: 22–36.

Gesner K, 1551. *Historiae animalium*. Vol I: *De Quadrupedibus viviparis*. Tiguri. Francofurti: 847–871.

Ghiglieri MP, 1987. Sociobiology of the great apes and the hominid ancestor. *J Human Evol* 16 (4): 319–357.

Ghoniem GM, Shaaban AM, Clarke MR, 1995. Irritable bladder syndrome in an animal model: A continuous monitoring study. *Neurourol Uroldynam* 14 (6): 657–665.

Gibbs CJ, Gajdusek DC, 1976. Studies on the viruses of subacute spongiform encephalopathies using primates, their only avaliable indicator. In: *First Inter-Amer Conf on Conservat and Utilizat of Amer Nonhuman Primates in Biomed Res*. PAHO Scient Publ # 317: 83–109.

Giddens WE, Jr, Combs CA, Smith OA, Klein ES, 1987. Spontaneous hypertension and its sequelae in woolly monkeys (*Lagothrix lagothricha*). *Lab Anim Sci* 37 (6): 750–756.

Gilmore HA, 1981. From Radcliffe-Brown to sociobiology: Some aspects of the rise of primatology within physical anthropology. *Amer J Phys Anthropol* 56: 387–392.

Gingerich PhD, 1985. Nonlinear molecular clocks and ape-man divergence times. In: *Past, Present, and Future of Hominoid Evolutionary Studies*. NY, Alan R Liss: 1–6 (Preprint).

Gingerich PhD, Uhen MD, 1994. Time of origin of primates. *J Human Evol* 27: 443–445.

Gladkova TD, 1966. [*Cutaneous Hand and Foot Patterns of Simians and Man*]. M, Nauka (Russian).

Glaser D, 1970. Uber Geschmacksleistungen bei Primaten. *Z Morphol Anthropol* 62 (3): 285–289.

Glaser D, Etzweiler F, Graf R, Neuner-Jehle N, Calame J-P and Mueller PM, 1994. The first odor threshold measurement in a non-human primate (*Cebuella pygmaea*; Callitrichidae) with a computerized olfactometer. *Advan Biosci* 93: 445–455.

Glaser D, Tinti J-M, Nofre C, 1996. Gustatory responses of non-human primates to dipeptide derivatives or analogues, sweet in man. *Food Chemistry* 56: 313–321.

Glaser HSR, 1986. Symmetric graphic patterns made by chimpanzee and child art. *Primate Rep* 14, July: 264–265.

Glaser HSR, 1996. The first two primate research stations. *Primate Rep* 45: 15–24.

Glickstein M, 1985. Ferrier's mistake. *Trends Neurosci* 8 (8): 341–344.

Goland RS, Stark RI, Wardlaw SL, 1990. Response to corticotropin-releasing hormone during pregnancy in the baboon. *J Clin Endocrinol Metabol* 70 (4): 925–929.

Goldman D, Giri PR, O'Brien SJ, 1987. A molecular phylogeny of the hominoid primates as indicated by two-dimensional protein electrophoresis. *Proc Natl Acad Sci USA* 84 (10): 3307–3311.

Goldman-Rakic PS, 1996. The neurobiology of cognition: Facts and concepts from the study of the prefrontal cortex in nonhuman primates. *IPS/ASP Congr Abstr:* # 246.

Goldschmid-Lange U, 1976. Beyiehungen der facialen Muskulatur zur Lippenregion bei Schimpanse, Gorilla und Mensch (Primates, Hominoidea). *Zool Anz* 187 (3–4): 272–287.

Goldsmith EI, 1968. Endangered nonhuman primates. *Science* 162 (3858): 1077.

Goldstein S, Chapman MJ, Laudat MH, 1978. Comparison of serum LDL in man and several animal species by immunological techniques. In: *Proteides of the Biological Fluids. Proc of the 25 Coll.* NY, Pergamon Press: 511–514.

Goldwasser E, McDonald J, Beru N, 1987. The molecular biology of erythropoietin and expression of its gene. In: *Mol and Cell Aspects Erythropoet and Erythropoies: Proc NATO Adv Res Worksh.* Windsheim: 11–21.

Goldzieher JW, Joshi S, Kraemer DC, 1974. Nonhuman primates in contraceptive research. In: *Pharmacological Models in Contraceptive Development.* Stockholm: 90–118.

Gomperts ED, Cantrell AC, Zail SS, 1971. Electrophoretic patterns of red-cell membrane protein from various mammalian species. *Brit J Haematol* 21 (4): 429–434.

Goncharenko EN, Kudryashov YB, 1996. [Problem of the chemical protection under action of chronic x-radiation]. *Radiatsionnaya Biologia i Radiekologia* 36 (4): 573–586 (Russian w/English sum).

Goncharov N, Katzija G, Todua T, 1980. Peripheral plasma levels of 12 steroids during pregnancy in the baboon (*Papio hamadryas*). *Europ J Obstetr Gynecol Reprod Biol* 11 (3): 201–208.

Goncharova ND, Goncharov NP, 1985. [Comparative characteristics of hormonal function of adrenal and sexual glands in human and *Papio hamadryas* sicking with hemoblastosis. *Exp Oncology*] 7 (3): 47–50 (Russian).

Goncharova ND, 1997. [Hormonal function of the adrenal glands and gonads in man and monkeys during aging]. *Zhurnal Evol Biokhimii i Fiziol* 33 (1): 44–51 (Russian).

Goncharow NP, Todua TN, Woronzow WI, Lomaja LP, Hobe G, Schubert K, 1978. Steroidstoffwechsel bei Primaten. XX. Ausscheidung von C_{21} – steroiden beim Urin bei Frauen und beim Pavian (*Papio hamadryas*) wahrend der Schwangerschawt. *Endokrinologie* 72 (2): 136–140.

Goodman JR, Wolf RH, Roberts JA, 1977. The unique kidney of the spider monkey (*Ateles geoffroyi*). *J Med Primatol* 8 (4): 232–236.

Goodman M, 1962. Immunochemistry of the primates and primates evolution. *Ann N. Y. Acad Sci* 102: 219–234.

Goodman M, 1963. Man's place in the phylogeny as reflected in serum proteins. In: *Classification and Human Evolution.* S. L. Washburn, ed. Chicago, Aldine: 204–234.

Goodman M, 1967. Effects of evolution on promate macromolecules. *Primates* 8: 1–22.

Goodman M, 1976. Toward a genealogical description of the primates. In: *Mol Anthropol.* New York – London: 321–353.

Goodman M, 1982. Biomolecular evidence on human origins from the standpoint of Darwinian theory. *Human Biol* 54 (2): 247–264.

Goodman M, 1991. Molecular evolution of the primates. In: *Primatology Today.* Amsterdam, Elsevier Science Publishers: 11–18.

Goodman M, 1992. Reconstructing human evolution from proteins. *The Cambridge Encyclopedia of Human Evolution.* Cambr Univ Press: 307–312.

Goodman M, 1996. Epilogue: A personal account of the origins of a new paradigm. *Mol Phylogeny Evol* 5 (1): 269–285.

Goodman M, Braunitzer G, Stangl A, Schrank B, 1983. Evidence on human origins from haemoglobins of African apes. *Nature* 303 (5917): 546–548.

Goodman M, Moore GM, 1971. Immunodiffusion systematics of the primates. I. The Catarrhini. *Syst Zool* 20 (1): 19–62.

Goodman M, Moore GW, Farris W, 1974. Primate phylogeny from the perspective of molecular systematics. *Transplant Proc* 6 (2): 17–22.

Goodman M, Moore GM, Farris W, Poulik E, 1970. The evidence from genetically informative macromolecules on the phylogenetic relationships of the chimpanzee. *The Chimpanzee* 2. Basel – New York: 318–360.

Goodman M, Olson CB, Beeher JE, Czelusniak J, 1982. New perspectives in the molecular biological analysis of mammalian phylogeny. *Acta zool fenn* 169: 19–35.

Goodman M, Porter CA, Czelusniak J, Page SL, Schneider H, Shoshani J, Gunnell G, Groves CP, 1998. Toward a phylogenetic classification of primates based on DNA evidence complemented by fossil evidence. *Mol Phylogen Evol* 9 (3): 585–598.

Goodwin WJ, 1972. Development of biomedical research in nonhuman primates in the United States. In: *Med Primatol 1972. Proc 3 Conf Exp Med Surg Primat – Lion 1972*. Part I. Basel, S. Karger: 15–29.

Goosen DJ, Dormehl IC, du Plessis DJ, Walters L, Grove HA, 1982. Non-human primates as models for the study of renal physiology and disease. *Intern J Primatol* 3 (3): 287.

Gordan MH, 1914. (About parotitis). *Rep Loc Covt. Bd New Ser*: # 96.

Gordon KD, 1984. Orientation of occlusal contacts in the chimpanzee, *Pan troglodytes verus*, deduced from scanning electron microscopic analysis of dental microwear patterns. *Archs Oral Biol* 29 (10): 783–787.

Gormus BJ, Xu K, Baskin GB, Martin LN, Bohm RP, Blanchard JL, Mack PA, Ratterree MS, McClure HM, Meyers WM, Walsh GP, 1995. Experimental leprosy in monkeys. I. Sooty mangabey monkeys: Transmission, susceptibility, clinical and pathological findings. *Leprosy Rev* 66 (2): 96–104.

Gould KG, 1983. Ovum recovery and in vitro fertilization in the chimpanzee. *Fertil Steril* 40 (3–4): 378–383.

Gouras P, Zrenner E, 1981. Color vision: A review from a neurophysiological perspective. In: *Progress in Sensory Physiol* 1. Berlin – Heidelberg – New York, Springer-Verlag: 139–179.

Goustard M, 1987. Les apprentissages et la connaissance chez les simiens et chez l'homme. In: *Zool Ethol et Paleontol*. Paris, Fondation Singer-Polignac Masson: 187–211.

Gower EC, 1990. The long-term retention of events in monkey memory. *Behav Brain Res* 38 (3): 191–198.

Goy RW, 1990. A comparative approach: Estrus or sexuality? *Congr IPS* 13 *Abstr*: 158.

Graham CE, Hodgen GD, 1979. The use of chimpanzee in reproductive biology. *J Med Primatol* 8 (5): 265–272.

Grant PR, Hoff CJ, 1975. The skin of Primates. Numerical taxonomy of primate skin. *Amer J Phys Anthropol* 42: 151–168.

Greenwald NS, 1991. Primate-bat relationships and cladistic analysis of morphological data. *Amer J Phys Anthropol* Suppl 12: 82.

Gregory WK, 1915. On the classification and phylogeny of the Lemuroidea. *Bull Geol Soc Amer* 26: 426–446.

Gregory WK, 1916. Studies on the evolution of the primates. *Bull Amer Museum Nat History* 35: 239–355.

Gregory WK, Hellman M, 1939. The dentition of the extinct South African man-ape *Australopithecus (Plesianthropus)*, Transvaalensis Broom. A comparative and phylogenetic study. *Ann Transvaal Museum* 29: 339–373.

Gribnau AAM, Geijsberts LGM, 1985. Morphogenesis of the brain in staged rhesus monkey embryos. *Advances Anat Embryol and Cell Biol* 91. Berlin etc, Springer-Verlag: 1–69.

Griffiths RR, Balster RL, 1979. Opioids: Similarity between evaluations of subjective effects and animal self-administration results. *Clin Pharmacol Ther* 25 (5, part 1): 611–617.

Groves CP, 1968. The classification of the gibbons (Primates, Pongidae). *Z Saugetier* 33 (4–5): 239–346.

Groves CP, 1970a. The forgotten leaf-eaters and the phylogeny of the Colobinae. In: *Old World Monkeys*. New York – London, Acad Press: 557–587.

Groves CP, 1970b. *Gorillas*. London, Arthur Baker.

Groves CP, 1972. Pylogeny and classification of primates. In: *Pathol of Simian Primates*. Part 1. Basel, S. Karger: 11–57.

Groves CP, 1984. A new look at the taxonomy and phylogeny of the gibbons. In: *The Lesser Apes: Evolutionary and Behavioral Biology*. Edinburg, Edinb Univ Press: 542–561.

Groves CP, 1986. Systematics of the great apes. In: *Syst, Evol, and Anat*. NY, Alan R Liss: 187–217.

Groves CP, 1987. Monkey business with red apes. (Rev: Schwartz JH. The Red Ape: Orang-utans and Human Origins. Boston, Houghton Mifflin, 1987, 352 pp). *J Human Evol* 16: 537–542.

Groves CP, 1988. The taxonomy of crested or concolor gibbons. *Austral Primatol* 3 (1): 33.

Groves CP, 1989. The model of modern major genera. In: *The Growing Scope of Human Biology (Proc Austr Soc Hum Biol)*. Nedlands, Western Australia: 25–36.

Groves CP, 1991. *A Theory of Human and Primate Evolution*. Oxford, Clarendon Press.

Groves CP, Eaglen RH, 1988. Systematics of the Lemuridae (Primates, Strepsirhini). *J Human Evol* 17 (5): 513–538.

Grunbaum ASF, Sherrington CS, 1904. Observations on the physiology of the cerebral cortex of the anthropoid apes. *Proc Roy Soc* 72: 152–155.

Guisto JP, Margulis L, 1981. Karyotypic fission theory and the evolution of Old World monkeys and apes. *Bio Systems* 13 (4): 267–302.

Gunderson VM, Rose SA, Grant-Webster KS, 1990. Cross-modal transfer in high- and low-risk infant pigtailed macaque monkeys. *Develop Psychol* 26 (4): 576–581.

Gupta RS, Chopra A, Stetsko DK, 1986. Cellular basis for the species differences in sensitivity to cardiac glycosides (digitalis). *J Cell Physiol* 127 (2): 197–206.

Gusdon JP, Leake NN, van Dyke AH, Atkins W, 1970. Immunochemical comparison of human placental lactogen and placental proteins from other species. *Amer J Obstetr Gynecol* (St Louis) 107 (3): 441–444.

Gyllensten UB, Erlich HA, 1989. Ancient roots for polymorphism at the HLA-DQα locus in primates. *Proc Natl Acad Sci USA* 86: 9986–9990.

Gysin J, 1990. L'utilisation des singes du nouveau monde *Saimiri* et *Aotus* dans le developpment d'un vaccin anti-paludique. *Pathol-Biol* 38 (3): 189–192.

Haaf T, Bray-Ward P, 1996. Redion specific YAC banding and painting probes for comparative genome mapping: Implications for the evolution of human chromosome 2. *Chromosoma* 104: 537–544.

Haber SN, Watson SJ, 1983. The comparison between enkephalin-like and dynorphin-like immunoreactivity in both monkey and human globus pallidus and substantia nigra. *Life Sci* 33 (1): 33–36.

Habgood PJ, 1989. An investigation into usefulness of a cladistic approach to the study of the origin of anatomically modern humans. *J Human Evol* 18 (4): 241–252.

Habteyesus A, Nordenskjold A, Bohman C, Eriksson S, 1991. Deoxynucleoside phosphorylating enzymes in monkey and human tissues show great similarities, while mouse deoxycytidine kinase has a different substrate specificity. *Biochem Pharmacol* 42 (9): 1829–1836.

Haekel E, 1866. *Generelle Morphologie der Organismen: Allgemeine Grundzuge der Organischen Formen-Wissenschaft, mechanisch Begrundet durch die von Charles Darwin reformiste Descendenz-Theorie*. 2 vols. Berlin, Georg Reisner.

Hahn RA, MacDonald BR, 1987. Primate myocardial and systemic hemodynamic responses to leukotriene D_4: Antagonism by LY 171883. *J Pharmacol Exp Ther* 242 (1): 62–69.

Hainsey BM, Hubbard GB, Leland MM, Brasky K, 1993. Clinical parameters of the baboons (*P. spp.*) and chimpanzees (*Pan troglodytes*). *Lab Anim Sci* 43: 236–243.

Hakkinen PJ, Witschi HP, 1985. Animal models. In: *Handbook of Experimental Pharmacology* 75. Berlin etc., Springer-Verlag: 95–114.

Halberstaedter L, von Prowazek S, 1907. Über Zelleinschlüsse parasitärer Natur beim Trachom. *Arn Gesundheitsamte* 126: 44–47.

Hales JRS, Rowell LB, King RB, 1979. Regional distribution of blood flow in awake heat-stressed baboons. *Amer J Physiol* 237 (6): H 705–H 712.

Hall J, Brésil H, Montesano R, 1985. O⁶-Alkylguanine DNA alkyltransferase activity in monkey, human and rat liver. *Carcinogenesis* 6 (2): 209–211.

Hallock MB, Worobey J, Self PA, 1989. Behavioural development in chimpanzee (*Pan troglodytes*) and human newborns across the first month of life. *Intern J Behav Develop* 12 (4): 527–540.

Halpern IM, Dong WK, 1986. D-phenylalanine: A putative enkephalinase inhibitor studied in a primate acute pain model. *Pain* 24 (2): 223–237.

Hamai M, 1994. Half-sibling relationship among wild chimpanzee in Mahale Mountains National Park, Tanzania. *Congr IPS* 15 *Abstr*: 412.

Hamilton WJ, Buskirk RE, Buskirk WH, 1975. Defensive stoning by baboons. *Nature* 256 (5517): 488–489.

Hammer C, 1996. Experimental models and progress in xenogeneic transplantation. In: *Organ Transplantation and Tissue Grafting*. INSERM/J. Libbey & Co Ltd: 1343–1352.

Hannothiaux M-H, Scharfman A, Lafitte J-J, Cornu L, Daniel H, Roussel P, 1987. Comparison of hydrolases, peroxidase and protease inhibitors in bronchoalveolar fluid from *Macacus cynomolgus* and human controls. *Compar Biochem Physiol* 88 B (2): 655–660.

Hansen B, 1996. Primate animal models on non-insulin-dependent diabetes mellitus. In: *Diabetes mellitus: A Fundamental and Clinical Text*. D. LeRoith, S. I. Taylor, J. M. Olevsky, eds. Philadelphia, Lippincott-Raven: 595–603.

Hansen BG, Jen Kai-Lin C, Koerker DJ, Goodner CJ, Wolfe RA, 1982. Influence of nutritional state on periodicity in plasma insulin levels in monkeys. *Amer J Physiol* 242 (3): R 255–R 260.

Hanson SR, Harker LA, 1987. Baboon models of acute arterial thrombosis. *Thromb Haemost* 58 (3): 801–805.

Hantraye P, Riche D, Maziere M, Isacson O, 1990. A primate model of Huntington's disease: Behavioral and anatomical studies of unilateral exitotoxic lesions of the caudate-putamen in the baboon. *Exp Neurol* 108: 91–104.

Harada ML, Scheider H, Scheider MPC, Sampaio I, Czelusniak J, Goodman M, 1995. DNA evidence on the phylogenetic systematics of the New World monkeys: Support the sister grouping of *Cebus* and *Saimiri* from two unlinked nuclear genes. *Mol Phylogen Evol* 4 (3): 331–349.

Hardy JD, Kurrus FE, Chavez CM *et al.*, 1964. Heart transplantation in man: Developmental studies and report of a case. *JAMA* 188: 1132.

Harlow HF, 1972. Love created – love destroyed, love regained. *Coloq Intern CRNS* # 198: 13–58.

Harlow HF, Harlow MK, 1962. Social deprivation in monkeys. *Sci Amer* # 203: 136–146.

Harrison MJS, 1988. A new species of guenon (genus *Cercopithecus*) from Gabon. *J Zool* 215 (3): 561–575.

Harrison RM, Phillippi PP, Roberts JA, 1995. Development of nonhuman primate models for the study of human reproductive pathophysiology. *Amer J Primatol* 36 (2): 127.

Harrison T, 1988. A taxonomic revision of the small catarrhine primates from the early Miocene of East Africa. *Folia Primatol* 50 (1–2): 59–108.

Harrison J, Wyner J, Tanner J, Barnicot N, Reynolds B, 1979. [*Biology of the Man*]. M, Mir (Russian).

Harwerth RS, Smith EL, Duncan GC, Crawford MLJ, von Noorden GK, 1986. Multiple sensitive periods in the development of the primate visual system. *Science* 232 (4747): 235–238.

Hasegawa M, 1990. Phylogeny and molecular evolution in primates. *Jap J Genet* 65 (4): 243–266.

Hasegawa M, 1991. Molecular phylogeny and man's place in Hominoidea. *J Anthropol Soc Nippon* 99 (1): 49–61.

Hasegawa M, Horai S, 1991. Time of the deepest root for polymorphism in human mitochondrial DNA. *J Mol Evol* 32: 37–42.

Hasegawa M, Yano T-a, Kishino H, 1984. A new molecular lock of mitochondrial DNA and the evolution of hominoids. *Proc Jap Acad* B 60 (4): 95–98.

Hasler T, Niederwieser A, 1986. Tetrahydrobiopterin-producing enzyme activities in liver of animals and man. In: *Chem and Biol Pteridines, 1986. Pterid Folic Acid Deriv. Proc 8 Intern Symp, Montreal, June 15–20, 1986*. Berlin – New York: 319–322.

Haude RH, Ray OS, 1968. The visual exploration method. In: *Use of Non-human Primates in Drug Evaluation*. Austin – London, Texas Univ Press: 265–282.

Hausfater G, 1974. History of three little-known New World populations of macaques. *Lab Primate Newslet* 13 (1): 16–18.

Hayasaka K, Gojobori T, Horai S, 1988. Molecular phylogeny and evolution of primate mitochondrial DNA. *Mol Biol Evol* 5 (6): 626–644.

Hayasaka K, Skinner C, Goodman M, 1992. Molecular phylogeny of three platyrrhine primates, capuchin monkey, spider monkey and owl monkey, as inferred from nucleotide sequences of the Ψ_η-globin gene. *J Human Evol* 23: 389–399.

Hayes VJ, Freedman L, Oxnard CE, 1990. The taxonomy of savannah baboons: An odonto-morphometric analysis. *Amer J Primatol* 22 (3): 171–190.

Haynes BF, Dowell BL, Henstley LL, Gore I, Metzgar RS, 1982. Human T cell antigen expression by primates T cells. *Science* 215 (4530): 298–300.

Healy KC, 1995. Fluorescence in situ hybridization reveals homologies among tarsier, galago and human karyotipes. In: *Creatures of the Dark: The Nocturnal Prosimians*. L. Alterman, G. Doyle, M. Azard, eds. NY, Plenum Press: 211–219.

Hearn J, 1994. The NIH – NCRR regional primate research centers program (USA). *Congr IPS 15*: 374.

Heberling RL, Kalter SS, 1974. Use of nonhuman primates in virus research. *Lab Anim Sci* 24 (1): 142–149.

Hedges TR, Zaren HA, 1969. Experimental papilledema: A study of cats and monkeys intoxicated with triethyl tin acetate. *Neurology* 19 (4): 359–366.

Heffner HE, 1987. Ferrier and the study of auditory cortex. *Arch Neurol* 44 (2): 218–221.

Heffner HE, Heffner RS, 1990. Role of primate auditory cortex in hearing. In: *Comparative Perception. II. Complex Signals*. NY, John Wiley and Sons: 279–310.

Heinz-Summer K, Greim H, 1981. Hepatic glutathione S-transferases: Activities and cellular localization in rat, rhesus monkey, chimpanzee and man. *Biochem Pharmacol* 30 (92): 1719–1720.

Hellekant G, Danilova V, 1996. Primate differences toward sweeteners. *IPS/ASP Congr Abstr*: # 651.

Hendrickson AE, Tigges M, 1985. Enucleation demonstrates ocular dominance columns in Old World macaque but not in New World squirrel monkey visual cortex. *Brain Res* 333 (2): 340–344.

Hendrickx AG, 1972a. A comparison of temporal factors in the embryological development of man, Old World monkeys and galagos, and craniofacial malformations induced by thalidomide and triamcinolone. In: *Med Primatol 1972*, part III. Basel, S. Karger: 259.

Hendrickx AG, 1972b. Early development of the embryo in nonhuman primates and man. *Acta Endocrinol* 71 (Suppl 166): 103–130.

Hendrickx AG, Dieter JA, Stewart DR, Tarantal AF, Overstreet JW, Lasley BL, 1996. Biomarkers of early fetal loss in the macaque: TCDD as a model compound of environmental toxicity. *IPS/ASP Congr Abstr*: # 444.

Hendrickx AG, Binkerd PE, 1990. Nonhuman primates and teratological research. *J Med Primatol* 19 (2): 81–108.

Hendrickx AG, Binkerd PE, Rowland JM, 1983. Developmental toxicity and nonhuman primates. Interspecies comparisons. In: *Issues and Reviews in Teratol* 1. NY, Plenum Publ Corp: 149–180.

Hendrickx AG, Peterson PE, 1997. Perspectives on the use of the baboon in embryology and teratology research. *Human Reprod Update* 3 (6): 575–592.

Heppner DG, Ballou WR, 1998. Malaria in 1998: Advances in diagnosis, drugs and vaccine development. *Curr Opinion in Infectious Diseases* 11 (5): 519–530.

Hershkovitz P, 1977. *Living New World Monkey (Platyrrhini)*. Chicago, Chicago Univ Press.

Hershkovitz P, 1981. Comparative anatomy of platyrrhine mandibular cheek teeth dpm_4, pm_4, m_1 with particular reference to those of *Homunculus* (Cebidae), and comments on platyrrhine origins. *Folia Primatol* 35 (2–3): 179–217.

Hershkovitz P, 1983. Two new species of night monkeys, genus *Aotus* (Cebidae, Platyrrhini): A preliminary report on *Aotus* taxonomy. *Amer J Primatol* 4 (3): 209–243.

Hershkovitz P, 1984. Taxonomy of squirrel monkeys, genus *Saimiri* (Cebidae, Platyrrhini): A preliminary report with description of a hitherto unnamed form. *Amer J Primatol* 6 (4): 257–312.

Hershkovitz P, 1990. *Titis, New World Monkeys of the Genus Callicebus (Cebidae, Platyrrhini): A Preliminary Taxonomic Review*. Chicago, Fieldiana, Zool New Ser 55 (Publ 1410).

Hess RF, Baker CL, Zrenner E, Schwarzer J, 1986. Differences between electroretinograms of cat and primate. *J Neurophysiol* 56 (3): 747–768.

Heyes CM, 1996. Theory of mind in monkeys and apes: The problem of test BIAS. *IPS/ASP Congr Abstr*: # 141.

Higgins PJ, Garlick RL, Bunn HP, 1982. Glycosylated hemoglobin in human and animal red cells. Role of glucose permeability. *Diabetes* 31 (9): 743–748.

Hill WCO, 1953. *Primates. Comparative Anatomy and Taxonomy*. I. *Strepsirrhini*. Edinburg Univ Press.

Hill WCO, 1969. The discovery of the chimpanzee. In: *The Chimpanzee*. 1. *Anatomy, Behavior and Diseases of Chimpanzees*. Basel – New York, S. Karger: 1–21.

Hines JW, Skidmore MG, 1994. *Final Report of the U.S. Primate Cardiovascular Experiment Floun on the Soviet Biosatellite Cosmos 1667. Final Report # NASA-TM–108803*. NASA Ames Res Center, Moffett, CA.

Hinshaw KC, Anderson JH, Rosenberg DP, 1981. Selected gynecological and obstetrical problems of nonhuman primates. *Amer Zoo Vet Ann Proc*: 125–127.

Hinton BT, Setchell BP, 1980. Micropuncture and microanalytical studies of the monkey and baboon epididymis and the ductus deferent. *Anthropol Contempor* 3 (2): 211.

Hiramotsu Y, Gikakis N, Gorman JH, III, Khan MMH, Hack CE, Velthuis HTE, Sun L, Marcinkiewicz C, Rao AK, Niewiarowski S, Colman RW, Edmunds LH, Jr, Anderson HL, III, 1997. A baboon model for hematologic studies of cardiopulmonary bypass. *J Lab Clin Med* 130 (4): 412–420.

Hisaoka KK, 1973. Nonhuman primates as a resourse in craniofacial research. *Symp 4 Congr Primatol, 1972* (Appendix) vol 3. Basel e. a: 258–261.

Hitchcock CR, 1969. Experimental surgery in primates and in standard laboratory animals: A comparative survey. *Ann N. Y. Acad Sci* 162: 392–403.

Hitzig E, 1874. *Untersuch. u. d. Gehirn*. Berlin.

Hixson JE, 1999. Use of baboons as an animal model to study heart disease. *Amer J Phys Anthropol* (Suppl 28): 153–154.

Hobson BM, 1970a. Comparison of chorionic gonadotropin in primates. *Proc Soc Endocrinol* 47: V–VI.

Hobson BM, 1970b. Excretion of gonadotropin by the pregnant baboon (*Papio cynocephalus*). *Folia Primatol* 12: 111–115.

Hobson WC, 1987. Overview: Laboratory experience. In: *Preclinical Safety of Biotechnology Products Intendent for Human Use*. NY, Alan R. Liss: 121–123.

Hobson W, Coulston F, Faiman C, Winter J, Reyes F, 1976. Reproductive endocrinology of female chimpanzee: A suitable model of humans. *J Toxicol Environm Health* 1 (4): 657–668.

Hobson WC, Graham CE, Rowell TH, 1991. National chimpanzee breeding program: Primate research institute. *Amer J Primatol* 24 (3–4): 257–263.

Hodgen GD, 1983. Surrogate embryo transfer combined with estrogen-progesterone therapy in monkeys. Implantation, gestation, and delivery without ovaries. *JAMA* 250, Oct: 2167–2171.

Hodges JK, Tarara R, Wangula C, 1984. Circulating steroids and the relationship between ovarian and placental secretion during early and mid pregnancy in the baboon. *Amer J Primatol* 7: 357–366.

Hoesch RM, Boyer TD, 1988. Purification and characterization of hepatic glutathione S-transferases of rhesus monkeys. A family of enzymes similar to the human hepatic glutathione S-transferases. *Biochem J* 251 (1): 81–88.

Hof PR, Nimchinsky EA, Perl DP, Ervin J, 1998. Identification of neural types in cingulate cortex that are unique to humans and great apes. *Amer J Primatol* 45 (2): 184–185.

Hoffstetter R, 1979. Controverses acteuelles sur la phylogenie et la classification des primates. *Bull Mem Soc d'anthropol de Paris* 6 (8): 305–332.

Hoffstetter R, 1982. Les primates Simiiformes (Anthropoidea). (Compréhension, phylogénie, histoire, biogeographique). *Ann paléontol Vertébr* 68 (3): 241–290.

Hofman MA, 1983. Energy metabolism, brain size and longevity in mammals. *Q Rev Biol* 58: 495–512.

Hohmann A, 1969. Fate of autogenous grafts, and processed heterogenous bone in the mastoid cavity of primates. *Laryngoscope* 79 (9): 1618–1646.

Holm S, 1978. Metabolism of 2, 4, 5, 2', 5'-pentachlorbiphenyl by lung and liver microsomes. *Arch Toxicol*, Suppl 1: 239–242.

Holmberg A, 1965. Schlemm's canal and the trabecular meshwork. An electron microscopic study of the normal structure in man and monkey (*Cercopithecus aethiops*). *Docum Ophtalmol* 19: 339–373.

Hoppius CE, 1760. Anthropomorphia, quae, praeside . . . Upsaliae.

Horai S, 1990. Molecular phylogeny and evolution hominoid mitochondrial DNA. *Congr IPS 13 Abstr*: 160.

Horai S, Hayasaka K, Kondo R, Tsugane K, Takahata N, 1995. Recent African origin of modern humans revealed by complete sequences of hominoid mitochondrial DNAs. *Proc Natl Acad Sci USA* 92: 532–536.

Horsburgh T, Cannon IK, Pitts RF, 1978. Glutamine synthetase in kidneys of monkey and man. *Compar Biochem Physiol* B 60 (4): 501–503.

Horsley V, Schafer EA, 1883. Experimental researches in cerebral physiology. *Proc Roy Soc* 36 (London): 437–442.

Hosaka S, Ozawa H, Tanzawa H, Kinitomo T, Nichols R, 1983. In vivo evalution of ocular inserts of hydrogel impregnated with antibiotics for trachoma therapy. *Biomaterials* 4, Octob: 243–248.

Howard AN, Blaton V, Vandamme D, van Landachoot N, Peeters H, 1972. Lipid changes in the plasma lipoproteins of baboons given an atherogenic diet. Part 3. A comparison between lipid changes in the plasma of the baboon and chimpanzee given atherogenic diets and those in human plasma lipoproteins of type II hyperlipoproteinaemia. *Atherosclerosis* 16: 257–272.

Howard CF, 1971. New technique for inducing diabetes mellitus in monkeys. *Primate News* 9 (3): 3–8.

Howard CF, Fang T-Y, 1988. Conributions of the insular amyloidotic islet lesion to the development of diabetes mellitus in *Macaca nigra*. In: *Nonhuman Primates Studies on Diabetes, Carbohydrate Intolerance, and Obesity*. NY, Alan R. Liss: 89–105.

Howard CF, Jr, Wolff J, van Bueren A, 1995. Effects of chronic insulin and tolbutamide administration on the progression of *Macaca nigra* towards diabetes. *Diabetes Res* 28 (4): 147–160.

Howard RJ, Kao V, 1981. Comparison of the surface membrane proteins of human and rhesus monkey (*Macaca mulatta*): Erythrocytes labelled with protein and glycoprotein probes. *Compar Biochem Physiol* B 70 (4): 767–774.

Hoyer BH, van de Velde NW, 1975. Relationships of some African monkeys as determined by their single-copy DNA sequence homologies. *Carnegie Inst Washington Year Book* 74, *1974–1975*. Baltimore, Md: 158–159.

Huang H-j, Zhang X-r, Chen Y-f, 1993. [Studies of the homology of chromosomes between human being and rhesus monkey with chromosomal in situ suppression hybridization]. *Acta Genet Sin* 20 (3): 193–200, 1 pl. (Chinese w/English sum).

Hubel D, 1990. [*Eye, Brain, Vision*]. M, Mir (Russian. Transl from English, 1987).

Hubel DH, Wiesel TN, 1968. Receptive fields and functional architecture of monkey striate cortex. *J Physiol* 195: 215–243.

Hubel DH, Wiesel TN, 1977. Functional architecture of macaque monkey visual cortex. *Proc Roy Soc London* 198: 1–59.

Hubel DH, Wiesel TN, 1982. [Central mechanisms of vision. In: [*Brain*].] M, Nauka: 167–197 (Russian. Transl from English).

Hume DM, Gayle WE, Williams GM, 1969. Cross circulation of patients in hepatic coma with baboon partners having human blood. *Surg Gynecol Obstetr* 128: 495–517.

Hurov JR, 1986. Terrestrial locomotion and back anatomy in vervet and patas monkeys. *J Phys Anthropol* 69: 217.

Huxley ThH, 1863. *Evidence as to Man's in Nature*. London, Williams & Norgate.

Imai M, Shibata T, Mineda T, Kubo K, 1975. *Lunula corneae* in man and primates. *J Human Evol* 4 (3): 259–266.

Imamura K, Bonfils A, Diani A, Uno H, 1998. The effect of topical RU 58841 (androgen receptor blocker) combined with minoxidil on hair growth in macaque androgenetic alopecia. *J Investigat Dermatol* 110 (4): 679.

Ironside JW, 1996. Review: Creutzfeldt-Jakob disease. *Brain Pathol* 6 (4): 379–388.

Isaacs A, Lindemann J, 1957. Virus interference. I. The interferon. *Proc Roy Soc* 147 B (927): 258–267.

Ishak B, Warter S, Dutrilaux B, Rumpler Y, 1988. Phylogenetic relation between Lepilemuridae and other lemuriform families. *Amer J Primatol* 15 (3): 275–280.

Ishizaka K, Ishizaka T, 1971. IgE and reaginic hypersensitivity. *Ann N. Y. Acad Sci* 190: 443–456.

Itakura Sh, 1987. Mirror guided behavior in Japanese monkeys (*Macaca fuscata*). *Primates* 28 (2): 149–161.

Itani J, Suzuki A, 1967. The social unit of chimpanzee. *Primates* 8 (4): 355–381.

Iwatsuki K, Iijima F, Yamagishi F, Chiba S, 1985. Effect of secretagogues on pancreatic exocrine secretion in the monkey and the dog. *Clin Exp Pharmacol Physiol* 12 (1): 47–51.

Jablonski NG, Pan YP, 1988. The evolution and palaeobiogeography of monkeys in China. In: *The Palaeoenvironment of East Asia from the Mid-Tertiary. Proc II Conf Oceanography, Palaeozool and Palaeoanthropol*: 849–867.

Jablonski O, Gossrau R, 1989. Epithels beim Marmoset. *Z mikrosk. -anatom Forsch* 103 (2): 190–220.

Jacobs GH, Neitz J, Crognale MA, Brammer GL, 1991. Spectral sensitivity of vervet monkeys (*Cercopithecus aethiops sabaeus*) and the issue of catarrhine trichromacy. *Amer J Primatol* 23 (3): 185–195.

Jacobs RL, Lux GK, Spielvogel RL, Eichberg JW, Gleiser CA, 1984. Nasal polyposis in a chimpanzee. *J Allergy Clin Immunol* 74 (1): 61–63.

Jacobsen L, Hamel R, eds, 1994, 1996. *International Directory of Primatology*. Wisconsin Reg Primate Res Ctr, Univ of Wisconsin – Madison.

Jaeger J-J, Chaimanee Y, Ducrocq S, 1998. Origin and evolution of Asian hominoid primates. Paleontological data versus molecular data. *C. R. Acad Sci Paris, Sciences de la vie/Live Sciences* 321 (1): 73–78.

Jalagoniya ShL, Sisoeva AF, Kuliev ZA, Antiya LE, Gvazava AM, Dzidzariya AT, Inal-Ipa ASh, Schoniya ZhB, Sochadze TN, Kokaya GYa, Shubladze TN, 1987. [Clinical and experimental study of intersystem relations in norm and pathology. *XV Congr of All-Union I. P. Pavlov Physiol Soc, Kishinev*, vol 2]. Leningrad, Nauka: 334–335 (Russian).

Janson HW, 1952. *Apes and Apes Lore in the Middle Ages and the Renaissance*. London, London Univ Press.

Jayaprakash D, Sharma K, Ansari AS, Lohiya NK, 1994. Testosterone ennathate, a single entity male contraceptive agent in langur monkey. *Europ Archs Biol* 105 (3–4): 103–110.

Jayo JM, Jayo MJ, Zanolli MD, Benedict L, 1988. Psoriasis in a cynomolgus monkey. *Lab Anim Sci* 38 (4): 512.

Jelinek J, Kappen A, Schonbaum E, Lomak P, 1984. A primate model of human postmenopausal hot flushes. *J Clin Endocrinol Metabol* 59 (6): 1224–1228.

Jemilev ZA, 1970. [Comparative of radio sensitivity of leucocytic chromosoma from peripheric blood of human and monkey *Macaca mulatta* on different phases of mitotic cycle]. *Genetica* 6 (15): 125–135 (Russian).

Jenner P, 1990. Models of neurological diseases and symptoms. *Psychopharmacol* 101 (Suppl): 27.

Jiang Z-h, Liu Z-g, Chen S-d, Zhou W-b, Cai J, Ni Z-m, Zhou C-f, 1995. [Fate of human fetal dopamine neurons transplanted into rhesus monkey model of Parkinson's disease: A tyrosine hydroxylase immunocytochemical study. *Acta Physiol Sin*] 47 (1): 31–37 (Chinese w/English sum).

Joffe TH, Dunbar RIM, 1998. Tarsier brain component composition and its implications for systematics. *Primates* 39: 211–216.

Johannessen JN, Chiueh CC, Burns RS, Markey SP, 1985. Differences in the metabolism of MPTP in the rodent and primate parallel differences in sensitivity to its neurotoxic effects. *Life Sci* 36 (3): 219–224.

Johanson DC, White TD, 1979. A systematic assessment of early African hominids. *Science* 202: 1321.

Johnsen DO, 1995. History. In: *Nonhuman Primates in Biomedical Research*. B. T. Bennett *et al.*, eds. San Diego, Acad Press: 1–14.

Johnson DE, Russel RG, 1996. Animal models of urinary tract infection. In: *Urinary Tract Infections: Molecular Pathogenesis and Clinical Management*. H. L. T. Mobley, J. W. Warren, eds. Herndon, ASM Press: 377–403.

Johnson RL, Kapsalis E, 1996. Is it likely that a rhesus female will experience menopause? *IPS/ASP Congr Abstr*: # 233.

Johnson ED, Johson BK, Silverstein D, Tukei P, Geisbert TW, Sanchez AN, Jahrling PB, 1996. Characterization of a new Marburg virus isolated from a 1987 fatal case in Kenya. *Archs Virol* (Suppl 11): 101–114.

Jolly A, 1972. Hour of birth in primates and man. *Folia Primatol* 18 (1–2): 108–121.

Jolly A, 1991. Conscious chimpanzee? A review of recent literature. In: *Cognitive Ethology: The Minds of Other Animals. Essays of Honor of D. R. Griffin.* Hillsdale, New Jersey: 231–252.

Jones CW, West MS, Hong DT, Jonasson O, 1984. Peripheral glomerular basement membrane thickness in the normal and diabetic monkey. *Lab Invest* 51 (2): 193–198.

Jones ML, 1980. Lifespan in mammals. In: *The Comparative Pathology of Zoo Animals.* Washington, D. C., Smithsonian Inst Press: 495–509.

Joosse MV, Wilson JR, Boothe RG, 1990. Monocular visual fields of macaque monkeys with naturally occurring strabismus. *Clin Vis Sci* 5 (2): 101–111.

Jordan C, 1982. Object manipulation and tool-use in captive pygmy chimpanzee (*Pan paniscus*). *J Human Evol* 11 (1): 35–39.

Jorgensen AL, Jones CA, Leth A, 1990. Evolution of alphoid DNA on gomologous pairs of NOR-bearing chromosomes in man and chimpanzee. *Abstr 41 Ann Meet Amer Soc Human Genet, Cincinnati, Ohio, 1990. Amer J Human Genet* 47 (3, Suppl): A 92.

Jorgenson RJ, 1982. The conditions manifesting taurodontism. *Amer J Med Genet* 11 (4): 435–442.

Joulian F, 1996. Comparing chimpanzee and early hominid techniques: Some contributions to cultural and cognitive questions. In: *Modelling the Early Human Mind.* P. Mllars and K. Gibson, eds. McDonald Inst. Monographs, Cambridge: 173–189.

Kaas JH, 1984. The organization of somatosensory cortex in primates and other mammals. In: *Somatosensory Mech. Proc Intern Symp, Stockholm, 1983.* New York – London: 51–59.

Kadanoff D, 1970. Zur Morphologie der Nervenend apparate in der Schleimhaut der Zuunge bei Affen im Vergleich zum Menschen. *Zeitschr mikrosk. -anat. Forschung* 82 (4): 445–460.

Kadanoff D, Chouchkov C, Michailova K, 1980. Die Evolution der eingekapselten sensiblen Nervenend apparate in dem unbehaarten Lippenteil der Saugetiere. *Zeitschr mikrosk. -anat. Forschung* 94 (5): 943–951.

Kalmus H, 1970. The sense of taste of chimpanzee and other primates. In: *The Chimpanzee* Vol 2: 130–141.

Kalter SS, 1969. Nonhuman primates in viral researches. *Ann N. Y. Acad Sci* 164 (art 1): 499–598.

Kalter SS, Heberling RL, 1990. Primate viral diseases in perspective. *J Med Primatol* 19 (6): 519–535.

Kalter SS, Heberling RL, Cooke AW, Barry JD, Tian PY, Northam WJ, 1997. Viral infection of nonhuman primates. *Lab Anim Sci* 47 (5): 461–467.

Kamback NC, 1972. Alcohol selection, learning performance, and emotionality in two subspecies of monkeys (*Macaca nemestrina*). *Proc Ann Convent Amer Psychol Ass* 17 (2): 841–842.

Kanazawa S, 1994. Recognition of facial expressions in humans and a Japanese monkey. *Congr IPS* 15: 420.

Kaplan DJ, Duncan CH, 1990. Novel short interspersed repeat in human DNA. *Nucl Acid Res* 18 (1): 192.

Kaplan JR, Adams MR, Anthony MS, Morgan TM, Manuck SB, Clarkson TB, 1995. Dominant social status and contraceptive hormone treatment inhibit atherogenesis in premenopausal monkeys. *Ateroscl Thrombos Vascul Biol* 15 (12): 2094–2100.

Kaplan RJ, Platt JL, Kwiatkowsi PA, Edwards NM, Xu H, Shah AS, Masroor S, Michler RE, 1995. Absence of hyperacute rejection in pig-to-primate orthotopic pulmonary xenografts. *Transplantation* 59 (3): 410–416.

Kappelman J., Kelley J, Pilbeam D, Sheikh KA, Ward S, Anwar M, Barry JC, Brown B, Hake P, Johnson NM, Raza SM, Shah SMI, 1991. The earliest occurrence of *Sivapithecus* from the middle Miocene Chinji Formation of Pakistan. *J Human Evol* 21: 61–73.

Karamyan AI, 1980. [*Ivan Michaylovich Sechenov and Evolutionary Neurophysiology*]. Leningrad, Nauka (Russian).

Karamyan AI, Sollertinskaya TN, 1990. [Evolution of the limbic brain. *Biol J Armenia*] 43 (10–11): 824–836 (Russian w/Armenian and English sum).

Karamyan AI, Sollertinskaya TN, Valuch TP, Ustoev M, Rizhakov MK, 1987. [Structural and functional evolution of limbic brain. *XV Congr of All-Union I. P. Pavlov Phisiol Soc, Kishinev,* vol 2]. Leningrad, Nauka: 532–533 (Russian).

Kasahara M, Klein D, Fan W, Gutknecht J, 1990. Evolution of the class II major histocompatibility complex alleles in higher primates. *Immunol Rev* # 113: 65–82.

Kasahara M, Klein D, Klain J, 1989. Nucleotide sequence of a chimpanzee DOB cDNA clone. *Immunogenet* 30 (1): 66–68.

Katsiya GV, Todua TN, Gorlushkin BM, Goncharov NP, 1986. [Study of the endocrine function of testis in experiments on sacred baboons (*Papio hamadryass*)]. *Vestn AMN SSSR* # 3: 42–45 (Russian).

Katz AI, 1972. [Primate intellectual behavior as a biological prerequisite of the instrumental activities. In: *XX Intern Psychological Congr, Tokio 1972*]. M: 83–86 (Russian).

Kaufman HE, 1971. Species differences in interferon stimulation: Minimal response in primates and man as compared to rodents. *J Din Invest* 50 (6): 53a.

Kaufman PL, Bito LZ, DeRousseau CJ, 1982. The development of presbyopia in primates. *Trans Ophtalmol Soc U.K.* 102: 323–326.

Kaufman PL, Erickson-Lamy KA, 1985. Effect of repeated anterior chamber perfusion on aqueous flow in the cynomolgus monkey. *Invest Ophtalmol Vis Sci* 26: 885–886.

Kaumanns W, Singh M, Beisenherz W, Schwitzer C, Knogge C, 2000. Bartaffen und ihr Lebensraum. *Zeitschrift des Kölner Zoo* 43 (4): 147–168.

Kaur J, Chakravarti RN, Chugh KS, Chuttani PN, 1968. Spontaneously occuring renal diseases in wild rhesus monkeys. *J Pathol Bacteriol* 95 (1): 31–36.

Kawai K, Chiba S, 1986. Vascular responses of isolated canine and simian femoral arteries and veins to α-adrenoceptor agonists. *Arch intern pharmacodyn et ther* 284 (2): 201–211.

Kawamura M, Tsuji N, Moriya N, Terashita Z-i, 1997. Effects of TAK–29, a novel GP IIb/IIIa antagonist, on arterial thrombosis in guinea pigs, dogs and monkeys. *Thrombos Res* 86 (4): 275–285.

Kawauchi M, Takeda M, Matsumoto J, Nakajima J, Furuse A, Oka T, Yoshitake T, 1996. Which are the target vessels of xeno- and allo-lung rejection? A primate single lung transplantation study. *Transplant Proc* 28 (3): 1416–1417.

Kay RF, 1982. *Sivapithecus simonsi*, a new species of Miocene hominoid, with comments on the phylogenetic status of the Ramapithecinae. *Intern J Primatol* 3 (2): 113–173.

Kay RF, 1990. The phyletic relationships of extant and fossil Pitheciinae (Platyrrhini, Anthropoidea). *J Human Evol* 19 (3): 175–208.

Kay RF, Thorington RW, Houde P, 1990. Eocene plesiadapiform shows affinities with flying lemurs not primates. *Nature* 345 (6273): 342–344.

Kay RF, Williams BA, 1994. Cladistics, computers, and character analysis. *Evol Anthropol* 3: 31–32.

Kaye B, Rance DJ, Waring L, 1985. Oxidative metabolism of carbazeran in vitro by liver cytosol of baboon and man. *Zenobiotica* 15: 237–242.

Kelley J, Pilbeam D, 1986. The dryopithecines: Taxonomy, comparative anatomy and phylogeny of Miocene large hominoids. In: *Syst, Evol Anat*. NY: 361–411.

Kemnitz JW, 1984. Obesity in macaques: Spontaneous and induced. *Adv Vet Sci Compar Med* 28: 81–114.

Kemnitz JW, Perelman RH, Engle MJ, Farrell PM, 1985. An experimental model for studies of fetal maldevelopment in the diabetic pregnancy. *Pediatr Pulmonol* 1 (Suppl): S 79–S 85.

Kemnitz JW, Zimbric ML, Thomson JA, O'Rourke CM, Uno H, 1996. Colorectal cancer in aged rhesus monkeys. *IPS/ASP Congr Abstr*: # 706.

Kemper TL, Moss MB, Hollander W, Prusty S, 1996. Cerebral microinfarction in a monkey model hypertension. *J Neuropathol Exp Neurol* 55 (5): 648.

Kendrick KM, Dixson AF, 1986. Anteromedial hypothalamic lesions block proceptivity but not receptivity in the female common marmoset (*Callithrix jacchus*). *Brain Res* 375 (2): 221–229.

Keough EM, Wilcox LM, Connoliy RJ, Hotte CE, 1981. Comparative ocular blood flow. *Compar Biochem Physiol* A 68 (2): 269–291.

Kessler MJ, 1989. Establishment of the Cayo Santiago colony. *Puerto Rico Health Sci J* 8 (1): 15–17.

Khan KNM, Kats AA, Fouant MM, Snook SS, McKearn JP, Alden CL, Smith PF, 1996. Recombinant human interleukin-3 induces extramedullary hematopoiesis at subcutateneosus injection sites in cynomolgus monkey. *Toxicologic Pathol* 24 (4): 391–397.

Khasawinah AM, Hardy CJ, Clark GC, 1989. Comparative ingalation toxicity of technical chlordane in rats and monkeys. *J Toxicol Environ Health* 28: 327–347.

Khassabov GA, 1973. [Comparative study of spastic activity in simians and rodents]. *J evol biochim i physiol* 9 (5): 502–511 (Russian).

Khassabov GA, 1978. [*Neurophysiology of the Relations of Great Hemispheres Cortex in Primates*]. M, Medicine (Russian).

Kiely PM, Crewther SG, Nathan J, Brennan NA, Efron N, Madigan M, 1987. A comparison of ocular development of the cynomolgus monkey and man. *Clin Vis Sci* 1 (3): 269–280.

Kimeldorf D, Hunt E, 1969. [*The Effect of Ionising Radiation on the Functions of Nervous System*]. M, Atomizdat (Russian).

Kimura T, 1990. Center of gravity of chimpanzee body in bipedal standing. *J Anthropol Soc Nippon* 98 (2): 213.

King BF, Wilson JM, 1983. A fine structural and cytochemical study of the rhesus monkey yolk sac: Endoderm and mesothelium. *Anat Rec* 205 (2): 143–158.

King FA, Yarbrough CJ, 1994. Kenya programs of the Yerkes Primate Research Center of Emory University. *Congr IPS 15*: 326.

King FA, Yarbrough CJ, Anderson DC, Gordon TP, Gould KG, 1988. Primates. *Science* 240: 1475–1482.

King M-C, Wilson AC, 1975. Evolution at two levels in human and chimpanzees. *Science* 188: 107–116.

Kingsley SR, 1988. Physiological development of male orang-utans and gorillas. In: *Orang-utan Biology*. NY, Oxford Univ Press: 123–131.

Kiorpes L, Boothe RG, Carlson MR, Alfi D, 1985. Frequency of naturally occurring strabismus in monkeys. *J Pediatr Ophtalmol Strab* 22 (2): 60–64.

Kirshenbaum AS, Bressler RB, Friedman MM, Metcalfe DD, 1988. Characterization of basophil-like cells derived from nonhuman primate bone marrow. *Intern Arch Allergy Appl Immunol* 87 (1): 1–8.

Kirshenblat YaD, 1971. [*General Endocrinology*]. Ed II. M, Vissh shkola (Russian).

Klein J, Gutknecht J, Fischer N, 1990. The major histocompatibility complex and human evolution. *Trends Genet* 6 (1): 7–11.

Kling AS, 1986. Neurological correlates of social behavior. Physiological – ethological approaches to ostracism. *Ethol Sociobiol* 7 (3–4): 175–186.

Klingbeil C, Hsu DH, 1999. Pharmacology and safety assessment of humanized monoclonal antibodies for therapeutic use. *Toxicologic Pathol* 27 (1): 1–3.

Klinger HP, Hamerton JL, Mutton D, Lang EM, 1963. The chromosomes of the Hominoidea. In: *Classification and Human Evolution*. S. L. Washburn, ed. Chicago, Aldine: 235–242.

Klots IN, Korzaya LI, Lapin BA, 1998. Telomere sizes and telomerase activity in normal lymphocytes of different monkey species. *Congr IPS 17 Abstr*: # 330.

Knobil E, 1981. The role of the central nervous system in the control of the menstrual cycle of the rhesus monkey. *Exp Brain Res* 42 (Suppl 3): 200–207.

Knobil E, 1995. [On the control of the menstrual cycle and ovulation]. *Contraception, tertilite, sexualite* 23 (12): 705–709 (French).

Knorre AG, 1969. [Current state of knowledge about early stages of normal embryonal development of man]. *Arch anat histol i embryol* 57 (8): 3–22 (Russian).

Köhler W, (1917) 1973. *Intelligenzprufungen an Menschenaffen*. Reprinting: Berlin, Springer-Verlag.

Kohne DE, Chiscon JA, Hoyer BH, 1972. Evolution of primate DNA sequences. *J Human Evol* 1: 627–644.

Koizumi H, 1989. Current topics of species specificity in drug toxicity – morphological aspect in monkeys. *J Toxicol Sci* 14 (4): 297.

Komuro A, Watanabe T, Miyoshi I, Hayami M, Tsujimoto H, Seiki M, Yoshida M, 1984. Detection and characterization of simian retroviruses homologous to human T-cell leukemia virus type I. *Virology* 138 (2): 373–378.

Kononova EP, 1962. [*The Frontal Region of the Cerebrum*]. Leningrad, Publish House of Acad Sci (Russian).

Kooksova MI, 1954. [On the order of the first dentition eruption in macaques rhesus]. *Bull exp biol i med* 38 (7): 69–72 (Russian).

Kooksova MI, 1972. [*The Hemopoiesis System of Simians in Norm and Pathology*]. M, Medicine (Russian).

Koontz MA, Hendrickson AE, 1987. Stratified distribution of synapses in the inner plexiform layer of primate retina. *J Compar Neurol* 263 (4): 581–592.

Koop BF, Siemieniak D, Slightom JL, Goodman M, Dunbar J, Wright PC, Simons EL, 1989. *Tarsius* delta- and beta-globin genes: Conversions, evolution, and systematic implications. *J Biol Chem* 264 (1): 68–79.

Koritnik D, Clarkson T, Adams M, 1986. Cynomolgus macaques as models for evaluating effects contraceptive steroids on plasma lipoproteins and coronary artery atherosclerosis. In: *Contraceptive Steroids: Pharmacology and Safety*. NY, Plenum Press: 303–319.

Korte R, Vogel F, Osterburg I, 1987. The primate as a model for hasard assessment of teratogens in human. *Arch Toxicol* Suppl 11: 115–121.

Korsch PC, Gillespie JR, Berry JD, 1979. Respiratory mechanics in normal bonnet and rhesus monkeys. *J Appl Physiol: Respirat Environ Exercise Physiol* 46 (1): 166–175.

Kotchetkova EI, (1962) 1973. [*Paleoneurology*]. M, Publish House of Moskow University (Russian).

Koto M, Yogo K, Takanashi H, Ozaki K, Matsu-Ura T, 1994. The gastrointestinal contractile activity and the response to motiline in conscious rhesus monkeys. *Congr IPS 15*: 224.

Kottler ML, Dang CD, Salmon R, Counts R, Degrelle H, 1990. Effect of testosterone on regulation of the level of sex steroid-binding protein mRNA in monkey (*Macaca fascicularis*) liver. *J Mol Endocrinol* 5 (5): 253–257.

Koyama S, Terada N, Shiojima Y, Takeuchi T, 1985. Some species differences in cardiovascular responses to intravenously injected leucine-enkephalin. *Experimentia* 41 (11): 1394–1396.

Kozey SA, 1975. [The nerves of the tongue in macaque rhesus fetals. In: *Actual Problems of Theoretical and Clinical Medicine*]. Minsk: 110–112 (Russian).

Kraemer GW, 1985. The primate social environment, brain neurochemical changes and psychopathology. *Trends Neurosci* 8 (8): 339–340.

Krell RD, 1979. Pharmacologic aspects of canine and rhesus monkey models of allergic asthma. In: *Proc Intern Symp 'Mechanisms of Airways Obstruction in Human Respiratory Disease'*. Capetown, Balkema: 117–137.

Krell RD, Williams JC, Giles RE, 1986. Pharmacology of aerosol leukotriene C_4- and D_4-induced alteration of pulmonary mechanics in anesthetized cynomolgus monkeys. *Prostaglandins* 32 (5): 769–780.

Krikorian A, Rahmani R, Bromet M, Bore P, Cano JP, 1989. Pharmacokinetics and metabolism of navelbine. *Seminars in Oncology* 16 (2, Suppl 4): 21–25.

Krukoff S, 1971. Rotation de la face autor du Nasion, par rapport à la base du crane, ches les hominidés. *C. r. Acad sci* D 272 (14): 1850–1853.

Krusko N, Dolhinow P, Anderson C, Bortz W, Kastlen J, Flesher K, Flood M, Howe R, Kelly A, Le Favour N, Leydorf C, Limbach C, Read E, 1986. Earthquake: Langur monkeys' response. *Lab Primate Newslet* 25 (1): 6–7.

Krygier G, Genko RJ, Mashimo PA, Hausmann E, 1973. Experimental gingivitis in *Macaca speciosa* monkeys: Clinical, bacteriological and histologic similarities to human gingivitis. *J Criodontol* 44 (8): 454–463.

Krylova RI, 1986. [*Pathologic Anatomy of some Infections and Tumours*. Synopsis of Thesis Doctor Diss]. M, Inst of Man Morphol (Russian).

Kuhl PK, 1988. Auditory perception and the evolution of speech. *Human Evol* 3 (1–2): 19–43.

Kuhn H-J, 1970. Primate centers of the world. *Z Fersuchtstierkunde* 2 (4): 265–266.

Kukuev LA, 1968. [*The Structure of Locomotory of Analyzer*]. Leningrad, Publish House of Acad Sci (Russian).

Kuntz RE, 1973. Models for investigation in parasitology. In: *Nonhuman Primates and Medical Research*. G. H. Bourne, ed. New York – London, Acad Press: 167–201.

Kuvaeva IB, 1976. [*The Metabolism in Organism and Intestinal Microflora*]. M, Medicine (Russian).

Laboratory animals: A turning point? *Inform Rep*, 1981, 30 (1): 8–9.

Ladigina-Kohts NN, 1935. [*Infant of Ape and Human Child*]. M, Publish House of USSR Acad Sci (Russian w/English great sum).

Lagarde D, 1990. Les primates comme modeles de l'etude du sommeil chez l'homme. *Pathol – biol* 38 (3): 214–220.

La Motte RH, Thalhammer JG, Robinson CJ, 1983. Peripheral neural correlates of magnitude of cutaneous pain and hyperalgesia: A comparison of neural events in monkey with sensory judgments in human. *J Neurophysiol* 50 (1): 1–26.

Landsteiner K, 1901. Ueber agglutinationserscheinungen normalen menschlichen blutes. *Wien Klin Wochenschr* 14: 1132–1134.

Landsteiner K, Miller C, 1925. Serological studies on the blood of the primates. II. The blood groups in anthropoid apes. *J Exp Med* 42: 853–862.

Landsteiner K, Popper E, 1908. Mikroskopische Präparate von einem menschlichen und zwei Affenrückemarken. *Wien Klin Wschr* 21: 1830.

Landsteiner K, Wiener AS, 1940. An agglutinable factor in human blood recognized by immune sera for rhesus. *Proc Soc Exp Biol Med* 43: 233.

Lane MA, 1996. Slowing aging by calorie restriction: Studies in rhesus monkeys. *IPS/ASP Congr Abstr*: # 750.

Langley JN, Sherrington CS, 1891. On pilo-motor nerves. *J Physiol* 12: 278–291.

Lapin BA, ed., 1967. [*Medical Primatology*]. Tbilisi (Russian).

Lapin BA, 1988. Permissibility and scientific grounds for the use of nonhuman primates in biomedical research. *Congr IPS 12* Brasilia: 127–141.

Lapin BA, 1994. History of Russian primatology. *Congr IPS* 15: 394.

Lapin BA, 1996. [Review and perspectives of medico-biological investigations on primates. *Bull exp biol i med*] 122 (9): 245–252.

Lapin BA, Dzhikidze EK, Fridman EP, 1987. [*A Handbook of Medical Primatology*]. M, Medicine (Russian w/English sum).

Lapin BA, Fridman EP, 1966. *Monkeys for Science*. M, APN.

Lapin BA, Fridman EP, 1983. [Primatology Medical. In: *Great Medical Encyclopedia*] 21. M, Medicine: 66 (Russian).

Lapin BA, Fridman EP, 1988. The central primatological institution of the USSR: The Institute of Experimental Pathology and Therapy. *Congr IPS 15*. Brasilia: 161–165.

Lapin BA, Yakovleva LA, 1960. [*Essays of Comparative Pathology of Simians*]. M, Medgiz (Russian; in English – 1964, USA; in German – 1964).

Lapin BA, Yakovleva LA, 1970. [On virus nature of leukosis in man]. *Vestn AMN SSSR* # 5: 60–71 (Russian w/English sum).

Lapin BA, Yakovleva LA, 1994. Pathology of primates. *Congr IPS* 15: 213.

Lapin BA, Yakovleva LA, Bukaeva IA, Indzhia LV, Kokosha LV, 1975. Induction of leukemia in monkeys with human leukemic blood. In: *Symposia of Congr IPS 4*. Vol 4. *Nonhuman Primates and Human Diseases*. Basel, S. Karger: 1–29.

Lapin BA, Yakovleva LA, Chikobava MG, 1996. Virus associated hemopoietic malignancies in primates. *IPS/APS Congr Abstr*: # 644.

Larsson H, Lund J, 1981. Metabolism of femoxetine. *Acta pharmacol et toxicol* 48 (5): 424–432.

Latanov AV, Leonova AY, Evtikhin DV, Sokolov EN, 1997. [Comparative neurobiology of colour vision in humans and animals]. *Z Vissh Nervn Deyateln im I. P. Pavlova* 47 (2): 308–319 (Russian).

Latanov AV, Polansky VB, Sokolov EN, 1991. [Fourth-measured spherical colour vision of simians]. *Z Vissh Nervn Deyateln im I. P. Pavlova* 41 (4): 636–646 (Russian).

Latimer B, Lovejoy CO, 1990. Metatarsophalangeal joints of *Australopithecus afarensis. Amer J Phys Anthropol* 83 (1): 13–23.

Laver HM, Robinson WG, Jr, Hansen BC, 1994. Spontaneously diabetic monkeys as a model for diabetic retinopathy. *Invest Ophtalmol Vis Sci* 35 (4): 1733.

Lawlor DA, Ward FE, Ennis PD, Jackson AP, Parham P, 1988. HLA-A and B-polymorphisms predate the divergence of humans and chimpanzees. *Nature* 335 (6187): 268–271.

Layton LL, 1965. Passive transfer of human atopic allergies into lemurs, lorises, pottos, and galagos: Possible primate-ordinal specificity of acceptance of passive sensitization by human atopic reagin. *J Allergy* 36 (6): 523–531.

Le XE, 1996. Endangered primate species in Vietnam. *IPS/ASP Congr Abstr*: # 399.

Leakey LSB, 1970. The relationship of African apes, man, and Old World monkeys. *Proc Natl Acad Sci USA* 67 (2): 746–748.

Lee Jo ES, Yoshida A, Brandt IK, 1979. Structural comparison of hexosaminidases in primates. *Compar Biochem Physiol* B 63 (4): 491–494.

Lee MML, Yeh MN, 1983. Fetal circulation of the placenta: A comparative study of human and baboon placenta by scanning electron microscopy of vascular casts. *Placenta* 4 (Spec Iss): 515–526.

Lehner NDM, Bullock BC, Clarkson TB, 1968. Squirrel monkeys (*Saimiri sciureus*) as models for the study of rickets. In: *Use of Nonhuman Primates in Drug Evaluation. A Symposium, San Antonio, 1967*. H. Vagtborg, ed., Austin, Texas Univ Press: 67–76.

Leibl H, Eder G, Pfeiler S, Eibl MM, 1986. IgG subclasses in chimpanzees. *Immunoglobulin Subclass Defic Intern Symp, Lund, 1985*. Basel e.a: 34–41.

Le May M, 1985. Asymmetries of the brains and skulls of nonhuman primates. In: *Cerebral Lateralization in Non-Human Species*. Orlando, Florida, Acad Press: 233–245.

Leomis MR, Rosenberg DP, 1980. Apparent psychogenic polydipsia in a rhesus monkey. *J Amer Vet Med Ass* 177 (9): 940–941.

Leone-Bay A, Ho K-K, Agarwal R, Baughman RA, Chaudhary K, DeMorin F, Genoble L, McInnes C, Lercara C, Milstein S, O'Toole D, Sarubbi D, Variano B, Paton DR, 1996. 4-[4-[(2-hydroxybenzoyl) amino]phenyl]butyric acid as a novel oral delivery agent for recombinant human growth hormone. *J Med Chem* 39 (13): 2571–2578.

Leroy E, Baize S, Wahl G, Egwang TG, Georges AJ, 1997. Experimental infection of a nonhuman primate with Loa loa induces transient strong immune activation followed by peripheral unresponsiveness of helper T cells. *Infect Immunity* 65 (5): 1876–1882.

Lester D, 1968. Nonhuman primates in research on the problem of alcohol. In: *Use of Nonhuman Primates in Drug Evaluation*. Austin – London, Texas Univ Press: 249–264.

Levenbook I, Dragunsky E, Chumakov K, Taffs R, Chernokhvostova Y, Nomura T, Hioki K, Gardner D, Cogan J, Asher D, 1997. Transgenic mice and a molecular assay as alternatives to the monkey neurovirulence test of oral poliovirus vaccine. *Develop Anim Vet Sci* 27: 1013–1016.

Levine RA, Hall RC, 1976. Cyclic AMP in secretin choleresis. Evidence for a regulatory role in man and baboons but not in dogs. *Gastroenterology* 70 (4): 537–544.

Levitina MW, 1982. [Cerebrosides and sulfocerebrosides of brain at Java macaque (*Macaca fascicularis*)]. *Z evol biochim i physiol* 18 (6): 573–576 (Russian).

Levitt MM, 1985. Dysesthesias and self-mutilation in humans and subhumans: A review of clinical and experimental studies. *Brain Res Rev* 10: 247–290.

Levy BM, 1980. Animal analogues for the study of dental and oral diseases. In: *16 IABS Congr: The Standardization of Animals to Improve Biomedical Research, Production and Control, San Antonio 1979*. Develop Biol Standard, vol 45. Basel, S. Karger: 51–59.

Levy EK, Levy DE, 1986. Monkey in the Middle: Pre-Darwinian evolutionary thought and artistic creation. *Persp Biol Med* 30 (1): 95–106.

Lewin R, 1987. My close cousin the chimpanzee. *Science* 238 (4825): 273–275.

Lewis DA, Campbell MJ, Foote SL, Morrison JH, 1986. The monoaminergic innervation of primate neocortex. *Human Neurobiol* 5 (3): 181–188.

Lewis JH, 1977. Comparative hematology rhesus monkeys (*Macaca mulatta*). *Compar Biochem Physiol* 56 A: 379–383.

Lewis RW, 1984. Benign prostatic hyperplasia in the nonhuman primate. In: *New Approaches to the Study of Benign Prostatic Hyperplasia*. NY, Alan R Liss: 235–255.

Leyton ASF, Sherrington CS, 1917. Observations of the excitable cortex of the chimpanzee, orang-utan, and gorilla. *Quart J Exp Physiol* 11: 135–222.

Li Choh Hao, Parkoff H, 1956. Preparation and properties of growth hormone from human and monkey pituitary glands. *Science* 124 (3233): 1293–1294.

Li W-H, Wolfe KH, Sourdis J, Sharp PM, 1987. Reconstruction of phylogenetic trees and estimation of divergence times under nonconstant rates of evolution. *Cold Spring Harbor Conf Cell Prolifer* 52: 847–856.

Lichtenstein LM, 1983. [Anaphylactic reactions. In: *Mechanisms of Immunopatholy*]. M, Medicine: 23–39 (Russian).

Lidow MS, Elsworth JD, Goldman-Racic PS, 1997. Down-regulation of the D1 and D5 dopamine receptors in the primate prefrontal cortex by chronic treatment with antipsychotic drugs. *J Pharmacol Exp Therapeutics* 281 (1): 597–603.

Lie TS, Ronschke A, Ukigusa M, Rommolsheim K, 1983. Indikation zur Behandlung des Leberversagenis mit Pavianleberperfusion. *Munch med Wochschr* 125 (12): 224–226.

Lieber CS, 1997. Ethanol metabolism, cirrhosis and alcoholism. *Clinica Chimica Acta* 257: 59–84.

Lieber CS, Decarli LM, 1989. Liquid diet techique of ethanol administration: 1989 update. *Alcohol and Alcoholism* 24 (3): 197–211.

Lieberman P, 1985. On the evolution of human syntactic ability. Its pre-adaptive bases – motor control and speech. *J Human Evol* 14 (7): 657–668.

Lieberman P, 1990. The evolution of human language. *Semin Speech Lang* 11 (2): 63–76.

Lin G, Wan C, Shao J, Liu F, 1989. [A comparative experimental study on the preliminary generalization of nonhuman primates. Generalization of objects, photographs and pictures in golden monkeys (*Rhinopithecus roxellanae*). In: *Progress in the Studies of Golden Monkey*]. China: 271–276 (Chinese w/English sum).

Lin SS, Weidner BC, Byrne GW, Diamond LE, Lawson JH, Hoopes CW, Daniels LJ, Daggett CW, Parker W, Harland RC, Davis RD, Bollinger RR, Logan JS, Platt JL, 1998. The role of antibodies in acute vascular rejection of pig-to-baboon cardiac transplants. *J Clin Invest* 101 (8): 1745–1756.

Linnankosli I, Lelnonen LM, 1985. Compatibility of male and female sexual behaviour in *Macaca arctoides*. *Z Tierpsychol* 70 (2): 115–122.

Linne C, von 1758. *Systema naturae* I, 10 ed. Stockholm.

Livingstone M, Hubel D, 1988. Segregation of form, color, movement, and depth: Anatomy, physiology, and perception. *Science* 240 (4853): 740–749.

Logothetis NK, Schall JD, 1990. Binocular motion rivary in macaque monkeys: Eye dominance and Tracking eye movements. *Vis Res* 30 (10): 1409–1419.

Lonsbury-Martin BL, Martin GK, 1988. Incidence of spontaneous otoacostic emissions in macaque monkeys: A replication. *Hear Res* 34: 313–318.

Lowenstein JM, Zihlman AL, 1984. Human evolution and molecular biology. *Persp Biol Med* 4 (27): 611–622.

Loy J, 1987. The sexual behavior of African monkeys and the question of estrus. In: *Comparative Behavior of African monkeys*. NY, Alan R. Liss: 175–195.

LPN (Laboratory Primate Newletter), 2000. Vol 39 (2): 22.

Lubach GR, 1996. Introduction: Primate psychoneuroimmunology. *Amer J Primatol* 39 (4): 203–204.

Luckett WP, ed., 1980. *Comparative Biology and Evolutionary Relationships of Tree Shrews*. NY, Plenum Press.

Lucotte G, Barriel V, Guerin P, Abbas N, Ruffie J, 1990. Rétro-transposition de la séquence humaine homologue à la p49f sur le chromosome Y au cours de l'évolution des singes anthropoides. *Biochem Syst Ecol* 18 (2–3): 199–204.

Lucotte G, Jouventin P, 1980. Distance électrophorétique entre le mandrill et le drill. *Ann génét* 23 (1): 46–48.

Lucotte G, Lefebvre J, 1981. Distances électrophorétiques entre l'homme, le chimpanzé (*Pan troglodytes*) et le gorille (*Gorilla gorilla*) basees sur la mobilité des enzymes erythrocytaires. *Human Genet* 57 (2): 180–184.

Lucotte G, Ruffié J, 1982. Variation électrophorétique et spéciation ches les différentes espêces de singes anthropoides. *Human Genet* 61 (4): 310–317.

Lue TF, Omraiya M, Takamura T, Schmidt RA, Tanagho EA, 1983. Animal models for penile erection studies. *Neurol Urodyn* 2 (3): 225–231.

Lueck CJ, Zeki S, Friston KJ, Deiber MP, Cope P, Cunningham VJ, Lammertsma AA, Kennard C, Frackowiak RSJ, 1989. The colour centre in the cerebral cortex of man. *Nature* 340 (6232): 386–389.

Luo W-y, Ying Y, Li Z-x, Fan C-m, Zhou J-h, Fang X-c, 1996. [Pharmacokinetics of contragestazol (DL-111 IT), a new non-steroid antifertility agent in monkeys. *Acta Pharm Sin*] 30 (6): 408–411 (Chinese w/English sum).

Luo Y, Taniguchi S, Kobayashi T, Niekrasz M, Cooper DK, 1995. Screening of baboons as potential liver donors for humans. *Xenotransplantation* 2 (4): 244–252.

Lynch JJ, Eskin TA, Merigan WH, 1986. Visual toxicity of 2, 5-hexanedione: Are effects different in felines and primates. *Toxicologist* 6 (1): 72.

Lynch M, 1991. Methods for the analysis of comparative data in evolutionary biology. *Evolution* 45 (5): 1065–1080.

Macleod MC, 1996. Multiple infanticide in blue monkeys: a successful reproductive strategy? *IPS/ASP Congr Abstr:* # 202.

MacPhee RDE, ed., 1993. *Primates and their Relatives in Phylogenetic Perspective.* NY, Plenum Press.

Macpherson EE, Montagna W, 1974. The mammary glands of rhesus monkeys. *J Invest Dermatol* 63 (1): 17–18.

Mac Vittie TJ, Dobson ME, Farese AM, Virca GD *et al.*, 1994. Comparative efficacy of soluble rhuIL-1 receptors type I and II in modulating the hemodynamic, metabolic and inflammatory cytokine response in a nonhuman primate model of endotoxin shock. *Exp Haematol* 22 (8): 683.

Madigan MC, Gillard-Crewther S, Kiely PM, Crewther DP, Brennan NA, Efron N, Holden BA, 1987. Corneal thickness changes following sleep and overnight contact lens wear in the primate (*Macaca fascicularis*). *Curr Eye Res* 6 (6): 809–815.

Magakyan GO, 1953. [Materials to study arterial pressure in monkeys. In: *Proc Inst Clin and Exp Cardiol of Acad Sci GeorgSSR*] 2. Tbilisi: 107–111 (Russian).

Magakyan GO, 1977. [An experimental study of pathogenesis of arterial hypertension and ischemic heart disease on monkeys]. *Vestn AMN SSSR* # 8: 20–24 (Russian).

Magrath D, Reeve P, 1993. On the role of the World Health Organization in the development of Sabin vaccines. *Biologicals* 21 (4): 345–348.

Mahaney MC, Sciulli PW, 1983. Hominid – pongid affinities: A multivavariate analysis of hominoid odontometrics. *Curr Anthropol* 24 (3): 382–386.

Mahaney MC, Sciulli PW, 1984. On the classification of *Homo. Curr Anthropol* 25 (1): 131–132.

Mahany T, Khirabadi BS, Gersten DM, Kurian P, Ledley RS, Ramwell PW. Studies on the affinity chromatography of serum albumins from human and animal plasmas. *Compar Biochem Physiol* B 68 (2): 319–323.

Maher WP, 1990. Palmar-digital arterial networks in fifty pairs of human fetal hands: Arteriographic models for clinical consideration. *Amer J Anat* 187: 201–210.

Mahoney CJ, 1985. Rev: Reproduction in New World Primates. New Models in Medical Science. J. P. Hearn, ed., 1983. *J Med Primatol* 14 (6): 363–370.

Mahoney WC, Nute PE, 1980. Amino acid sequence of the hemoglobin α-chain from a baboon (*Papio cynocephalus*): A product of gene fusion? *Biochemistry* 19: 1529–1534.

Mai LL, 1985. Chromosomes and the taxonomic status of the genus *Tarsius*: Preliminary results. *J Human Evol* 14 (3): 229–240.

Maina JN, 1987. The morphology and morphometry of the adult normal baboon lung (*Papio anubis*). *J Anatomy* # 150, Feb: 229–245.

Maita T, Aizawa N, Kuwahara M, Fukushima K, Matsuda G, Goodman M, 1979. Amino acid sequences of α- and β-chains of adult hemoglobin of the patas monkey (*Erythrocebus patas*). *J Chem Soc Pak* 1 (1): 5–13.

Malley A, Baecher L, Burger D, 1971. The role of complement in allergen-reagin mediated histamine release from monkey lung tissue. *Proc Soc Exp Biol Med* 136 (2): 341–343.

Maloney AG, Schmucker DL, Vessey DS, Wang RK, 1986. The effects of aging of the hepatic microsomal mixed-function oxidase system of male and female monkeys. *Hepatology* 6 (2): 282–287.

Malyukova IV, Nikitin VS, Uvarova IA, Sylakov VL, 1990. [Comparative-physiological study of generalization function in primates. *J evoluzion biochim i physiol* 26 (6): 801–810 (Russian w/English sum).

Manuelidis L, Wu JC, 1978. Homology between human and simian repeated DNA. *Nature* 276 (5683): 92–94.

Maples WR, 1972. Systematic reconsideration and a revision of the nomenclature of Kenya baboons. *Amer J Phys Anthropol* 36 (1): 9–19.

Marcario JK, Raghavan R, Joag SV, Foresman LL *et al.*, 1997. Correlation of behavioral, physiological, virological and neuropathological variables associated with SIV infection in rhesus macaques. *Soc Neurosci Abstr* 23 (Pt 1): 464.

Marfurt CF, Echtenkamp SF, Jones MA, 1989. Origins of the renal innervation in the primate, *Macaca fascicularis*. *J Auton Nerv Syst* 27: 113–126.

Markaryan DS, Adjigitov FI, 1980. [Using monkeys for estimation of mutagenous and cancerogenic activity of chemical substances in surroundings of human inhabits. In: *Biological Characteristics of Animals and Extrapolation to Man of Experimental Data: Materials of All-Union Conference*] M: 300–302 (Russian).

Marks J, Diamond JM, 1988. Relationships of human to chimps and gorillas. *Nature* 334: 654–658.

Marks J, 1991. What's old and new in molecular phylogenetics? *Amer J Phys Anthropol* 85 (2): 207–219.

Marshall JT, Jr, Marshall ER, 1976. Gibbons and their territorial songs. *Science* 193 (4249): 235–237.

Marshall J, Sugardjito J, 1986. Gibbon systematics. In: *Systematics, Evolution and Anatomy*. NY: 137–185.

Martin L, 1986. 10 relationships aming extant and extinct great apes and humans. In: *Major Topics in Primate and Human Evolution*. B. Wood, L. Martin, P. Andrews, eds. Cambridge Univ Press: 162–187.

Martin MA, Hoyer BH, 1967. Adenine plus thymine and guanine plus cytosine enriched fractions of animal DNAs as indicators of polynucleotide homologies. *J Mol Biol* 27: 113–129.

Martin RD, 1985. Primates: A definition. Symp: 'Major Topics in Primate Evolution'. *J Anatomy* 140 (3): 509.

Martin RD, 1990. *Primate Origins and Evolution*. London, Chapman & Hall.

Martin RD, 1993. Primate origins: Pluggig the gaps. *Nature* 363: 223–234.

Martin SL, Zimmer EA, Davidson WS, Wilson AC, 1981. The untranslated regions of β-globin mRNA evolve at a functional rate in higher primates. *Cell* 25 (3): 737–741.

Mason RG, Reed MS, 1971. Some species differences in fibrinolysis and blood coagulation. *J Biomed Mater Res* 5: 121–128.

Mason WA, 1968. Scope and potential of primate research. *Sci and Psychol* 12: 101–112.

Mason WA, 1990. Premises, promises, and problems of primatology. *Amer J Primatol* 22 (2): 123–138.

Matano S, 1987. A volumetric comparison of the vestibular nuclei in primates. *Folia Primatol* 47 (4): 189–203.

Mathew J, Balasubramanian AS, 1984. Anionic forms of brain arylsulfatase B: Evidence for a phosphorylated form in man and monkey. *Developm Neurosci* 6 (4–5): 278–284.

Matsuda G, 1985. Phylogeny of primates inferred from α-hemoglobin sequences. *Proc Jap Acad* B 61 (8): 359–362.

Matsumiya G, Gundry SR, Nehlsen-Cannarella S, Fagoaga OR, Morimoto T, Arai S, Fukushima N, Zuppan CW, Bailey LL, 1997. Serum interleukin-6 level after cardiac xenotransplantion in primates. *Transplant Proc* 29 (1–2): 916–919.

Matsuzawa T, 1985. Colour naming and classification in a chimpanzee (*Pan troglodytes*). *J Human Evol* 14 (3): 283–291.

Matsuzawa T, 1990. Spontaneous sorting in human and chimpanzee. In: *'Language' and Intelligence in Monkeys and Apes. Comparative Developmental Perspectives*. NY, Cambr Univ Press: 451–468.

Mayr E, 1965. Classification and philogeny. *Amer Zool* # 5: 165–174.

Mayr E, 1971. [*The Principles of Zoological Systematics*]. M, Mir (Russian).

Mazue G, Richez P, 1982. Problems in utilizing monkeys in toxicology. In: *Animals in Toxicology Research*. I. Bartosek *et al.*, eds. NY, Raven Press: 147–162.

Mazur A, 1985. A biosocial model of status in face-to-face primate groups. *Soc Forces* 64 (2): 377–402.

McConathy W, Weech P, 1981. Antigenic homologies among primates in studying human apolipoproteins. *Arteriosclerosis* 1: 54.

McConkey EH, 1997. The origin of human chromosome 18 from a human/ape ancestor. *Cytogenet a. Cell Genet* 76 (3–4): 189–191.

McDermid EM, Ananthakrishnan R, 1972. Red cell enzymes and serum proins of *Cercopithecus aethiops* (South African green monkey). *Folia Primatol* 17 (1–2): 122–131.

McDermott WC, 1938. *The Ape in Antiquity*. Baltimore, J. Hopkins Press.

McEvoy SM, Maeda N, 1988. Complex events in the evolution of the haptoglobin gene cluster in primates. *J Biol Chem* **263** (30): 15740–15747.

McGraw AP, Sim AK, 1972. Clinical biochemistry of the baboon. *J Compar Pathol* **82**: 193–200.

McGrew WC, 1987. Tools to get food: The substants of Tasmanian aborigines and Tanzanian chimpanzees compared. *J Anthropol Res* **43** (3): 247–258.

McGrew WC, 1992. *Chimpanzee Material Culture Implications Human Evolution*. Cambridge, Cambr Univ Press.

McGrew WC, Marchant LF, Nishida T, eds, 1986. *Great Ape Societies*. Cambridge, Cambr Univer Press.

McGuire MT, 1974. *The St. Kitts Vervet. (Contribs Primatol* 1). Basel, S. Karger.

McGullough B, Wackwitz R, Johanson WG, 1979. Total lung capacity of baboons and humans determined by planimetry of radiographs. *Lab Anim Sci* **29** (1): 61–67.

McHenry HM, 1984. The common ancestor. A study of the postcranium of *Pan paniscus, Australopithecus*, and other hominoids. In: *Pygmy Chimpanzee: Evol Biol Behav*. New York – London: 201–230.

McIntosh GH, Lawson CA, Rodgers SE, Lloyd JV, 1985. Haemotological characteristics of the common marmoset (*Callithrix jacchus jacchus*). *Res Vet Sci* **38** (1): 109–114.

McIntosh GH, McMurchie EJ, James M, Lawson CA, Bulman FH, Charnock JS, 1987. Influence of dietary fats on blood coagulation and prostaglandin production in the common marmoset. *Arteriosclerosis* **7** (2): 159–165.

McKenna JJ, 1986. An anthropological perspective on the sudden infant death syndrome (SIDS): The role of parental breathing cues and speech breathing adaptations. *Med Anthropol* **10** (1): 9–92.

McKenna MC, 1975. Toward a phylogenetic classification of the Mammalia. In: *Phylogeny of Primates. Multidisciplinary Approach*. New York – London: 21–45.

McKinney WT, Moran EC, Kraemer GW, Prange AJ, 1980. Long-term chlorpromazine in rhesus monkeys: Production of dyscinesias and changes in social behavior. *Psychopharmacologia* (Berlin) **72** (1): 35–39.

McMahan MR, 1982. Complement components C3, C4, and Bf in six nonhuman primate species. *Lab Anim Sci* **32** (1): 57–59.

Mechnikov I. I., 1887 (1950). [On the struggle of phagocytes in relapsing fever. In: *Academ Collection of I. I. Mechnikov's Works*], vol 6. M: 91–101 (Russian).

Mechnikov I. I. (Metchnikoff Elie), 1903. *Études sur la nature humains*. Paris.

Mechnikov I. I., 1903–1905 (1959). [Experiments on infection syphilis to simians. In: *Academ Collection of I. I. Mechnikov's Works*], vol 10: 347–387 (Russian).

Mechnikov I. I., 1909. [The survey of main success achievements of science on microbes for 1909. In: *Academ Collection of I. I. Mechnikov's Works*], vol 9. M: 136–139 (Russian).

Mechnikov I. I. (with A. Bezredka), 1910 (1955). [Experimental enteric fever. In: *Academ Collection of I. I. Mechnikov's Works*], vol 9. M: 255–256 (Russian).

Mechnikov I. I., (1910) 1943. [Darwinism and medicine. In: *About Darwinism*]. Moscow-Leningrad (Russian).

Mechnikov I. I. (Metchnikoff E. E.), 1915. Causerie de E. E. Metchnikoff. *Ann de l'Inst Pasteur* **29** (8): 364–368.

Mechnikov I. I. (Metchnikoff E) et Roux Em, 1903. Études expérimentales sur la Syphilis (Premiére memoire). *Ann de l'Inst Pasteur* **18** (12).

Mehta S, Nain CK, Relan NK, Kalsi HK, 1980. Energy metabolism of the brain in malnutrition. A experimental study in young rhesus monkeys. *J Med Primatol* **9** (6): 335–342.

Mehtali M, Imler JL, Sorg T, Pavirani A, 1995. [Gene therapy for inherited and acquired human diseases]. *Ann d'endocrinol* **56** (6): 571–574 (French w/English sum).

Meier B, Albignac R, Peyrieras A, Rumpler Y, Wright P, 1987. A new species of *Hapalemur* (Primates) from South East Madagascar. *Folia Primatol* **48** (3–4): 211–215.

Meireles CMM, Czelusniak J, Schneider MPC, Muniz JAPC, Brigido MC, Ferreira HS, Goodman M, 1999. Molecular phylogeny of Ateline New World monkeys (Platyrrhini, Atelinae) based on gamma-globin gene sequences: Evidence that *Brachyteles* is the sister group of *Lagothrix*. *Mol Phylogenet and Evol* **12** (1): 10–30.

Meireles CMM, Schneider MPC, Sampaio MGC, Schneider N, Slightom JL, Chi-hua Ch, Neiswanger K, Gumucio DL, Czelusniak J, Goodman M, 1995. Fate of a reduntant γ-globin gene in the atelid clade of New World monkeys: Implications concerning fetal globin gene expression. *Proc Natl Acad Sci USA* 92: 2607–2611.

Melendez LV, Hunt RD, Daniel MD, Blake BJ, Garcia FG, 1971. Acute lymphocytic leukemia in owl monkeys inoculated with *Herpesvirus saimiri*. *Science* 171 (3976): 1161–1163.

Mello NK, Bree MP, Ellingboe J, Mendelson JH, Harvey KL, 1984. Lack of acute alcohol effects on estradiol and luteinizing hormone in female macaque monkey. *Pharmacol Biochem Behav* 20 (2): 293–299.

Mello NK, Mendelson JH, Kelly M, Diaz-Migoyo N, Sholar JW, 1997. The effects of chronic cocaine self-administration on the menstrual cycle in rhesus monkeys. *J Pharmacol Exp Therapeut* 281 (1): 70–83.

Melnick DJ, Ashley MV, Williams A, Absher R, 1989. Evolutionary relationships and molecular divergence among Asian macaques. *Amer J Phys Anthropol* 78 (2): 270–272.

Melnik LA, 1978. [Comparative quantitative characteristics of cell structures of 44 and 45 cortical fields in man and chimpanzee brain. In: *Studies on Localization and Organization of Cerebral Functions on Contemporary Stage*]. M, Institute of Brain, Acad Med Sci USSR: 104–106 (Russian).

Melnikov RA, Barabadze EM, 1968. [*Malignant Tumors in Simians*]. Leningrad, Medgiz (Russian).

Melsen B, Melsen F, Rolling I, 1977. Dentin formation rate in human teeth. *Calcified Tissue Res* 24 (Suppl): 16.

Mendelow B, Grobicki D, de la Hunt M, Marcus F, Metz J, 1980. Normal cellular and humoral immunologic parameters in the baboon (*Papio ursinus*) compare to human standards. *Lab Anim Sci* 30 (6): 1018–1021.

Mendoza CE, Hatina GV, 1971. Comparative starch-gel electrophoresis of liver esterases from seven species. *Compar Biochem Physiol* B 39 (3): 483–488.

Menini C, Silva-Barrat C, 1990. Interet du singe *Papio papio* pour l' etude de l'epilepsie. *Pathol-biol* 38 (3): 205–213.

Menini C, Silva-Barrat C, 1998. The photosensitive epilepsy of the baboon. A model of generalized reflex epilepsy. *Advan Neurol* 75: 29–47.

Menzel EW, 1984. Spatial cognition and memory in captive chimpanzee. In: *Biol Learning*. Berlin e.a: 509–531.

Mercier-Bodard C, Renoir JM, Baulieu E-E, Emiliozzi R, Duval D, 1981. Human sex steroid binding plasma protein. In: *Hormones in Normal and Abnormal Human Tissues* 2. Berlin – New York, Walter de Gruyter: 25–48.

Mering TA, 1990. [Structural prerequisites of the temporal ensuring of nervous activity]. *Uspechi Physiol Nauk* 21 (4): 103–106 (Russian).

Merker H-J, Sames K, Csato W, Heger W, Neubert R, 1988. The embryology of *Callithrix jacchus*. In: *Non-Human Primates – Developmental Biology and Toxicology*. Berlin, Ueberreuter-Wissenschafts Verlag: 217–242.

Metivier H, Nolibe D, Masse R, Lafuma J, 1974. Excretion and acute toxicity of 239 PuO$_2$ in baboons. *Health Phys* 27 (5): 512–514.

Meyers WM, Gormus BJ, Walsh GP, Baskin GB, Hubbard GB, 1991. Naturally acquired and experimental leprosy in nonhuman primates. *Amer J Trop Med Hyg* 44 (4, Suppl): 24–27.

Mezey E, 1989. Animal models for alcoholic liver disease. *Hepatology* 9 (6): 904–905.

Micha JP, Quimby F, 1984. Baboon cervical colposcopy, histology, and cytology. *Gynecol Oncol* 17 (3): 308–313.

Michejda M, 1980. Growth standard in the skeletal age of rhesus monkeys (*Macaca mulatta*), chimpanzee (*Pan troglodytes*) and man. In: *16 IABS Congr: The Standardization of Animals to Impuve Biomed Res and Control, San Antonio 1979. Develop Biol Stand*, vol 45. Basel, S. Karger: 45–50.

Michel J-B, Gueltier C, Philippe M, Galen F-X, Cowol P, Menard J, 1987. Active immunization against renin in normotensive marmoset. *Proc Natl Acad Sci USA* 84 (12): 4346–4350.

Miczek KA, Weerts EM, 1987. Seizures in drug-treated animals. *Science* 235 (4793): 1127.

Miller DA, 1977. Evolution of primate chromosomes: Man's closest relative may be the gorilla, not the chimpanzee. *Science* 198: 1116–1124.

Miller DA, Sharma V, Mitchell AR, 1988. A human-derived probe, p82H, hybridizes to the centromeres of gorilla, chimpanzee, and orangutan. *Chromosoma* 96 (4): 270–274.

Miller ER, Simons EL, 1996. Age of the first cercopithecoid, Prohylobates tandyi, Wadi Moghara, Egypt. *Amer J Phys Anthropol* Suppl 22: 168–169.

Miller J, Donaldson J, 1976. Primate model for studies of clinical problems of middle ear functions. *Ann Otol Rhinol Laryngol* 85 (2, Suppl 25): 1–8.

Miller YE, 1987. Monoclonal antibody Tf-1, Tf2 anti-transferrin. *Monoclon Antibody News* 5 (4): 17.

Milton K, Demment MW, 1988. Digestion and passage kinetics of chimpanzees fed high and low fiber diets and comparison with human data. *J Nutrit* 118: 1082–1088.

Miminoshvili DI, Magakyan GO, Kokaya GY, 1956. [Experience in modelling of experimental hypertony and coronary insufficiency on simians. In: *Theoretical and Practical Questions of Medicine and Biology in Experiments on Monkeys*]. M, Medgiz: 85–97 (Russian).

Miranda OC, Bito LZ, Kaufman PL, DeRosseau CJ, Raley SE, 1986. Ocular development and aging. 2. Different basic patterns of lenticular growth in cats and rhesus monkeys and the lack of discernible effect of sunlight on lenticular aging in Rh Ms. *Invest Ophtalmol Vis Sci* 27 (3): 215.

Misra S, Tyagi K, Chatterjee RK, 1997. Experimental transmission of nocturnally periodic Wuchereria bancrofti to Indian leaf monkey (*Presbytis entellus*). *Exper Parasitol* 86 (2): 155–157.

Mitchell AR, Gosden JR, 1978. Evolutionary relatioships between man and the great apes. *Sci Progr* (Oxford) 65: 273–293.

Mitchell GM, Frykman GK, Morrison WA, O'Brien BM, 1986. The nature and extent of histopathologic injury in human avulsed arteries and veins and in experimentally avulsed monkey arterias. *Practic a. Reconstr Surg* 78 (6): 801–810.

Mitchell RW, 1996. Videotaped evidence of self-recognition in primates. *IPS/ASP Congr Abstr*: # 307.

Mitchell RW, 1999. Scientific and popular conceptions of the psychology of great apes from the 1790s to the 1970s: de ja vu all over again. *Primate Report* 53: 3–118.

Mitchell SC, Waring RH, Wilson WL, Idle JR, Autrup H, Harris CC, Ritchie JC, Crothers MJ, Sieber SM, 1986. Sulphoxidation of S-carboxymethyl-L-cysteine in the rhesus monkey (*Macaca mulatta*), cynomolgus monkey (*Macaca fascicularis*), African green monkey (*Cercopithecus aethiops*) and the marmoset (*Callithrix jacchus*). *Compar Biochem Physiol* B 84 (2): 143–144.

Mittermeier RA, Konstant WR, 1998. Primate conservation: A retrospective and a look into the 21st Century survival of the World's most endangered primates. *Congr IPS* 17 *Abstr*: # 144.

Mivart St-G, 1873. On *Lepilemur* and *Cheirogaleus*, and the zoological rank of the Lemuroidea. *Proc Sci Meet Zool Soc London*: 484–510.

Miyamoto MM, Goodman M, 1990. DNA systematics and evolution of primates. *Annu Rev Ecol Syst* 24: 197–220.

Miyamoto MM, Koop BF, Slightom JL, Goodman M, Tennant MR, 1988. Molecular systematics of higher primates: Genealogical relations and classification. *Proc Natl Acad Sci USA* 85 (20): 7627–7631.

Miyamoto MM, Slightom JL, Goodman M, 1987. Phylogenetic relations of humans and African apes from DNA sequences in the Ψ_η-globin region. *Science* 238 (4825): 369–373.

Mladinich CRJ, Collins BR, Huang SW, 1987. Penicillin allergy in a stump-tailed macaque (*Macaca arctoides*). *J Zoo Med* 18 (2–3): 108–109.

Modell W, 1968. Malaria and victory in Vietnam. *Science* 162 (3860): 1346–1352.

Monaco WA, Wormington CM, 1990. The rhesus monkey as an animal model for age-related maculopathy. *Optometry Vis Sci* 67 (7): 532–537.

Monkey colonies for research on aging. *Res Resour Rep*, 1982, 6 (9): 14.

Monkeys, 1980. NY, Workman Publishing.

Montagna W, 1968. Why study primates? *Primate News* 6 (5): 3–5.

Montagna W, 1980. The skin of primates and origin of human skin. *Anthropol Contempor* 3 (2): 242–243.

Montagna W, 1985. The evolution of human skin. *J Human Evol* 14 (1): 3–22.

Montagna W, Uno H, 1968. The phylogeny of baldness. In: *Phylogenesis and Ontogenesis*. Basel, S. Karger: 9–24.

Montagu MFA, 1941. Knowledge of ancients regarding the ape. *Bull Hist Med* 10 (4): 525–543.

Montano LM, Selman M, Hong E, 1985. Differentes efectos de la adrenalina sobre el musculo liso traqueal de *Erythrocebus patas*: predominio de los receptores adrenergicos alfa. *Arch Invest Med* (Mex) 16 (2): 169–174.

Moore GW, Barnabas J, Goodman M, 1973. A method for constructing maximum parsimony ancestral amino acid sequences on a given network. *J Theoret Biol* 38: 459–485.

Moore LJ, Machlan LA, Lim MO, Yergey AL, Hansen JW, 1985. Dynamics of calcium metabolism in infancy and childhood. 1. Methodology and quantificatio in the infant. *Pediatr Res* 19 (4): 329–334.

Moor-Jankowski J, 1978. Primate models for man. *Pharm Med Future, 3 Intern Meet, Pharm Phys*, Brussels, # 79: 211–225.

Moor-Jankowski J, Wiener A, 1971. Blood groups of primates: Their contribution to taxonomy and phylogenetics. In: *Med Primatol 1970. Proc 2 Conf Exp Med Surg Primates, New York 1969*. Basel, S. Karger: 939–946.

Moor-Jankowski J, Wiener AS, Socha WW, Gordon EB, Davis JH, 1973. A new taxonomic tool: Serological reactions in cross-immunized baboons. *J Med Primatol* 2 (2): 71–84.

Mootnick AR, Haimoff EH, 1986. Species and sub-species identification of gibbons for use by zoological parks and breeding facilities. *Amer J Primatol* 10 (4): 420.

Mori Y, Iesato K, Ueda S, Wakashin Y, Wakashin M, Okuda K, 1981. Purification and immunilogical characterization of human liver-specific F antigen. *Clin Chem Acta* 112: 125–134.

Moriyama A, Takahashi K, 1980. Studies on distribution of acid proteases in primate lungs and other tissues by diethylaminoethylcellulose chromatography. *J Biochem* 87: 737–743.

Morris R, Morris D, 1966. *Men and Apes*. London, Hutchinson & Co.

Mouri T, 1990. Macaque phylogenies viewed from cranial nerve foramina. *J Anthropol Soc Nippon* 98 (2): 201.

Muchmore E, Potter JL, Moor-Jankowski J, Goldsmith EI, 1971. Teaching program in medical primatology for medical students. In: *Med Primatol 1970. Proc 2 Conf Exp Med Surg Primates*, New York 1969. Basel, S. Karger: 986–992.

Mufson EJ, Kordover JH, 1998. Nerve growth factor and its receptors in the primate forebrain: Alterations in Alzheimer's disease and potential use in experimental therapeutics. In: *Neuroprotect Signal Transductio*. M. P. Mattson, ed. Totowa, NJ, Humanan Press: 23–59.

Muggenburg BA, Hahn FF, Bowen JA, Bice DE, 1982. Flexible fiberoptic bronchoscopy of chimpanzee. *Lab Anim Sci* 32 (5): 534–537.

Murayama J-I, Utsumi H, Hamada A, 1989. Amino acid sequence of monkey erythrocyte glycophorin MK. Its amino acid sequence has a striking homology with that of human glycophorin A. Biochim et biophys acta. *Protein Struct and Mol Enzymol* 999 (3): 273–280.

Murphy GM, Zollman PE, Greaves MV, Winkelmann RK, 1987. Symptomatic dermographism (*Factitious urticaria*) -passive transfer experiments from human to monkey. *Brit J Dermatol* 116 (6): 801–804.

Musher DM, Schell RF, Knox JM, 1976. The immunology of siphilis. *Intern J Dermatol* 15 (8): 566–576.

Myers RE, 1969. The clinical and pathological effects of asphyxation in the fetal rhesus monkey. In: *Diagnosis and Treatment of Fetal Disorder*. K. Adamsons, ed. NY: 226–249.

Nadler RD, 1986. Great ape sexual behavior: Human implications? *Proc 7 World Congr Sexology*. P. Kothri, ed. Bombay, Vakil & Sons: 45–49.

Nadler RD, 1994a. Robert M. Yerkes and the early studies of primate sexual behavior. *Congr IPS 15*: 394.

Nadler RD, 1994b. The chimpanzee: A useful model for the human in research on hormonal contraception. In: *The Role of the Chimpanzee in Research*. G. Eder, E. Kaiser, F. A. King, eds. Basel, Karger: 56–67.

Nadler RD, 1996. Robert M. Yerkes and the early studies of primate sexual behavior. *Primate Rep* 45: 65–77.

Nadler RD, Dahl JF, 1989. Sexual behavior of the great apes. In: *Perspectives in Primate Biology* 3. New Delhi: 37–49.

Nadler RD, Dukelow WR, 1996. The first symposium on the history of primatology. Congr IPS 15, 3–8 August 1994, Kuta, Indonesia. *Primate Rep* 45: 3–4.

Nadler RD, Graham CE, Collins DC, Gould KG, 1979. Plasma gonadotropins, prolactin, gonadal steroids, and genital swelling during the menstrual cycle of lowland gorillas. *Endocrinology* 105 (1): 290–299.

Nadler RD, Graham CE, Gosselin RE, Collins DC, 1985. Serum levels of gonadotropins and gonadal steroids including testosterone during the menstrual cycle of the chimpanzee (*Pan troglodytes*). *Amer J Primatol* 9 (4): 273–284.

Nadler RD, Phoenix CH, 1991. Male sexual behavior: Monkeys, men, and apes. In: *Understanding Behavior. What Primate Studies tell Us about Human Behavior*. NY, Oxford Univ Press: 152–189.

Nagel U, 1971. Social organization in a baboon hybrid zone. *Proc 3 Intern Congr Primatol, Zurich 1970*. Vol 3. Basel, S. Karger: 48–57.

Naithani VK, Steffens GJ, Tager HS, Buse G, Rubenstein AH, Steiner DF, 1984. Isolation and amino-acid sequence determination of monkey insulin and proinsulin. *Hoppe-Seyler's Z Physiol Chem* 365, Mai: 571–575.

Nakajima T, Miyazaki S, Kogure T, Furukawa K, 1986. Expression and distribution of human blood group antigens in nonhuman primates. *Japan J Human Genet* 31 (2): 177–178.

Napier JR, Napier PH, 1967. *A Handbook of Living Primates*. London, Acad Press.

Napier JR, Napier PH, 1985. *The Natural History of the Primates*. Cambridge, MA, MIT Press.

Naquet R, 1973. L'epilepsie photosensible du *Papio papio*. Un modele de l'epilepsie photosensible de l'homme. *Arch ital biol* 111 (3–4): 516–526.

Naquet R, Silva-Barrat C, Menini C, 1995. Reflex epilepsy in the *Papio papio* baboon, particularly photosensitive epilepsy. *Ital J Neurologic Sci* 16 (1–2): 119–125.

Nash LT, Bearder SK, Olson TR, 1989. Synopsis of *Galago* species characteristics. *Intern J Primatol* 10 (1): 57–79.

National Primate Plan, 1978. Prepared by Interagency Primate Steering Committee. Washington, US Department of Health, Education and Welfare, 81 pp.

Neborskiy AT, Belkaniya GS, 1986. [Electrodermal activity in man and simians. *Space Biol Aviacosm Med* (Moscow)] 20 (3): 61–68 (Russian).

Nei M, 1978. The theory of genetic distance and evolution of human races. *Japan J Human Genet* 23: 341–369.

Neoh SH, Jahoda DM, Rowe DS, Voller A, 1973. Immunoglobin classes in mammalian species identified by cross-reactivity with antisera to human immunoglobulin. *Immunochemistry* 10 (12): 805–813.

Nesturch MF, 1960. [*Primatology and Anthropogenesis (Simians, Prosimians and Descent of Man)*]. M, Medgiz (Russian).

Nesturch MF, 1970. [*The Descent of Man*]. M, Nauka (Russian, German, Spanish).

Neubert D, Heger W, Klug S, Merker H-J, 1990. Marmosets as a convenient primate species for studies in reproductive biology and toxicology. *Teratology* 41 (5): 581.

Neubert R, Helge H, Neubert D, 1996. Nonhuman primates as models for evaluating substance-induced changes in the immune system with relevance for man. In: *Experim Immunotoxicol*. R. J. Smialowicz, M. P. Holsapple, eds. Boca Raton, CRC Press: 63–98.

Neubert R, Hinz N, Thiel R, Neubert D, 1996. Down-regulation of adhesion receptors on cells of primate embryos as a probable mechanism of the teratogenic action of thalidomide. *Life Sci* 58 (4): 295–316.

Newcomer CE, Fox JG, Taylor RM, Smith DE, 1984. Seborrheic dermatitis in a rhesus monkey (*Macaca mulatta*). *Lab Anim Sci* 34 (2): 185–187.

Newell-Morris L, 1981. Potentialities for research in primate dermatoglyphics. *Amer J Phys Anthropol* 54 (2): 258.

Newman JD, 1988. Primate hearing mechanisms. In: *Comparative Primate Biol*. Vol 4. *Neuroscience*. NY, Alan R. Liss: 469–499.

Nickells MW, Atkinson JP, 1990. Characterization of CR1- and membrane cofactor protein-like proteins of two primates. *J Immunol* 144 (1): 4262–4268.

Nicolle C, Comte C, Conseil E, 1909. [Experimentale transmission of Typhus exanthematicus by the body louse]. *C. r. Acad sci* (Paris) 149: 486–489 (French).

Nicolle C, Conseil E, 1911. Reproduction experimentale de la rougeole ches le Bonnet chinois. Virulence du sang des malades 24 heure avant le debut de l'eruption. *C. r. Acad sci* (Paris) 153: 1522–1524.

Niemitz C, 1984. Taxonomy and distribution of the genus *Tarsius* Storr, 1780. In: *Biol of Tarsiers*. Stuttgart – New York, Fischer Verlag: 1–16.

Niemitz C, 1985. Der Koboldmaki. Evolutionsforschung an einem Primaten. *Naturwiss Rdsch* 38 (2): 43–49.

Niemitz C, Nietsh A, Warter S, Rumpler Y, 1991. *Tarsius dianae*: A new primate species from central Sulawesi (Indonesia). *Folia Primatol* 56 (2): 105–116.

Nikitenko MF, 1967. [On the brain types of mammals in relation with their evolution and way of life. In: *Ecology of Mammalia and Birds*]. M, Nauka: 273–285 (Russian).

Nimchinsky EA, Gillissen E, Allman JM, Pere DP, Ervin JM, Hof PR, 1999. A neuronal morphologic type unique to humans and great apes. *Proc Nat Acad Sci USA* 96 (9): 5268–5273.

Nishida T, 1968. The social group of wild chimpanzees in the Mahali mountains. *Primates* 9: 167–224.

Nishida T, 1996. Mahale chimpanzee studies: Past, present and future. *IPS/ASP Congr Abstr*: # 046.

Nishida T, 1998. Primate culture and primate conservation. *Congr IPS* 17 *Abstr*: # 161.

Nishida T, Kawanaka K, 1985. Within-droup cannibalism by adult male chimpanzees. *Primates* 26 (3): 274–284.

Nishikawa I, Kawanishi G, Cho F, Honjo S, Hatakeyama T, Wako H, 1976. Chemical composition of cynomolgus monkey milk. *Exp Anim* 25 (4): 253–264.

Nishio H, Gibbs PEM, Minghetti PP, Zielinski R, Dugaiczyk A, 1995. The chimpanzee alpha-fetoprotein-encoding gene shows structural similarity to that of but distinct differences from that of human. *Gene* 16 (2): 213–220.

Nissen HW, 1944. Yerkes laboratory of primate biology. *Anim Kingdom* 47 (6): 12–19.

Nobrega FG, Ozols J, 1971. Amino acid sequences of tryptic peptides of cytochromes B_5 from microsomes of human, monkey, porcine, and chicken liver. *J Biol Chem* 216 (6): 1706–1717.

Noireau F, 1987. HIV transmission from monkey to man. *Lancet* # 8548: 1498–1499.

Nooij FJM, Borst JG, van Meurs GJE, Jonker M, Balner H, 1986. Differentiation antigens on rhesus monkey lymphocytes. 1. Identification of T cells bearing CD3 and CD8, and of subset of CD8-bearing cells. *Europ J Immunol* 16 (8): 975–979.

Norley S, Kurth R, 1997. Simian immunodeficiency virus as a model of HIV pathogenesis. *Springer Semin Immunopathol* 18 (3): 391–405.

Normile D, 2001. Gene expression differs in human and chimp brains. *Science* 292: 44–45.

Nosawa R, Shotake T, Kawamoto Y, Tanabe Y, 1982. Electrophoretically estimated genetic distance and divergence time between chimpanzee and man. *Primates* 23 (3): 432–443.

Nussenblatt RB, Gery I, 1996. Experimental autoimmune uveitis and its relationship to clinical ocular inflammatory disease. *J Autoimmunity* 9 (5): 575–585.

Nute PE, Mahoney WC, 1979. Complete sequence of the β-chain from the fetal hemoglobin of the baboon, *Papio cynocephalus*. *Haemoglobin* 3: 399–410.

Nute PE, Mahoney WC, 1980. Complete primary structure of the β-chain from the hemoglobin of a baboon, *Papio cynocephalus*. *Haemoglobin* 4 (2): 109–123.

Nuttall GHF, 1902. The new biological test for blood in relation to zoological classification. *Proc Roy Soc London* 80, April: 150–155.

Nuttall GHF, 1904. *Blood Immunity and Blood Relationship*. Cambridge, Cambr Univ Press.

O'Conor GT, 1969. Cancer – a general review. In: *Using Primates in Med Res. Part II. Rec Compar Research*. Basel – New York, S. Karger: 9–22.

Ochoa-Zarzosa A, Carapia A, Hernandez-Lopez L, Mayagoitia L, 1994. Female behavioral variations in relation to the menstrual cycle in stumptail macaques. *Congr IPS* 15: 231.

Oehler R, Sharpe LT, 1989. Dark adaption and increment threshold in rhesus monkey and man. *Exp Brain Res* 75 (3): 664–668.

Offenbacher S, Alexander SP, McClure H, Strobert E, Orkin JL, van Dyke TE, 1977. Cross-sectional study of the prevalence and severity of periodontal disease in rhesus monkeys. *J Dent Res* 66 (Spec iss): 232.

Ogata M, Kotani H, Koshimizu K, Magaribuchi T, 1981. Isolation and serological characterization of ureaplasmas from nonhuman primates. *Japan J Vet Sci* 43 (4): 521–529.

Ohgaki H, Hasegawa H, Kusama K, Morino K, Matsukura H, Sato S, Maruyama K, Sugimura T, 1986. Induction of gastric carcinomas in nonhuman primates by N-ethyl-N'-nitro-N-nitrosoguanidine. *J Natl Cancer Inst* 77 (1): 179–186.

Ohmori S, Kitada M, Hamada A, Igarashi T, Ueno K, Kanakubo Y, Kitagawa H, 1988. Purification and same properties of NADPH-cytochrome P-450 reductase from crab-eating monkey liver microsomes. *Biochem Int* 17 (2): 249–256.

Ohnishi K, 1991. A tentative evolutionary tree of mammalian orders constructed by hennigian comparison of the amino acid sequences of alpha-crystallin A chain, myoglobin, and hemoglobin alpha chain. *Sci Rep Nigata Univ* # 28: 19–31.

Ohshima T, Takayasu T, Maeda H, Nagano T, 1988. Immunohistochemical studies on ABO (H)- and Lewis-activities in the primate tissues and cells. *Japan J Legal Med* 42 (1): 13–20 (Japanese w/English sum).

Ohye C, 1979. Primate model of Parkinsonian motor symptoms. *Advances Neurol Sci* 23 (5): 949–955 (Japanese w/English sum).

Okada M, Okada T, Kawaguchi N, Sato Y, Ide K, 1988. Use of latex sensitized with human hemoglobin in species identification. *Japan J Legal Med* 42 (2): 169–172 (Japanese w/English sum).

Oksche A, Liesner R, Tigges J, Tigges M, 1984. Intraepithelial inclusions resembling human Biondi bodies in the choroid plexus of an aged chimpanzee. *Cell Tissue Res* 235: 467–469.

Okuyama Y, Morino A, 1997. Pharmacokinetics of prulifloxacin. 3 communication: Metabolism in rates, dogs and monkeys. *Arzneim-Forsch//Drug Res* 47 (1–3): 293–298.

Olenev SN, 1987. [*The Construction of Brain*]. Leningrad, Medicine (Russian).

Ordog T, Chen MD, O'Byrne KT, Goldsmith JR, Connaughton MA, Hotchkiss J, Knobil E, 1998. On the mechanism of lactational anovulation in the rhesus monkey. *Amer J Physiol* 274 (4, Pt 1): E 665–E 676.

Orihel TC, 1971. Primates as models for parasitological research. In: *Med Primatol 1970*. Basel, Karger: 772–782.

Orjechovskaya NS, 1983. [Cytoarchitectonics of the frontal cortical areas in *Papio hamadryas* and *Macaca mulatta* ontogeny. *Arch anat, histol i embryol*] (Strasburg) 84: 32–33 (Russian).

Orlov OYu, 1972. [On the evolution of color vision in vertebrates. In: *Problems of Evolution*]. Vol 2: 69–94.

Oshigova AP, 1981. [*Morphological Bases of Visual Analyzer Evolution of Primates*] Synopsis of Thesis Doctor Diss. M, Moscow Lomonosov Univ (Russian).

Oshigova AP, 1982. Morphology of the visual system of primates in an evolutionary context. *Intern J Primatol* 3 (3): 321.

Overman WH, Bachevalier J, Schuhmann E, Ryan P, 1996. Cognitive gender differences in very young children parallel biologically based cognitive gender differences in monkeys. *Behav Neurosci* 119 (4): 673–684.

Overman WH, Ormsby G, Mishkin M, 1990. Picture recognition vs. picture discrimination learning in monkeys with medial temporal removals. *Exp Brain Res* 76 (1): 18–24.

Owiti GE, Tarara RP, Hendrickx AG, 1989. Fetal membranes and placenta of the African green monkey (*Cercopithecus aethiops*). *Anat and Embriol* 179 (6): 591–604.

Oxbury JM, 1970. Prefrontal lesions in man and monkey. *Brain Res* 24: 555.

Oxnard CE, 1983. Anatomical, biomolecular and morphometric views of the primates. *Progr Anat* 3. Anat Ass, Great Britain and Ireland: 113–142.

Oxnard CE, 1984. *The Order of Man. A Biomathematic Anatomy of the Primates*. New Haven – London, Yale Univ Press.

Palatnik M, Rowe A, 1984. Duffy and Duffy-related human antigens in primates. *J Human Evol* 13 (2): 173–179.

Palfi S, Conde F, Riche D, Brouillet E, Dautry C, Mittoux V, Chibois A, Peschanski M, Hantraye P, 1998. Fetal striatal allografts reverse cognitive deficits in a primate model of Huntington disease. *Nature Medicine* 4 (8): 963–966.

Panov EN, 1983. [*Animals Behavior and Ethological Structure of Populations*]. M, Nauka (Russian).

Pare M, Munoz DP, 1996. Saccadic reaction time in the monkey: Advanced preparation of oculo-motor programs is primarily responsible for express saccade occurrence. *J Neuropathol* 76 (6): 3666–3681.

Parham P, 1996. Major histocompatibility complex (MHC) class I molecules and the immune system. *IPS/ASP Congr Abstr:* # 599.

Parker DC, Morishima M, Koerker DJ, Gale CC, Goodrer CJ, 1972. Pilot study of growth hormone release in sleep of the chair-adapted baboon: Potential as model of human sleep release. *Endocrinology* 91 (6): 1462–1467.

Passingham RE, Ettlinger G, 1974. A comparison of cortical functions in man and other primates. *Intern Rev Neurobiol* 16. NY e.a: 233–299.

Patterson F, 1983. Child, chimp and gorilla. A coparison of early language samples. *Gorilla* 7 (1): 2–3.

Patterson FGP, Cohn RH, 1994. Self-recognition and self-awareness in lowland gorillas. In: *Self-Awareness in Animals and Humans: Developmental Perspectives.* Parker ML *et al.*, eds. Cambridge, Cambr Univ Press: 273–290.

Patterson FGP, Holts CL, 1989. A review of project Koko: Language acquisition by two lowland gorillas. Part I. *Gorilla* 12 (2): 2–5.

Patterson F, Tanner J, Mayer N, 1988. Pragmatic analysis of gorilla utterances. *J Pragmat* 15: 35–54.

Patterson R, Harris KE, 1997. Models of asthma. In: *Allergic Diseases*, 5 edit. Patterson R., Grammer L., Greenberger P., eds. Philadelphia, Lippincott-Raven Publishers: 41–48.

Paule MG, Forrester TM, Maher MA, Cranmer JM, Allen RR, 1990. Monkey versus human performance in the NCTR operant test battery. *Neurotoxicol Teratol* 12 (5): 503–507.

Paules RS, Propst F, Dunn KJ, Blair DG, Kaul K, Palmer AE, Vande Woude GF, 1988. Primate c-mos proto-oncogene structure and expression: Transcription initiation both upstream and within the gene in a tissue-specific manner. *Oncogene* 3: 59–68.

[*Pavlov's Wednesdays*], vols I, II, III, 1949. Moscow – Leningrad, Acad Sci USSR (Russian).

Pearson PL, 1973. The uniqueness of the human karyotype. *Nobel Symp* 23: 145–151.

Pekow CA, Weller RE, Schulte SJ, Lee SP, 1995. Dietary induction of cholesterol gallstones in the owl monkey: Preliminary findings in a new model. *Lab Anim Sci* 45 (6): 657–662.

Peng Y, Ye Z, Zhang Y, Liu R, 1989. Observations on the position of genus *Rhinopithecus* in phylogeny. In: *Progress in the Studies of Golden Monkey.* China: 11–19 (Chinese w/English sum).

Pennisi E, 1997. Monkey virus DNA found in rare human cancers. *Science* 275 (5301): 748–749.

Perera KL, Handunnetti SM, Holm I, Longacre S, Mendis K, 1998. Baculovirus merozoite surface protein 1 C-terminal recombinant antigens are highly protective in a natural primate model for human *Pasmodium vivax* malaria. *Infect Immunity* 66 (4): 1500–1506.

Perez Arellano JL, Barrios Gonzalez MN, Martin Dominguez T, Sanchez Benitez de Soto ML, Jimenez Lopez A, 1992. Experimental models of hypersensitivity pneumonitis. *J Investigational Allergol Clin Immunol* 2 (4): 219–228.

Perlman F, 1969. Allergy. *Primate News* 7 (8): 4–11.

Permadi D, Tumbelaka LI, Yusuf TL, 1994. Reproductive pattern of *Tarsius* spp. in the captive breeding. *Congr IPS* 15: 421.

Perrett DI, Mistlin AJ, Chitty AJ, Smith PAJ, Potter DD, Broennimann PK, Harries M, 1988. Special-ized face processing and hemispheric asymmetry in man and monkey: Evidence from single unit and reaction time studies. *Behav Brain Res* 29: 245–250.

Perry TL, Yong VW, Hansen S, Jones K, Bergeron C, Foulks JA, Wright JM, 1987. Alpha-tocopherol and beta-carotene do not protect marmosets against the dopaminergic neurotoxicity of N-methyl-4-phenyl-1, 2, 3, 6-tetra-hydropyridine. *J Neurol Sci* 81: 321–331.

Persson GR, Engel LD, Whitney CW, Weinberg A, Moncla BJ, Darveau RP, Houston L, Braham P, Page RC, 1994. *Macaca fascicularis* as a model in which to assess the safety and efficacy of a vaccine for periodontitis. *Oral Microbiol Immunol* 9 (2): 104–111.

Peters JH, Berridge BJ, Chao WR, Cummings JG, Lin SC, 1971. Amino acid patterns in the plasma of Old and New World primates. *Compar Biochem Physiol* 39: 639–647.

Petersen SE, Robinson DR, 1986. Damage to parietal cortex produces a similar deficit in man and monkey. *Invest Ophtalmol Vis Sci* 27 (3, Suppl): 18.

Petrov NN, Krotkina NA, Vadova AV, Postnikova ZA, 1951. [*Dynamics of Origin and Development of the Cancer Growth in Experiments on Simians*]. M, Medgiz (Russian).

Pettigrew JD, 1991. Primate relations as revealed by brain characters: The 'flying primate' hypothesis. *Amer J Phys Anthropol* Suppl 12: 142–143.

Phan DT, Benczur M, Gidali J, Feher I, Harsany V, Nemes L, Petranyi G, Hollan SN, 1989. *Cercopithecus aethiops* monkey as a reliable model for in vitro study of T cell depletion of bone marrow with Campath-1 plus complement. *Folia Haematol* (DDR) 116 (1): 97–106.

Phoenix CH, Chambers KC, 1986. Aging and primate male sexual behavior. *Proc Soc Exp Biol Med* 183 (2): 151–162.

Pieper WA, Skeen MJ, McClure HM, Bourne GH, 1972. The chimpanzee as an animal model for investigating alcoholism. *Science* 176 (4030): 71–73.

Pierce EA, Dame MC, Bouillon R, van Baelen H, DeLuca HF, 1985. Monoclonal antibodies to human vitamin D-binding protein. *Proc Natl Acad Sci USA* 82 (24): 8429–8433.

Pierpont GL, DeMaster EG, Reynold S, Pederson J, Cohn JN, 1985. Ventricular myocardial catecholamines in primates. *J Lab Clin Med* 106 (2): 205–210.

Pike LM, Carlisle A, Newell C, Hong S-B, Musich PR, 1986. Sequence evolution of rhesus monkey alphoid DNA. *J Mol Evol* 23 (2): 127–137.

Pilbeam D, 1984. [The origin of hominoids and hominids]. *V Mire Nauki* (M) # 5: 38–48 (Russian).

Pilbeam D, 1985. Patterns of hominoid evolution. In: *Ancestors: The Hard Evidence*. NY, Alan R. Liss: 51–59.

Pilbeam D, 1986. Distinguished lecture: Hominoid evolution and hominoid origins. *Amer Anthropologist* 88 (2): 295–312.

Pilbeam D, 1996. Genetic and morphological records of the Hominoidea and hominid origins: A synthesis. *Mol Phylogenet Evol* 5 (1): 155–168.

Pilleri G, 1984. Movements related to the preservation of the species and social instincts appearing as nueurological symptoms in degenerative dementias. *Brain Pathol*, vol 1. Bern: 39–69.

Pindak FF, Gardner WA, Mora de Pindak M, Abee CR, 1987. Detection of hemagglutinins in cultures of squirrel monkey investinal trichomonads. *J Clin Microbiol* 25 (4): 609–614.

Ploog D, 1972. Kommunikation in Affengesellschaften und deren Bedeutung für die Verständigungsweisen des Menschen. *Neue Anthropologie* 2 (2): 98–178.

Poggio GF, 1995. Mechanisms of stereopsis in monkey visual cortex. *Cerebral Cortex* 3, May/Jun: 193–204.

Pohl CR, Knobil E, 1982. The role of the central nervous system in the control of ovarian function in higher primates. *Ann Rev Physiol*, vol 44. Palo Alto, Calif: 583–593.

Pohl P, 1983. Central auditory processing. V. Ear advantages for acoustic stimuli in baboons. *Brain Lang* 20 (1): 44–53.

Polites HG, Melchior GW, Castle CK, Marotti KR, 1986. The primary structure of cynomolgus monkey apolipoprotein A-I deduced from the cDNA sequence: Comparison to the human sequence. *Gene* 49: 103–110.

Pollock JI, Mullin RJ, 1987. Vitamin C biosynthesis in prosimians: Evidence for the anthropoid affinity of *Tarsius*. *Amer J Phys Anthropol* 73 (1): 65–70.

Pond CM, Mattacks CA, 1987 (1988). The anatomy of adipose tissue in captive *Macaca* monkeys and its implications for human biology. *Folia Primatol* 48 (3–4): 164–185.

Pope E, Dresser B, Chin N, Liu J, Loskutoff N, Behnke E, Brown C, McRae M, Sinoway C, Campbell M, Cameron C, Owens O, Johnson C, Evans R, Cedars M, 1996. Birth of a western lowland gorilla (*Gorilla gorilla gorilla*) following in vitro fertilization and embryo transfer. *IPS/ASP Congr Abstr*: N 006.

Pope NS, Gould KG, Anderson DC, Mann DR, 1989. Effects of age and sex on bone density in the rhesus monkey. *Bone* 10: 109–112.

Porter CA, Sampaio I, Schneider H, Schneider MPC, Czelusniak J, Goodman M, 1995. Evidence on primate phylogeny from E-globin gene sequences and flanking regions. *J Mol Evol* 40: 30–55.

Post PW, Szabo G, Keeling ME, 1975. A quantitative and morphological study of the pigmentary system of the chimpanzee with the light and electron microscope. *Amer J Phys Anthropol* 43 (3): 435–443.

Poulsen K, Haber E, Burton J, 1976. On the specificity of human renin. Studies with peptide inhibitors. *Biochim et biophys acta* 452 (2): 533–537.

Povinelli DJ, Boysen ST, Nelson KE, 1989. Can chimpanzees guess what others know? *Amer J Primatol* 18 (2): 161.

Povinelli DJ, Rulf AB, Landau KR, Bierschwale DT, 1993. Self-recognition in chimpanzees (*Pan troglodytes*): Distribution, ontogeny, and patterns of emergence. *J Compar Physiol* 107: 347–372.

Power M, 1991. *The Egalitarians – Human and Chimpanzee: An Anthropological View of Social Organization.* Cambridge, Cambr Univ Press.

Prasad MRN, Diczfalusy E, 1983. New contraceptives for men. What are the prospects? *Intern J Androl* 6 (4): 305–309.

Prathap K, 1975. Diet-induced aortic atherosclerosis in Malaysian long-tailed monkeys (*Macaca irus*). *J Pathol* 115, March: 163–174.

Premack AJ, Premack D, 1972. Teaching language to an ape. *Sci Amer* 227 (4): 92–99.

Premack D, 1983. Animal cognition. *Ann Rev Psychol.* Vol 34. Palo Alto, Calif: 351–362.

Premack D, 1984. Possible general effects of language training on the chimpanzee. *Human Develop* 27 (5–6): 268–281.

Premaratne S, May ML, Nakasone CK, McNamara JJ, 1995. Pharmacokinetics of endotoxin in a rhesus macaque septic shock model. *J Surg Res* 59 (4): 428–432.

Preobrajenskaya NS, 1974. [On some aromorphic changes in the brain of mammals]. *Uspechi sovremen biol* 78 (1): 157–167 (Russian).

Preston NW, Stanbridge TN, 1976. Mouse or man? Which are pertussis vaccines to protect. *J Hygiene* 76 (2): 249–256.

Preuschoft H, 1996. History of the International Primatology Society: The first sixteen years. *Primate Rep* 45: 5–13.

Preuschoft H, Chivers DJ, eds, 1993. *Hands of Primates.* NY, Springer–Verlag.

Preuschoft H, Hayama S, Gunther MM, 1989. Curvature of the lumbar spine as a consequence of mechanical necessities in Japanese macaques trained for bipedalism. *Folia Primatol* 50 (1–2): 42–58.

Preuschoft S, 1996. Behaviour phylogeny of primate 'laughter' and 'smile'. *Primate Rep* 44, Jan: 38.

Price DL, Martin LJ, Sisodia SS, Walker LC, Voytko ML, Wagster MV, Cork LC, Koliatsos VE, 1994. The aged nonhuman primate: A model for the behavioral and brain abnormalities occurring in aged humans. In: *Alzheimer Disease.* R. D. Terry, R. Katzman, K. L. Bick, eds. NY, Raven Press: 231–245.

Prichard RW, 1968. Some human diseases for which animal models are needed. In: *Animal Models for Biomedical Research. Proc Symp Natl Acad Sci USA.* Washington, D.C.: 157–167.

Pride MW, Bailey CR, Muchmore E, Thanavala Y, 1998. Evaluation of B and T-cell responses in chimpanzees immunized with Hepagene (R), a hepatitis B vaccine containing pre-S1, pre-S2 and S gene products. *Vaccine* 16 (6): 543–550.

Primate supply endangered. India cuts rhesus exports. 1974. *Natl Soc Med Res Bull* 25 (4): 1–2.

Prince AM, 1981. The use of chimpanzees in biomedical research. In: *Trends in Bioassay Methodology: In vivo, In vitro and Mathematical Approaches* (NIH publ # 82–2382). Washington, D.S.: 81–98.

Proctor P, 1970. Similar functions of uric acid and ascorbate in man? *Nature* 228 (5274): 868.

Promislov MSh, 1974. [On low molecular substances which are specific to CNS]. In: *Progress in Neurochemistry. Rep of 6 All-Union Conf of Neurochem].* Leningrad, Nauka: 174–179 (Russian).

Prost JH, 1979. Origin of bipedalism. *Amer J Phys Anthropol* 50 (3): 472.

Prouty LA, Buchanan PD, Pollitzer WS, Mootnick AR, 1983. A presumptive new hylobatid subgenus with 38 chromosomes. *Cytogenet Cell Genet* 35: 141–142.

Pryce C, Scott L, Schnell C, eds, 1997. *Handbook: Marmosets and Tamarins in Biological and Biomedical Research.* Salisbury, UK, DSSD Imagery.

Rangel DM, Bruckner WL, Byfield JE, Adomian GE, Dinbar A, Fonkalsrud EW, 1970. Enzymatic and ultrastructural evaluation of hepatic preservation in primates. *Arch Surg* 100, March: 284–286.

Rao LV, Singh O, Talwar GP, 1988. Immunological cross-reactivity of antibodies with species chorionic gonadotropin is a critical requirement for efficacy testing of human gonadotropin vaccines in sub-human primates. *J Reprod Immunol* 13: 53–63.

Rao PN, Davis FM, Wagner RD, 1985. Evolutionary significance of a monoclonal antibody specific for chromosomes of primates. *J Cell Biol* 101 (5, Part 2): 75.

Rapaport DH, Provis J, 1996. Preface: Why retina? Why primate? *Persp Developmental Neurobiol* 3 (3): 143–145.

Rapaport DH, Racic P, LaVail MM, 1996. Spatiotemporal gradients of cell genesis in the primate retina. *Persp Developmental Neurobiol* 3 (3): 147–159.

Rasmussen DT, 1990. The phylogenetic position of *Mahgerita stevensi*: Protoanthropoid or lemuroid? *Intern J Primatol* 11 (5): 439–469.

Rathbun WB, 1986. Glutathione synthesis in evolution: An Achilles' heel of human and other Old World simian lenses. *Ophtalmol Res* 18 (4): 236–242.

Raviola E, Wiesel TN, 1985. An animal model of myopia. *New Engl J Med* 312: 1609–1615.

Ravkina LI, 1972. [Experimental allergic encephalomyelitis as a model of human demyelising diseases. In: *Actual Problems of Virology and Prophylaxis of Viral Diseases. Theses of XVII Scient Session of Inst*]. M: 230 (Russian).

Ray J, 1693. *Sinopsis Methodica Animalium Quagrupedum et Serpenti Genaris*. Londini.

Rearden A, 1986. Evolution of glycophorin A in the hominoid primates studied with monoclonal antibodies, and description of a sialoglycoprotein analogous to human glycoprotein B in chimpanzee. *J Immunol* 136 (7): 2504–2509.

Redman CM, Lee S, Ten Bokkel HD, Rabin BI, Johnson CL, Oyen R, Marsh WL, 1989. Comparison of human and chimpanzee Kell blood group systems. *Transfusion* 29 (6): 486–490.

Redmond TM, Wiggert B, Robey FA, Chader GJ, 1986. Interspecies conservation of structure of inter-photoreceptor retinoid-binding protein. Similarities and differences as adjudged by peptide mapping and N-terminal sequencing. *Biochem J* 240 (1): 19–26.

Redshaw ME, 1989. A comparison of neonatal behaviour and reflexes in the great apes. *J Human Evol* 18 (23): 191–200.

Reed CA, 1963. Rev: Simpson GG. 'Principles of Animal Taxonomy'. NY, Columbia Univ Press, 1961, XII, 247 pp. *Amer J Phys Anthropol* 21 (2): 241–243.

Refino CJ, Modi NB, Bullens S, Pater C, Lipari MT, Robarge K, Blackburn B, Beresini M, Weller T, Steiner B, Bunting S, 1998. Pharmacokinetics, pharmacodynamics and tolerability of a potent, non-peptidic, GP IIb/IIIa receptor antagonist following multiple oral administrations of a prodrug form. *Thrombos Haemostas* 79 (1): 169–176.

Regional Primate Research Centers – The Creation of a Program, 1968. NIH, Bethesda.

Reimann KA, Waite BCD, Lee-Parritz DE, Lin W-y, Uchanska-Ziegler B, O'Connell MJ, Letvin NL, 1994. Use of human leukocyte-specific monoclonal antibodies for clinically immunophenotyping lymphocytes of rhesus monkeys. *Cytometry* 17 (1): 102–108.

Reindel JF, Dominick MA, Bocan TMA, Gough AW, McGuire EJ, 1994. Toxicologic effects of a novel acyl-CoA: Cholesterol acyltransferase inhibitor in cynomolgus monkeys. *Toxicologic Pathol* 22 (5): 510–518.

Reite M, Pegram GV, Stephens LM, Bixler EC, Lewis OL, 1969. The effect of reserpine and monoamine oxidase inhibitors on paradoxical sleep in the monkey. *Psychopharmacologia* (Berlin) 14: 12–17.

Reite M, Kaemingk K, Boccia ML, 1989. Maternal separation in bonnet monkey infants: Altered attach-ment and social support. *Child Develop* 60: 473–480.

Relf MV, 1996. Xenotransplantation of baboon bone marrow cells: A historical review of the protocols as a possible treatment modality for HIV/AIDS. *J Ass Nurses AIDS Care* 7 (5): 27–35.

Ren RM, 1996. The snub-nosed monkeys (*Rhinopithecus*) of China. *IPS/ASP Congr Abstr*: N 398.

Repin YuM, Startsev VG, 1975. [The mechanism of selective lesion of the cardiovascular system in the psycho-emotional stress]. *Vestn Acad Med Nauk SSSR* # 8: 71–76 (Russian).

Reynolds HH, 1969. Nonhuman primates in the study of toxicological effects on the central nervous system: A review. *Ann N. Y. Acad Sci* 162 (1): 617–629.

Reynolds TR, 1985. Stresses on the limbs of quadrupedal primates. *Amer J Phys Anthropol* 67 (4): 351–362.

Reynolds V, 1967. *The Apes. The Gorilla, Chimpanzee, Orangutan, and Gibbon – their History and their World.* NY.

Reynolds V, Reynolds J, 1995. Riding of the backs of apes. In: *Ape, Man, Apeman: Changing Views since 1600. Evaluative Proc Symp.* Netherlands, Leiden Univ: 395–405.

Rice DC, 1987. Primate research: Relevance to human learning and development. *Develop Pharmacol Ther* 10: 314–327.

Richt JA, Herzog S, Rott R, Oldach D, Pyper JM, Clements JE, Narayan O, 1994. Molecular biology and neuropathogenesis of Borna disease virus. *Neurolog Disease Ther* 27: 679–686.

Rimmelzwaan GF, Baars M, van Amerongen G, van Beck R, Claas ECJ, Osterhaus ADME, 1996. Protective immunity against influenza in a monkey model after experimental infection or vaccination with a new candidate vaccine. In: *Options for the Control of Influenza III.* L. E. Brown, A. W. Hampson, R. G. Webster, eds. Amsterdam, Elsevier: 653–660.

Rivers TM, 1927. Varicella in monkeys. Nuclear inclusions produced by varicella virus in the testicles of monkeys. *J Exp Med* 45: 961–968.

Rivkees SA, 1997. Developing circadian rhythmicity: Basic and clinical aspects. *Pediatr Clinics North Amer* 44 (2): 467–487.

Robbins MM, 1996. Male mating patterns in multimale mountain gorilla groups. *IPS/ASP Congr Abstr*: # 456.

Roberts A, 1996. Comparison of cognitive function in human and non-human primates. *Cognitive Brain Res* 3: 319–327.

Roberts ED, Baskin GB, Watson E, Henk WG, Shelton TC, 1984. Calcium pyrophosphate deposition disease (CPDD) in nonhuman primates. *Amer J Pathol* 116 (2): 359–361.

Roberts JA, Kaack MB, Baskin G, Svenson SB, 1993. Prevention of renal scarring from pyelonephritis in nonhuman primates by vaccination with a synthetic *Escherichia coli* serotype O8 oligosaccharide-protein conjugate. *Infect a. Immunity* 61 (12): 5214–5218.

Roberts JA, Kaack MB, Martin LN, 1995. Cytokine and lymphocyte activation during experimental acyte pyelonephritis. *Urol Res* 23 (1): 33–38.

Robinson JG, Ramirez JC, 1982. Conservation biology of Neotropical primates. In: *Mammalian Biology in South America.* M. A. Mares, N. N. Gonoways, eds. Pittsburg, Univ of Pittsb: 329–344.

Robinson PA, Hawkey C, Hammond GL, 1985. A phylogenetic study of the structural and functional characteristics of corticosteroid binding globulin in primates. *J Endocrinol* 104 (2): 251–257.

Roçh-Ramel F, Schali G, 1981. Renal excretion of urate in mammals. *Fortschr Urol und Nephrol* 16: 36–42.

Rodieck RW, 1988. The primate retina. In: *Comparative Primate Biology.* Vol 4: *Neurosciences.* NY, Alan R. Liss: 203–278.

Rogers J, Kammerer CM, Parry P, Morin P, 1996. Development and applications of a genetic linkage map of the baboon. *IPS/ASP Congr Abst*: # 605.

Rogers J, Mahaney MC, Witte SM, Nair S, Newman D, Wedel S, Rogriguez LA, Rice KS, Slifer SH, Perelygin A, Slifer M, Palladino-Negro P, Newman T, Chambers K, Joslyn G, Parry P, Morin PA, 2000. A genetic linkage map of the baboon (*Papio hamadryas*) genome based on human microsatellite polymorphisms. Genomics 67 (3): 237–247.

Rogers LJ, Kaplan G, 1994. Handedness in orang-utans. *Congr IPS* 15: 206.

Roginski YaYa, 1969. [*Contemporary Problems of Anthropogenesis*]. M, Znanie (Russian).

Rohles FH, 1969. The impact of Robert Yerkes on chimpanzee research. In: *Proc 2 Intern Congr Primatol, Atlanta 1968.* Vol 2. Basel – New York, S. Karger: 11–15.

Rojas FJ, Moretti-Rojas I, Balbaceda JP, Asch RH, 1989. Regulation of gonadotropin-stimulable adenylyl cyclase of the primate corpus luteum. *J Steroid Biochem* 32 (18): 175–182.

Romer AS, 1971. *The Vertebrate Body.* 4th ed. Philadelphia, Saunders.

Romera-Hererra AE, Lehmann H, Joysey KA, Friday AE, 1976. Evolution of myoglobin amino acid sequences in primates and other vertebrates. In: *Molecular Anthropology: Genes and Proteins in the Evolutionary Ascent of the Primates.* M. Goodman, R. E. Tashian, J. H. Tashian, eds. New York – London, Plenum Press: 289–300.

Rosa MGP, Gattass R, Fiorani M, Jr, 1988. Complete pattern of ocular dominance stripes in V1 of a New World monkey, *Cebus apella. Exp Brain Res* 72 (3): 645–648.

Rose LM, Richards T, Alvord EC, Jr, 1994. Experimental allergic encephalomyelitis (EAE) in nonhuman primates: A model of multiple sclerosis. *Lab Anim Sci* 45 (5): 508–512.

Rosenberger AL, 1984. Fossil New World monkeys dispute the molecular clock. *J Human Evol* 13 (8): 737–742.

Rosenberger AL, Setoguchi T, Shigehara N, 1990. The fossil record of callitrichine primates. *J Human Evol* 19 (1–2): 209–236.

Roskosz T, 1984. The arteries of the brain base in baboon, *Papio cynocephalus*. In: *Ann Warsaw Agr Univ SGGW-AR Vet Med* # 12: 1–16.

Roth RH, Bacopoulos NG, Bustos G, Redmond DE, 1980. Antipsychotic drugs: Differential effects on dopamine neurons in basal ganglia and mesocortex following chronic administration in human and non-human primates. In: *Long-term Effects of Neuroleptics* (*Adv Biochem Psychopharmacol*, vol 24). NY, Raven Press: 513–520.

Roussel F, 1985. Marquage des cellules endotheliales vasculaires animales par la lectine d' *Evonymus europaeus*. *C. r. Soc biol* 179 (6): 795–800.

Rowe AW, 1995. Primate models: Cryopreservation and transfusion of red cells. *Cryobiology* 32 (6): 547.

Rozsa AJ, Molinary HH, Grinspan JD, Kenshalo DR, 1985. The primate as a model for the human temperature-sensing system: 1. Adapting temperature and intensity of thermal stimuli. *Somatosen Res* 2 (4): 303–314.

Rubin E, Lieber CS, 1973. Experimental alcoholic hepatitis: A new primate model. *Science* 182 (4113): 712–713.

Ruch TC, 1941. *Bibliographia Primatologica*. New Haven, Yale Med Libr.

Ruch TC, 1959. *Diseases of Laboratory Primates*. Philadelphia – London, W. B. Saunders Co.

Ruch TC, 1966. [Primatological information 1941–1966. In: *Biology and Pathology of Simians, Studies of Human Diseases in Experiments on Simians. Materials of Intern Symp in Sukhumi*]. Tbilisi: 104 (Russian).

Ruch TC, Fulton JF, 1942. Scientific literature: Growth of primate literature since 1800. *Science* 95 (2454): 47–48.

Ruffie J, Moor-Jankowski J, Socha W, 1982. Immunogenetic evolution of primates. *Main Lectures of the 8 Congr Intern Prim Soc, Florence 1980*. A. B. Chiarelli, R. S. Corruccini, eds. Berlin – Heidelberg – New York, Springer-Verlag: 28–34.

Rumbaugh DM, 1973. The importance of nonhuman primate studies of learning and related phenomena for understanding human cognitive development. In: *Nonhuman Primates and Med Res*. G. H. Bourne, ed. New York – London, Acad Press: 415–429.

Rumbaugh D, 1995. Primate language and cognition: Common ground. *Soc Res* 62 (3): 711–730.

Rumbaugh DM, Hopkins WD, Washburn DA, Savage-Rumbaugh ES, 1989. Lana chimpanzee learns to count by 'HUMATH': A summary of a videotaped experimental report. *Psychol Rec* 39 (4): 459–470.

Rumbaugh DM, Washburn DA, Richardson WK, 1996. Selection of behavioral tasks and development of software for evaluation of rhesus monkey behavior during spaceflight. *Supplemental Status Report* # *NAG-2-438*. Georgia State Univ Language Res Ctr, Atlanta.

Rumennik L, Vincent GP, Schwam E, Sepinwall J, 1989. Flumazenil (RO 15–1788) improved learning and memory in mice but not in monkeys. *Abstr Soc Neurosci* 15: 1172.

Rumpler Y, Dutrillaux B, 1978. Chromosomal evolution in Malagasy lemurs. III. Chromosome banding studies in the genus *Hapalemur* and the species *Lemur catta*. *Cytogenet Cell Genet* 21 (4): 201–211.

Rumyantsev PP, 1982. [*Cardiomyocytes in Processes of Reproduction. Differentiating and Regeneration*]. Leningrad, Nauka (Russian).

Rupniak NMJ, Iversen SD, 1987. Primate models of senile dementia. In: *Cognitive Neurochemistry*. NY, Oxford Univ Press: 57–72.

Russel IS, Pereira SC, 1981. Visual neglect in rat and monkey: An experimental model for study of recovery of function following brain damage. *Develop Neurosci* 13: 209–238.

Ruvolo M, Disotell TR, Allard MW, Brown WM, Honeycutt RL, 1991. Resolution of African hominoid trichotomy by use of a mitochondrial gene sequence. *Proc Natl Acad Sci USA* 88: 1570–1574.

Ruvolo M, Smith T, 1986. Phylogeny and DNA-DNA hybridization. *Mol Biol Evol* 3 (3): 285–289.

Rylands AB, 1994a. [Black-faced lion tamarin *Leontopithecus caissara* Lorini & Persson, 1990]. In: *Livro*

Vermelho dos Mamiferos Brasileiros Ameacados de Extincao. G. A. B. da Fonseca *et al.*, eds. Belo Horizonte, Fundacao Biodiversitas: 73–81 (Portuguese).

Rylands AB, 1994b. [Black lion tamarin *Leontopithecus chrysopygus* Mikan, 1823]. In: *Livro Vermelho dos Mamiferos Brasileiros Ameacados de Extincai.* G. A. B. da Fonseca *et al.*, eds. Balo Horizonte, Fundacao Biodiversitas: 97–107 (Portuguese).

Rylands AB, 1996. The endangered lion tamarins of Brazil's Atlantic forest. *IPS/ASP Congr Abstr:* # 391.

Rylands AB, Caram AAA, 1998. Old World primates – new species and subspecies. *Oryx* 32 (2): 88–89.

Sabin AB, 1985. Oral poliovirus vaccine: History of its development and use and current challenge to eliminate poliomyelitis from the world. *J Infect Dis* 151 (3): 420–436.

Sabin AB, 1993. Reflections on the qualitative and quantative aspects of neurovirulence of different polioviruses. *Biologicals* 21 (4): 317–320.

Saitou N, 1991a. Phylogeny of extant hominoids reconstructed from molecular data. In: *Primatol Today.* A. Ehara *et al.*, eds: 627–630.

Saitou N, 1991b. Reconstruction of molecular phylogeny of extant hominoids from DNA sequences data. *Amer J Phys Anthropol* 84 (1): 75–85.

Sakura O, Matsuzawa T, 1991. Flexibility of wild chimpanzee nut-cracking behavior using stone hammers and anvils: an experimental analysis. *Ethology* 87 (3–4): 237–248.

Salvignol I, Blancher A, Calvas R, Clayton J, Socha WW, Colin Y, Ruffie J, 1994. Molecular genetics of chimpanzee Rh-related genes: Their relationship with the R-C-E-F blood group system, the chimpanzee counterpart of the human rhesus system. *Biochem Genet* 32 (5–6): 201–221.

Sands SF, Wright AA, 1982. Monkey and human pictorial memory scanning. *Science* 216 (4552): 1333–1334.

Sapin MR, Charin GM, 1985. [The structure of the spleen in baboon ontogenesis]. *Arch anat histol i embryol* 88 (4): 65–69 (Russian).

Sarich VM, 1968. The origin of the hominids: An immunological approach. In: *Perspectives on Human Evolution.* S. L. Washburn, P. C. Jay, eds. New York *et al.*, Holt, Rinehart and Winston: 94–119.

Sarich VM, 1993. Mammallian systematics: Twenty-five years among their albumins and transferrins. In: *Mammal Phylogeny: Placentals.* F. S. Szalay, M. J. Novacek, M. C. McKenna, eds. Secaucus, NJ, Springer-Verlag: 103–114.

Sarich VM, Cronin JE, 1976. Molecular systematics of the primates. In: *Molecular Anthropol: Genes and Proteins in the Evolutionary Ascent of the Primates.* M. Goodman, R. E. Tashian, J. H. Tashian, eds. NY, Plenum Press: 141–170.

Sarich VM, Wilson AC, 1967. Immunological time scale for hominid evolution. *Science* 158 (3805): 1200–1203.

Sarmiento EE, 1988. Anatomy of the hominoid wrist joint: Its evolutionary and functional implications. *Intern J Primatol* 9 (4): 281–345.

Sarmiento RF, 1975. The stereoacuity of macaque monkey. *Vis Res* 15 (4): 493–498.

Sato P, Udenfriend S, 1978. Scurvy-prone animals, including man, monkey, and guinea pig, do not express the gene for gulonolactone oxidase. *Arch Biochem Biophys* 187 (1): 158–162.

Satoh K, Kawada M, Wada Y, Taira N, 1984. Different responses of coronary circulation to autonomic drugs and autacoids in the monkey and the dog. *Japan J Pharmacol* 36 (Suppl): 271.

Satoh K, Yamashita S, Maruyama M, Taira N, 1982. Comparison of the responses of the simian and canine coronary circulations to autonomic drugs. *J Cardiovasc Pharmacol* 4: 820–828.

Savage TS, Wyman J, 1847. Notice of external characters and habits of *Troglodytes gorilla*, a new species of orang from Gabon River . . . Osteology of the same . . . *Bost J Natur Hist* 5: 417–443.

Savage-Rumbaugh ES, 1986. *Ape Language: From Conditioned Response to Symbol.* NY, Columbia Univ Press.

Savage-Rumbaugh ES, Levin R, 1994. *Kanzi: At the Brink of the Human Mind.* NY, J. Wiley and Sons.

Savage-Rumbaugh ES, Murphy J, Sevcik RA, Rumbaugh DM, Brakke KE, Williams S, 1993. *Language Comprehension in Ape and Child.* Monographs of the Society for Research in Child Development. Ser # 233, vol 58.

Savage-Rumbaugh ES, Sevcik RA, Hopkins WD, 1988. Symbolic cross-modal two species of chimpanzees. *Child Develop* 59: 617–625.

Savage-Rumbaugh S, Shanker SG, Taylor TJ, 1998. *Apes, Language, and the Human Mind*. NJ, Oxford Univ Press.

Sawada I, Schmid CW, 1986. Primate evolution of the α-globin gene and its Alu-like repeats. *J Mol Biol* 192: 693–709.

Sayer AM, Littlefield LG, DuFrain RJ, Richter CB, 1980. Mutagen-induced chromosome lesions in lymphocytes of *Saguinus oedipus* tamarins. A possible genetic marker for animals at risk for colon cancer. *Envir Mut Soc* 2: 207.

Scallet AC, McKay D, Bailey JR, Ali SF, Paule MG, Slikker W, Rayford PL, 1989. Meal-induced increase in plasma gastrin immunoreactivity in the rhesus monkey (*Macaca mulatta*). *Amer J Primatol* 18 (4): 315–319.

Schaller G, 1988. *The Year of the Gorilla*. Chicago, Chic Univ Press.

Schapovalov AI, 1975. [*Neurons and Synapses of the Supraspinal Motor Systems*]. Leningrad, Nauka (Russian).

Schastny A, 1972. [*The Complicated Forms of Ape Behavior*]. Leningrad, Nauka.

Schellekens H, De Reus A, van der Meide PH, 1984. The chimpanzee as a model to test the side effects of human interferons. *J Med Primatol* 13 (5): 235–245.

Schellekens H, Niphuis H, Buijs L, van der Rapp PD, Hochkeppel HK, Heeney JL, 1996. The effect of recombinant human interferon alpha B/D compared to interferon alpha 2b on SIV infection in rhesus macaques. *Antivir Res* 32 (1): 1–8.

Scherbakova OP, 1937. [Materials to study the twenty-four-hours periodicity of physiological processes in higher mammals. Report I. *Bull Exp Biol Med*] 4 (4): 335–337 (Russian).

Schlemmer RF, Jr, Davis JM, 1986. A primate model for the study of hallucinogens. *Pharmacol Biochem Behav* 24 (2): 381–392.

Schlemmer RF, Jr, Young JE, Davis JM, 1996. Stimulant-induced disruption of non-human primate social behavior and the psychopharmacology of schizophrenia. *J Psychopharmacol* 10 (1): 67–76.

Shevtchenko Yu, 1971. [*Evolution of the Brain Cortex of the Primates and Man*]. M, Moscov Univ Publish House (Russian).

Schmidt GA, 1969. [Embryological prerequisites for anthropogenesis. In: *Transactions of the 7 All-Union Congress of Anat, Histol & Embryol*]. Tbilisi: 455–457 (Russian).

Schmidt H, 1991. Woher kam die moderne Mensch? Wie breitete et sich uber die Erde aus? Teil 1. *Prax Naturwiss Biol* 40 (6): 42–47.

Schmidt LH, 1961. Chemotherapy of tuberculosis. *Progress Rep on Grant E-509 (Covering the Period Novemb 1, 1952, through Decemb 31, 1960)*.

Schmidt S, Neubert R, Schmitt M, Neubert D, 1996. Studies on the globulin-E system of the common marmoset in comparison with human data. *Live Sci* 59 (9): 719–730.

Schmitt T, Graur D, Tomiuk T, 1990. Phylogenetic relationships and rates of evolution in primates: Allozymic data from catarrhine and platyrrhine species. *Primates* 31 (1): 95–108.

Schneider H, Schneider MPC, Sampaio I, Harada ML, Stanhope M, Czelusniak J, Goodman M, 1993. Molecular phylogeny of the New World monkeys (Platyrrhini, Primates). *Mol Phyl Evol* 2 (3): 225–242.

Schoenemann PT, 1989. Comparison of intraspecific craniometric variability in *Homo*, *Pan*, and *Gorilla*. *Amer J Phys Anthropol* 78 (2): 298.

Schrier AM, 1969. Psychology and psychiatry. In: *Using Primates in Med Res. Part II. Recent Compar Res*. Basel – New York, S. Karger: 28–40.

Schrier AM, Brady PM, 1987. Categorization of natural stimuli by monkeys (*Macaca mulatta*): Effects of stimulus set size and modification of exemplars. *J Exp Psychol: Anim Behav Process* 13 (2): 136–143.

Schroeder WA, de Simone J, Shelton JB, Shelton JR, Espinueva Z, Hall L, Zwiers D, 1983. Changes in the γ-chain heterogeneity of hemoglobin F of the baboon (*Papio cynocephalus*) postnatally and after partial switching to hemoglobin F production by various stimuli. *J Biol Chem* 258 (5): 3121–3125.

Schulster D, Burstein S, Cooke BA, 1976. *Molecular Endocrinology of the Steroid Hormones*. London, Wiley.

Schultz AH, 1966. Changing views on the nature and interrelations of the higher primates. *Yerkes Newslet* 3 (1): 15–29.

Schultz AH, 1969. *The Life of Primates*. NY, Universe Books,

Schultz W, 1988. MPTP-induced parkinsonism in monkeys: Mechanism of action, selectivity and pathophysiology. *Genet Pharmacol* 19 (2): 153–161.

Schumann P, Touzani O, Young AR, Verard L, Morello R, MacKenzie ET, 1996. Effects of indomethacin on cerebral blood flow and oxigen metabolism: A positron emission tomographic investigation in the anaesthetized baboon. *Neurosci Lett* 220 (2): 137–141.

Schwaier A, Weis H, 1982. *Tupaia*. A new animal model for gallstone research. *Exp Biol Med* 7: 206.

Schwartz H, Chu I, Villeneuve DC, Benoit FM, 1987. Metabolism of 1, 2, 3, 4-, 1, 2, 3, 5-, and 1, 2, 4, 5-tetrachlorobenzene in the squirrel monkey. *J Toxicol Environ Health* 22 (3): 341–350.

Schwartz JH, 1984. The evolutionary relationships of man and orangutans. *Nature* 308 (5959): 501–505.

Schwartz JH, 1986. Primate systematics and a classification of the order. In: *Comparative Primate Biology*. Vol 1. *Systematics, Evolution and Anatomy*. NY, Alan R. Liss: 1–41.

Schwartz JH, 1988. History, morphology, paleontology, and evolution. In: *Orang-utan Biology*. NY, Oxford, Oxf Univ Press: 69–89.

Schwartz SS, 1967. [Current problems of evolution theory]. *Voprosi phylosoph* # 10: 143–153 (Russian).

Schwartzberg M, Unger J, Weindl A, Landa W, 1990. Distribution of neuropeptide Y in the prosencephalon of man and cotton-head tamarin (*Saguinus oedipus*): Colocalization with somatostatin in neurons of striatum and amygdala. *Anat and Embryol* 181 (2): 157–166.

Scott GBD, 1979. The comparative pathology of the primate colon. *J Pathol* 127 (2): 65–72.

Scott GBD, 1980. The primate caecum and appendix vermiformis: A comparative study. *J Anat* 131 (3): 549–563.

Scott GBD, 1992. *Comparative Primate Pathology*. NY, Oxford Univ Press.

Scott TR, Gize BK, 1987. The effect of physiological condition on taste in rats and primates. In: *Umami: Basic Taste. Phisiol Biochem, Nutr, Food Sci*. New York – Basel: 409–437.

Seevers MH, 1969. Psychopharmacological elements of drug dependence. *JAMA* 206 (6): 1263–1266.

Seljeskog EL, Hitchcock CR, Groover ME, Haglin JJ, Strobel CJA, 1963. Isosorbide dinitrate during acute coronary occlusion. Effects in the baboon. *JAMA* 183 (3): 210–213.

Semenov LF, 1957. [Acute radial diseases by irradiation of different parts of organism]. In: *Rasschirennoe zasedanie Buro OMBN AMN SSSR*. Sukhumi: 116–118 (Russian).

Semenov LF, 1967. [*Prophylaxis of Acute Radial Disease in Experiment*]. Leningrad, Medicine (Russian).

Semenov LF, Yakovleva LA, 1965. [Comparative characteristics of radial disease in different species of mammals including primates]. *Vestn AMN SSSR* # 11: 50–57 (Russian).

Sengpiel F, Troilo D, Kind PC, Graham B, Blakemore C, 1996. Functional architecture of area 17 in normal and monocularly deprived marmosets (*Callithrix jacchus*). *Vis Neurosci* 13 (1): 145–160.

Senut B, le Eloch-Prigent P, 1982. Etude tomodensitometriqye d'un coude entier chez *Pan*: comparison avec *Homo*. *C. r. Acad sci*. Ser 3, vol 294 (22): 1035–1040.

Seth PK, Seth S, Saxena MB, 1983. Chromosomal homology in *Macaca mulatta*, *Papio papio* and man. In: *Perspectives in Primate Biology*. New Delhi, Today and Tomorrow's Printers and Publishers: 161–165.

Seuanez HN, Lachtermacher M, Canavez F, Moreira MAM, 1997. The human chromosome 3 gene cluster ACY1-CACNA1D-ZNF64-ATP2B2 is evolutionary conserved in *Ateles paniscus chamek* (Platyrrhini, Primates). *Cytogenet Cell Genet* 77 (3–4): 314–317.

Shannon RP, Shen Y-t, 1996. The coronary vasoconstrictor effects of cocaine are more intense in conscious baboons than in dogs. *J Amer Coll Cardiol* 27 (2, suppl A): 263 A.

Shaparenko KK, 1935. [Linnaeus and theologists. *Sovet Botany*] # 5: 156 (Russian).

Shapiro LJ, Jungers WL, 1988. Back muscle function during bipedal walking in chimpanzee and gibbon: Implications for the evolution of human locomotion. *Amer J Phys Anthropol* 77 (2): 201–212.

Shapiro M, Cohen N, 1980. Normal erythrocytic σ-aminolevulinic acid dehydratase activity in ten nonhuman primate species. *J Med Primatol* 9: 304–308.

Sharples K, Plowman GD, Rose TM, Twardzik DR, Purchio AI, 1987. Cloning and sequence analysis of simian transforming growth factor-β cDNA. *DNA* 6 (3): 239–244.

Shaw ST, Elsahw SJ, Moyer DL, 1972. Menstrual blood quantitation in the rhesus monkey: An experimental tool for improving intrauterine contraceptive devices (IUDS). *Fertil Steril* 23 (4): 257–263.

Sheffield WD, Squire RA, Strandberg DH, 1981. Cerebral venus thrombosis in the rhesus monkey. *Vet Pathol* 18 (3): 326–334.

Shen Y-t, Vatner SF, 1996. Differences in myocardial stunning following coronary artery occlusion in conscious dogs, pigs, and baboons. *Amer J Physiol* 270 (4, pt 2): H 1312–H 1322.

Shepard AD, Connolly RJ, Callow AD, Ramberg-Laskaris K, Foxall TL, Libby P, Keough EM, O'Donnell TF, 1984. Endothelial cell seeding of small-caliber synthetic vascular prostheses in the primate: Sequential indium 111 platelet studies. *Surg Forum* 35: 432–434.

Sherrington ChS, 1889. On nerve-tracts degenerating secondarily to lesions of the cortex cerebri. *J Physiol* # 10: 429–432.

Sherrington ChS, 1906 (1969). [*The Integrative Action of the Nervous System*]. Leningrad, Nauka (Russian).

Shestopalova SK, 1968. [*To the Physiological Analysis of Salivation in Hamadryas Baboons*. Theses of Candidate Diss]. M (Russian).

Shevtsova ZV, Lapin BA, 1987. [A comparative aspect in the studies of viral infections in monkeys and man]. *Vestn AMN SSSR* # 10: 70–74 (Russian w/English sum).

Shively CA, Grant KA, Ehrenkaufer RL, Mach RH, Nader MA, 1997. Social stress, depression, and brain dopamine in female cynomolgus monkeys. *Ann N. Y. Acad Sci* 807: 574–577.

Shoshani J, Groves CP, Simons EL, Gunnell GF, 1996. Primate phylogeny: Morphological vs molecular results. *Mol Phylog Evol* 5 (1): 102–154.

Shroyt IG, 1961. [*Experimental Measles*]. Kishinev (Russian).

Sibley CG, Ahlquist JE, 1984. The phylogeny of the hominoid primates, as indicated by DNA-DNA hybridization. *J Mol Evol* 20: 2–15.

Sibley CG, Ahlquist JE, 1987. DNA hybridization evidence of hominoid phylogeny: Results from an expanded data set. *J Mol Evol* 26: 99–121.

Sibley CG, Comstock JA, Ahlquist JE, 1990. DNA hybridization evidence of phylogeny: A reanalysis of the data. *J Mol Evol* 30 (3): 202–236.

Siddall RA, 1978. The use of marmosets (*Callithrix jacchus*) in teratological and toxicological research. In: *Marmosets in Exp Med* 10. Basel e.a.: 215–244.

Siddiqui WA, 1977. An effective immunization of experimental monkeys against a human malaria parasite, *Plasmodium falciparum*. *Science* 197 (4301): 388–389.

Siddons RC, 1974. Experimental nutritional folate deficiency in the baboon (*Papio cynocephalus*). *Brit J Nutr* 32 (3): 579–587.

Silverman S, Morgan JP, 1980. Thoracic radiography of the normal rhesus macaque (*Macaca mulatta*). *Amer J Vet Res* 41 (10): 1704–1719.

Simon M, Green H, 1989. Involucrin in the epidermal cells of subprimates. *J Invest Dermatol* 92 (5): 721–724.

Simonds PS, 1962. The Japan Monkey Center. *Curr Anthropol* 3 (3): 7–11.

Simons EL, 1987. New faces of Aegyptopithecus from the Oligocene of Egypt. *J Human Evol* 16 (3): 273–289.

Simons EL, 1990. Discovery of the oldest known anthropoidean skull from the Paleogene of Egypt. *Science* 247 (4950): 1567–1569.

Simons EL, 1992. The fossil history of primates. In: *The Cambridge Encyclopedia of Human Evolution*. S. Jones, R. Martin, D. Pilbeam, eds. Cambridge, Camb Univ Press: 199–208.

Simons EL, 1993. New endocasts of *Aegyptopithecus*: Oldest well-preserved record of the brain in Anthropoidea. *Amer J Sci* 239 A: 383–390.

Simons EL, 1995. Skulls and anterior teeth of *Catopithecus* (Primates: Anthropoidea) from the Eocene and anthropoid origins. *Science* 268: 1885–1888.

Simons EL, Covert HH, 1981. Paleoprimatological research over the last 50 years: Foci and trends. *Amer J Phys Anthropol* 56 (4): 373–382.

Simons EL, Rasmussen DT, 1991. The generic classification of Fayum Anthropoidea. *Intern J Primatol* 12 (2): 163–178.

Simons EL, Rumpler Y, 1988. Eulemur: New generic name for species of *Lemur* other than *Lemur catta*. *C. r. Acad. sci.* Ser 3, 307 (9): 547–551.

Simpson GG, 1945. The Principles of Classification and a Classification of Mammals. *Bull Amer Mus Nat Hist* 45: 1–350.

Simpson GG, 1962a. Primate taxonomy and recent studies of nonhuman primates. *Ann N. Y. Acad Sci* 102 (1): 497–514.

Simpson GG, 1962b. *Principles of Animal Taxonomy*. NY, Columb Univ Press.

Simpson GG, 1971. William King Gregory (1876–1970). *Amer J Phys Anthropol* 35 (2): 158.

Sinanović O, Chiba S, 1988. Different antagonistic effects of bunazosin and ketanserin on the norepinephrine-induced vasoconstriction in isolated, perfused canine and simian skeletal muscle arteries. *Gap g. Pharmacol* 46 (2): 200–203.

Sinclair RJ, Burton H, 1991. Tactile discrimination of gratings: Psychophysical and neural correlates in human and monkey. *Somatosens a. Motor Res* 8 (3): 241–248.

Sinnott JM, Owren MJ, Petersen MR, 1987. Auditory frequency discrimination in primates: Species differences (*Cercopithecus, Macaca, Homo*). *J Compar Psychol* 101 (2): 126–131.

Sinosich MJ, 1986. Evolution of pregnancy-associated plasma protein-A. In: *Pregnancy Proteins Anim Proc – Intern Meet, Copenhagen, 1985*. Berlin – New York: 269–280.

Sipes IG, Slocumb ML, Perry DF, Carter DE, 1980. 4, 4'-Dichlorobiphenil: Distibution, metabolism, and excretion in the dog and the monkey. *Toxicol Appl Pharmacol* 55 (3): 554–563.

Sirianni JE, 1985. Nonhuman primates as models for human craniofacial growth. In: *Nonhuman Primates Models for Human Growth and Development*. NY, Alan R. Liss: 94–124.

Slavin RG, Fischer VW, Levine EA, Tsai CC, Winzenburger P, 1978. A primate model of allergic bronchopulmonary aspergillosis. *Intern Arch Allergy Appl Immunol* 56: 325–333.

Slonim AD, 1952. [*Animal Warmth and Its Regulation in Organism of Mammals*]. Moscow – Leningrad (Russian).

Slonim AD, 1986. [*Evolution of Termoregulation*]. Leningrad, Nauka (Russian).

Smeets DFCM, van de Klundert FAJM, 1990. Common fragile sites in man and three closely related species. *Cytogenet Cell Genet* 53 (1): 8–14.

Smith AP, Wolthuis OL, 1983. H16 as an antidote to soman poisoning in rhesus monkey respiratory muscles in vitro. *J Pharmac a. Pharmacol* 35 (3): 157–160.

Smith BH, Tompkins RL, 1993. Toward a life history of the Hominidae. *Ann Rev Anthropol* 24: 257–279.

Smith CJ, Norman RL, 1987. Circadian periodicity in circulating cortisol is absent after orcidectomy in rhesus macaques. *Endocrinology* 121 (6): 2186–2191.

Smith RL, Caldwell J, 1976. Drug metabolism in non-human primates. In: *Drug Metabolism – from Microbe to Man*. London, Taylor and Francis: 331–356.

Smouse PE, Li W-H, 1987. Likelihood analysis of mitochondrial restriction-cleavage patterns for the human-chimpanzee-gorilla trichotomy. *Evolution* 41 (6): 1162–1176.

Soave O, 1981. Viral infections common to human and nonhuman primates. *J Amer Vet Med Ass* 179 (12): 385–388.

Socha WW, 1986. Blood groups of nonhuman primates. In: *Systematics, Evolution and Anatomy*. NY: 299–333.

Socha WW, Gordon EB, Saltzman M, Wiener SA, 1971. The use of anti-globulin inhibition test for serological studies in primates (seroprimatology). *Proc 3 Intern Congr Primatol – Zurich 1970*, vol 2. Basel, S. Karger: 70–80.

Socha WW, Marboe CC, Michler RE, Rose EA, Moor-Jankowski J, 1987. Primate animal model for the study of ABO incompatibility in organ transplantation. *Transplant Proc* 19 (6): 4448–4455.

Socha WW, Moor-Jankowski J, 1979. Blood groups of anthropoid apes and their relationship to human blood groups. *J Human Evol* 8 (4): 453–465.

Socha WW, Moor-Jankowski J, 1980. Chimpanzee R-C-E-F blood group system: A counterpart of the human Rh-Hr blood groups. *Folia Primatol* 33 (3): 172–188.

Socha WW, Moor-Jankowski J, 1989. Primate animal model for xenotransplantation. Serological criteria of donor-recipient selection. In: *Xenographt 25*. M. Hardy, ed. Amsterdam, New York, Oxford, Excerpta Med: 51–57.

Socha WW, Moor-Jankowski J, Blancher A, 1994. M-N and M-N-related blood groups of nonhuman primates. *Curr Primatol 1. Ecol Evol.* B. Thierry, J. R. Anderson, J. J. Roeder, H. Herrenschmidt, eds. Strasbourg, Univ Louis Pasteur: 341–348.

Socha WW, Moor-Jankowski J, Ruffié J, 1983. The B^P graded blood group system of baboon: Its relationship with macaque red cell antigens. *Folia Primatol* 40 (3): 205–216.

Socha WW, Ruffie J, 1983. Blood Groups of Primates. *Theory, Practice, Evolutionary Meaning.* Monographs in Primatol, vol 3. NY, Alan R. Liss.

Socha WW, Wiener AS, Moor-Jankowski J, 1972. Homologues of the human A-B-O-blood groups in non-human primates. *Med Primatol 1972. Proc 3 Conf Exp Med Surg Prim, Lyons 1972.* Part 1. Basel, S. Karger: 390–397.

Solomon S, Leung K, 1972. Steroid hormones in non-human primates during pregnancy. *Acta Endocrinol* 71 (Suppl 166): 178–190.

Soma H, 1983. Notes in the morphology of the chimpanzee and orangutan placenta. *Placenta* 4 (3): 279–290.

Somit A, 1990. Humans, chimps, and bonobos. The biological bases of aggresion, war, and peacemaking. *J Conflict Resolut* 34 (3): 553–582.

Sonnefeld P, van Bekkum DW, 1979. The effect of whole-body irradiation skeletal growth in the rhesus monkeys. *Radiology* 130 (3): 789–791.

Sopelak VM, Hodgen GD, 1984. Techniques for research in fetal and neonatal primates. In: *Animal Models in Fetal Medicine.* NY, Perinatal Press: 161–181.

Sorokin YN, 1970. Representation of nuclear topics in VINITI's abstracts journals. *Handl Nucl Inform Proc Symp.* Vienna: 385–391.

Spencer F, 1995. Pithekos to Pithecanthropus: An abbreviated review of changing scientific views on the relationship of the anthropoid apes to Homo. *Ape, Man, Apeman: Changing Views since 1600.* Leiden Univ: 13–27.

Spencer J, Finn T, Isaacson PG, 1986. A comparative study of the gut-associated lymphoid tissue of primates and rodents. *Virchows Arch* B 51 (6): 509–519.

Sperk E, 1896. *Ocures completes*, II, Paris: 614.

Srivastava A, Mohnot SM, Southwick CH, 1998. Status of golden langurs in India. *Congr IPS 17 Abstr*: # 086.

Stahlmann R, Neubert R, 1995. Value of non-human primate data. *Human Exp Tox* 14 (1): 146–147.

Stankov NS, 1958. [The life and scientific works of Karl Linneas]. In: *Karl Linneas.* M, Acad Sci USSR Publish House: 7–77 (Russian).

Stanyon R, 1989. Implications of biomolecular data for human origin with particular reference to chromosomes. In: *Proc 2 Intern Congr Human Paleontol.* Milan, Ja ca Book: 35–44.

Startsev VG, 1971. [*Modelling of Neurogenic Diseases of Man in Experiments on Monkeys*]. M, Medicine (Russian).

Startsev VG, 1972. [*Neurogenic Gastric Achylia*]. Leningrad, Nauka (Russian).

Startsev VG, 1980. [Comparative characteristics of some digestive functions in man, simians and dogs in norm and pathology. In: *Biological Characteristics of Laboratory Animals and Extrapolation of Experimental Data to Man*]. M, Acad Med Sci USSR: 244–245 (Russian).

Startsev VG, 1991. [Emotional stress: The actual problem of medicine and physiology]. Sukhumi, '*Abkhazian Univ*', Apr 5.

Startsev VG, Chirkova SK, Chirkov AM, Startsev SV, Butnev VYu, 1987. [Neurogenic arterial hypertension in monkeys: Neurohormonal mechanisms, prevention, treatment]. *Vestn Acad Med Nauk SSSR* # 10: 83–88 (Russian).

Stebbins WC, 1975. Hearing of the anthropoid primates: A behavioral analysis. In: *The Nervous System*, vol 3. *Human Communication and Its Disorders.* NY, Raven Press: 113–124.

Stegink LD, Moss J, Printen KJ, Soon CE, 1980. D-methionine utilization in adult monkeys fed diets containing DL-methionine. *J Nutr* 110 (6): 1240–1246.

Steklis HD, Gerald CN, Madry S, 1996. The mountain gorilla – conserving an endangered primate in conditions of extreme political instability. *IPS/ASP Congr Abstr*: # 394.

Stephens LC, King GK, Ang KK, Shultheiss TE, Peters LJ, 1986. Surgical and microscopic anatomy of parotid and submandibular salivary glands of rhesus monkeys (Macaca mulatta). *J Med Primatol* 15: 105–119.

Stephens VC, 1996. Reproductive parameters of the baboon menstrual cycle. *IPS/ASP Congr Abstr*: # 073.

Stevens VC, Sparks SJ, Powell JE, 1970. Levels of estrogens, progestogens and luteinizing hormone during the menstrual cycle of the baboon. *Endocrinology* 87 (4): 658–666.

St George JA, Nishio SJ, Cranz DL, Plopper CG, 1986. Carbohydrate cychemistry of rhesus monkey tracheal submucosal glands. *Anat Rec* 216 (1): 60–67.

Stinson CH, Sben DM, Burbacher TM, Mohamed MK, Mottet NK, 1989. Kinetics of methyl mercury in blood and brain during chronic exposure in the monkey *Macaca fascicularis*. *Pharmacol Toxicol* 65 (3): 223–230.

Stokes A, Bauer JH, Hudson ND, 1928. Experimental transmission of yellow fever to laboratory animals. *Amer J Trop Med* # 8: 103–164.

Stone WH, Treichel RCS, VandeBerg JL, 1987. Genetic significance of some common primate models in biomedical research. In: *Animal Models: Assessing the Scope of Their Use in Biomedical Research.* NY, Alan R. Liss: 73–93.

Stout LC, Folse DS, Meier J, Crosby WM, Kling R, Williams GR, Price WE, Geyer JR, Padula R, Whorton E, Kimmelstiel P, 1986. Quantitative glomerular morphology of the normal and diabetic baboon kidney. *Diabetologia* 29 (10): 734–740.

Struthers BJ, MacDonald JR, Dahlgren RR, Hopkins DT, 1983. Effects on monkey, pig and rat pancreas of soy products with varying levels of trypsin inhibitor and comparison with the administration of cholecystokinin. *J Nutr* 113 (1): 86–97.

Sturman JA, 1990. Taurine deficiency. In: *Taurine: Functional Neuro-Chemistry, Physiology, and Cardiology.* NY, Willey-Liss: 385–395.

Sturman JA, Wen GY, Wisniewski HM, Neuringer MD, 1984. Retinal degeneration in primates raised on a synthetic human infant formula. *Intern J Developm Neurosci* 2 (2): 121–129.

Su Y-g, Lu M-y, Pan B-h, Pan H-p, 1988. Examination of the diseases in genital organs of 100 female rhesus monkeys. *J Exp Anim Sci* 8 (3): 176, Suppl 2pp. (Chinese w/English sum).

Suda T, Takahashi N, Shinki T, Yamaguchi A, Tanioka Y, 1986. The common marmoset as animal model for vitamin D-dependent rickets, type II. In: *Steroid Hormone Resistance.* NY, Plenum Publish Co: 423–435.

Sugiyama Y, 1985. The brush-stick of chimpanzees found in South-West Cameroon and their cultural characteristics. *Primates* 26 (4): 361–374.

Sullivan HR, Hanasono GK, Miller WM, Wood PG, 1987. Species specificity in the metabolism of nabilone. Relationship between toxicity and metabolic routes. *Xenobiotica* 17 (4): 459–468.

Sulman FM, 1983. Primate circadian rhythms. *Bioscience* 33 (7): 445–449.

Sulser F, Vetulani J, Mobley PL, 1978. Mode of action of antidepressant drugs. *Biochem Pharmacol* 27 (3): 257–261.

Sumner DR, Morbeck ME, Lobick JJ, 1989. Apparent age-related bone loss among adult female Gombe chimpanzees. *Amer J Phys Anthropol* 79 (2): 225–234.

Sun FF, Czuk CI, Taylor BM, Crittenden NI, Stout BK, Johnson HG, 1989. Biochemical and functional differences between eosinophils from guinea pig, primate and man. *FASEB J* 3 (3): A 909.

Susman RL, 1989. Ape affinites. (Rev: Orang-utan Biology. J. H. Schwartz, ed. NY, Oxford Univ Press, 1988). *Science* 244 (4906): 859–860.

Suzman JM, 1981. The internal architecture of So. African plio-pleistocene hominid pelves and femora. *Amer J Phys Anthropol* 54 (2): 282.

Suzuki H, Kawamoto Y, Takenaka O, Munechika I, Hori H, Sakurai S, 1994 Phylogenetic relationships among Homo sapiens and related species based on restriction site variations in rDNA spacers. *Biochem Genet* 32 (7–8): 257–269.

Swan KG, Reynolds DG, 1972. The splanchnic circulation in shock, species differences between subhuman primate and canine models. In: *Book of Abstr 3 Conf Exp Med Surg in Primates.* Lyon: 138.

Swindler DR, 1985. Nonhuman primate dental development and its relationship to human dental development. In: *Nonhuman Primate Model for Human Growth and Development.* NY, Alan R. Liss: 67–94.

Swindler DR, 1987. Rev: Alternatives to animal use in research, testing, and education . . . *Amer J Phys Anthropol* 74 (2): 276–277.

Swindler DR, Erwin J, eds, 1986. *Systematics, Evolution and Anatomy (Compar Primate Biol*, vol 1). NY, Alan R. Liss.

Szalay FS, Dagosto M, 1988. Evolution of hallucial grasping in the primates. *J Human Evol* 17 (1–2): 1–33.

Szalay FS, Delson E, 1979. *Evolutionary History of the Primates.* NY, Acad Press.

Szeinberg A, Zoreff E, Golan R, 1969a. Immunochemical gel diffusion study of relationships between erythrocyte catalase of various species. *Biochim et biophys acta* 188 (2): 287–294.

Szeinberg A, Zoreff E, Golan R, 1969b. Immunoelectrophoretic-chemical investigation of erythrocyte aspartate aminotransferase. *Life Sci* 8 (18): 943–948.

Szostakiewicz-Sawicka H, Grzybiak B, Kubasik A, Kubasik-Juraniec J, Prejzner-Morawska A, Dudziak M, 1980. Some features of the morphology and internal organs in *Ateles. Anthropol Contemporary* 3 (2): 281.

Szostakiewicz-Sawicka H, Grzybiak M, Prejzner-Morawska A, 1981. Some features of the morphology of internal organs in *Ateles. Arch ital anat e embriol* 86 (3): 271–276.

Symposium discussion, 1987. In: *Preclinical Safety of Biotechnology Products intended for Human Use.* NY, Alan R. Liss: 189–206.

Takagi SF, 1986. Studies of the olfactory nervous system of the Old World monkey. *Progr Neurobiol* 27 (3): 195–249.

Takano N, Lever MJ, Lambertsen CJ, 1979. Acid-base curve nomogram for chimpanzee blood and comparison with human blood characteristics. *J Appl Physiol: Respir Environ a. Exercise Physiol* 46 (2): 381–386.

Takemoto H, 1996. Comparative anatomy of the tongue muscles of humans and chimpanzees. *IPS/ASP Congr Abstr*: # 418.

Takenaka A, Takenaka O, Ohuchi M, Nakamura S, Takahashi K, 1986. Complete amino acid sequence of γ-chain of fetal hemoglobin of Japanese macaque (*Macaca fuscata*). *Haemoglobin* 10 (1): 1–13.

Takenaka A, Takenaka O, 1996. Processed P117 gene in primates. *IPS/ASP Congr Abstr*: # 487.

Takenaka O, 1985. Molecular evolution of non-human primates – hemoglobins of macaque monkeys. *Japan J Med Sci Biol* 38 (1): 40–43.

Talwar GP, 1983. [*Immunology of Contraception*]. M, Medicine (Russian, transl from English).

Tamraz JC, Rethore M-O, Iba-Zizen M-T, Lejeune J, Cabanis EA, 1987. Contribution of magnetic resonance imaging to the knowledge of CNS malformations related to chromosomal aberrations. *Human Genet* 76 (3): 265–273.

Tanaka M, 1995. Object sorting in chimpanzees (*Pan troglodytes*): Classification based on physical identity, complementarity, and familiarity. *J Compar Psychol* 109 (1): 151–161.

Taniguchi S, Cooper DKC, 1997. Clinical xenotransplantation: Past, present and future. *Ann Roy Coll Surg Engl* 79: 13–19.

Tappen NC, 1960. Some potentials of a Primate program for medical research. *Bull Tulane Univ Med Faculty* 20: 17–22.

Tartabini A, 1991. Mother infant cannibalism in thick-tailed bushbabies (*Galago crassicaudatus umbrosus*). *Primates* 32 (3): 379–383.

Teller DY, 1981. The development of visual acuity in human and monkey infants. *Trends Neurosci* 4: 21–24.

Teller DY, 1983. Measurement of visual acuity in human and monkey infants: the interface between laboratory and clinic. *Behav Brain Res* 10 (1): 15–23.

Teofilovski-Parapid G, Bogdanovic D, Minic L, Rankovic A, Vela A, 1988. Dominance of coronary arteries in green monkey (*Cercopithecus aethiops*). *Acta biol et meol exp (SFRY)* 13 (1): 47–50.

Terasawa E, 1996. Symposium: 'What is New in Neuroendocrine Research in Non-Human Primates?' *IPS/ASP Congr Abstr*: # 254.

Terry RD, Katzman R, Bick KL, eds, 1994. *Alzheimer Disease.* NY, Raven Press.

't Hart BA, Bontrop RE, 1998. The relevance of arthritis research in non-human primates. *Brit J Rheumatol* 37 (3): 239–242.

The Carnegie Monkey Colony, 1925–1959. Carnegie Institution of Washington. *Year Book* 58: 415–419.

Thenius E, 1981. Remerkungen zur taxonomischen unf stammesgeschictlichen position der Gibbons (Hylobatidae, Primates). *Z. f. Saugetierkunde* 46 (4): 232–241.

Thiranagama R, Chamberlain AT, Wood BA, 1989. Valves in superficial limb veins of humans and nonhuman primates. *Clin Anat* 2: 135–145.

Thomas ET, Ricordi C, Contreras JL, Hubbard WJ, Jiang XL, Ecknoff DE, Cartner S, Bilbao G, Neville DM Jr, Thomas JM, 1999. Reversal of naturally occuring diabetes in primates by immunodivided islet xenografts without chronic immunosuppression. *Transplantation* 67 (6): 846–854.

Thomas RK, Phillips JA, Young CD, 1990. Numerousness judgments by *Homo sapiens* in a species-comparative context. *Bull Psychon Soc* 28 (6): 500.

Thomas S, Dave PK, 1985. Experimental scoliosis in monkeys. *Acta Orthop Scand* 56: 43–46.

Thorington RW, Anderson S, 1984. Primates. In: *Orders and Families of Recent Mammals of the World*. S. Anderson, J. K. Jones, eds. NY, John Viley & Sons: 187–217.

Thorndike EL, 1901. The mental life of the monkeys. *Psychol Monogr* 3 (5): 1–57.

Tich NA, 1970. [*Prehistory of Society. Comparative-Psychologic Study*]. Leningrad, Len Univ Publish House (Russian).

Tine JA, Lanar DE, Smith DM, Wellde BT, Schultheiss P, Ware LA, Kauffman EB, Wirtz RA, de Taisne C, Hui GSN, Chang SP, Church P, Hollingdale MR, Kaslow DC, Hoffman S, Guito KP, Ballou WR, Sadoff JC, Paoletti E, 1996. NYVAC-Pf7: A poxvirus-vectored, multiantigen, multistage vaccine candidate for *Plasmodium falciparum* malaria. *Infect Immunity* 64 (9): 3833–3844.

Titenko AM, 1991. [Marburg hemorrhage fever]. *J microbiol epidemiol i immunol* # 5: 67–71 (Russian).

Toda N, 1983. Alpha adrenergic receptor subtypes in human, monkey and dog cerebral arteries. *J Pharmacol Exp Ther* 226 (3): 861–868.

Toda N, 1985. Reactivity in human cerebral artery: species variation. *Fed Proc* 44 (2): 326–330.

Toda N, Okamura T, 1990. Beta adrenoreceptor subtype in isolated human, monkey and dog epicardial coronary arteries. *J Pharmacol Exp Ther* 253 (2): 518–524.

Tomasello M, Call J, 1997. *Primate Cognition*. NY, Oxford Univ Press.

Tomiinson RW, Schwartz DW, 1988. Perception of the missing fundamental in nonhuman primates. *J Acoust Soc Amer* 84 (2): 560–565.

Tonooka R, 1996. Leaf selectivity and development in leaf-folding behavior by wild chimpanzees at Bossou. *ISP/ASP Congr Abstr*: # 593.

Toro-Goyco E, Cora EM, Kessler MJ, Martinez-Corrion M, 1986. Induction of experimental myasthenia gravis in rhesus monkeys: A model for the study of the human disease. *Puerto Rico Health Sci J* 5 (1): 13–18.

Traverso LW, Gomez RR, 1982. Hemodynamic measurements after administration of aprotinin and/or heparin during pancreatic cell autotransplantation in the dog, pig, and monkey. *Ann Surg* 195 (4): 479–485.

Tripathi BJ, Geanon JD, Tripathi RC, 1987. Distribution of tissue plasminogen activator in human and monkey eyes. An immunohistochemical study. *Ophtalmology* 94 (11): 1434–1438.

Troilo D, Judge SJ, 1993. Ocular development and visual deprivation myopia in the common marmoset (*Callithrix jacchus*). *Vis Res* 33: 1311–1324.

Trown PW, Willis RJ, Kam JJ, 1986. The preclinical development of roferon[R] – A. *Cancer* 57 (8): 1648–1656.

Tsutsumi H, Katsumata Y, Sato K, Tamaki K, Yada S, 1985. Antigenic relationships between human and non-human primate plasma protein studied by indirect hemagglutination inhibition test. *Act Crim Japan* 51 (4): 149–156.

Tsutsumi H, Nakamura S, Sato K, Tamaki K, Katsumata Y, 1989. Comparative studies on an antigenicity of plasma proteins from humans and apes by ELISA: A close relationship of chimpanzee and human. *Compar Biochem Physiol* 94 (4): 647–649.

Tsutsumi H, Sato K, Htay HH, Orajima H, Katsumata Y, 1988. Immunological studies of the antigenic properties of plasma proteins of prosimian, insectifore, and rodent. *Res Pract Forens Med* 31: 61–64 (Japanese w/English sum).

Tsutsumi H, Sato K, Htay HH, Tamaki K, Okajima H, 1988. Identification of human urinary stains by enzyme-linked immunosorbent assay (ELISA) using rabbit anti-human uromucoid. *Act Crim Japan* 54 (4): 163–168 (Japanese w/English sum).

Tulp N, 1641. *Observationum medicarum, libri tres.* Amsterdam, Apud, Ludovidum Elzevirium: 274–279.

Turanov NM, Stoodnicin AA, Zalkan PM, Ievleva EA, Zvetkova GM, Akimov VG, 1973. [Experience of experimental study of lepra]. *Vestn dermatol i venerol* # 1: 9–15 (Russian).

Turleau C, Creau-Goldberg N, Cochet C, de Grouchy J, 1983. Gene mapping of the gibbon. Its position in primate evolution. *Human Genet* 64 (1): 65–72.

Turleau C, de Grouchy J, Klein M, 1973. Phylogenie chromosomique de l'homme et des primates. (Essai de reconstitutions du caryotype de l'ancetre commun). *Cahiers de medecine* 14 (5): 367–379.

Tuttle RH (Editorial), 1998. Global primatology in a new millennium. *Intern J Primatol* 19 (1): 1–12.

Tyson E, 1699. *Orang-Outang, sive Homo Silvestris: Or the Anatomy of a Pigmie Compared with That of a Monkey, and Ape, and a Man.* London, T. Bennet and D. Brown.

Udaev NA, Afinogenova CA, Goncharov NP, Katzija GV, 1977. [The ways of biosynthesis of corticosteroids in adrenal glands of monkeys macaque rhesus]. *Problem endocrinol* 23 (5): 92–96 (Russian).

Ueda S, 1991. Human genome and human evolution. In: *Evolution of Life: Fossils, Molecules and Culture.* Tokyo e.a., Springer-Verlag: 429–438.

Ueda S, Matsuda F, Honjo T, 1988. Multiple recombinational events in primate immunoglobulin epsilon and alpha genes suggest closer relationship of humans to chimpanzees than to gorillas. *J Mol Evol* 27: 77–83.

Ueda S, Watanabe Y, Saitou N, Omoto K, Hayashida H, Miyata T, Hisajima H, Honjo T, 1989. Nucleotide sequences of immunoglobulin-epsilon pseudogenes in man and apes and their phylogenetic relationships. *J Mol Biol* 205 (1): 85–90.

Understanding aging, 1990. *Wisconsin Reg Primate Res Ctr* 1 (3): 1–2.

Uno H, Ye F-f, Imamura K, Obana N, Allen-Hoffman L, Pan H-j, 1997. Dose-dependent and long term effects of RU58841 (androgen receptor blocker) on hair growth in the bald stumptailed macaque. *J Invest Dermatol* 108 (4): 651.

Utami SS, 1994. Meat eating behavior of adult female orangutans. *Congr IPS* 15: 336.

Urmancheeva TG, 1980. [The bases to extrapolation of experimenatal data to man in study of organization mechanisms of higher nervous action in simians. In: *Biological Characteristics of Laboratory Animals and Extrapolation of Experimental Data to Man.* M, Acad Med Sci USSR]: 211–212 (Russian).

Urmancheeva TG, 1986. [Structure-functional peculiarities of central nervous system and use of simians in study of behavior regulation mechanisms]. *Vestn Acad Med Nauk SSSR* # 3: 85–88 (Russian).

Urmancheeva TG, Dyakonova IN, Osipova IA, 1966. [Model of subcortical syndrome in monkeys. In: *Deep Structures of Human Brain in Norm and Pathology*]. Moscow-Leningrad, Nauka: 172–176 (Russian).

Vagtborg H, 1967. Foreword. In: *The Baboon in Medical Research*, vol II. *Proc Second Intern Symp 'Baboon and Its Use as an Exp Animal'.* H. Vagtborg, ed. Austin, London, Univ Texas Press.

Vaitukaitis JL, 1994. Importance of primate models in U.S. biomedical research efforts and the implications for involvement in international collaborative programs. *Congr IPS* 15: 325.

Vajda J, Branston NM, Ladds A, Symon L, 1985. A model of selective experimental ischaemia in the primate thalamus. *Stroke* 16 (3): 493–501.

Valeri CCR, Valeri DA, Gray A, Leavy PD, Contreras TJ, Linberg JR, 1983 Rhesus macaque red blood cells frozen with 40% glycerol and stored at −80°C. *Amer J Vet Res* 44 (9): 1786–1788.

Valin A, Voltz C, Naquet R, Lloyd KG, 1991. Effects of pharmacological manipulation on neurotransmitter and other amino acid level in the CSF of the Senegalese baboon *Papio papio. Brain Res* 538 (1): 15–23.

Van Alphen MMA, de Rooij DG, 1986. Depletion of the seminiferous epithelium of the rhesus monkey, *Macaca mulatta*, after X-irradiation. *Brit J Cancer* 53 (7, Suppl): 102–104.

Van Bekkum DW, Balner H, 1968. Development of primate centers in the European community. *Experimentation Anim* 1 (2): 109–112.

Van Buul PPW, 1983. X-ray-induced chromosomal aberrations in germ cells of the rhesus monkey. In: *Biol Effect Low-Level Radiat. Proc Intern Symp, 1983.* Vienna: 625–626.

Van Camp K, van Sande M, 1988. Lactate dehydrogenase isoenzymes in prostate and testis of wild animals and some histological remarks. *Urol Res* 16 (5): 373–375.

Van der Waaij D, van der Waaij BD, 1990. The colonization resistance of the digestive tract in different animal species and in man: A comparative study. *Epidemiol a. Infect* 105: 237–243.

Van Hooff JARAM, 1981. Facial expressions. In: *The Oxford Companion to Animal Behavior*. D. Mc Farland, ed. Oxford Univ Press: 165–176.

Van Hooff JARAM, 1989. Laughter and humour, and the 'duo-in-uno' of nature and culture. In: *The Nature of Culture*. Bohum, W. Germany. BPX Publishers: 120–149.

Van Lawick-Goodall J, 1967. *My Friends the Wild Chimpanzees*. Washington, DC, Natl Geograph Soc.

Van Lawick-Goodall J, 1968. *The Behaviour of Free Living Chimpanzees in the Gombe Stream Reserve*. Ann Behav Monogr 1: 161–311.

Van Lawick-Goodall J, 1971. Exemples de comportement agressif dans un groupe de chimpanzes vivant en liberte. *Rev Intern Sci Soc* 23 (1): 102–111.

Van Schaik CP, Knott C, 1998. Orangutan culture? *Amer J Anthropol* Suppl 26: 223.

Van Stee EW, Back K, 1969. Short-term inhalation exposure to bromotrifluoromethane. *Toxicol Appl Phermacol* 15 (1): 164–174.

Van Vark GN, Bilsborough A, Schaafsma W, 1990. Affinity, hominoid evolution and creationism. New computer methods weaken the creationists position. *Human Evol* 5 (5): 471–482.

Van Winkle DM, Romson JL, Feigl EO, 1988. Cholinergic coronary vasodilatation in dogs and baboon. *FASEB J* 2 (4): 495.

Van Zyl JJW, 1971. Report on the activities of the University of Stellenbosch primate colony. *Med Primatol 1970*. Basel, Karger: 976–979.

Van Zyl JJM, Murphy GP, Klerk JN, 1968. Baboons in transplation. Research at the University of Stellenbosch. *Lab Prim Newsl* 7 (4): 17–20.

Vancatova M, Jerabkova Z, Firsov LA, 1996. The ape's picture-making activity. *IPS/ASP Congr Abstr*: # 321.

VandeBerg JL, 1987. Historical perspective of genetic research with nonhuman primates. *Genetica* 74 (1–2): 7–14.

Verhaegen M, 1995. Aquatic ape theory, speech origins, and brain differences with apes and monkeys. *Med Hypothes* 44: 409–413.

Ver Hoeve JN, Kim CBY, Danilov Y, Spear PD, 1996. Visual acuity in aging rhesus monkeys. *IPS/ASP Congr Abstr*: # 647.

Verjee ZHM, Damji A, 1974. Some properties of red phosphate dehydrogenase from the East African baboon. *Intern Biochem* 5: 41–54.

Vesalius A, (1543) 1954. *De Humani corporis fabrica* (Latin) Basel. (Russian: M, Acad Sci USSR Publ House).

Viallet F, Trouche E, Beaubaton D, Nieoullon A, Legaliet E, 1981. Bradykinesia following unilateral lesions restricted to the substantia nigra in the baboon. *Neurosci Lett* 24 (1): 97–102.

Vickers JH, 1983. FDA regulatory use of nonhuman primates. *Lab Anim* April: 28–34.

Vilensky JA, 1988. Neural control of locomotion in primates: Comparisons with cats. *Amer J Phys Anthropol* 75 (2): 283–284.

Villarroya S, Merand B, Scholler R, 1987. Primate specific sialogly-coprotein of sperm head plasma membrane defined by an anti-carbohydrate monoclonal antibody. *Pathol Biol* 35 (8): 1155–1159.

Vincent JW, Falkler WA, Craig JA, 1983. Comparison of serological reaction of typed *Fusebacterium nucieatum* strains with those of isolated from humans, canines, and a *Macaca mulatta*. *J Clin Microbiol* 17 (4): 631–635.

Vinogradov-Volzhinski DV, Shargorodskaya VA, 1976. [*Epidemic Parotitis*]. M, Medicine (Russian).

Virchow R, 1899. [*Successes of Contemporary Science and Their Relation to Medicine and Surgery*]. M, Snegirev (Russian).

Visalberghi E, 1997. Success and understanding in cognitive tasks: A comparison between *Cebus apella* and *Pan troglodytes*. *Intern J Primatol* 18 (5): 811–830.

Visalberghi E, Fragaszy DM, 1990. Do monkeys ape? In: *Language and Intelligence in Monkeys and Ape: Comparative Developmental Perspectives*. Cambridge, Cabr Univ Press: 247–274.

Visalberghi E, McGrew WC, 1997. *Cebus* meets *Pan*. *Intern J Primatol* 18 (5): 677–681.

Volkova OV, 1983. [*Functional Morphology of Female Reproductive System*]. M, Medicine (Russian).

Von Sanderson JT, Steinbacher G, 1957. *Knaurs Affenbuch*. München-Zürich, Droemersche Verlagsansalt T. Knaur Nachf.

Voronin LG, Firsov LA, 1967. [Study of higher nervous action of anthropoid simians in Soviet Union]. *J visshey nervn deyatelnosti* 17 (5): 834–846 (Russian).

Voytko ML, 1998. Nonhuman primates as models for aging and Alzheimer's disease. *Lab Anim Sci* 48 (6): 611–617.

Voytonis NYu, 1949. [*Prehistory of the Intellect*]. Moscow-Leningrad (Russian).

Wagner JD, Thomas MJ, Williams JK, Zhang L, Greaves KA, Cefalu WT, 1998. Insulin sensitivity and cardiovascular risk factors in ovariectomized monkeys with estradiol alone or combined with nomegestrol acetate. *J Clin Endocrinol Metabol* 83 (3): 896–901.

Wahl G, Georges AJ, 1995. Current knowledge on the epidemiology, diagnosis, immunology, and treatment of loiasis. *Trop Med and Parasitol* 46: 287–291.

Waisman HA, Harlow HF, 1965. Experimental phenylketonuria in infant monkeys. A high phenylalanine diet produces abnormalities simulating those of the hereditary disease. *Science* 147 (3659): 685–695.

Walcroft, 1983. Primates – requirements by the pharmaceutical industry. *J Med Primatol* 12 (1): 1–7.

Walker A, 1976. Splitting times among hominoids deduced from the fossil record. In: *Mol Anthropol*. New York – London: 63–77.

Walker A, Teaford M, 1989. *The hunt for Proconsul. Sci Amer* 260 (1): 76–82.

Walker LC, Cork LC, Price DL, 1988. Age-associated brain abnormalities in nonhuman primates. *Amer J Primatol* 14 (4): 450.

Wallen K, Davis-DaSilva M, Mann DR, Gaventa S, Lovejoy JC, 1984. Suppression of ovulation and sexual behavior in group-living rhesus by GNRh treatment. *Biol Reprod* 30 (Suppl 1): 76.

Wallow IHL, Lund O-E, Gabel V-P, Birngruber R, Hillenkamp F, 1974. A comparison of retinal argon laser lesions in man and in cynomolgus monkey. *Albrecht v Graefe's Arch Klin Exp Ophtalmol* 189: 159–164.

Walston J, Lowe A, Silver K, Yang Y-f, Bodkin NL, Hansen BC, Shuldiner AR, 1997. The beta-3-adrenergic receptor in the obesity and diabetes prone rhesus monkey is very similar to human and contains arginine at codon 64. *Gene* 188 (2): 207–213.

Warfvinge J, 1986. Dental pulp inflammation: Experimental studies in human and monkey teeth. *Dental J* (Göteborg) 39: 2–36.

Washburn DA, Hopkins WD, Rumbaugh DM, 1989. Video-task assessment of learning and memory in macaques *(Macaca mulatta)*: Effects of stimulus movement on performance. *J Exp Psychol Anim Behav Process* 15 (4): 393–400.

Washburn SL, 1960. Tools and human evolution. *Sci Amer* 203: 63–75.

Washburn SL, 1982. Human evolution. *Persp Med* 4 (25): 583–602.

Wasmoen TL, Benirschke K, Gleich GJ, 1987. Demonstration of immunoreactive eosinophil granule major basic protein in the plasma and placentae of non-human primates. *Placenta* 8 (3): 283–292.

Waters DJ, Bostwick DG, Murphy GP, 1998. Conference summary: First International Workshop on Animal Models of Prostate Cancer. *Prostate* 36 (1): 47–48.

Watts ES, 1985. Adolescent growth and development of monkeys, apes and humans. In: *Nonhuman Primate Models for Human Growth and Development*. NY, Alan R. Liss: 41–65.

Weber A, 1997. Survey on the use of primates in the European Pharmaceutical Industry. *Primate Rep* 49, Oktober: 11–18.

Weber H, Madoerin M, 1977. The use of primates in pharmacology and toxicology. In: *Use of Nin-Human Primates in Biomedical Research. Intern Symp, New Delhi, India, 1975*. New Delhi, Ind Natl Acad Sci: 380–390.

Weber M, (1904) 1936. [*Primates. Anatomy, Systematics and Paleontology of Lemurs, Tarsiers and Simians*. Moscow – Leningrad, State Publ House Biol Med Literat] (Russian, transl from German).

Webster BR, Paice JC, Gale CC, 1970. Evidence for immunological identity of human and simian (baboon) TSH in serum. *Nature* 227: 712–714.

Weesner KM, Kaplan K, 1987. Hemodynamic echocardiographic evaluation of the stump-tailed macaque: A potential nonhuman primate model pulmonary vascular disease. *J Med Primatol* 16 (3): 185–202.

Wein AM, 1970. [*Wakefulness and Sleep*]. M, Nauka (Russian).

Weiss B, Laties VC, 1967. Comparative pharmacology of drugs effecting behavior. *Fed Proc* 26 (4): 1146–1156.

Weitzel V, Groves CP, 1985. The nomenclature and taxonomy of the colobine monkyes of Java. *Intern J Primatol* 6 (4): 399–409.

Wemelsfelder F, Dolins F, 1994. Different models of animal awareness and their consequences for understanding chronic suffering in captive primates. *Congr IPS* 15: 358.

Wen D, Boissel J-PR, Tracy TE, Gruninger RH, Czelusniak J, Goodman M, Bunn HF, 1993. Erythropoetin structure-function relationships: High degree of sequence homology among mammals. *Blood* 82 (5): 1507–1516.

Wenk GL, Pierce DJ, Struble RG, Price D, Cork LC, 1989. Age-related changes in multiple neurotransmitter systems in the monkey brain. *Neurobiol Aging* 10 (1): 11–19.

Wester RC, Mc Master J, Bucks DAW, Bellet EM, Maiback HI, 1989. Percutaneous absorption in rhesus monkeys and estimation of human chemical exposure. *Amer Chem Soc Symp Series* 382: 152–157.

Westergaard GC, 1988. Lion-tailed macaques (*Macaca silenus*) manufacture and use tools. *J Compar Psychol* 102 (2): 152–159.

Westergaard GC, Fragaszy DM, 1987. The manufacture and use of tools by capuchin monkeys. *J Compar Psychol* 101 (2): 159–168.

Westergaard GC, Kuhn HE, Suomi SJ, 1998. Bipedal posture and hand preference in humans and other primates. *J Compar Psychol* 112 (1): 55–64.

Wheeler MD, Styne DM, 1988. The nonhuman primate as a model of growth hormone physiology in the human being. *Endocrine Rev* 9 (2): 213–246.

White RJ, Wolin LR, Massopust LC, Taslitz N, Verdura J, 1971. Primate cephalic transplantation: Neurogenic separation, vascular association. *Transplant Proc* 3 (1): 602–604.

Whitehair L, Gay W, 1981. The seven Primate Research Centers. *Lab Animal*, Nov-Dec: 28.

Whitehead VM, Kamen BA, Beaulieu D, 1987. Levels of dihydrofolate reductase in livers of birds, animals, primates, and man. *Cancer Drug Deliv* 4 (3): 185–189.

Whiten A, Byrne RV, eds, 1997. *Machiavellian Intelligence II: Extensions and Evaluations*. Cambridge, UK, Cambr Univ Press.

Whiten A, Goodall J, McGrew WC, Nishida T, Reynolds V, Sugiyama Y, Tutin CEG, Wrangham RW, Boesch C, 1999. Cultures in chimpanzees. *Nature* 399 (6737): 682–685.

Whitney RA, 1995. Taxonomy. In: *Nonhuman Primates in Biomedical Research: Biology and Management*. NY, Acad Press: 33–47.

WHO (World Health Organization), 1971. [*Use of Nonhuman Primates for Bio-Medical Purposes*]. *Khronica WOZ* 25 (7) Geneva – Moscow (Russian).

WHO, 1973. [*Clinical Immunology*, Report 496]. Geneva – Moscow, Medicine (Russian).

WHO, 1975. *Guidelines of Evolution of Drug for Use in Man*. Tech Dept Ser # 563. Geneva.

WHO, 1976. [*Laboratory animals for medico-biological studies. 29 Session of All-World Assembly of Public Health*]. Geneva – Moscow, Khronica WOZ 30 (12): 595–607 (Russian).

WHO, 1981. *Report of an Informal Consultation on the WHO International Primate Resources Programme. BLG/PRI/81.1*, Geneva, 9–12 Novemb.

WHO, 1986. [*Viral Diseases Transmitted from Anthropoda and Rodents: Rep of WHO Sci Group*]. M, Medicine (Russian).

WHO, 1988. [Sense of nonhuman primates for elaboration vaccines against malaria: Memorandum of the WHO Meeting]. *Bull WHO* 66 (6): 35–45 (Russian).

WHO, 1991. [*WHO Expert Committee of Drug Dependences. Report 25*]. M, Medicine (Russian).

Wickings EJ, 1996. Sexual maturation in male lowland gorillas: A comparison with chimpanzees. *IPS/ASP Congr Abstr*: # 032.

Wiener AS, 1975. Blood groups of nonhuman primates: The development of new concepts and their applications. *J Med Primatol* 4 (6): 377–378.

Wiener AS, Gordon EB, Moor-Jankowski J, Socha WW, 1972. Homologues of the human M-N blood types in gorillas and other non-human primates. *Haematologia* 6 (4): 419–432.

Wiener AS, Gordon EB, Socha WW, Moor-Jankowski J, 1970. Further observations on the immunological relationships among serum globulins of man and other primates, revealed by serological inhibition test. *Intern Arch Allergy Appl Immunol* **39** (4): 368–380.

Wiener AS, Moor-Jankowski J, 1969. The A-B-O blood groups of baboons. *Amer J Phys Anthropol* **30** (1): 117–122.

Wiener AS, Moor-Jankowski J, Gordon EB, 1964. Blood group antigens and cross-reacting antibodies in primates including man. II. Studies of the M-N types of orangutans. *J Immunol* **93** (1): 101–105.

Wiener AS, Socha WW, Moor-Jankowski J, 1974. Homologues of the human ABO blood groups in apes and monkeys. *Haematologia* **8** (2): 195–216.

Wilks JM, Hodgen GD, Ross GT, 1979. Endocrine characteristics of ovulatory and anovulatory menstrual cycles in the rhesus monkey. In: *Human Ovulation. Mechanisms. Prediction, Detection and Induction.* Amsterdam *et al.*, North-Holland Publ Co: 205–218.

Williams DD, Bowden DM, 1984. A nonhuman primate model for the osteopenia of aging. In: *Comparative Pathobiology of Major Age-related Diseases: Current Status and Research Frontiers.* NY, Alan R. Liss: 207–219.

Williams JK, Adams MR, Kaplan JR, 1994. Use of cynomolgus monkeys as models of coronary vasospasm and thrombosis. *Congr IPS* **15**: 291.

Williams JK, Anthony MS, Clarkson TB, 1991. Coronary heart disease in rhesus monkeys with atherosclerosis. *Amer J Primatol* **24** (2): 140.

Williams JK, Kim YD, Adams MR, Chen M-F, Myers AK, Ramwell PW, 1994. Effects of estrogen on cardiovascular responses of premenopausal monkeys. *J Pharmacol Exp Terapeut* **271** (2): 671–676.

Williams JK, Wagner JD, Li Z, Golden DL, Adams MR, 1997. Tamoxifen inhibits arterial accumulation of LDL degradation products and progression of coronary artery atherosclerosis in monkeys. *Arterioscler Vascul Biol* **17** (2): 403–408.

Williams RT, 1976. Future developments. In: *Drug Metabolism – from Microbe to Man.* London, Taylor and Francis: 1–3.

Williams SA, Goodman M, 1989. A statistical test that supports a human/chimpanzee clade based on noncoding DNA sequence data. *Mol Biol Evol* **6** (4): 325–330.

Willoughby J, Glover V, Sandler M, Reynolds G, Baron B, 1987. Monoamine oxidase B distribution in the human, marmoset and rat brain. *Brit J Pharmacol* **92** (Suppl): 681.

Wilner P, 1997. Animal models of addiction. *Human Psychopharmacol* **12** (Suppl 2): S 59–S 68.

Wilson AC, Zimmer EA, Prager EM, Kocher TD, 1989. Restriction mapping in the molecular systematics of mammals: A retrospective salute. In: *The Hierarchy of Life: Molecules and Morphology in Phylogenetic Analysis. Proc from Nobel Symposium* **70**. NY, Elsevier Sci Publishers: 407–419.

Wilson IG, 1972. Use of primates in teratologic investigations. In: *Med Primatol 1972. Proc 3 Conf Exp Med Surg Primates, Lyon 1972.* Part 3 Basel, S. Karger: 286–295.

Wilson JG, 1978. Feasibility and design of subhuman primate studies. In: *Handbook of Teratology.* **4.** *Research Procedures and Data Analysis.* NY, Plenum Press: 255–273.

Wilson MC, Bedford JA, Kibbe AH, Sam JA, 1978. Comparative pharmacology of norcocain in *M. mulatta* and *M. fascicularis. Pharmacol Biochem Behav* **9** (1): 141–145.

Wilson V, Jeffreys AJ, Barrie PA, Boseley PG, Slocombe PM, Easton A, Burke DC, 1983. A comparison of vertebrate interferon gene families detected by hybridization with human interferon DNA. *J Mol Biol* **166** (4): 457–475.

Winfield AF, 1971. Comparative metabolism of Abbott 25794 in dogs, monkeys and man. *Fed Proc* **30**: 392.

Wissler RW, 1996. Atheroarteritis: A combined immunological and lipid imbalance. *Intern J Cardiol* **54** (Suppl): S 37–S 49.

Wissler RW, Vesselinovitch D, Hughes R, Turner D, Frazier L, 1983. Arterial lesions and blood lipids in rhesus monkeys fed human diets. *Exp Mol Pathol* **38** (1): 117–136.

Wolf DP, Stouffer RL, Brenner RM, eds, 1993. *In vitro Fertilization and Embryo Transfer in Primates.* NY, Springer-Verlag.

Wolpoff MH, 1982. Ramapithecus and hominid origins. *Curr Anthropol* 23 (5): 501–522.

Wood DH, Yochmowitz MG, Hardy KA, Salmon VL, 1986. Animal studies of life shortening and cancer risk from space radiation. *Adv Space Res* 6 (11): 275–283.

Wrandham RW, McGrew WC, de Waal FBM, Heltne PG, eds, 1994. *Chimpanzee Cultures.* Cambridge, MA, Harvard Univ Press.

Wright AA, Sperling HG, Mills SL, 1987. Researches on a unilaterally blue-blinded rhesus monkey. *Vis Res* 27 (9): 1551–1564.

Wright J, Johnson DR, Peterson DA, Wolfe LG, Deinhardt FW, Maschgan ER, 1982. Serology and lymphocyte surface markers of great apes maintained in a zoo. *J Med Primatol* 11 (2): 67–76.

Wu X, Lee CC, Muzny DM, Caskey CT, 1989. Urate oxidase: Primary structure and evolutionary implications. *Proc Natl Acad Sci USA* 86: 9412–9416.

Wyatt RJ, Morihisa JM, Nakamura RK, Freed WJ, 1986. Transplanting tissue into the brain for function: Use in a model of Parkinson's disease. In: *Neuropeptides in Neurologic and Psychiatric Disease.* NY, Raven Press: 199–208.

Wyss AR, Flynn JJ, 1995. 'Anthropoidea': A name, not an entity. In: *Evolutionary Anthropology.* NY, Wiley – Liss: 187–188.

Yakovleva LA, 1968. [*Comparative Reseach of Radiation Disease and Its Consequences on the Model of Radiation Disease of Monkeys*]. M, Medicine (Russian).

Yakovleva LA, Indzhiya LV, Chikobava MG, Schatzl H, Lapin BA, 1997. [Morphoimmunological pheno-types of baboon T-cell non-Hodgkin's malignant lymphomas (T-NHL) and their connection with T-lymphotropic retrovirus STLV-1]. *Arch patol* 59 (1): 19–25 (Russian w/English sum).

Yamashita M, Miura T, Ibuki K, Takehisa J, Chen J-I, Ido E, Hayami M, 1997. Phylogenetic relation-ships of HTLV-1/STLV-1 in the world. *Leukemia* 11 (Suppl 3): 50–51.

Yan R-q, Su J-j, Huang D-r, Gan Y-c, Yang C, Huang G-h, 1996. Human hepatitis B virus and hepatocellular carcinoma. I. Experimental infection of tree shrews with hepatitis B virus. *J Cancer Res and Clin Oncol* 122 (5): 283–288.

Yankell SL, 1985. Oral disease in laboratory animals: Animal models of human dental disease. In: *Veterinary Dentistry.* Philadelphia, Saunders Co: 281–288.

Yerkes RM, 1915. Maternal instinct in a monkey. *J Anim Behav* 5: 403–405.

Yerkes RM, 1916. Provision for the study of monkeys and apes. *Science* 43: 231–234.

Yerkes RM, 1925. *Almost Human.* New York – London.

Yerkes RM, 1943. *Chimpanzee. A Laboratory Colony.* London.

Yerkes RM, Yerkes AW, 1929. *The Great Apes.* New Haven, Yale Univ Press.

Yole DS, Reid GDF, Wilson RA, 1996. Protection against *Schistosoma mansoni* and associated immune responses induced in the vervet monkey *Cercopithecus aethiops* by the irradiated cercaria vaccine. *Amer J Trop Med and Hygiene* 54 (3): 265–270.

Yoshida T, 1983. Serum gonadotropin levels during the menstrual cycle and pregnancy in the cynomolgus monkey. *Japan J Med Sci Biol* 36 (4): 231–236.

Yoshioka H, Katsume Y, Akune H, 1982. Experimental central serous chorioretinopathy in monkey eyes. Fluorescein angiographic findings. *Ophtalmologica* 185 (3): 168–178.

Young FA, 1973. The distribution of myopia in man and monkey. *Doc Ophthalmol Proc* Ser 28: 5–11.

Young FA, Farrer DN, 1971. Visual similarities of nonhuman and human primates. *Med Primatol 1970.* Proc 2nd Conf Exp Med & Surg. Basel, S. Karger: 316–328.

Young MD, 1973. Monkeys and malaria. In: *Nonhuman Primates and Med Res.* New York – London, Acad Press: 17–24.

Younis AI, Rooks B, Khan S, Gould KG, 1998. The effects of antifreeze peptide III (AFP) and insulin transferrin selenium (ITS) on cryopreservation of chimpanzee (*Pan troglodytes*) spermatozoa. *J Androl* 19 (2): 207–214.

Yunis JJ, Prakash O, 1982. The origin of man: A chromosomal pictorial legacy. *Science* 215 (4539): 1525–1530.

Yurovskaya VZ, 1982. [Peculiarities of the fastening of the upper extremity muscles of primates in the comparative light]. *Voprosi anthropol* # 70: 38–53 (Russian w/English sum).

Yurovskaya VZ, 1983. [On locomotion of anthropomorphic ancestor of man according to data of comparative myology]. *Voprosi anthropol* # 72: 42–49 (Russian w/English sum).

Yurovskaya VZ, 1991. [Some morphological aspects of locomotor behavior of primates. In: *Behavior of Primates and Problems of Anthropogenesis*]. M, Moscow Soc Investigat of Nature: 67–78 (Russian).

Zacchei AG, Weidner L, 1974. The physiological disposition of a new antiarrhythmic agent, α, α-dimethyl-4-(α, α, β, β-tetrafluorophenethyl) benzylamine (I) in several species. *Fed Proc* 33 (3): 475.

Zaiceva NS, 1976. [*Trachoma*]. M, Medicine (Russian).

Zamir M, Medeiros JA, 1982. Arterial branching in man and monkey. *J Genet Physiol* 79 (3): 353–360.

Zdrodovskiy PF, 1961. [*Problems of Infection and Immunity*]. M, Medgiz (Russian).

Zelinski-Wooten MB, Stouffer RL, Molskness TA, Wolf DP, 1994. Repeated follicular stimulation in rhesus monkeys with recombinant human gonadotropins. *Amer J Primatol* 33 (3): 253–254.

Zeveloff SI, Boyce MS, 1982. Why human neonates are so altricial. *Amer Natur* 120 (4): 537–542.

Zhdanov VM, Lvov DK, 1984. [*The Evolution of Pathogens of Infection Diseases*]. M, Medicine (Russian).

Zhou D, Sun YL, Vacek I, Ma P, Sun AM, 1994. Normalization of diabetes in cynomolgus monkeys by xenotransplantation of microencapsulated porcine islets. *Transplant Proc* 26 (3): 1091.

Zihlman AL, 1979. Pygmy chimpanzee morphology and the interpretation of early hominids. *South Afr J Sci* 75 (April): 165–168.

Zihlman AL, 1984. Pygmy chimps, people, and the pundits. *New Sci* 104 (1430): 39–40.

Zihlman AL, 1989. Common ancestors and uncommon apes. *Human Origins*. Oxford, Clarendon Press: 81–105.

Zihlman AL, 1990. Knuckling under: Controversy over hominid origins. In: *From Apes to Angels: Essays on Anthropology in Honor of P. V. Tobias*. Wiley – Liss: 185–196.

Zilles K, Armstrong E, Moser KH, Schleicher A, Stephan H, 1989. Gyrification in the cerebral of primates. *Brain Behav Evol* 34 (3): 143–150.

Zimbric ML, Uno H, Thomson JA, Kemnitz JW, 1996. Cerebral beta-amyloidosis (Alzheimer's lesion) in aged captive macaques. *IPS/ASP Congr Abstr*: # 702.

Zimmermann E, Ehresmann P, Zietemann V, Radespiel U, Randrianambinina B, Pakotoarison N, 1997. A new primate species in North-Western Madagaskar: The golden-brown mouse lemur (*Microcebus ravelobensis*). *Primate Eye* 63: 26–27.

Zucker H, Flurer CJ, 1989. The protein requirement of adult marmosets: Nitrogen balances and net protein utilization of milk proteins, soy protein, and amino acid mixtures. *Z Ernahrungswiss* 28 (2): 142–148.

Zuckerkandl E, 1963. Perspectives in molecular anthropology. In: *Classification and Human Evolution*. S. L. Washburn, ed. Chicago, Aldine: 243–272.

Zuckerkandl E, Pauling L, 1965. Evolutionary divergence and convergence in proteins. In: *Evolving Genes and Proteins*. V. Bryson, H. J. Vogel, eds. NY, Acad Press: 97.

Zuckerman S, 1963. Laboratory monkeys and apes from Galen onwards. *Proc Conf Res with Primates*. Beaverton, Oregon: 3–7.

Index

Added to the page number, f denotes a figure; t denotes a table.

Printed and bound by CPI Group (UK) Ltd, Croydon, CR0 4YY

23/10/2024

01778226-0007